CATALYSIS OF ORGANIC REACTIONS

CHEMICAL INDUSTRIES

A Series of Reference Books and Textbooks

Consulting Editor

HEINZ HEINEMANN

73. *Clathrate Hydrates of Natural Gases: Second Edition, Revised and Expanded,* E. Dendy Sloan, Jr.
74. *Fluid Cracking Catalysts,* edited by Mario L. Occelli and Paul O'Connor
75. *Catalysis of Organic Reactions,* edited by Frank E. Herkes

ADDITIONAL VOLUMES IN PREPARATION

The Chemistry and Technology of Petroleum, Third Edition, Revised and Expanded, James G. Speight

Synthetic Lubricants and High-Performance Functional Fluids, Second Edition: Revised and Expanded, Leslie R. Rudnick and Ronald L. Shubkin

CATALYSIS OF ORGANIC REACTIONS

edited by

Frank E. Herkes

E. I. DuPont de Nemours & Co.
Wilmington, Delaware

MARCEL DEKKER, INC.   NEW YORK · BASEL · HONG KONG

Library of Congress Cataloging-in-Publication Data

Catalysis of organic reactions / edited by Frank E. Herkes
    p. cm.— (Chemical industries; v. 75)
    A compendium of papers originating from papers and posters presented at the
17th Conference on Catalysis of Organic Reactions sponsored by the Organic
Reactions Catalysis Society.
    Includes bibliographical references and index.
    ISBN 0-8247-1929-8 (alk. paper)
    1. Organic compounds—Synthesis—Congresses. 2. Catalysis—Congresses.
I. Herkes, Frank E. II. Organic Reactions Catalysis Society. III. Series.
QD262.C35533 1998
547'.2—dc21   *QD262*   98-29982
                           . C35533   CIP
                           1998

This book is printed on acid-free paper.

Headquarters
Marcel Dekker, Inc.
270 Madison Avenue, New York, NY 10016
tel: 212-696-9000; fax: 212-685-4540

Eastern Hemisphere Distribution
Marcel Dekker AG
Hutgasse 4, Postfach 812, CH-4001 Basel, Switzerland
tel: 44-61-261-8482; fax: 44-61-261-8896

World Wide Web
http://www.dekker.com

The publisher offers discounts on this book when ordered in bulk quantities. For more information, write to Special Sales/Professional Marketing at the headquarters address above.

Current printing (last digit)
10 9 8 7 6 5 4 3 2 1

PRINTED IN THE UNITED STATES OF AMERICA

Preface

Catalysis of Organic Reactions is a compendium of 59 chapters originating from contributed papers and posters presented at the 17th Conference on Catalysis of Organic Reactions sponsored by the Organic Reactions Catalysis Society (ORCS). This self-contained reference documents recent and novel developments in the study of catalysis as it relates to organic synthesis and its application in industrial processes. Over the years, the ORCS Conference has provide a forum for chemists and engineers from chemical and pharmaceutical industries and from colleges and universities to present and discuss their work on the use and application of catalysis in organic synthesis.

Written by more than 200 internationally renowned experts from industry and academia, *Catalysis of Organic Reactions* examines the wide area of homogeneous and heterogeneous catalysis for industrial and pharmaceutical chemicals. The focus of this reference is on the latest developments in asymmetric synthesis and hydrogenation, heterogeneously and homogeneously catalyzed hydrogenations and oxidation, environmental catalysis, solid acid-base catalysis, isomerization, amination, dehydrogenation, hydroformylation catalysts, catalyst selection and optimization and a variety of other topics.

The 17th Conference recognizes the generous support of Searle, which, with the sponsorship of Mark Scaros, published the conference preprints. Thanks also go to the banquet sponsors, Activated Metals-PMC, Calsicat Division of Mallinckrodt Chem., Degussa, Engelhard, W.R. Grace & Co.-Conn.-Grace Davison, and Johnson Matthey, and the corporate sponsors, Air Products & Chemicals, Autoclave Engineers, Dow Chemical, , Eastman Kodak, E. I. DuPont , Elf-Atochem, Hercules, Novartis-Basel, Merck Co., Monsanto, Parr Instruments, Parke-Davis Pharmaceutical, Phillips Petroleum, Uniroyal Chemical, and Witco OSI Specialities.

I wish to express my appreciation and thanks to all the contributors for their superb efforts in the preparation and presentation of papers and posters and to the session Chairs for their introduction of the speakers. I personally want to thank the Executive and Editorial Boards of the Organic Reactions Catalysis Society for their time, encouragement and help in putting this conference together. Additional thanks go to all the referees who gave some of their time to review the manuscripts. Finally, I want to thank my wife, Mary, for her loving support and encouragement throughout the two years in preparation for the conference.

Frank E. Herkes

Acknowledgments

The Organic Reaction Catalysis Society gratefully acknowledges the generous support of the following companies. The support of each company was instrumental in allowing our Society to fund the Seventeenth Conference and in particular the graduate students, post-graduates and academicians participating in the conference.

Air Products and Chemicals
Autoclave Engineers
Calsicat Division of Mallinckrodt Chemical Inc.
Degussa Corporation
Dow Chemical Co.
Eastman Kodak Co.
E. I. DuPont de Nemours & Co.
Elf-Atochem NA
Engelhard Corp
Hercules Inc.
Johnson Matthey Corp.
Novartis - Basel
Merck Corp.
Monsanto Corp.
Parke-Davis Pharmaceutical
Phillips Petroleum Co.
Precious Metals Corp. and Activated Metals
Uniroyal Chemical
Witco OSI Specialities
W. R. Grace & Co. - Conn. - Grace Davison

Our refreshments during the morning and afternoon meeting breaks were provided by the **Parr Instrument Company**. We wish to acknowledge and thank them for this generous and valuable service.

The **Organic Reactions Catalysis Society** also wishes to thank **G. D. Searle** for their generous help and support in publishing the preprints for this Conference.

Contents

Contents

Contents

Contributors

S. Abro, Laboratoire de catalyse, UMR CNRS, Poitiers, France

David J. Ager, NSC Technologies, Mt Prospect, Illinois

Christopher S. Alexander, University of British Columbia, Vancouver, Canada

E. Angelescu, University of Bucharest, Bucharest, Romania

E. Armbruster, Lonza Spa, Naters, Switzerland

J. N. Armor, Air Products and Chemicals Inc., Allentown, Pennsylvania

Mezzie L. Ash, Dow Chemical Co., Midland, Michigan

E. Auer, Degussa AG, Research and Applied Technology Chemical Catalysts and Zeolites, Hanau, Germany

Robert L. Augustine, Seton Hall University, South Orange, New Jersey

Alfons Baiker, Swiss Federal Institute of Technology, Zürich, Switzerland

M. Banciu, Polytechnic University of Bucharest, Bucharest, Romania

J. Barrault, Laboratoire de catalyse, UMR CNRS, Poitiers, France

A. G. M. Barrett, Imperial College of Science, Technology and Medicine, London, England

Mihály Bartók, Hungarian Academy of Sciences, József Attila University, Szeged, Hungary

Peter Baumeister, Novartis Service AG, Basel, Switzerland

Richard P. Beatty, E. I. DuPont Co., Wilmington, Delaware

Matthias Beller, Technische Universität München, Garching, Germany

I. P. Belomestnykh, Russian Academy of Sciences, Moscow, Russia

C. Bendic, University of Bucharest, Bucharest, Romania

T. Beregszászi, Hungarian Academy of Sciences, József Attila University, Szeged, Hungary

David E. Bergbreiter, Texas A&M University, College Station, Texas

Donald Bethell, University of Liverpool, Liverpool, England

John J. Birtill, Air Products (Chemicals) Teesside Ltd., Wilton, Middlesbrough, England

Donna G. Blackmond, Max-Planck-Institut für Kohlenforschung, Mülheim an der Ruhr, Germany

Hans-Ulrich Blaser, Novartis Service AG, Basel, Switzerland

Michael Bodmer, Swiss Federal Institute of Technology, Zürich, Switzerland

I. Borbáth, Hungarian Academy of Sciences, Budapest, Hungary

D. C. Braddock, Imperial College of Science, Technology and Medicine, London, England

B. Briffa, Eastman Kodak Co., Rochester, New York

John A. Brinkman, Seton Hall University, South Orange, New Jersey

Brenda L. Case, Texas A&M University, College Station, Texas

M. Campanati, Dip. Chimica Industr. e Materiali, Bologna, Italy

L. Carmona, Société Française Hoechst, Cuise-Lamotte, France

N. Carmona, Institut de Recherches sur la Catalyse du CNRS, Villeurbanne, France

G. L. Castiglioni, Lonza Spa, Scanzorosciate BG, Italy

Mark Chamberlain, Air Products (Chemicals) Teesside Ltd., Wilton, Middlesbrough, England

B. D. Chandler, University of Minnesota, Minneapolis, Minnesota

Chih-Yuan Chuang, State University of New York at Stony Brook, Stony Brook, New York

J-M. Clacens, Laboratoire de catalyse, UMR CNRS, Poitiers, France

S. Coman, University of Bucharest, Bucharest, Romania

Ian Costello, Air Products (Chemicals) Teesside Ltd., Wilton, Middlesbrough, England

K. Cottin, Laboratoire de catalyse, UMR CNRS, Poitiers, France

A. A. Davydov, University of Alberta, Edmonton, Canada

W. N. Delgass, Purdue University, West Lafayette, Indiana

Robert A. DeVries, Dow Chemical Co., Midland, Michigan

A. Donato, University of Reggio Calabria, Italy

M. Dubovik, TDA Research Inc., Wheat Ridge, Colorado

Martin Eichberger, Technische Universität München, Garching, Germany

A. Eisenstadt, IMI (TAMI) Institute for R&D Ltd., Israel

François Fajula, Matériaux Catalytiques et Catalyse, Montpellier, France

Anatoly B. Fasman, San Francisco, California

F. Fazzini, Dip. Chimica Industr. e Materiali, Bologna, Italy

Károly Felföldi, Hungarian Academy of Sciences, József Attila University, Szeged, Hungary

F. Figueras, Institut de Recherches sur la Catalyse du CNRS, Villeurbanne, France

Annie Finiels, Matériaux Catalytiques et Catalyse, Montpellier, France

Michela Finiguerra, Centro C.N.R. MISO, Bari, Italy

L. A. Fiorella, Eastman Kodak Co., Rochester, New York

K. Fodor, Technical University of Budapest, Budapest, Hungary

M. E. Ford, Air Products and Chemicals Inc., Allentown, Pennsylvania

G. Fornasari, Dip. Chimica Industr. e Materiali, Bologna, Italy

Justine G. Franchina, Texas A&M University, College Station, Texas

A. Freund, Degussa AG, Research and Applied Technology Chemical Catalysts and Zeolites, Hanau, Germany

V. Z. Fridman, Reflex, Inc., Brooklyn, New York

Diane E. Froen, NSC Technologies, Mt Prospect, Illinois

C. Fumagalli, Lonza Spa, Scanzorosciate BG, Italy

P. Gallezot, Institut de Recherches sur la Catalyse du CNRS, Villeurbanne, France

S. Galvagno, University of Messina, Messina, Italy

C. Gangemi, University of Messina, Messina, Italy

Michele Gargano, Centro C.N.R MISO, Bari, Italy

Laurent Gilbert, Rhône-Poulenc Industrialisation, Saint-Fons, France

K. M. Gitis, Russian Academy of Sciences, Moscow, Russia

K. G. Griffin, Johnson Matthey PMD-Chemicals, Royston, England

S. P. Griffiths, University of Hull, Hull, England

M. Gross, Degussa AG, Research and Applied Technology Chemical Catalysts and Zeolites, Hanau, Germany

T. F. Guidry, Louisiana State University, Baton Rouge, Louisiana

John Hall, Air Products (Chemicals) Teesside Ltd, Wilton, Middlesbrough, England

Fredrick E. Hancock, ICI Katalco, Billingham, Teeside, England

R. Hartung, Degussa AG, Research and Applied Technology Chemical Catalysts and Zeolites, Hanau, Germany

Stuart Hayden, Merck & Co., Rahway, New Jersey, and Seton Hall University, South Orange, New Jersey

M. Hegedüs, Hungarian Academy of Sciences, Budapest, Hungary

Craig L. Hill, Emory University, Atlanta, Georgia

M. Hillebrand, University of Bucharest, Bucharest, Romania

M. G. Hitzler, University of Nottingham, Nottingham, England

Thomas Q. Hu, Pulp and Paper Research Institute of Canada, Vancouver, Canada

Ian J. Huntingdon, ICI Katalco, Billingham, Cleveland, England

Naseem A. Hussain, ICI Katalco, Billingham, Cleveland, England

Graham L. Hutchings, University of Wales Cardiff, Cardiff, Wales

G. V. Isaguliants, Russian Academy of Sciences, Moscow, Russia

Donna M. Iula, State University of New York at Stony Brook, Stony Brook, New York

S. David Jackson, ICI Katalco, Billingham, Cleveland, England

Roland Jacquot, Rhône-Poulenc Industrialisation, Saint-Fons, France

Brian R. James, University of British Columbia, Vancouver, Canada

P. Johnston, Johnson Matthey PMD-Chemicals, Royston, England

Cs. Keresszegi, Hungarian Academy of Sciences, József Attila University, Szeged, Hungary

Frank King, ICI Katalco, Billingham,Teeside, England

John F. Knifton, Shell Chemical Co., Houston, Texas

T. A. Koch, E. I. DuPont Co., Wilmington, Delaware

W. Koo-amornpattana, University of Birmingham, Edgbaston, Birmingham, England

Scott A. Laneman, NSC Technologies, Mt Prospect, Illinois

Christopher Langham, University of Liverpool, Liverpool, England

R. Larsen, Merck & Co., Inc., Rahway, New Jersey

C. LeBlond, Merck & Co., Inc., Rahway, New Jersey

Darren F. Lee, University of Liverpool, Liverpool, England

Zhaoyang Li, State University of New York at Stony Brook, Stony Brook, New York

Yun-Shan Liu, Texas A&M University, College Station, Texas

Z. Liu, Purdue University, West Lafayette, Indiana

B. Lücke, Institut für Angewandte Chemie Berlin-Adlershof, Berlin, Germany

D. B. MacQueen, TDA Research, Inc., Wheat Ridge, Colorado

Tamas Mallat, Swiss Federal Institute of Technology, Zürich, Switzerland

T. A. Manz, Purdue University, West Lafayette, Indiana

J. L. Margitfalvi, Hungarian Academy of Sciences, Budapest, Hungary

R. K. Mariwala, TDA Research, Inc., Wheat Ridge, Colorado

Andreas Martin, Institut für Angewandte Chemie Berlin-Adlershof, Berlin, Germany

T. Máthé, Hungarian Academy of Science, Hungary

G. Mattioda, Centre de Recherches et d'Applications, Stains, France

M. Messori, Dip. Chimica Industr. e Materiali, Bologna, Italy

C. Milone, University of Messina, Messina, Italy

S. Minicò, University of Messina, Messina, Italy

Arpád Molnár, Hungarian Academy of Sciences, József Attila University, Szeged, Hungary

Claude Moreau, Matériaux Catalytiques et Catalyse, Montpellier, France

M. Ebrahimi-Moshkabad, University of Birmingham, Edgbaston, Birmingham, England

M. G. Musolino, University of Messina, Messina, Italy

Victor L. Mylroie, Eastman Kodak Co., Rochester, New York

M. Bennani Naciri, Laboratoire de Matériaux Catalytiques, Montpellier, France

G. Neri, University of Reggio Calabria, Italy

Iwao Ojima, State University of New York at Stony Brook, Stony Brook, New York

C. Orella, Merck & Co., Inc., Rahway, New Jersey

M. A. Osypian, Eastman Kodak Co., Rochester, New York

Philip C. Bulman Page, Loughborough University, Loughborough, Leicestershire, England

P. Panster, Degussa AG, Research and Applied Technology Chemical Catalysts and Zeolites, Hanau, Germany

Walter Partenheimer, E. I. DuPont Co., Wilmington, Delaware

V. I. Pârvulescu, University of Bucharest, Bucharest, Romania

A. Perrard, Institut de Recherches sur la Catalyse du CNRS, Villeurbanne, France

A. Petride, Institut of Organic Chemistry, Bucharest, Romania

Paola Piaggio, University of Wales Cardiff, Cardiff, Wales

O. Piccolo, Chemi Spa, Balsamo MI, Italy

L. H. Pignolet, University of Minnesota, Minneapolis, Minnesota

A. Pistone, University of Messina, Messina, Italy

M. Poliakoff, University of Nottingham, Nottingham, England

Y. Pouilloux, Laboratoire de catalyse, UMRS CNRS, Poitiers, France

G. L. Price, Louisiana State University, Baton Rouge, Louisiana

S. N. Thomas-Pryor, Purdue University, West Lafayette, Indiana

D. Ramprasad, Air Products and Chemicals Inc., Allentown, Pennsylvania

S. N. R. Rao, I² Technology, Dallas, Texas

Nicoletta Ravasio, Centro C.N.R. SSSCMTBO, Milano, Italy

S. Raymahasay, University of Birmingham Edgbaston, Birmingham, England

Sylvie Razigade, Matériaux Catalytiques et Catalyse, Montpellier, France

C. Rehren, Degussa AG, Research and Applied Technology Chemical Catalysts and Zeolites, Hanau, Germany

S. K. Ross, Thomas Swan & Co. Ltd., Country Durham, England

L. I. Rubinstein, University of Minnesota, Minneapolis, Minnesota

R. Ruiz, Universidad de Granada, Cordoba, Spain (on leave)

A. B. Schabel, University of Minnesota, Minneapolis, Minnesota

S. K. Sengupta, E. I. DuPont Co., Wilmington, Delaware

S. Sharma, University of Birmingham, Edgbaston, Birmingham, England

Roger Sheldon, Delft University of Technology, Delft, The Netherlands.

Urs Siegrist, Novartis Service AG, Basel, Switzerland

Chris Sly, Robinson Brothers Ltd., West Bromwich, West Midlands, England

F. Smail, University of Nottingham, Nottingham, England

G. V. Smith, Southern Illinois University, Carbondale, Illinois

R. Song, Southern Illinois University, Carbondale, Illinois

John R. Sowa, Jr., Seton Hall University, South Orange, New Jersey

Michel Spagnol, Rhône-Poulenc Industrialisation, Saint-Fons, France

G. Srinivas, TDA Research, Inc., Wheat Ridge, Colorado

Martin Studer, Novartis Service AG, Basel, Switzerland

Steven L. Suib, University of Connecticut, Storrs, Connecticut

Y.-K Sun, Merck & Co., Inc., Rahway, New Jersey

S. Szabó, Hungarian Academy of Sciences, Budapest, Hungary

Gerda Szakonyi, Albert Szent-Györgyi University, Szeged, Hungary

T. Tacke, Degussa AG, Research and Applied Technology Chemical Catalysts and Zeolites, Hanau, Germany

A. Tagliani, Dip. Chemica Industr. e Materiali, Bologna, Italy

E. Tálas, Hungarian Academy of Sciences, Budapest, Hungary

Setrak K. Tanielyan, Seton Hall University, South Orange, New Jersey

E. Tfirst, Hungarian Academy of Sciences, Budapest, Hungary

D. Tichit, Laboratoire de Matériaux Catalytiques, Montpellier, France

J. Toler, Thomas Swan & Co. Ltd., Country Durham, England

A. Tompos, Hungarian Academy of Sciences, Budapest, Hungary

Béla Török, Hungarian Academy of Sciences, József Attila University, Szeged, Hungary

Harald Trauthwein, Technische Universität München, Garching, Germany

A. Tungler, Technical University of Budapest, Budapest, Hungary

J. Tyrrell, Southern Illinois University, Carbondale, Illinois

Maria Tzamarioudaki, State University of New York at Stony Brook, Stony Brook, New York

A. Vaccari, Dip. Chimica Industr. e Materiali, Bologna, Italy

L. F. Valente, Eastman Kodak Co., Rochester, New York

J. C. Vallejos, Centre de Recherches et d'Applications, Stains, France

C. Vanhove, Laboratoire de catalyse, UMR CNRS, Poitiers, France

L. Vishwakarma, Eastman Kodak Co., Rochester, New York

Paul C. Vosejpka, Dow Chemical Co., Midland, Michigan

F. J. Waller, Air Products and Chemicals, Inc., Allentown, Pennsylvania

J. Wang, Merck & Co., Inc., Rahway, New Jersey

Jin-Yun Wang, University of Connecticut, Storrs, Connecticut

Y. Wang, Southern Illinois University, Carbondale, Illinois

Yu Wang, University of British Columbia, Vancouver, Canada

Derrick J. Watson, BP Chemicals, Hull, England

Peter B. Wells, University of Hull, Hull, England

M. G. White, Georgia Institute of Technology, Atlanta, Georgia

S. Wieland, Degussa AG, Research and Applied Technology Chemical Catalysts and Zeolites, Hanau, Germany

David J. Willock, University of Wales Cardiff, Cardiff, Wales

Robert Wilson, Air Products (Chemicals) Teesside Ltd., Wilton, Middlesbrough, England

J. M. Winterbottom, University of Birmingham, Edgbaston, Birmingham England

Guan-Guang Xia, University of Connecticut, Storrs, Connecticut

Yuan-Gen Yin, University of Connecticut, Storrs, Connecticut

L. Zhang, University of Birmingham, Edgbaston, Birmingham, England

Xuan Zhang, Emory University, Atlanta, Georgia

1998 Paul N. Rylander Plenary Lecture.
New Applications for Inorganic Solid Acid Catalysis

John F. Knifton

Shell Chemical Company, Houston, Texas

Abstract

The effectiveness of inorganic solid acid catalysis for three industrially important syntheses: selective phenol/acetone cogeneration via cumene hydroperoxide cleavage, methyl *tert*-butyl ether production from *tert*-butanol, and isobutene separation from mixed C-4 streams, is described. Preferred classes of catalysts include acidic montmorillonite clays and heteropoly acids impregnated into clay and Group IV oxide structures.

Introduction

Solid acid catalysis finds ever expanding usage as the chemical industry seeks more selective and more environmentally benign process technology [1]. In this plenary lecture we describe the use of inorganic solid catalysts for three industrially-important syntheses:

- Phenol/acetone cogeneration via cleavage of cumene hydroperoxide (CHP)(eq 1).
- Methyl *tert*-butyl ether (MTBE) production from *tert*-butanol (eq 2).
- Isobutene separation from mixed C-4 streams (eq 3).

Classes of inorganic solid acid catalysts found to be most effective for these syntheses include acidic montmorillonite clays and heteropoly acids impregnated into clay or Group IV oxide structures.

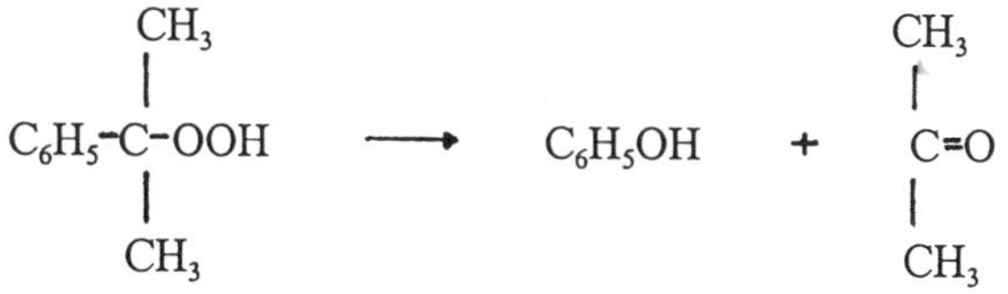

$$C_6H_5\text{-}\underset{\underset{CH_3}{|}}{\overset{\overset{CH_3}{|}}{C}}\text{-}OOH \longrightarrow C_6H_5OH + \underset{\underset{CH_3}{|}}{\overset{\overset{CH_3}{|}}{C}}=O \qquad (1)$$

$$CH_3\text{-}\underset{\underset{CH_3}{|}}{\overset{\overset{CH_3}{|}}{C}}\text{-}OH + CH_3OH \longrightarrow CH_3\text{-}\underset{\underset{CH_3}{|}}{\overset{\overset{CH_3}{|}}{C}}\text{-}OCH_3 + H_2O \qquad (2)$$

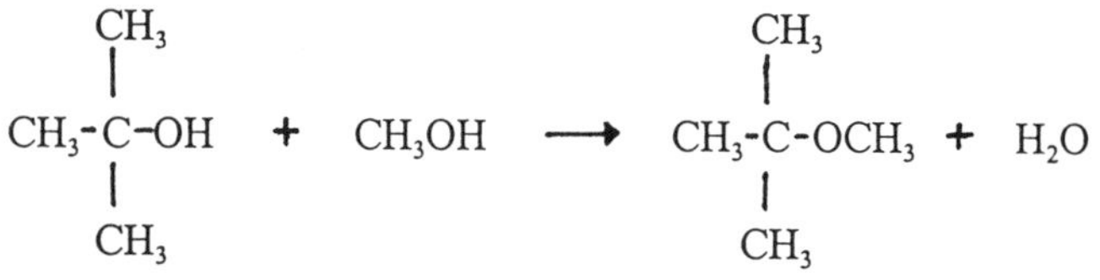

1

$$\diagup \!\!\!\!\diagdown C{=}CH_2 \ + \ \begin{array}{c} CH_2OH \\ | \\ CH_2OH \end{array} \ \longrightarrow \ \begin{array}{c} CH_2O{-}C\diagup\!\!\!\!\diagdown \\ | \\ CH_2OH \end{array} \ \longrightarrow \ \begin{array}{c} CH_2OH \\ | \\ CH_2OH \end{array} \ + \ \diagup\!\!\!\!\diagdown C{=}CH_2 \quad (3)$$

Phenol/Acetone Cogeneration

The cumene peroxidation process for phenol/acetone cogeneration accounts for more than 90% of the phenol produced commercially worldwide [2]. The principal elements of the process include cumene oxidation with air to give cumene hydroperoxide, cleavage of the concentrated CHP to phenol plus acetone, followed by further purification via downstream fractionation. The cleavage step is rapid and highly exothermic (Δ H_{298} = -52.9 Kcal/g/mole CHP). The commercial acid catalyst most commonly used is concentrated (98%) sulfuric acid, although other suitable catalysts include strong organic acids such as sulfonic acids, sulfur dioxide dissolved in cumene, and hydrogen ion-exchanged resins [3].

The mineral acid catalyzed process for cumene hydroproxide cleavage, first commercialized in the 1950's, has certain intrinsic disadvantages in today's 'clean technology' environment, problems that include the disposal of salt by-products, the potential for acid entrainment into the phenol/acetone purification train, the operating and maintenance costs associated with downstream acid removal steps, and fourthly, the metallurgical costs for those kettles, towers, etc., handling acidic streams. Clearly there are potentially significant cost savings that could accrue from the development of suitable solid acid catalysts for reaction (1).

We have previously disclosed four classes of novel solid acid catalysts for the selective cleavage of cumene hydroperoxide to phenol plus acetone. They include:

• Mineral acid-treated montmorillonite clays [4].
• Montmorillonite clays modified with heteropoly acids or certain Lewis acids [5].
• Group IV oxide supports impregnated with heteropoly acids [6].
• Fluorophosphoric acid and HF-treated oxides [7].

Table 1 illustrates the performances of a typical, commercially-available, acidic montmorillonite clay in CHP cleavage under continuous, plug-flow, conditions.

Base line conditions are detailed in ex. 1, variations in operating temperature, pressure and feed rate are illustrated in ex. 2-6. Due to the very exothermic nature of the cumene hydroperoxide cleavage reaction, the 78.5 wt % CHP feed stock (also containing 16.5 wt % cumene, 4.7 wt % 2-phenol-2-propanol and 0.4 wt % acetophenone) was diluted to ca. 23% concentration with acetone/cumene/phenol mixture (64:24:92 weight ratio). In each experimental series the cumene hydroperoxide conversion levels remain quantitative over the range of conditions considered, even at high feed rates, i.e. LHSV's of 10 (ex. 2-5). The estimated phenol yield in base line ex. 1 is 99 mole %, the estimated acetone yield is 97 mole %. Material balances are consistently > 99%. Acetophenone, α-methylstyrene and 4-cumylphenol concentrations rise only slightly during peroxide decomposition, the 2-phenyl-2-propanol fraction in the crude CHP is almost quantitatively converted, and most notably, we see no evidence of mesityl oxide formation with the montmorillonite clay catalysts. At higher operating temperatures (80°C, ex. 3) the phenol yields drop to 94 mole %; pressure appears to have little influence on yield structure (typical phenol yields in ex. 4 and 5 are 97 and 95%, respectively). While the reactor system was jacketed and cooled with Dowtherm, the exothermic heat led to temperature profiles along the catalyst bed of up to 25°C above set point at LHSV 1 and as much as 45°C at LHSV 10. Even so, the clay Grade F24 maintained good performance for at least 1000 hours of continuous service, with no need for catalyst regeneration or replacement [4]. The recovered granular clay was discolored with heavy organics, but generally in excellent physical condition.

Further significant improvements in the performances of the acidic montmorillonite clays were discovered by impregnating these layer structures with various Lewis and Bronsted acids, particularly with heteropoly acids such as 12-tungstophosphoric acid and 12-molybdophosphoric acid [5]. However, these improvements were most evident with respect to the modified clay's initial activity for CHP cleavage.

Data in Figure 1 illustrate the performance of clay Grade F62, 1/16" diameter, extrudates modified with 12-tungstophosphoric acid. The evaluation was completed at three space velocities. Even at the highest feed rates, the calculated phenol and acetone yields exceeded 95 mole %. Reproducibility between batch-to-batch samples of this heteropoly acid-modified clay was excellent.

A variety of heteropoly acids impregnated into Group IV oxide supports have also been found to be useful catalysts for phenol/acetone cogeneration. Data in

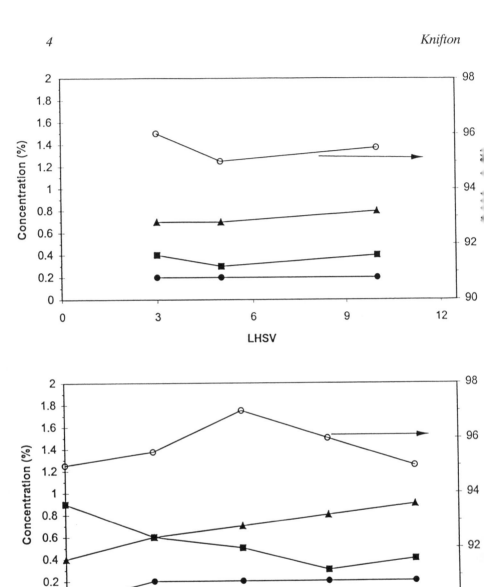

FIGURES 1 and 2: Phenol/acetone cogeneration using 12-tungstophosphoric
acid-on-clay Grade F62 and 12-molybdophosphoric acid-on-titania.
Designations: phenol, o; α-methylstyrene, ▲; 4-cumylphenol, ■; acetophenone, ●.
Operating conditions: 60°C, 20 bar.

Figure 2 are illustrative of the performances of 12-molybdophosphoric acid (Mo-P)-on-titania in the same continuous, plug-flow, reactor described *supra.* Equivalent performance was realized over five days of uninterrupted service and at higher feed rates. On the fourth day of operation, for example, effluent sample analyses show essentially complete cumyl hydroperoxide conversion with estimated phenol and acetone yields of 96 and 99 mole %, respectively. However, the slightly lower activity of the 12-molybdophosphoric acid-on-titania at higher LHSV's - in comparison with, for example, 12-tungstophosphoric acid-on-montmorillonite clay - is evidenced by the fact that at LHSV of ten, the cumyl hydroperoxide conversion level drops to ca. 97%. 2-Phenyl-2-propanol conversion to α-methylstyrene and cumylphenol are also acid-catalyzed reactions [8] and each class of solid acid catalyst is very effective in this regard (see Figures 1 and 2). On the other hand, none of the solid acids tested in this work gave evidence of acetone dehydration to mesityl oxide (e.g. Table 1), which is normally also an acid catalyzed process. Possibly the mesityl oxide is undergoing rapid further condensation to undetected, higher MW products. Formation of small, but measurable, quantities of acetophenone is consistent with free-radical pathways being important with the solid acid catalyzed CHP cleavage [9]. Overall, however, the product distributions summarized in Table 1 and Figures 1 and 2 are consistent with a classical acid-catalyzed CHP cleavage mechanistic scheme [10]. The significant differences in product distribution and catalyst performance between the preferred solid acids versus traditional mineral acids include significantly higher phenol yields and selectivities, no mesityl oxide, lower acetophenone production, and less heavy end oligomers.

In conclusion, the performances of the acidic clay/heteropoly acid-modified clay catalysts, having demonstrated long life characteristics, in comparison with traditional mineral-acid processes for CHP cleavage, provide both significant technological improvements, as well as potentially important process cost savings. These technology/cost improvements include:

- Measurably higher yields of phenol per pass.
- No by-product salt disposal problems or costs.
- Simpler downstream processing - since there is no longer a need for acid removal and recovery equipment.
- Less stringent metallurgy requirements - particularly in the CHP cleavage and phenol/acetone recovery sections.

The economic benefits to using this solid acid technology for CHP cleavage have been reviewed and highlighted elsewhere [11].

Table 1
Phenol/Acetone Cogeneration Using Mineral Acid-Treated Montmorillonite Clays, Continuous Studies[a]

Ex.	Clay Catalyst	Temp. (°C)	Pressure (bar)	Feed rate (LHSV)	Sample	Feed / Product Composition (wt. %)								
						Acetone	Mesityl Oxide	Cumene	Phenol	α-Methyl Styrene	Aceto-phenone	2-Phenyl-2-Propanol	4-Cumyl-Phenol	Cumene Hydro-Peroxide
1	Grade F24	60	20	1	F[b]	24.3	--	13.8	36.6	--	0.1	1.4	0.1	23.2
					P[b]	33.3	--	14.0	50.8	0.4	0.2	--	0.7	--
2	Grade F24	50	20	10	F	24.6	--	13.7	35.7	--	0.1	1.6	--	23.9
					P	33.9	--	13.7	50.2	0.8	0.3	--	0.7	--
3	Grade F24	80	20	10	F	24.7	--	13.8	35.5	--	0.2	1.5	--	24.0
					P	34.2	--	13.8	49.5	0.9	0.3	--	0.7	--
4	Grade F24	60	65	10	F	24.9	--	13.7	35.7	--	0.1	1.6	--	24.0
					P	34.1	--	13.7	49.8	0.9	0.3	--	0.6	--
5	Grade F24	60	7	10	F	24.6	--	13.6	35.2	--	0.1	1.6	--	24.5
					P	33.8	--	13.8	49.9	0.9	0.3	--	0.6	--
6	Grade F24	60	20	3	F	24.4	--	14.1	37.2	--	0.1	1.5	--	22.7
					P	33.4	--	14.1	50.5	0.9	--	0.2	0.3	--

[a] Continuous syntheses conducted using cumene hydroperoxide (78.5%) diluted to ca. 23% with acetone/cumene/phenol (64:24:92) mix.

[b] F, Feed composition; P, crude effluent product composition, determined by GLC analyses.

MTBE from *tert*-Butanol
Both supported heteropoly acids and montmorillonite clays are also very effective inorganic solid acid catalysts for the industrially-important generation of methyl *tert*-butyl ether from *tert*-butanol/methanol mixtures (eq 2). In comparison with traditional hydrogen ion-exchange resins, inorganic solid acids - with their greater thermal stability - have the intrinsic advantages of:

1. Allowing etherification at higher operating temperatures (> 140°C) where rates of MTBE formation are considerably more attractive [12].
2. Optionally producing isobutylene as an attractive coproduct [13].

Typical syntheses of MTBE from *tert*-butanol (tBA)/methanol (1:2 molar) liquid feeds are illustrated in Table 2 for a variety of oxide-impregnated 12-molybdophosphoric acid (Mo-P) and 12-tungstophosphoric acid (W-P) catalyst compositions. Continuous, plug-flow, studies over a range of etherification temperatures (100-180°C), yield MTBE with moderate to high levels of *tert*-butanol conversion even at LHSV's of 4-8 (ex. 7-8). While MTBE molar selectivities may reach 90 mole % at 100 C, increasing temperatures lead to greater isobutene (i-C_4) by-product make, so that by 180°C, the crude effluents comprise two-phase liquid mixtures - a heavier, aqueous MTBE-MeOH rich phase and an isobutene-rich (25-40% i-C_4), lighter phase (ex. 8)[13].

The preparation, calcination, and analysis of each oxide-impregnated heteropoly acid catalyst has been detailed elsewhere [12].

Isobutene Separation
In recent years there has been renewed interest in technology for efficiently separating butenes present in crude C-4 streams generated via steam and catalytic cracking operations. For isobutene, there are currently three important separation processes, including: (1) an extraction route using a mineral acid to isolate the isobutene, (2) the dehydration of *tert*-butanol, and (3) the cracking of methyl *tert*-butyl ether [14].

We have developed a 2-step procedure for separating isobutene from mixed C-4 hydrocarbon streams, such as Raffinate-1, using inorganic solid acid catalysis. The procedure involves initial etherification of the C-4 mixture with a suitable diol, such as ethylene glycol (eq 3) or 1,2-propylene glycol, to give the corresponding glycol mono *tert*-butyl ethers, followed by deetherification of said mono ethers at higher temperatures to yield pure isobutene plus regenerated glycol. Suitable classes of solid acid catalysts once again include

Table 2
Methyl tert-Butyl Ether Synthesis Using Oxide-Supported Heteropoly Acid Catalysts[a]

Ex.	Catalyst	Temp. (°C)	LHSV	Sample	----- Feed / Product Composition (wt. %) -----					tBA Conv. (%)	MTBE Selectivity (mole%)
					MTBE	i-C$_4$	MeOH	tBA	H$_2$O		
7	Mo-P / TiO$_2$			F[b]			47.5	52.4			
		100	1.0	P[b]	23.2	2.2	38.5	30.2	6.0	42	87
		120	1.0	P	35.6	5.0	32.5	16.1	10.6	69	82
		150	1.0	P	36.2	7.6	32.1	11.7	11.6	78	75
		150	4.0	P	34.8	6.5	33.5	14.4	10.7	73	77
8	W-P/TiO$_2$			F			47.1	52.7			
		100	1.0	P	28.6	1.4	37.0	26.1	7.0	50	90
		150	1.0	P	36.9	6.7	32.5	12.2	11.6	77	74
		180	1.0	P[c]	24.0	8.4	43.4	5.8	18.2	[c]	[c]
		150	8.0	P	26.1	4.0	37.5	23.4	8.8	56	75
9	W-P/SiO$_2$			F			46.4	53.5			
		150	1.0	P	45.9	8.7	27.9	10.0	7.5	81	88
10	W-P/Al$_2$O$_3$			F			45.9	54.0			
		150	1.0	P	39.7	6.2	31.8	12.8	9.4	76	81

[a] Continuous syntheses: 20 bar.

[b] F, Feed composition; P, crude effluent product composition, determined by GLC analyses.

[c] Two-phase product liquid, analysis of heavier phase, lighter phase is rich in isobutene (i-C$_4$).

acidified montmorillonite clays and heteropoly acids dispersed on Group IV oxides [15,16].

The scope of both the glycol mono *tert*-butyl ether synthesis and the isobutene regeneration is illustrated in Figures 3 and 4, using 1,2-propylene glycol as coreactant and 12-tungstophosphoric acid-on-titania as catalyst. Both syntheses steps are conducted in plug-flow continuous reactors [14]. The Raffinate-1 feedstock typically contains 12.7% isobutene. After etherification the heavier propylene glycol *tert*-butyl ethers (PGTBE) - rich glycol phase (Figure 3) is separated from the lighter Raffinate-1 phase and passed separately through a similar continuous reactor also charged with 12-tungstophosphoric acid-on-titania. Typically, the etherification step is conducted in the range 80-120°C. Figure 3 illustrates the estimated isobutene conversion levels in the Raffinate-1 phase and the typical concentrations of PGTBE product in the heavier glycol-rich phase. PGTBE deetherification was examined at higher temperatures (ca. 120-180°C) and the influence of these higher temperatures upon the levels of PGTBE conversion and the concentrations of liberated isobutene are illustrated in Figure 4. Here there is an excellent parallel between the level of PGTBE undergoing deetherification and the total isobutene liberated.

The chemical and dynamic nature of the 12-tungstophosphoric acid species adsorbed on the formed titania utilized in these etherification/deetherification chemistries has been examined by ^{31}P and ^{1}H solid state NMR spectroscopy, as well as diffuse reflectance FT-IR spectroscopy. Analyses of the MAS ^{31}P data indicates there are at least five phosphorus species on the fresh and used catalyst surfaces. These include a bulk salt phase, two weakly-bound intact Keggin unit species, a strongly bound partially fragmented Keggin unit and a pure phosphate phase formed by complete fragmentation of the Keggin unit to tungstate and phosphate [17]. ^{1}H NMR suggests that adsorption of the 12-tungstophosphoric acid on the titania surface leads to an increase in the acidity of the Keggin unit acid protons (11.4 ppm). With increasingly effective adsorption of the Keggin ion there is an accompanying dispersion in the types of acid sites present on the surface, along with possibly a decrease in the mobility of the acidic protons on the surface [17].

Related changes in oxide-supported heteropoly acid structure have been identified during their usage in MTBE production (eq 2) as well as CHP cleavage to phenol/acetone.

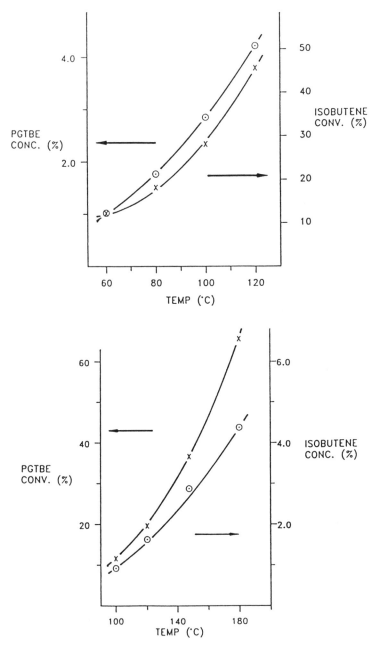

FIGURES 3 and 4: Production of propylene glycol *tert*-butyl ethers (PGTBE) from Raffinate-1 plus 1,2-propylene glycol and the liberation of isobutene from crude PGTBE using 12-tungstophosphoric acid-on-titania catalysts.

References
1. K. Tanabe. *Applied Catal.*, **113**, 147 (1994).
2. Stanford Research Institute Report No. 22C, March 1991, pp. 2-1.
3. Chem Systems Report No. 88-6, March 1990, pp. 64.
4. J. F. Knifton, U.S. Patent 4,870,217 to Texaco Chemical Company (1989).
5. J. F. Knifton, U.S. Patent 4,898,987 to Texaco Chemical Company (1990).
6. J. F. Knifton, U.S. Patent 4,898,995 to Texaco Chemical Company (1990).
7. J. F. Knifton and N. J. Grice, U.S. Patent 4,876,397 to Texaco Chemical Company (1989).
8. P. L. Beltrame, P. Carniti, A. Gramba, O. Cappellazo, L. Lorenzoni, and G. Messina. *Ind. Eng. Chem. Res.*, **27**, 4 (1988).
9. G. Lauterbach, W. Pritzkow, T. D. Tien, and V. Voerckel. *J. f. prakt. Chemie,* **330**, 933 (1988).
10. W. Kiessling, I. Krafft, K. K. Moll, and K. Pelzing. *Chem. Tech. (Leipzig),* **23**, 423 (1971).
11. Chem Systems PERP Report 91-8, March 1993, pp. 1.
12. J. F. Knifton, U.S. Patent 4,827,048 to Texaco Chemical Company (1989).
13. J. F. Knifton, U.S. Patent 5,157,161 to Texaco Chemical Company (1992).
14. J. F. Knifton. *Applied Catal. A* , **109**, 247 (1994).
15. J. F. Knifton, U.S. Patent 5,146,041 to Texaco Chemical Company (1992).
16. J. F. Knifton, U.S, Patent 5,177,301 to Texaco Chemical Company (1993).
17. J. C. Edwards, C. Y. Thiel, B. C. Benac, and J. F. Knifton, presented by J. F. Knifton, Am. Chem. Soc. Southwest Regional Meeting, Houston, 1996.

Selective Esterification and Etherification of Glycerol Over Solid Catalysts

J. BARRAULT , Y. POUILLOUX, C. VANHOVE , K. COTTIN, S. ABRO , J-M. CLACENS

Laboratoire de catalyse en chimie organique, UMR 6503, ESIP, 40, avenue du recteur Pineau, 86022 Poitiers cedex, France.

Abstract : The selective preparation of linear di- or triglycerol in the presence of solid catalysts is an interesting procedure for valorizing glycerol. Polyglycerols or polyglycerol esters can be used for the manufacture of non-ionic tensioactives for detergents or of additives to lubricants. A study of the etherification of glycerol carried out in the presence of different resins showed that diglycerol was the main product. However, the dehydration of glycerol to acrolein over acid catalysts was observed. The direct etherification of glycerol into di- or triglycerol was less selective over homogeneous and heterogeneous base catalysts (oxides, hydroxides or carbonates). Polyglycerols were also formed in the presence of base zeolites but formation of the main by-product, acrolein, was not observed. Indeed, di- and triglycerol were selectively formed over NaX or Cs exchanged zeolites.

1. Introduction

Glycerol is mainly a natural product issued from the methanolysis of vegetable oils . In Europe, due to the increasing use of methyl esters as fuel additives, one can expect an increase of glycerol production which could become a cheaper raw material for chemistry (1). For example, polyglycerols (PG) and especially polyglycerols-esters (PGEs) are becoming very important among the new products in the manufacture of tensioactive , lubricants , cosmetics , food additives, ... Indeed, PGEs could exhibit multifunctional properties if it was possible to control : i) the length of the polyglycerols chain , ii) the degree of esterification and iii) the fatty acid molecular weight. These reactions are quite interesting objectives for shape - selective catalytic processes.

In our laboratory, we are involved in a general program concerning the selective transformation of fatty acids (or of methyl esters) and glycerol issued from sunflower oil (2,3). We have studied the esterification of glycerol with oleic acid in the presence of solid catalysts (4). Monoglycerides are selectively formed over cationic exchange resins under mild conditions. Polyglycerols and polyglycerol esters as well as acrolein, issued from glycerol dehydration, were obtained as main by-products. The modification of the pseudo-pore size of the cationic exchange resin improved the selectivity to (PG + PGEs) but acrolein was always produced over these acid catalysts (4).

To control the selective formation of diglycerol, triglycerol or polyglycerols, we investigated first the selective etherification of glycerol over solid catalysts :

Polyglycerols are generally obtained i) from the etherification of glycidol with polyol or glycerol in the presence of acid or base catalysts (5,6) or ii) from the reaction between the epichlorohydrin and glycerol over base catalysts such as NaOH or Na_2CO_3 (7,8). The direct polymerization of glycerol can be performed at high temperature but cyclic by-product or dehydrated compounds can be formed (9-12). Generally, homogeneous catalysts used in the polymerization of glycerol are corrosive, difficult to separate from the reaction medium and lead to the formation of excessive wastes (salts). Garti et al. classified different base catalysts according to their activity in glycerol etherification (13) : K_2CO_3 > Li_2CO_3 > Na_2CO_3 > KOH > NaOH > CH_3ONa > $Ca(OH_2)$ > LiOH > $MgCO_3$ > MgO > CaO > $CaCO_3$.

The objective of our work was to find new solid catalysts in order to control the selectivity of the reaction, di- or triglycerol being the compound we aim to obtain. We present in this paper the results obtained with solid base catalysts in the direct polymerization of glycerol without solvent. We compare the behavior of different catalysts (ion exchange resin or zeolite) with usual catalysts, in particular, Na_2CO_3. The influence of the zeolite structure and the basicity of exchanged alkaline catalysts are discussed.

2. Experimental

2.1. Catalytic tests.
The polymerization reaction was carried out at atmospheric pressure in a glass-reactor equipped with a mechanical stirrer. The water formed during the reaction was eliminated using a Dean-Stark apparatus.
The reaction was performed at 240-260°C for 8 hours with 50g of glycerol under nitrogen, the weight percentage of catalysts was around 2-4%.

supplied by SGE (L = 25 m, ID = 0.22 mm, thickness of film = 0.10 μm). The molar percentage of each compound was determined by using standardization methods with methyl laurate as an internal reference.

2.2. Catalysts

The catalysts used in this study were homogeneous bases (KOH, Na_2CO_3, La_2O_3), zeolites exchanged or not with alkaline elements and cationic exchange resins. The characteristics of the zeolites are reported on Table 1.

Table 1 : Characteristics of zeolites

Zeolite	Si/Al	pore size (Å)	specific surface area (m^2/g)
HZSM5	18.0	5.5	430
HY ZF110	2.7	7.4	800
HY ZF 510	15.0	7.4	800
NaX	1.2	13.0	780
Mordenite	5.0	7	330

The exchange reaction was performed with a nitrate salt solution for 24 hours at room temperature. The zeolites were then calcined under air at 500°C for 12 hours.

3. Results

3.1. Comparison between usual base catalysts

In the first part of our study, the polymerization of glycerol over different base catalysts (carbonate, hydroxide and oxide) was studied. Blank tests showed that, between 240 and 260°C, the glycerol was not transformed into polyglycerol. Figure 1 shows that Na_2CO_3 and KOH are very active since the conversion of glycerol is around 90% after 9 hours reaction. In agreement with previous results reported in the literature, we can observe that carbonate is more active than hydroxide and oxide.

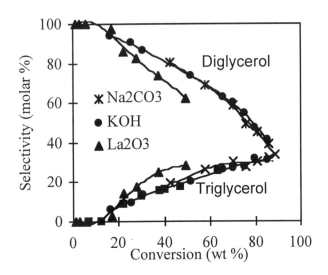

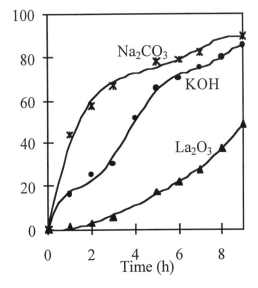

Figure 1 : Polymerization of glycerol in the presence of usual base catalysts. Variation of the conversion with reaction time. Temperature : 240°C, catalyst : 4 (wt%)

Figure 2 : Polymerization of glycerol in the presence of usual base catalysts. Selectivity to diglycerol and triglycerol versus glycerol conversion. Temperature : 240°C, catalyst : 4 (wt %).

The main products of the reaction were diglycerol and triglycerol but the formation of cyclic diglycerol, tetra- and pentaglycerol was also observed. Figure 2 which presents the molar selectivity to di- and triglycerol shows that the selectivity is quite independent of the type of catalyst. The selectivity to diglycerol decreases with the conversion. At the end of the reaction, an equimolar ratio of di- and triglycerol is obtained and the formation of higher polyglycerols is also detected. As reported in the literature, these catalyts are not selective for the formation of low molecular weight polyglycerols.

3.2. Cationic exchange resins
In our previous works, we showed that polyglycerols could be formed in the presence of cationic exchange resins (4). Different resins (characteristics reported in Table 2) were studied.

Table 2 : Characteristics of cationic exchange resins used in the polymerization of glycerol

Resin	Supplier	Type	Crosslinking level[a] (%)	Acidity (meq H^+/g)	Particle size (mm)
K1481	Bayer	gel	8	4.8	powder < 0,05
Amberlyst 31	Rohm & Haas	gel	4	4.8	1.2 to 1.3
Amberlyst 16	Rohm & Haas	macroporous	12	5.0	1.2 to 1.6

[a] % Divinylbenzene (DVB) added to the polymer matrix.

Table 3 shows that if resins have a similar activity, the selectivity varies with the catalyst structure. Indeed, the macroporous resin (Amberlyst 16) gives mainly diglycerol whereas the resins with a gel structure favor the formation of triglycerol. It is well known that the polymer resins can swell in the presence of a polar medium. Both the pore size and the swelling properties, depending on the crossliking level, can explain these results. Owing to the acid sites of the catalyst there was a significant formation of acrolein :

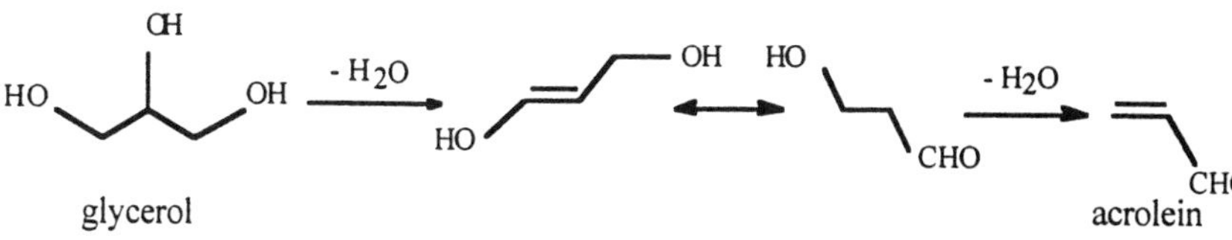

glycerol   acrolein

Table 3 : Etherification of glycerol in the presence of cationic exchange resins

Resins	Conversion	Selectivity (%)	
	(%)	Diglycerol	Triglycerol
Amberlyst 16	40	85	15
Amberlyt 31	35	75	25
K 1481	35	75	25

Temperature : 140°C, Catalyst : 5 (wt %)

3.3. Zeolites

To limit the formation of acrolein, the polymerization of glycerol was performed in the presence of different sodium zeolites. Figure 3 shows that the glycerol conversion varies with the zeolite structure. Indeed, NaX is the best zeolite since the glycerol conversion attains 70% after 9 hours reaction. NaHZSM5 and NaHY zeolites are not totally exchanged with Na. Therefore, even if the acido-base properties of zeolites are modified, the formation of acrolein is not suppressed, especially in the presence of NaHZSM5.

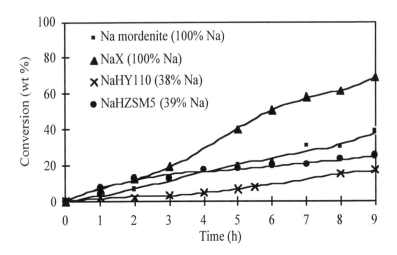

Figure 3 : Etherification of glycerol in the presence of Na-zeolites. Variation of the conversion with reaction time. Temperature : 260°C, catalyst :(4 %).

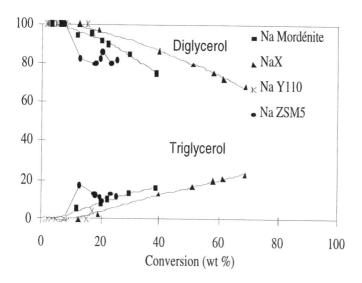

Figure 4 : Etherification of glycerol in the presence of Na-zeolites. Selectivity to di- and triglycerol versus glycerol conversion. Temperature : 260°C, catalyst : 4 % (weight).

Diglycerol is the main compound produced over NaX zeolite (Figure 4). Apparently the use of a catalyst with a controlled porosity therefore increases the selectivity to di- and triglycerol. On the contrary, Na-Mordenite and NaZSM5 favor the formation of triglycerol and in this case, it has been suggested that owing to the small pore size of Mordenite or ZSM5 the polymerization of glycerol occured on their external surface. It must also be remarked that in the presence of NaX, there is no dehydration of glycerol.

3.3. Alkaline exchanged zeolite
To determine the influence of alkaline elements, the HY ZF110 zeolite (Si/Al = 2.7) was exchanged with a constant atomic percentage of alkaline elements (20%). The reaction was performed at 260°C with 4 (wt %) of catalyst.

a) Effect of alkaline elements
Table 4 shows that the conversion of glycerol increases with the basicity of the alkaline element : CsY > KY > NaY. This results is in agreement with previous works proposed by Barthomeuf et al. (15). Indeed, the basicity of the alkaline element increases with the size of the ionic radius, cesium being the most basic. Furthermore, the introduction of an alkaline element modifies the hydrophilic properties of the zeolite and consequently the adsorption rate of glycerol.

On the other hand, the selectivity to di- and triglycerol is not modified by an agent added to the zeolite.

Table 4 : Etherification of glycerol in the presence of a HY ZF110 zeolite exchanged with alkaline elements.

	Conversion (wt %)		
Time (h)	alkaline element Na	K	Cs
3	3.5	5.2	36.2
5	7	7.6	58.3
9	16	21.5	79.9
Content (wt %))	4.5	7.6	34.1

Temperature : 260°C, catalyst : 4 (wt %)

b) Effect of the exchange level

Figure 5 shows that the conversion of glycerol increases with the exchange level of cesium. We can see that the protonic HY zeolite is not active under the conditions of the reaction. At a low exchange level, the reduced activity of the catalyst can be explained by i) a strong adsorption of glycerol on acid sites, ii) a catalyst deactivation due to the formation of coke (polymerisation of acrolein) or iii) a limitation of the accessibility of the base sites to glycerol especially those located in the sodalite cages of the zeolite. When the cesium percentage increases, the sites located on the external surface can be exchanged and increase the basicity of the catalyst. On the other hand, whatever the amount of the cesium it has no effect on the selectivity to di- and triglycerol.

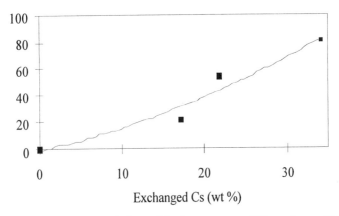

Figure 5 : Etherification of glycerol in the presence of Cs exchanged Y zeolite. Effect of exchange level.

c) Effect of the Si/Al ratio of the CsHY zeolite

To study the effect of the Si/Al ratio, two zeolites differing by their Si/Al ratio were exchanged (same exchange level). Table 5 shows that the conversion of glycerol decreases with the Si/Al ratio. Owing to the variation of the number of available aluminum atoms, for a same exchange level, the weight percentage of Cs in the HY ZF110 is higher than the one observed with the HY ZF510. Moreover, the variation of the Si/Al ratio modified the hydrophilic properties of the zeolites and thus, the adsorption rate of glycerol ; one can expect a variation of the total reaction rate. Whatever the Si/Al ratio, the selectivity to di- and triglycerol is comparable.

Table 5 : Etherification of glycerol in the presence of Cs-HY zeolite. Effect of the Si/Al ratio.

	Si/Al	Conversion (%)		Selectivity
		2 (h)	7 (h)	(20% conv.)
HY ZF110	2.7	21	71	94
HY ZF510	15.0	3	17	91

Temperature : 260°C, catalyst : 4 (wt %)

d) Effect of the structure of Cs-zeolites

The etherification reaction of glycerol is greatly dependent on the hydrophilic character of the catalyst but it can be also affected by the zeolite structure. We compared three Cs exchanged zeolites with different structures (exchange level : 50%). Table 6 shows that the conversion of glycerol is similar (around 80%) over CsNaX and CsHY but only of 30% over CsZSM5. The order of activity - CsNaX ≈ CsHY > CsZSM5 - is quite in agreement with the one already proposed by Barthomeuf (16) or Kaliaguine (17) concerning the basicity of these catalysts

Moreover, the structure of the zeolite does not modify the selectivity to di- and triglycerol (Figure 6). It seems that the reaction occurs at the surface of the catalyst. Indeed, at a constant exchange level, the Cs weight content is different and can influence the activity of the catalyst.

Table 6 : Etherification of glycerol in the presence of Cs-zeolites. Effect of the zeolite structure.

	Conversion (wt %)		
Time (h)	CsZSM5	CsY110	CsNaX
3	18.2	36.2	31.1
5	20.4	58.3	60.2
8	27.5	75.9	77.3
9	27.8	79.9	80.9

Temperature : 260°C, catalyst : 4 (wt %), Cs exchanged level : 50%.

Barrault et al.

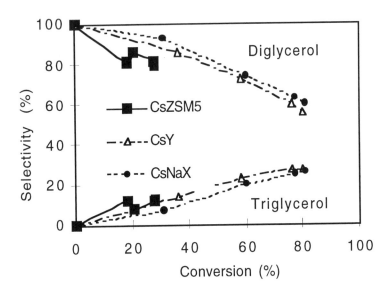

Figure 6 : Etherification of glycerol in the presence of alkaline exchanged Y zeolites. Effect of the zeolite structure on the selectivity to di- and triglycerol..

4. Conclusions

The etherification of glycerol in the presence of solid catalysts shows that the activity and the selectivity depend on the nature of the catalyst. Indeed, it appears that a carbonate such as Na_2CO_3 is more active than a hydroxide or an oxide but the selectivity to diglycerol or triglycerol is very low. Although the NaX and CsHY zeolites are less active at the beginning of the reaction, the conversion reaches 80% after 8h. Moreover, base zeolites catalyzed the etherification reaction of glycerol without any formation of acrolein. As for the selectivity, the Na_2CO_3, catalyst which is the most active, is less selective to diglycerol than the base zeolite. Indeed, the selectivity to diglycerol is higher than 90% over a CsHY or a NaX zeolite. In this particular case, it seems that the porosity of the catalyst slightly increases the selectivity of the reaction, especially when the glycerol conversion is under 80%. We have also shown what is the effect of alkaline elements on the activity and on the selectivity ; the best results are obtained with the most basic promoter.

5. Acknowledgments

The authors thank the organizers of the "AGRICE" program and the European Community "FAIR program" for their financial support.

6. References

1. A. J. Kaufman, R. J. Ruebush, Proceedings of the world conference on oleochemicals into 21st century, T.H. Applewhite Ed., American Oli Chemist Society (1991) 10-25.
2. A. Piccirilli, Y. Pouilloux, J. Barrault, J. Mol. Cat., accepted for publication.
3. X. Caillault, Y. Pouilloux, J. Barrault, J. Mol. Cat. A: 103 (1995) 117-123.
4. S. Abro, Y. Pouilloux and J. Barrault, 4th Symposium on Heterogeneous Catalysis and Fine Chemicals, Basel, 8-12 september 1996.
5. Dow Chemical, WO, 91/10368 (1991).
6. Kawai, JP 0,680,600 (1992).
7. Solvay, DE 3,842,692 (1988) ; DE 4,228,147 (1992).
8. Kashima Kemikaru, JP 05,117,182 (1993).
9. Onidol, EP 518,765 (1993).
10. Henkel, DE 4,029,323 (1993).
11. Nippon oils and fats, JP 61,238,749 (1986).
12. E. Jungermann, Cosm. Sci. Tech. Serv. , 11 (1991) 97-112.
13. N. Garti, A. Aserin, J.A.O.C.S., September (1981) 878-883.
14. M.R. Sahasrabudhe, J.A.O.C.S., 44 (1966) 376-378.
15. D. Barthomeuf, J. Phys. Chem., 88 (1984) 42.
16. D. Barthomeuf, Proceedings of Catalysis on solids acids and bases, 14-15 march, Berlin.
17. M. Huang and S. Kaliaguine, J. Chem. Soc., Faraday Trans, 88 (5) (1992), 751-758.

Catalytic Aziridination of Alkenes Using Microporous Materials

Christopher Langham[2], Paola Piaggio[1], Donald Bethell[2], Darren F. Lee[2], Philip C. Bulman Page[3], David J. Willock[1], Chris Sly[4], Frederick E. Hancock[5], Frank King[5] and Graham J. Hutchings[1].

[1]Department of Chemistry, University of Wales Cardiff, PO Box 912, Cardiff, CF1 3TB, UK.

[2]Leverhulme Centre for Innovative Catalysis, Department of Chemistry, University of Liverpool, Liverpool, L69 3BX, UK.

[3]Department of Chemistry, Loughborough University, Loughborough, Leicestershire, LE11 3TU, UK.

[4]Robinson Brothers Ltd., Phoenix Street, West Bromwich, West Midlands, B70 OAH, UK.

[5]ICI Katalco, R&T Division, PO Box 1, Billingham, Teeside, TS23 1LB, UK.

Abstract

Copper exchanged zeolite Y is shown to be an efficient catalyst for the aziridination of alkenes employing (N-(p-tolylsulfonyl)imino)phenyliodinane, PhI=NTs, as the nitrogen source.

In contrast to many homogeneous aziridinations high yields can be achieved without the use of an excess of alkene. We have shown that the reaction is truly heterogeneous by recycling the catalyst and by demonstrating the size-specificity of the zeolite. Water is found to act as an inhibitor and therefore dry solvents and reagents are required. Deactivation caused by water is easily reversed by a simple drying procedure prior to reuse of the catalyst.

Introduction

For some time zeolites have been extensively used as heterogeneous catalysts for the production of gasoline and related petrochemicals. These commercial applications have been based on both the shape selective properties of the zeolite porous framework and the cation exchange properties (1). In recent years, however, there has been a significant increase in interest in using microporous materials as high area heterogeneous catalysts for many organic processes (1). For example, the alkylation of aromatic compounds (2) and the hydrolysis of disaccharides (3) have received recent attention. In addition, we have shown that modification of zeolite Y with chiral sulphoxides leads to the formation of a high activity enantioselective acid catalyst (4-6). An area that has started to receive attention is the development of catalysts that effect heteroatom transfer to alkenes, for example epoxidation (7), cyclopropanation (8) and aziridination (9). Heterogeneously catalysed epoxidation has been extensively studied since the discovery of the titanium silicalite TS-1 and the observation that alkene epoxidation could be achieved with TS-1 using hydrogen peroxide as the oxygen donor under ambient conditions (11,12). To date, there remains considerable interest in using related systems to effect enantioselective epoxidation in a manner analogous to the Sharpless epoxidation procedure which utilize homogeneous titanium catalysts (13) but no success has so far been reported. In contrast, the analogous nitrogen atom transfer process, aziridination, has not been well studied. Aziridination is an important process in organic chemistry. Some natural products contain aziridines as subunits and their biological activity is influenced by the stereochemistry of the aziridine moiety. Aziridines are also useful as chiral synthons in the construction of compounds containing nitrogen functionalities (14). Evans *et al* (9) have described the use of soluble copper salts as efficient catalysts for the aziridination of alkenes, employing (*N*-(*p*-tolylsulfonyl)imino)phenyliodinane, PhI=NTs, as the nitrogen source. In this paper we present our initial results on the use of Cu^{2+}-exchaged zeolite Y as a heterogeneous catalyst for the aziridination of alkenes.

Experimental

The Cu-exchanged zeolite was prepared by conventional ion-exchange methods using an aqueous $Cu(OAc)_2$ solution, the concentration of which was calculated to obtain the required exchange level. The exchanged zeolite was recovered by filtration, washed with distilled water at least fifteen times and dried at 110°C in an oven. Finally the zeolite was calcined at 550°C prior to use. The degree of ion exchange was determined by atomic adsorption spectroscopy. All reactions were performed in MeCN with 5.0 equiv. of alkene at 25°C using a stirred batch reactor unless otherwise noted. All aziridines were isolated by flash chromatography and isolated yields are quoted.

Results And Discussion

The aziridination of alkenes described by Evans(9) was particularly appealing for development of a heterogeneous analogue, since it can be catalysed by a number of copper salts, e.g. copper triflate, at room temperature to give the corresponding aziridines in high yields (Figure 1).

Evans *et al.* (9) found that N-(p-toluenesulfonyl)imino)phenyliodinane (PhI=NTs) is a suitable nitrogen source for this reaction and that the aziridination occurs at room temperature and pressure to give high yields for a range of alkenes. Our approach involved the use of cation exchanged zeolites as catalysts since it is well known (10) that several zeolites can undergo ion-exchange reactions with metals, potentially providing an ideal heterogeneous catalyst for this process.

Before attempting to carry out this reaction with microporous materials we used computer simulation studies to confirm that the aziridine formed during the reaction would be able to diffuse through the pores of the zeolite structure. On the basis of these studies, zeolite Y was chosen for investigation since this zeolite has a pore structure comprising 12Å supercages connected by 8Å pores and the target aziridines as well as the proposed substrates could all readily be accommodated within this structure.

The aziridination of a range of alkenes was investigated using CuHY as catalyst and the results are shown in Table 1. Comparison of the homogeneous catalysed reaction using copper triflate (Cu(OTf)$_2$) as catalyst (9) indicates that the use of

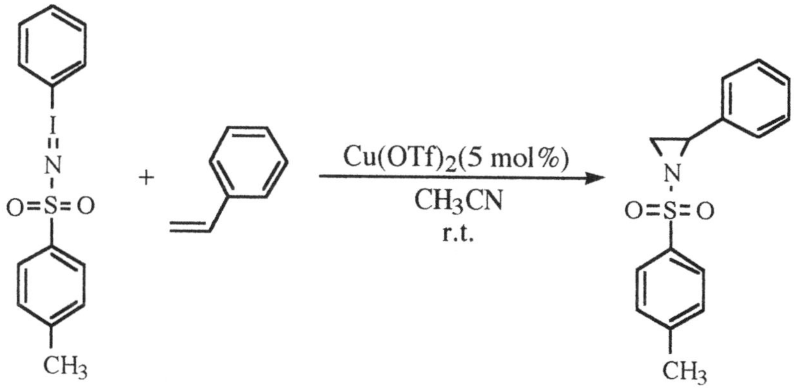

Figure 1. Homogeneous Aziridination of Alkenes

copper-exchanged zeolites is equally successful and comparable yields are achieved in both the homogeneous and heterogeneous systems.

Initially, CuHY was screened in the aziridination of styrene (Table 1, entries 1-3), since this substrate affords good yields of aziridine when the homogeneous catalyst $Cu(OTf)_2$ is used as the catalyst. Using a five-fold excess of styrene the desired N-tosylaziridine was obtained in 90% yield (entry 1). To our knowledge this is the first example of a heterogeneously catalysed aziridination reaction.

It was noted by Evans (9) that in the homogeneously catalysed reaction, the yield of aziridine significantly decreased when only 1.0 equiv. of styrene was employed, due to the competing breakdown of the PhI=NTs reagent. This decrease was found to be less significant using the heterogeneous catalyst, where 87% yield was obtained with 1.0 equivalent of styrene (Table 1, entry 2) as compared to a yield of 90% when 5.0 equivalents of styrene were used (Table 1, entry 1).

Table 1. CuHY catalysed Aziridination of Representative Alkenes

entry	alkene [a]	Cu mol%	Yield % [b]
1	styrene	25	90 (92)
2	styrene [c]	25	87 (35)
3	styrene	5	62
4	α-methylstyrene	25	33
5	p-chlorostyrene	25	76
6	p-methyl styrene	25	66
7	cyclohexene	25	50 (60)
8	methyl cinnamate	25	84 (73)
9	*trans*-stilbene	25	0 (52)
10	*trans*-2-hexene	25	44

[a] *All reactions were performed in MeCN with 5.0 equiv. of alkene at 25C unless otherwise noted.* [b] *Isolated yield of aziridine based on 1.0 equiv. of PhI=NTs.* [c] *One equivalent of alkene was used. Values in parentheses indicate yields obtained from homogeneous reactions.*

In order to confirm that this process was truly heterogeneous, after reaction the catalyst was removed by filtration, another aliquot of reactants, without the catalyst, was added to the solvent and the reaction monitored. No further product was observed. Futher, the removed catalyst was reused with fresh reagents and solvent, and the zeolite demonstrated similar activity to when it was used fresh.

To demonstrate the catalytic efficacy of CuHY, the reaction was carried out using 5.0 mol% of CuHY (rather than 25 mol%), which resulted in a yield of 62% of the aziridine (Table 1, entry 3). It is often observed that heterogeneous catalysts react more slowly than their homogeneous analogues under comparable reaction conditions. We consider this to be the case in our system, and when only 5.0 mol% of catalyst is present the competing breakdown of the PhI=NTs reagent becomes more significant than if the reaction were proceeding more rapidly.

CuHY was found to be successful in catalysing the aziridination of a range of alkenes (Table 1). It is observed that the catalyst gives best results with phenyl-substituted alkenes and lower yields are observed with cyclohexene and *trans*-hex-2-ene.

Interestingly, in the aziridination of *trans*-stilbene the product, *trans*-N-(p-tolylsulfonyl)-2,3-diphenylaziridine (I) could not be observed, and the zeolite turned black during reaction. This was thought to be the due to the relatively bulky aziridine product being too large to be accommodated within the small pores of CuHY. However, the aziridination of *trans*-β-methyl cinnamate, to form *trans*-N-(p-tolylsulfonyl)-2-(carbomethoxy)-3-phenylaziridine (II), inside CuHY proceeded in 84% yield, despite being similar in structure type to the *trans*-stilbene product (figure 2).

We used molecular modelling to investigate the ease with which these two aziridines could be placed into the pore structure of zeolite Y. These calculations suggest that the N-tosylaziridine formed from *trans*-β-methyl cinnamate can adopt a conformation which can easily be accommodated in the pores of zeolite Y. The aziridine derived from *trans*-stilbene can be constructed within the supercages but is indeed too bulky to diffuse through the connecting windows of the zeolite. We consider this to be a crucial piece of evidence since it shows that the aziridination reaction with the CuHY catalyst occurs within the intracrystalline space. Furthermore, this exciting result illustrates the potential for a heterogeneous catalyst to possess size-specificity to a precise degree.

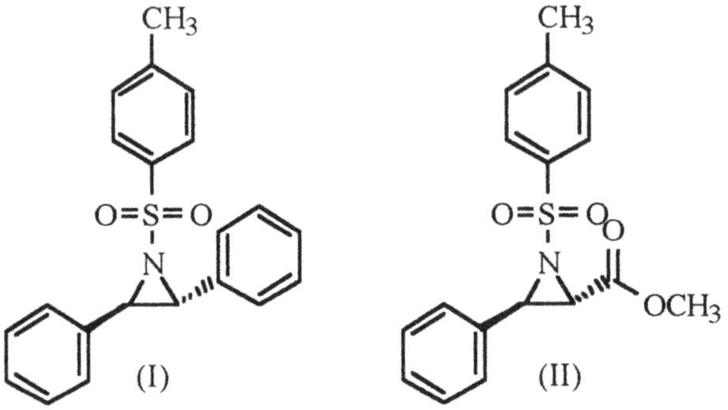

trans-*N*-(*p*-Tolylsulfonyl)   trans-*N*-(*p*-Tolylsulfonyl)-2
-2,3-diphenylaziridine   -(carbomethoxy)-3-phenylaziridine

Figure 2.

Such a property could be exploited by constructing zeolites with a range of pore sizes, and could also be developed to achieve regio-selectivity in a reagent containing two or more double bonds.

Investigation Of Alternative Cations

A range of alternative cation-exchanged zeolite Y catalysts were investigated for the aziridination of styrene and the results are shown in Figure 3.

In most cases when zeolite Y was exchanged with other metals, the yield of aziridine was lower than that obtained if no catalyst was present. We considered this to be due to the presence of other metals which catalyze the breakdown of the PhI=NTs reagent. Copper is therefore the preferred metal for the heterogeneous aziridination reaction, as for the homogeneous reaction (9).

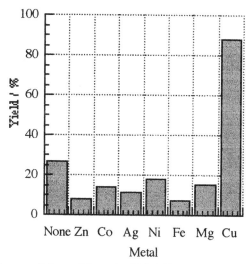

Reaction conditions: 1.0 equiv. PhI=NTs added to 5-fold excess styrene and metal-exchanged zeolite Y (25 mol%) in CH_3CN at room temperature. Yields determined by HPLC.

Figure 3. Effect of metal-exchanged zeolite as catalysts for the aziridination of styrene

Effect Of Water On Reaction

Zeolites are known to be affected by the adsorption of water and it therefore appeared possible that traces of water might act as an inhibitor for the aziridination reaction. The effect of water on the aziridination reaction was investigated for both the homogeneous and heterogeneous systems.

Initially we repeated the homogeneous reaction, in total absence of water, and with acetonitrile containing water (1.0%) and the results are shown in Figure 4. It is apparent that in the absence of water the reaction proceeds readily to completion in less than 30 minutes. When the water-containing solvent is used, however, the rate of reaction is significantly reduced. After five hours the reaction is complete, as determined by the consumption of the PhI=NTs reagent, albeit with a reduction in yield to only 50%.

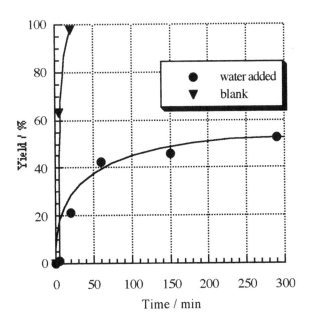

Reaction conditions: 1.0 equiv. PhI=NTs added to 5-fold excess styrene and Cu(OTf)₂ (5.0 mol%) in CH₃CN at room temperature. Yields determined by HPLC.

Figure 4. Effect of water added to homogeneous aziridination of styrene

The effect of water is more marked in the heterogeneous reaction. We have already shown that the rate of the heterogeneous reaction is generally lower than the homogeneous system, and this effect can be observed in Figure 5 - the heterogeneously catalysed reaction takes about twice as long to reach completion as does the homogeneous system (Figure 4).

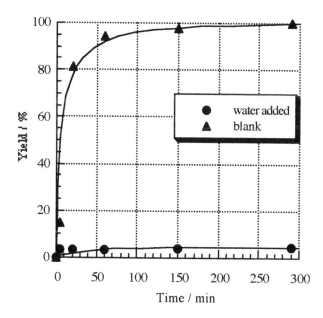

Reaction conditions: 1.0 equiv. PhI=NTs added to 5-fold excess styrene and Cu-
exchanged zeolite Y (25 mol%) in CH₃CN at room temperature. Yields
determined by HPLC.

Figure 5. Effect of water added to heterogeneous aziridination of styrene

Use of a solvent containing water (1.0%), however, has a significant effect on
the heterogeneously catalyst and the reaction proceeds very slowly, such that
after 5 hours only a 5.0% yield was observed.

To demonstrate further the effect of water on catalyst activity further the effect of
water adsorption during catalyst aging was studied. CuHY was calcined, then left
exposed to moist air for two hours, after which time thermal gravimetric
analysis showed the catalyst to have absorbed 20% weight of water from the
atmosphere.

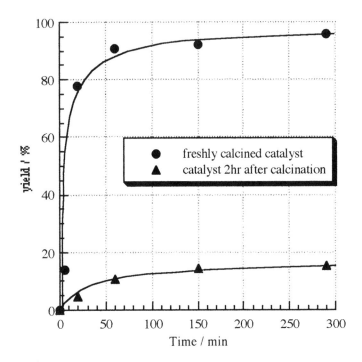

Reaction conditions: 1.0 equiv. PhI=NTs added to 5-fold excess styrene and metal-exchanged zeolite Y (25 mol%) in CH_3CN at room temperature. Yields determined by HPLC.

Figure 6. Effect of catalyst aging on heterogeneous aziridination of styrene

This is equivalent to 50% of the amount of water added to acetonitrile solvent in the previously described studies. The freshly calcined CuHY and the aged CuHY were then tested as catalysts for the aziridination of styrene and the results are shown in Figure 6. As expected both rate of reaction and overall yield are decreased significantly for the catalyst left in air, with a maximum yield of 17% being observed. These studies show that dry solvents must be used when CuHY is used as a catalyst for the aziridination of alkenes

Catalyst Reusability

The main advantage of a heterogeneous process is the ease with which the catalyst can be recovered and reused. Investigations into the reusability of the CuHY catalyst have been very encouraging, with only a small drop in activity after multiple re-use being observed. A series of reactions have been carried out using the same catalyst. In each case the reaction was carried out until complete disappearance of the PhI=NTs reagent was observed and the catalyst was recovered by filtration, washed with fresh solvent then added to a fresh reaction mixture of styrene, PhI=NTs and acetonitrile. The yield was found to decrease only slightly after each reuse. This we believe to be due to absorption of water into the zeolite which inhibit the reaction and not due to loss of copper which we have found is minimal under our reaction conditions (following thirteen reactions with the same catalyst only 8% of the total copper had leached out of the zeolite). We have found that full catalytic activity can be restored if the catalyst is dried before being reused.

Conclusions

In this study we have shown that the copper-exchanged zeolite CuHY is an effective catalyst for the aziridination of alkenes. To our knowledge this is the first example of heterogeneously catalysed aziridination reaction. The high catalytic activity and reusability of this modified zeolite makes it a superior catalyst to the homogenous analogue. In addition, have shown examples of the size specificity of this catalyst, and this suggests great potential for stereo- and regioselectivity that can be explored with these catalysts. The catalyst is inhibited by the competitive adsorption of water and therefore it is essential that dry solvents are used.

References

(1) W. Hölderich, M. Hesse and F. Neumann, *Angew. Chem. Int. Ed. Engl.*, **27**, 226 (1988).
(2) R. Ganti and S. Bhatia, *Stud. Surf. Sci. Catal.*, **98**, 167 (1995).
(3) C. Buttersack and D. Laketic, *Stud. Surf. Sci. Catal.*, **98**, 190 (1995).
(4) S. Feast, D. Bethell, P.C.B. Page, F. King, C.H. Rochester, M.R.H. Siddiqui, D.J. Willock, and G.J. Hutchings, *J. Chem. Soc., Chem. Comm.*, 2409 (1995).
(5) S. Feast, D. Bethell, P.C. Page, M.R.H. Siddiqui, D.J. Willock, G.J.Hutchings, F. King and C.H. Rochester, *Stud. Surf. Sci. Catal.*, **101**, 211 (1996).
(6) D.J. Willock, S. Feast, G.J. Hutchings, P.C.B. Page and D. Bethell, *Topics Catal.*, **3**, 77 (1996).
(7) K.A.Jorgenson, *Chem. Rev.*, **89**, 431 (1989).

(8) M.P. Doyle, *Chem. Rev.*, **86**, 919 (1986).

(9) D.A. Evans, M.M. Faul and M.T. Bilodeau, *J. Am. Chem. Soc.*, **116**, 2742 (1994).

(10) K. Klier, *Langmuir*, 4, 13-25 (1998)

(11) G.J. Hutchings, D.F. Lee and A.R. Minihan, *Catal. Lett.*, **33**, 369 (1995).

(12) G.J. Hutchings, D.F. Lee and A.R. Minihan, *Catal. Lett.*, **39**, 83 (1996).

(13) T. Katsuki and K.B. Sharpless, *J. Am. Chem. Soc.*, **102**, 5974 (1980).

(14) A. Padwa and A. D. Woolhouse, In "Comprehensive Heterocyclic Chemistry," Ed. by W. Lwowski. Pergamon, Oxford, Vol. 7, p. 47 (1984).

Selective Aldolisation of Acetone into Diacetone Alcohol Using Hydrotalcites as Catalysts

F. Figueras[1], D. Tichit[2], M. Bennani Naciri[2], and R. Ruiz[2,3]

[1]Institut de Recherches sur la Catalyse du CNRS, 2 avenue A. Einstein, 69626 Villeurbanne Cedex, France.

[2] Laboratoire de Matériaux Catalytiques et Catalyse en Chimie Organique, UMR 5618 ENSCM, 8 rue de l'Ecole Normale, 34296-Montpellier Cedex 5, France.

[3] on leave from Departamento de Quimica Organica, Universidad de Granada, Avenida San Alberto Magno, E-14004 Cordoba, Spain.

Abstract:
The aldolisation of acetone has been investigated at 273 K on a series of hydrotalcite-like materials synthesized by classical routes and activated in different conditions. The activity increases noticeably after decarbonation, but a simple thermal treatment at 673 K yields a catalyst of low activity with a selectivity of about 4% to mesityl oxide (MO) formed by dehydration of diacetone alcohol (DAA). MO is a strong inhibitor of the reaction. This undesired reaction can be decreased by a controlled rehydration at room temperature of the a solid first decarbonated at 723 K. This rehydration decreases the basic strength according to calorimetric determinations of CO_2 adsorption. An excess of water inhibits the reaction. Trace amounts of chlorine retained by the solid are detrimental for catalytic activity. They can be eliminated by exchange with carbonates. Activity goes through a maximum for a Mg/Al ratio of about 3. Thermodynamic equilibrium is then reached at 273 K in less than 1 h. These catalysts appear to have possible applications in the field of basic catalysis.

Introduction

The substitution of liquid acids and bases by solids as catalysts in organic reactions is a requisite for a better preservation of environment (1, 2). It is also a very difficult task because industrial processes have often been optimised to such extend that the performances are excellent. In the case of acids, many solid acids are known, but in the case of bases fewer solid bases are available. Among them hydrotalcites (HDT) are interesting because they can give wide pore solids which would be able to accept the substrates of organic chemistry: numerous studies on the physico chemical properties and thermal stability of various HDT have been reported (3-5). Upon thermal decomposition, a highly active homogeneous mixed oxide is obtained from these materials at about 723 K which is potentially

pore solids which would be able to accept the substrates of organic chemistry: numerous studies on the physico chemical properties and thermal stability of various HDT have been reported (3-5). Upon thermal decomposition, a highly active homogeneous mixed oxide is obtained from these materials at about 723 K which is potentially a basic catalyst for a variety of organic transformations such as aldol condensations (6), olefin isomerisation (7), nucleophilic halide exchange (8), alkylation of diketones (9), epoxidation of activated olefins with hydrogen peroxide (10) or Claisen-Schmidt condensation (11).

The aldol condensation of acetone yields diacetone alcohol, which is both a Cl free solvent, and an intermediate for the synthesis of mesityloxide and isophorone. The aldol condensation has numerous applications in the synthesis of fine chemicals classically catalyzed by bases (12). At the laboratory scale many catalysts have been proposed including alumina (13, 14), zeolites (15), sepiolite (16), and hydrotalcites (11, 17, 18). It has to be remarked that if high conversions can be reached with these solids, the yields are generally modest. In a former work (17), the condensation of acetone with benzaldehyde was investigated at 383 K: the observation of a good Hammett correlation showed that aldolisation was indeed base catalysed on hydrotalcites , and that traces of chlorine left on the solid strongly inhibited the catalytic properties by blockage of the stronger basic sites.

In all these former studies, aldolisations were performed at relatively high temperatures: 673 K for the addition of formaldehyde to acetone (6), 573 K for the condensation of acetone (19), and 383 K for the condensation of benzaldehyde with ketones (11, 17, 18). Thermodynamic data are scarce for this type of reaction, but it appears that the aldolisation of acetone is an exothermic reaction (20), in which yield decreases with increasing temperature. The work at 383 K gives low yields in aldol, and the main product is benzalacetone obtained by dehydration of the aldol. The reaction was then performed here in an ice bath i.e. 273 K.

The second aspect is the procedure of activation of the solid : it is a general belief that the strongest basic sites, produced by evacuation at high temperature, are oxygen anions of low coordination. The procedures of activation of basic solids have therefore been oriented at the formation of these sites, which would correspond to Lewis basic sites. We investigate here solids which have suffered rehydration, and in which the Lewis basic sites have been converted to hydroxyls or Brønsted sites. A recent report indeed shows that these solids are good catalyst for the aldolisation of acetone (21).

Experimental

Preparation of hydrotalcites

The hydrotalcites were prepared by coprecipitation at 343 ± 5 K as described previously (17, 18). An aqueous solution containing 0.3 mol of $MgCl_2.6H_2O$, 0.1 mole of $AlCl_3.6H_2O$ (for a ratio Mg/Al = 3), and one containing 0.8 mol of NaOH and 0.02 mol of Na_2CO_3, were slowly mixed under vigorous stirring, maintaining the pH between 8 and 10. The mixture was heated at this temperature for 15 hours under stirring. The precipitate was washed several times until the solution was chloride free ($AgNO_3$ test). The products were dried at 313 K. Using the same procedure, different samples were prepared by mixing the required amounts of Mg and Al salts to change the Mg/Al ratio. They contain both Cl^- and CO_3^{2-} as charge compensating anions, as established from the chemical analysis data reported in Table 1. These samples will be referred to as HT(x)NE (non exchanged).

Table 1. Compositions of the samples

Sample	Mg/Al in the solid	Composition
HT(2.5)NE	2.1	$Mg_{0.67} Al_{0.32}(OH)_2(CO_3^{2-})_{0.1} Cl_{0,07} \cdot 0.59H_2O$
HT(2.5)E-1	2.4	$Mg_{0.71} Al_{0.29}(OH)_2(CO_3^{2-})_{0.16} Cl_{0,0005}$
HT(2.5)E-2	2.5	$\cdot 0.30H_2O$
		$Mg_{0.71} Al_{0.29}(OH)_2(CO_3^{2-})_{0.16} Cl_{0,0002}$
		$\cdot 0.32H_2O$
HT(3)NE	3	$Mg_{0.75} Al_{0.25}(OH)_2(CO_3^{2-})_{0.1} Cl_{0,11} \cdot 0.52H_2O$
HT(3)E-1	3	$Mg_{0.75} Al_{0.25}(OH)_2(CO_3^{2-})_{0.15} Cl_{0,004}$
HT(3)E-2	3	$\cdot 0.55H_2O$
		$Mg_{0.75} Al_{0.25}(OH)_2(CO_3^{2-})_{0.16} Cl_{0,001}$
		$\cdot 0.68H_2O$

Procedure of anion exchange

The exchange of Cl^- anions was performed by contacting the hydrotalcite with a solution of Na_2CO_3 : 2 g of HT(x)NE were dispersed in a $1.5 \; 10^{-3}$ M solution of sodium carbonate and stirred at 343 K for 2 h. After filtration the solids were washed and dried at 313 K. These samples will be referred to as

HT(x)E-1. In order to achieve a higher degree of exchange, a second exchange was carried out on HT(x)E-1 to obtain HT(x)E-2. The compositions and surface areas of the samples are reported in Table 1 and 2.

Activation of the catalysts
    Catalysts were first activated by calcining to 723 K in a flow of air. The temperature was raised at the rate of 10 K per minute to reach 723 K and maintained for 8 h. The solid was then cooled in dry nitrogen and rehydrated at room temperature under a flow of nitrogen gas saturated with water vapour. The flow of wet nitrogen of 6 L/h was maintained for a specified period depending on the amount of catalyst to be rehydrated (Table 3),.

Table 2 . Anionic Content and Specific areas of catalysts.

Mg/Al	Samples	Specific area $m^2 g^{-1}$	CO_3^{2-} (% of charge)	Cl^- (% of charge)
	HT(2.5)NE	313	74.30	25.7
2.5	HT(2.5)E-1	167	99.85	0.15
	HT(2.5)E-2	122	99.93	0.07
	HT(3)NE	235	62.80	37.2
3	HT(3)E-1	120	98.54	1.46
	HT(3)E-2	122	99.69	0.31
-	MgO	28	-	-
-	NaY	740	-	-

Characterization
    X-ray powder diffraction patterns were recorded with a CGR Theta 60 instrument using $CuK_{\alpha 1}$ radiation. DTA and TGA analyses were carried out in a Setaram microbalance operated under a flow of dry nitrogen at a heating rate of 5 K/min. The adsorption of CO_2 was followed both by calorimetry and measurements of uptake using the experimental apparatus described in the literature (22). After outgassing the sample at 673 K, CO_2 was introduced in small increments. The heat of adsorption of CO_2 can be taken as a measure of the basic strength of the adsorption sites, and the distribution in strengths of basic sites was obtained using the differential heat of adsorption as a function of coverage.

Measurements were also performed on a sample rehydrated at room temperature by contact with water vapour, then outgassed in vaccuum at 373 K. Before characterization, the samples were first calcined to decompose the carbonates and possible carbon deposits.

Catalysis

The kinetics of aldolisation were followed in a 3 necked 50 ml round bottomed flask, cooled to 273 K in an ice bath and equipped with a magnetic stirrer. 15 mL of acetone (RP Normapur) were introduced. After 15 min of cooling, the freshly activated catalyst (0.5 g) was introduced and the progress of the reaction followed by gas chromatography using a DB2 column. For the analysis of the reaction mixture, small samples (30 μL) were taken with a syringe at different reaction times, diluted in 30 μL of dichloroethane to which decane (60 μL) was added, as internal standard for quantitative GC analysis. In a few experiments dichloroethane was used as solvent for the reaction.

RESULTS

1) Synthesis and characterization of hydrotalcites

The structural formulas given in Table 1 were established from the chemical analysis of the solids and thermogravimetric analysis. The surface areas and Cl contents are presented in Table 2. The X-ray diffraction patterns of hydrotalcites are reported in Fig. 1. These solids show typical crystalline hydrotalcite patterns. After calcination at 723 K, their structure is that of mixed oxide of the MgO type and after hydration the original lamellar structure was reformed. From the physico-chemical characterizations the original solids used here appear as pure hydrotalcites, which reproduce the solids described in the literature. However, after rehydration the anions compensating the structure are OH instead of carbonates for the hydrotalcite structure, therefore the solid shows now a meixnerite-like structure.

<u>Basicity from CO_2 adsorption</u>

The results are reported in Figure 2, for two hydrotalcite samples just calcined or calcined then rehydrated. The calcined sample is more basic than the rehydrated one, both in strength and number of sites. The higher amount of CO_2 adsorbed and the higher heat of interaction show that surface oxygens have a higher basicity than OH groups. Therefore water inhibits the stronger basic sites.

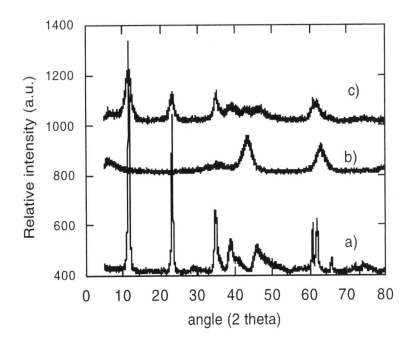

Fig. 1: XRD spectra of HT(2.5) after a): synthesis, b) calcination at 773 K, and c) rehydration at room temperature in a stream of nitrogen saturated with water vapor.

2) Reaction kinetics

The conversion of acetone on hydrotalcites yields diacetone alcohol, mesi-tyl oxide and traces of triacetone alcohol which were not quantified here. A simple kinetic study of the reaction was determined on a HDT (2.89) sample activated at 723 K, in order to select the best conditions of reaction. Aldolisation of acetone to diacetone alcohol is an exothermothermic reaction, and the yield which can be reached at equilibrium decreases then with temperature. This reaction was performed at different temperatures up to 300 K, with the result that the selectivity to mesityloxide increases with the reaction temperature. Consequently both thermodynamic and kinetic reasons lead to the conclusion that the reaction would preferably be run at 273 K.

selectivity to mesityloxide increases with the reaction temperature. Consequently both thermodynamic and kinetic reasons lead to the conclusion that the reaction would preferably be run at 273 K.

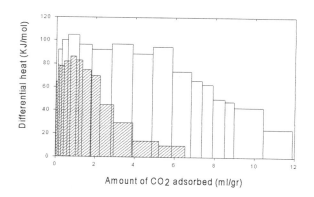

Fig.2. Differential heats of adsorption of carbon dioxide over HDT(2.7) either calcined at 773 K (open bars), or calcined then rehydrated at room temperature (grey bars).

The rate is proportional to the weight of catalyst in the range investigated, and a rate equal to $3 \ 10^{-6}$ mol. $\text{sec}^{-1}.\text{g}^{-1}$ was measured using pure acetone (concentration: 0.013 mol.cm^3) and a powder with particle size of about 0.01 cm.

The criterion of Weisz (23) :

$$\frac{dN}{dt} \frac{1}{C_0} \frac{R^2}{D_{\text{eff}}} = \Phi$$

in which C_0 is the concentration of reactant (mol/mL), R (cm) the radius of the catalyst particles, dN/dt (mol/s/g) the rate and D_{eff} (cm^2/s) the diffusion coefficient, permits to estimate the importance of internal diffusion. The low value of the Weisz modulus $\Phi=10^{-2}$ ensures that neither the rates of reaction nor the selectivities are perturbed by diffusion.

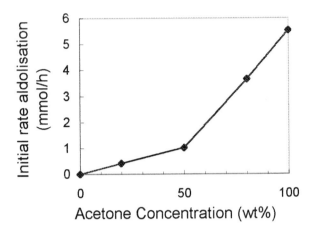

Fig.3. Influence of the concentration of acetone on the initial rate of reaction of aldol formation using HDT(2.89)E-2

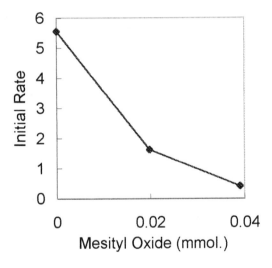

Fig.4. Influence of the addition of mesityloxide on the initial rate of reaction of aldol formation, using HDT(2.89) E-2.

The influence of the concentration of acetone was studied using dichlorethane as solvent . The initial rate of reaction is plotted against the concentration of acetone in Fig. 3. The reaction follows a second order which is consistent with a bimolecular mechanism of condensation.

Mesityloxide (MO) is a secondary product of the reaction, and its kinetic influence was determined by introducing it with the substrate. As illustrated in Fig .4, MO is a strong inhibitor of the rate. Therefore a non-selective catalyst cannot reach high conversions because of this inhibition by the product.

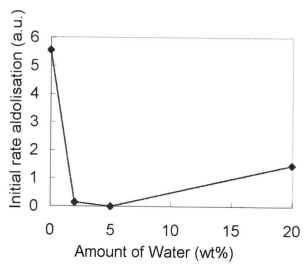

Fig.5. Influence of the addition of water to the feed, on the initial rate of formation of diacetone alcohol, using HDT(2.89)E-2 as catalyst.

The other secondary product of reaction, water, shows a more complex influence: marked inhibition of the initial rate, but increase of activity after a sufficiently long time, suggesting that new catalytic sites had been formed (Fig. 5). MO is formed by dehydration, which is believed to be an acid catalysed reaction due to the exposed Mg^{2+} cations . This reaction should be inhibited by organic bases like pyridine. Fig. 6 proves that this is indeed the case: the addition of about 2 % of pyridine to the charge somewhat decreases somewhat the selectivity for MO at low conversions, and decreases the rate by a factor of about 2.

In conclusion of this brief kinetic study, it appears that the reaction is inhibited by mesityloxide and water, and that high yields in diacetone alcohol, which is the desired product cannot be reached by changing the reaction conditions only. The activation of the solid has to be modified.

3) Influence of the conditions of activation of the solid.

Strong basic sites only appear after decarbonation at high temperature, and a much lower activity is observed on samples calcined below 673 K. Constantino and Pinnavaia (24) recently reported a good activivity for the conversion of 2-methyl-3butyne-2-ol below the decomposition temperature of HDT (i.e. < 523 K). Indeed the application of similar conditions of activation of the catalysts for our reaction gave no conversion at all. Hence we conclude that the reaction investigated here requires a much higher basic strength than that needed for reaction of 2-methyl-3butyne-2-ol, thus a much higher decomposition temperature of the carbonates.

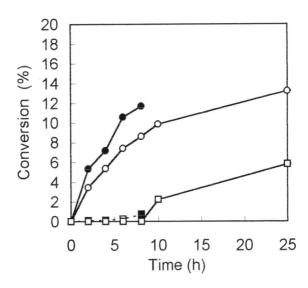

Fig.6. Kinetics of formation of diacetone alcohol with HDT(2.89)E-2, in the absence of pyridine (black points), and in presence of 2wt% of pyridine (open points) : circles for diacetone alcohol and squares for mesityloxide.

An important factor affecting the basicity is the presence of trace amounts of chlorine. Cl- anions are detrimental for activity as illustrated in Fig: 7.

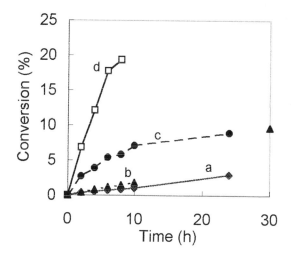

Fig. 7. Comparison of the conversion of acetone to diacetone alcohol on for two hydrotalcite samples : HT (2.85) non exchanged (curve a) and exchanged (curve c), HT(4.68) non exchanged (b) and exchanged (d).

Since water showed this effect of formation of new active sites, the rehydration of the solid was investigated and appears to be the main parameter for activity. Selectivity is also improved since only traces of mesityloxide are then formed. For the same weight of solid (0.5 g before activation) the effect of activation conditions on the initial rate of reaction are summarised in Table 3.

Table 3. Effect of the conditions of rehydration on the initial rate of aldolisation.

Sample	Rehydration time (h)	Rehydration Temp (K)	Initial rate (mmol. min^{-1})
HT(2.7) A	17	373	0.114
HT(2.7) B	16	323	0.225
HT(2.7) J	8	298	1.003
HT(2.7) H	5	298	1.117
HT(2.7) E	2	298	2.038

A well activated catalyst reaches thermodynamic equilibrium in less than one hour. This is mainly due to the suppression of the formation of mesityloxide which is a strong inhibitor of the reaction. An excess of water, i.e. a too high rehydration, has a negative effect, as found before for the calcined sample. The results of catalysis can be compared with those of calorimetry. The latter show lower basicity of the surface after rehydration, both in strength and number of sites, while kinetic results indicate that the initial rate of reaction, i.e. in the absence of mesityloxide, is much higher on the hydrated samples. Surface areas have not been determined systematically on rehydrated samples, but it has been observed that rehydration decreases the surface area of calcined HDT(2.5) to about 50 m^2/g, therefore the activity is also higher for the rehydrated catalyst when calculated per unit surface area. This result suggests that aldolisation is rather specific to OH^-, and works worse on oxygens of higher basic strength.

4) Comparison of different samples.
Magnesia and hydrotalcites of different Mg/Al ratios, activated by the same procedure are compared for the aldolisation of acetone in Fig. 8. Activity goes through a maximum for a Mg/Al ratio of 3 and hydrotalcites appear much more active than magnesia.

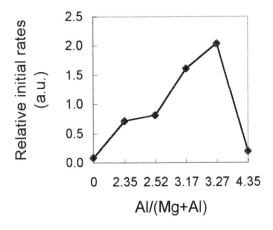

Fig.8. Effect of the Al content on the initial rate of formation of aldol for samples calcined and rehydrated.

In conclusion, a good catalyst for aldolisation reactions can be obtained by combining decarbonation and controlled rehydration of hydrotalcites. Even if rehydration suppresses the stronger basic sites of the solid, it increases the activity and selectivity of aldolisation of acetone and Claisen Schmidt reaction.

Aknowledgements: the financial support of ELF-Atochem-France is warmly aknowledged.

References

1. L. Krumenacker and S.Ratton, L'actualité Chimique, **26** (1986).
2. W. F Hölderich and H. Van Bekkum, Stud. Surf. Sci. Catal. **58**, 664 (1991).
3. F Rey and V.Fornes, *J. Chem. Soc. Faraday Trans.* **88**, 2233 (1992)
4. W.T Reichle,.S.Y.Kang, and D.S. Everhardt. *J. Catal.* **101**, 352 (1986)
5. G.W .Brindley, and S. Kikkawa, *Clays & Clay Minerals.* **28**, 87 (1980)
6. E. Suzuki and Y. Ono, *Bull. Chem. Soc. Jpn.* **61**, 1008 (1988)
7. W. T Reichle, *J. Catal.* **94**, 547 (1985)
8. E. Suzuki, M. Okamoto, and Y. Ono, *J. Mol. Catal.* **61**, 283 (1990)
9. C. Cativiela, F. Figueras, J.I. Garcia, J. A. Mayoral, and M.M. Zurbano, *Synth. Commun.* **25**, 1745 (1995).
10. C. Cativiela, F. Figueras, J.M. Fraile, J.I. Garcia, J. M. Mayoral, *Tetrahedron Lett.* **36**, 4125 (1995).
11. M.J. Climent, A. Corma, S. Iborra and J. Primo, *J. Catal.***151**, 60 (1995)
12. J. March, *Advanced Organic Chemistry*, 4th Edition, J. Wiley &Sons, 938 (1992)
13. J. Muzart, *Synth.Commun.* **15**, 285 (1985)
14. J. Muzart, *Synthesis.* **1**, 60 (1982)
15. A. Corma and R.M. Martin -Aranda, *J.Catal.* **130**, 130 (1991)
16. A. Corma, V. Fornes, R. M. Martin -Aranda, H. Garcia and J. Primo , *Appl. Catal.* **59**, 237 (1990)
17. D. Tichit, M. H. Lhouty, A. Guida, B. Chiche, F. Figueras, A. Auroux, D. Bartalini and E. Garrone, *J. Catal.* **151**, 50 (1995)
18. A. Guida, M.H. Lhouty, D. Tichit, F. Figueras, and P. Geneste, *Applied Catal.* **394**, 1 (1997).
19. W.T. Reichle, *US Patent* 4,458,026 (1984) to Union Carbide
20. E.G Craven., J. Appl. Chem. **13** , 71 (1963)
21. R. Tessier, D. Tichit, F. Figueras and J. Kervenal, *French Pat.* n°95 00094 to Atochem. (1995)
22. P. C. Gravelle, *Adv. Catal.* **22**, 191 (1972)
23. P. B. Weisz , *Adv. Catal.* **13**, 137 (1962).
24. V. R. L. Constantino and T. J. Pinnavaia, *Catal. Lett.* **23**, 361 (1994).

Selective Hydroxymethylation of Guaiacol to Vanillic Alcohols in the Presence of H-Form Mordenites

Claude Moreau, François Fajula, Annie Finiels and Sylvie Razigade

*Matériaux Catalytiques et Catalyse en Chimie Organique
UMR 5618 CNRS-ENSCM, ENSC Montpellier, France*

Laurent Gilbert, Roland Jacquot and Michel Spagnol

*Rhône-Poulenc Industrialisation
CRIT-Carrières Saint-Fons, France*

Abstract

Hydroxymethylation of guaiacol with aqueous formaldehyde, yielding 3-methoxy-4-hydroxy benzylic alcohol (vanillic alcohol) and 2-hydroxy-3-methoxy benzylic alcohol, was performed in a batch mode in the presence of various dealuminated H-form mordenites. A high selectivity to hydroxymethylated products ($\approx$ 80 to 90 %) was achieved in the presence of a H-form mordenite with a Si/Al ratio of 18, at 338 K and with a high formaldehyde/guaiacol ratio. Under these reaction conditions, the formation of vanillic alcohol (3-methoxy-4-hydroxy benzylic alcohol) is favored over that of 2-hydroxy-3-methoxy benzylic alcohol, by a factor of about 5-6.

1. INTRODUCTION

Introduction of a CH_2OH group into an aromatic or heteroaromatic compound using formaldehyde as the hydroxymethylating agent only occurs under the effect of activating functional groups, e.g. hydroxyl or alkoxy groups [1]. However, the selectivity to hydroxymethylated products is generally poor because the latter react rapidly enough with the starting material to give bisarylmethane derivatives.

Several attempts were recently made to replace homogeneous catalysts by solid acids such as zeolites in order to gain a better selectivity to the hydroxymethylated product from the specific properties of these microporous materials. Indeed, it was reported that hydroxymethylation of phenol yields nearly selectively para and ortho hydroxymethylphenols in the presence of H-form zeolites [2], but these results could not be reproduced. In fact, under the reported operating conditions, hydroxymethylation of phenolic compounds goes generally to bisarylmethane derivatives [3,4].

In this paper, we present our results on the selective hydroxymethylation of guaiacol to 3-methoxy-4-hydroxy benzylic alcohol (**1**) and 2-hydroxy-3-methoxy benzylic alcohol (**2**) using dealuminated H-form mordenites (Scheme 1). These catalysts were already shown to be highly selective in reactions capable of forming condensation products [5,6].

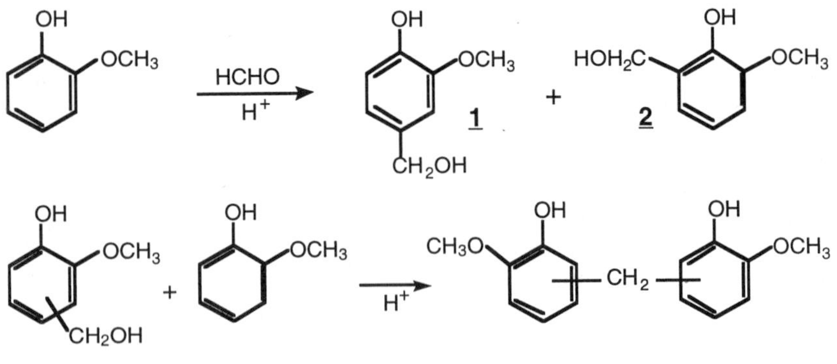

Scheme 1. Reaction scheme for hydroxymethylation of guaiacol.

2. EXPERIMENTAL

A. Catalysts

Dealuminated H-form mordenites were obtained from Institut Français du Pétrole, Zeocat and PQ Zeolites. They were generally used without further activation. The reaction was carried out using water as solvent and activation procedures consisting in thermal treatments had little if any influence on catalytic behavior. For all dealuminated zeolites, total and framework Si/Al ratios ranging from 10 to 108 were given as identical.

B. Procedure

Experiments were carried out in a 0.1 l stirred glass reactor working in a batch mode. Except otherwise mentioned, the general procedure was as follows: 0.25 g of catalyst was stirred for 30 min in the presence of 12 ml of formalin (37 wt % aqueous solution of formaldehyde), at 358 K. The agitation speed was 700 rpm. Zero time was set when 1 ml of guaiacol was added to the suspension of catalyst in formalin .

C. Analyses

Analyses were performed by HPLC using a Shimadzu LC-6A pump equipped with a UV spectrophotometer SPD-6A detector at 280 nm and controlled by a PC software package (ICS). A RP18 column was used (Merck) and the mobile phase was an acetonitrile/water mixture (70/30 by volume) at 0.7 ml/min.

Product identification was performed by gas chromatography on a GC-MS apparatus (HP 5890A/5970A) with an OV-1 type glass capillary column and controlled by a PC software package (HP G1034C).

D. Kinetic modelling

The initial rate constants were deduced from the experimental plots of concentrations vs time by curve fitting using the AnaCin software [7].

3. RESULTS AND DISCUSSION

A. Identification of products

The major products identified from combined gas chromatography and mass spectrometry analysis or comparison with reference samples were 3-methoxy-4-hydroxy benzylic alcohol (**1**) and 2-hydroxy-3-methoxy benzylic alcohol (**2**). Small amounts of the corresponding methyl ethers and acetals were observed, as due to the presence of methanol (15 %) in the starting formalin solution. They were rapidly hydrolyzed to their parent alcohols during the course of the reaction. Finally, traces of dihydroxymethylated compounds and bisarylmethane derivatives were also observed.

B. Catalytic tests

Under the operating conditions described above, the reaction was not limited by external or internal diffusion (see below). Indeed, it has been already shown in a closely related reaction, namely hydroxymethylation of furane derivatives [8], that external diffusion limitations could be ruled out on the basis of studies of the influence of agitation speed and catalyst weight and by the determination of activation energy. For the internal diffusion, quantitativedata on the diffusivity of benzylic alcohols within the pores of a zeolite are not available. It is generally accepted, according to Weisz equation, that internal diffusion is negligible for values of diffusion coefficients between 10^{-6} and 10^{-8} cm^2/s in the liquid phase.

Experimental results obtained over a series of dealuminated H-mordenites after 30 and 60 minutes of reaction are reported in Table 1. The effect of the extent of dealumination on the conversion of guaiacol is further illustrated in Figure 1.

The plots of the conversion of guaiacol, after 30 and 60 minutes reaction, as a function of the atomic Si/Al ratio (Fig.1), show that the activity is determined by the acicity up to a Si/Al ratio of c.a. 50. The maximum conversion for a Si/Al ratio of 18 corresponds to the better compromise between the strength and the number of sites, in agreement with Barthomeuf proposals [9].

Table 1. Influence of Si/Al ratio on guaiacol conversion, selectivity to vanillic alcohol (**1**) and isomer (**2**), and para/ortho selectivity at different reaction times (12 ml of formalin, 0.250 g of catalyst, 1 ml of guaiacol, 358 K).

Si/Al ratio	guaiacol conversion (%)		selectivity to alcohols **1** and **2** (%)		para/ortho ratio*	
	30 min	60 min	30 min	60 min	30 min	60 min
10	6	9	25	25	5.3	4.1
15	12	19	49	56	5.1	5.2
18	18	30	48	50	5.0	4.3
40	11	18	35	37	7.7	4.5
49	10	19	46	46	3.6	4.3
100	51	69	35	34	4.0	3.6
108	53	74	33	30	5.6	5.0

* The para/ortho ratio refers to para or ortho orientation of the reaction with respect to the OH group of guaiacol, yielding **1** and **2**, respectively.

For higher Si/Al ratios (18 <Si/Al ≤50), although the strength of acidic sites remains nearly constant, their number decreases. The further increase in activity for Si/Al ratios higher than 50 can only result from the increase of the hydrophobicity of the catalysts. Such a behavior has been already observed in the reaction of hydroxymethylation of furfuryl alcohol [8] and we have then proposed that it would result from competitive adsorption, on the catalyst surface, between formaldehyde, present as the hydrate form, and the starting alcohol. The latter becomes preferentially adsorbed on hydrophobic zeolites.

Concerning the selectivity of the reaction, it is seen in Table 1 that high selectivities to monohydroxymethylated products were found for H-mordenites with Si/Al ratios of 15 and 18 and, to a lesser extent, with a more dealuminated catalyst (Si/Al=49). It is also worth mentioning here that only mordenites with a bidimensional porosity with a single channel (6.5 x 7.0 Å) accessible to organic molecules are suitable catalysts in such reactions involving a consecutive kinetic network and formation of reactive intermediates capable of reacting with the starting material. Indeed, zeolites with a large pore apertures and large cavities (Faujasite-type Y zeolite for instance) were found much less selective.

Finally, the para vs ortho selectivity of the orientation of the reaction to 3-methoxy-4-hydroxy benzylic alcohol (vanillic alcohol, **1**) and 2-hydroxy-3-

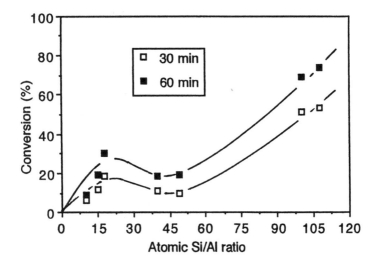

Figure 1. Plot of guaiacol conversion against atomic Si/Al ratio of H-mordenites at 30 and 60 minutes reaction times.

methoxy benzylic alcohol (**2**) did not seem to be influenced by conversion or dealumination level, particularly at 60 minutes reaction time where a value of 4 to 5 is found for the para/ortho (**1/2**) selectivity ratio.

The feasibility of the reaction was thus demonstrated, and we shall focus now on the different reaction parameters able to improve the selectivity to alcohols **1** and **2** in the presence of the dealuminated H-mordenite with a Si/Al ratio of 18.

C. H-mordenite (Si/Al=18)

Effect of catalyst weight and reaction temperature

The effect of catalyst weight on the conversion of guaiacol, the selectivity to 3-methoxy-4-hydroxy benzylic alcohol (**1**) and 2-hydroxy-3-methoxy benzylic alcohol (**2**), and the para/ortho ratio is reported in Tables 2 and 3.

Table 2. Influence of reaction time on guaiacol conversion, selectivity to vanillic alcohol (**1**) and isomer (**2**), and para/ortho selectivity (12 ml of formalin, 0.250 g of catalyst, 1 ml of guaiacol, 358 K).

time min	guaiacol conversion (%)	selectivity to alcohols **1** and **2** (%)	para/ortho ratio*
30	18	47	5.0
60	28	51	4.3
120	42	57	4.2

*See footnote in Table 1.

Table 3. Influence of reaction time on guaiacol conversion, selectivity to vanillic alcohol (**1**) and isomer (**2**), para/ortho selectivity (12 ml of formalin, 0.500 g of catalyst, 1 ml of guaiacol, 358 K).

time min	guaiacol conversion (%)	selectivity to alcohols **1** and **2** (%)	para/ortho ratio*
30	32	58	4.8
60	49	64	4.8
120	73	60	4.4

*See footnote in Table 1.

Plots of the conversion of guaiacol as a function of catalyst weight at two temperatures are given in Figure 2 which also includes data from Tables 4 and 5. They show a direct dependence of conversion on catalyst weight suggesting uniformity in the catalyst surface acidity and operation in a non diffusion-limited regime.

A net increase in the selectivity to 3-methoxy-4-hydroxy benzylic alcohol (**1**) and 2-hydroxy-3-methoxy benzylic alcohol (**2**) is also observed when increasing catatlyst weight, particularly at reaction times of 30 and 60 minutes. The tendency would be then to increase catalyst weight and decrease reaction temperature as far as the guaiacol conversion is maintained at a reasonable level.

This is illustrated in Table 4 from the experimental results obtained at a lower temperature (338 K).

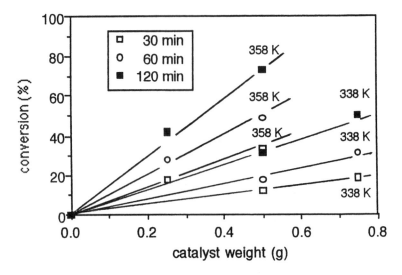

Figure 2. Plot of guaiacol conversion against catalyst weight at 338 and 358 K, and at 30, 60 and 120 minutes reaction times.

From the plots of guaiacol conversion against time, it was possible to calculate the initial reaction rates and the activation energies for the hydroxymethylation step ($\approx$ 50 kJ/mol) and for the condensation step ($\approx$ 99 kJ/mol). These values are close to the ones observed in hydroxymethylation of furan derivatives [8,11] and confirm that the reaction was not controlled by external diffusion.

Table 4. Influence of reaction time on guaiacol conversion, selectivity to vanillic alcohol (**1**) and isomer (**2**), and para/ortho selectivity (12 ml of formalin, 0.500 g of catalyst, 1 ml of guaiacol, 338 K).

time min	guaiacol conversion (%)	selectivity to alcohols **1** and **2** (%)	para/ortho ratio*
30	12	71	4.9
60	18	82	4.8
120	32	75	4.7

*See footnote in Table 1.

In addition, the selectivity to hydroxymethylated products was, as expected from a consecutive reaction mechanism, higher at low temperatures.

Effect of the formaldehyde to guaiacol ratio

Another important parameter to be considered in this reaction is the effect of the relative concentrations of formaldehyde and guaiacol reactants. The results obtained for formaldehyde/guaiacol molar ratios from 17.6 to 189.4 are given in Table 5.

It is seen from Table 5 that there was no fundamental change in guaiacol conversions by increasing the formaldehyde/guaiacol ratio. However, higher selectivities (80 to 85 %) to 3-methoxy-4-hydroxy benzylic alcohol (**1**) and 2-hydroxy-3-methoxy benzylic alcohol (**2**), and higher para/ortho ratios, 5 to 6, could be reached when the reaction was carried out in the presence of a large excess of formalin. Nevertheless, it should be noted that a slight decrease in selectivity was observed at same reaction time and reactant ratio, when the weight of catalyst was increased (Table 4 and Table 5, set *1*). Such a behavior is not unexpected for a consecutive reaction mechanism because the corresponding conversions are different.

Table 5. Influence of formaldehyde/guaiacol ratio on guaiacol conversion, selectivity to vanillic alcohol (**1**) and isomer (**2**), and para/ortho selectivity at different reaction times (0.750 g catalyst, 338 K).

set number	time min	guaiacol conversion (%)	selectivity to alcohols **1** and **2** (%)	para/ortho ratio*
1	30	19	70	5.3
	60	31	69	4.1
	120	50	66	4.8
2	30	13	75	6.3
	60	24	78	6.2
	120	41	72	6.2
	240	60	82	5.7
3	30	9	77	5.5
	60	15	91	5.7
	120	30	78	5.6
	240	45	89	5.5
4	30	14	83	4.7
	60	25	71	7.6
	120	39	81	5.9
	240	59	85	6.0

*See footnote in Table 1.

1: 12 ml of formalin, 1ml of guaiacol (formaldehyde/guaiacol molar excess = 17.6).

2: 12.75 ml of formalin, 0.25 ml of guaiacol (formaldehyde/guaiacol molar excess = 75.2).

3: 20 ml of formalin, 0.75 ml of guaiacol (formaldehyde/guaiacol molar excess = 146.8).

4: 12.9 ml of formalin, 0.1 ml of guaiacol (formaldehyde/guaiacol molar excess = 189.4).

Effect of the rate of addition of guaiacol

Another way to simulate the effect of a large molar excess of formaldehyde versus guaiacol is to add guaiacol progressively to the reaction medium. An experiment was thus performed by adding 5 aliquots of guaiacol (0.2 ml each) at 1 h intervals. The results obtained using such a procedure are given in Table 6.

Table 6. Effect of slow addition of guaiacol on guaiacol conversion, selectivity to vanillic alcohol (**1**) and isomer (**2**), and para/ortho selectivity (12 ml of formalin, 0.750 g of catalyst, 5 x 0.2 ml of guaiacol, 338 K).

time min	guaiacol conversion (%)	selectivity to alcohols **1** and **2** (%)	para/ortho ratio*
60	33	55	5.0
120	33	72	5.1
180	35	78	4.5
240	33	89	4.9
300	33	98	5.0

*See footnote in Table 1.

These results may be considered as a simulation of a continuous process and it can be seen that slightly higher selectivities to 3-methoxy-4-hydroxy benzylic alcohol (**1**) and 2-hydroxy-3-methoxy benzylic alcohol (**2**), up to ≈ 90 %, can now be obtained without important consequences on the para/ortho selectivity. The high formaldehyde/guaiacol ratio and, as a consequence, the presence of a large excess of water, helps hydroxymethylated intermediates to desorb rapidly, thus avoiding the formation of side reactions.

4. CONCLUSION

The objective of this work was to find an alternative route to 3-methoxy-4-hydroxy benzylic alcohol (vanillic alcohol, **1**) and 2-hydroxy-3-methoxy benzylic alcohol (**2**) using solid acid catalysts. The success of such a task was largely dependent on the operating conditions and on the knowledge of basic principles concerning both reaction mechanisms and catalysis by microporous materials.

For hydroxymethylation of guaiacol by aqueous formaldehyde, a H-form mordenite with a Si/Al ratio of 18 featuring a strong acidity and a moderate hydrophobicity allows the reaction to proceed with a high selectivity to 3-methoxy-4-hydroxy benzylic alcohol (vanillic alcohol, **1**) and 2-hydroxy-3-methoxy benzylic alcohol (**2**). The selectivity to these alcohols, as well as the para vs ortho orientation of the reaction, have been improved by working at low temperature in the presence of a large excess of formaldehyde and using appropriate catalyst loadings [12].

ACKNOWLEDGEMENTS

IFP, PQ Zeolites and Zeocat are gratefully acknowledged for providing catalysts samples and Jérôme Lecomte for fruitful discussions.

REFERENCES

1. J. Mathieu and J. Weill-Raynal, in *Formation of C-C bonds, Vol. I, Introduction of a Functional Carbon Atom*, George Thieme Verlag, Stuttgart, 1973.
2. T. Kiyora, Japanese Patent JP 63307835 (1988).
3. M.J. Climent, A. Corma, H. Garcia and J. Primo, *J. Catal.*, 130 (1991) 138.
4. M.H.W. Burgers and H. van Bekkum, *Stud. Surf. Sci. Catal.*, 78 (1993) 567.
5. R. Durand, P. Geneste, J. Joffre and C. Moreau, *Stud. Surf. Sci. Catal.*, 78 (1993) 647.
6. C. Moreau, R. Durand, C. Pourcheron and S. Razigade, *Industrial Crops and Products*, 3 (1994) 85.
7. J. Joffre, P. Geneste, A. Guida, G. Szabo and C. Moreau, *Stud. Phys. Theor. Chem.*, 71 (1990) 409.
8. J. Lecomte, A. Finiels, P. Geneste and C. Moreau, *J. Mol. Catal.*, in press.
9 D. Barthomeuf, *Mater. Chem. Phys.*, 17 (1987) 49.
10. N.Y. Chen, *J. Phys. Chem.*, 80 (1976) 60.
11. O. Boulet, R. Emo, P. Faugeras, J. Jobelin, F. Laporte, J. Lecomte, C. Moreau, M.C. Neau, G. Roux, J. Simminger and A. Finiels, French Appl. 95/13827 (1195).
12. C. Moreau, A. Finiels, S. Razigade, L. Gilbert and F. Fajula, PCT Int. Appl. WO 96/00778 (1996).

The Reaction of Methanol to Hydrocarbons: Effect of Adding Cu^{2+} to ZSM-5 Using $[Cu(en)_2]^{2+}$ by Ion Exchange

S. N. R. Rao[1] and M. G. White*

Focused Research Program in Surface Science and Catalysis, School of Chemical Engineering, Georgia Institute of Technology; Atlanta, GA, 30332-0100 (USA)

[1]*Present address: I[2] Technology, 1603 LBJ Freeway, Suite 780, Dallas TX, 75234 (USA)*

*Corresponding author

Abstract

Two series of catalysts were prepared for which Cu^{2+} ions were introduced into the zeolite ZSM-5. For one series, an aqueous solution of $Cu(OAc)_2$ was the sources of the Cu^{2+} ions; and for the other series, $Cu(en)_2(ClO_4)_2$ dissolved in acetonitrile was the source of the Cu^{2+} ions. These catalysts were tested for reactivity and selectivity in the methanol to hydrocarbon reaction. The selectivity of this reaciton to C_8-C_{10} hydrocarbons was favored by the catalyst prepared using $Cu(en)_2(ClO_4)_2$ as the source of Cu^{2+} ions.

Introduction

White (1) and Van Der Voort, *et al.* (2, 3) discussed the uses of metal complexes to synthesize supported metal oxide adsorbents and catalysts which demonstrated unusual thermal stability against sintering. Central to this technology was the strong interaction between the metal complex and the support which was the result of (a) hydrogen-bonding interactions between hydroxyl groups of the support and base functions of the metal complex, (b) ion exchange between surface protons and the cation complex, and (c) ligand exchange of the surface oxide anions for the anionic ligands of the complex. By this technology, the support surface was decorated with a thin layer of the metal complexes, and the ligands were removed by controlled heating to produce a surface oxide which was well-dispersed on the support oxide. Metal ion dispersions as high as 100% were common-place for systems such as Cu^{2+}/silica, TiO_x/silica, VO_x/silica and these dispersions were stable against heating in air to 450°C.

Moreover, the catalytic properties of these dispersions were sensitive to the metal ion ensemble size for some reactions. White (1) showed that

Cu^{2+}/silica catalysts were very selective to the ethanol product, 98-100%, of the acetaldehyde hydrogenation over catalysts synthesized from a single layer of $Cu(acac)_2$/silica. However, other catalysts showed significant yields of the coupling product, ethyl acetate, from the same reaction when dinuclear or multinuclear Cu complexes were used as the precursors on silica (1, 3-6). Thus, the product yields of the acetaldehyde hydrogenation reaction over the Cu/silica catalysts could be altered drastically with the change in the nuclearity of the Cu metal complex precursor. We have demonstrated the beneficial effect of this technology for the methanol oxidation reaction catalyzed over VO_x/silica which was developed from silica decorated with a monolayer of $VO(acac)_2$ (3). It appeared prudent to extend this technology to create *zeolitic* solids with *novel* and *controlled* catalytic properties.

The zeolite known as ZSM-5 is characterized by channels having an aperture size of 10-oxygen members and 10-T atoms; where T is the metal cation (*i. e.*, Si^{4+} or Al^{3+}). The regular alternating arrangement of TO_4 which creates this "10-ring" aperture results in an aperture size of 5-6 Å. This zeolite is rich in silicon having a SiO_2/Al_2O_3 ratio > 10. The framework describes one set of sinusoidal channels (5.1 x 5.5 Å, minor x major diameters) intersecting at right angles to a set of straight channels (5.3 x 5.6 Å). The crystal structure is characterized by an orthorhombic unit cell belonging to the space group Pnma with the unit cell constants: a = 20.1 Å, b = 19.9 Å, and c = 13.4 Å. The void volume is 0.10 cc/g. The zeolite may be synthesized according to the method of Rollmann (7) or Masuda (8). Rollmann's method requires up to 24 days to complete the crystallization; whereas, Masuda's modification reduces the crystallization time to 2-4 days.

Transition metal ions have been ion exchanged into zeolites from aqueous salts of the tmi (9 and many others); however, the ion exchange technique from aqueous solutions does not offer a high degree of control on the morphology when the tmi forms multiple species in aqueous solution as a result of dynamic equilbria. Attempts to control the morphology of the tmi in zeolites include the "ship-in-the-bottle synthesis" technique. Examples of this technique include Schiff base complexes (10) and noble metal carbonyls (11). If the pristine complex is active for the desired reaction, then this approach offers a means to prepare a well-characterized catalyst and derive the benefits offered by the structure directing properties of the zeolite. However, these benefits are often forfeited if the ligands must be removed to render the catalyst active as the metal residue will migrate from the pores at some reaction temperatures.

Other researchers devised another means to immobilize metal complexes into the zeolite by using a metal complex as a template. Bruce reported on the use of $[Co^{3+}(en)_3]^{3+}$ as a template for the crystallization of zeolite A (12) to produce microporous solid having high loadings of Co^{3+}. Later, Bruce (13-15) showed how novel, AlPO-layered structures were grown using other $[Co^{3+}(L)_3]^{3+}$ metal complexes as the template where L is trimethylenediamine or diethylenetriamine. We show here the use of stable metal complexes as ion exchange agents into ZSM-5 using an aprotic solvent.

Experimental methods and results

A. Catalysts synthesis

Samples of Na-ZSM-5 were synthesized according to the procedure of Masuda, *et al.* (8) and the template, tetrapropylammoniumbromide, was removed by heating in air to 450°C. One series of Cu/ZSM-5 catalysts, designated CuZ5LHx (x = 4, 5), were prepared using aqueous solutions of copper acetate as the precursor after Iwamoto, *et al.* (16). This series of catalysts represents the state-of-the-art technology. Other catalysts were prepared using the perchlorate salt of copper(II) ethylenediamine $[Cu(en)_2]^{2+}$ (50 x 10^{-3} M solution) as the ion exchange species in an aprotic solvent (acetonitrile, 200 cm^3) with 10 g of Na-ZSM-5. This solution was stirred for 16 h at room temperature before filtering onto a Buchner funnel. The solid retained on the filter was washed 4 times with fresh acetonitrile and dried at room temperature. The catalysts were stored in this condition in stoppered bottles until needed.

This second series of catalysts, desinged CuZ5En2x (x = 1, 2) enables a test of the new technology for preparing ion exchanged zeolites using stable cationic metal complexes as the source of the metal ion. Copper ethylenediamine is known to be very stable even in strong acid solutions, thus, we expect that it will remain intact during the ion exchange process. Molecular modeling shows that $Cu(en)_2^{2+}$ fits easily into the channels of ZSM-5 and moreover, simple energy minimization calculations demonstrated that the cation complexes sited in these channels where two or more framework Al^{3+} ions were located (17). We analyzed for the perchlorate anion (18) to determine if any $Cu(en)_2^{2+}$ remained in the solid without exchanging for host cations.

All samples were pretreated before characterization and reaction studies. This pretreatment was oxidation in O_2/He at 500°C for 1 h followed by heating in He only at 500°C for 1 h. Thus no template was present in any of the samples after this treatment.

Table 1 Copper metal loading and bulk Si/Al ratio

Sample	Cu (wt%)	Si/Al mol/mol	Cu Exch (%)	Solvent
CuZ5LH4	1.81	16	72	Water
CuZ5LH5	0.78	113	179	Water
CuZ5En21	1.45	16	58	Acetonitrile
CuZ5En22	1.91	16	99	Acetonitrile

B.   Elemental analyses

The catalysts were characterized by Galbraith Laboratories, Inc. (Knoxville, TN) for metal loading and Si/Al ratio. Table 1 shows the copper metal loading, the bulk Si/Al ratio, percentage Cu exchange, and type of solvent used in the ion exchange. None of the samples prepared from $[Cu(en)_2]^{2+}$ contained the perchlorate anion which suggests that all of the $[Cu(en)_2]^{2+}$ cations were ion exchanged. One may observe Cu exchage capacities which exceed 100%. This anomalous looking report is a result of the procedure for calculating the exchange capacity which assumes that the Cu ion species bears a 2+ charge, and that all of the Al in the sample is framework Al^{3+}. If either of these assumptions is not true, then it is possible to report Cu exchange capacities which exceed 100%. This result is often noticed when the Si/Al ratio in the sample is very large.

C.   Powder XRD

A standard obtained from Texaco Research and Development (Port Arthur, TX) was also examined by powder x-ray diffraction (PXRD), Fig. 1-A, to confirm the phase(s) of the unknowns. A ZSM-5 sample, prepared in our laboratories, was examined for crystal structure by PXRD to determine the phase(s), Fig. 1-B. These two samples, Figure 1 A-B, show the same PXRD thus confirming the structure of the zeolites we synthesized in our laboratory. One more sample containing Cu^{2+} ions (CuZ5LH4, Fig. 1-C) showed the same PXRD pattern as the zeolites prior to ion exchange.

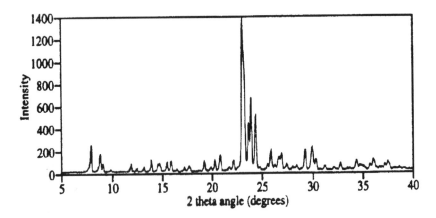

Figure 1-A Powder X-Ray Diffraction of Standard ZSM-5

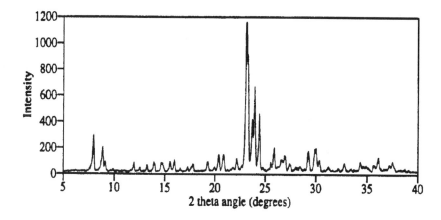

Figure 1-B Powder X-Ray Diffraction of Synthesized ZSM-5

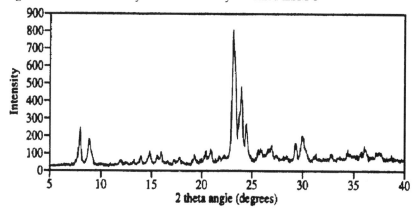

Figure 1-C Powder X-Ray Diffraction of Cu/ZSM-5

D.   FTIR of catalyst titrated with CO.

The oxidation state of the copper ion controls the activity/selectivity of some reactions and therefore is an important catalyst property for supported Cu catalysts. We examined the infrared spectra of two catalysts (CuZ5LH4 and CuZ5En21) which were titrated with CO to characterize the copper species in the catalysts. FTIR studies were conducted in a DRIFTS cell in which the samples were evacuated and/or exposed to various gases. The samples were pretreated first by heating in air for 4 hours at 500°C and then evacuated at 500°C for 2 hours, cooling to room temperature. Carbon monoxide was introduced at room temperature and 20 Torr. The samples were heated in this atmosphere to 450°C and cooled to 40°C, then evacuated. This step should reduce the Cu ions to the monovalent state (Cu^{1+}). Thus, we used the IR spectrum of chemisorbed CO before and after this reduction to determine the types of sites present in the Cu/ZSM-5 samples.

The sample prepared by the aqueous ion exchange of Cu(II) acetate (CuZ5LH4) showed four peaks in the IR spectrum before (solid line) and after the reduction (dashed line) step (Table 2, Figure 2-A). The catalyst prepared from the $Cu(II)(en)_2$ cation (CuZ5En21) showed the same four peaks before reduction (Fig 2-b); however, the peak at 2135 cm^{-1} was weaker than the corresponding peak in the CuZ5LH4 sample. After reduction in CO, the sample prepared from the $Cu(OAc)_2$ precursor showed an IR spectrum of the chemisorbed CO which was similar to that before reduction. However, reduction of the sample prepared from the $[Cu(en)_2]^{2+}$ cation caused the IR CO spectrum to change dramatically. The peak at 2157 cm^{-1} decreased as a result of the reduction treatment, whereas the peak at 2135 cm^{-1} was larger in the reduced sample than in the oxidized sample.

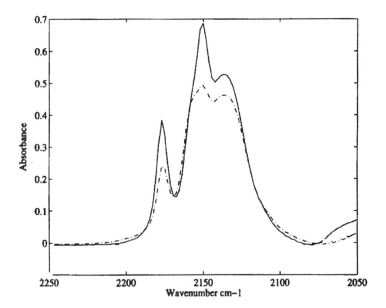

Figure 2-A FTIR of CO Adsorbed on Cu/ZSM-5 Sample (CuZ5LH4)
Dashed line is for sample reduced, solid line is for sample before reduction.

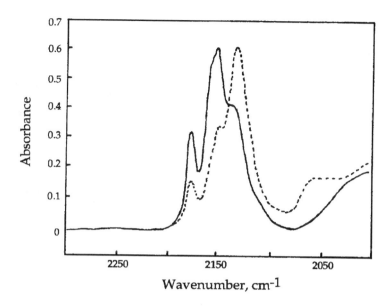

Figure 2-B FTIR of CO Adsorbed on Cu/ZSM-5 Sample (CuZ5En21)
Dashed line is for sample reduced, solid line is for sample before
reduction.

Table 2 Summary of IR Peaks in CO/Cu/ZSM-5

peak locations: CuZ5LH4		peak locations: CuZ5En21		
Oxidized	Reduced	Oxidized	Reduced	Assignments
2157(m)*	2157(s)*	2157(s)*	2157(w)*	Cu^+CO: copper species located near isolated Al ions in framework-- Type I site (19, 20)
2150(s)	2150(s)	2150(s)	2150(m)	symmetric and
2177(s)	2177(s)	2177(s)	2177(m)	asymmetric $Cu^+(CO)_2$ --Type III site (19)
2135(m)	2135(s)	2135(m)	2135(s)	Cu^+CO: copper species located near two Al ions in the framework -- Type II site (20)

Notes
Lower case letters in parentheses indicate intensity of peak: (s) = strong; (m) =
medium; and (w) = weak. Numbers in brackets show references to assignments.
*The peak at 2157 cm^{-1} is a shoulder to the 2150 cm^{-1} peak

Data from the literature of peak assignments (Table 2) permit us to relate the changes in the IR spectrum to the changes in surface sites. The type I site has been identified as a cuprous ion located near an isolated Al^{3+} ion; whereas the type II site is a cuprous ion located near two Al^{3+} ions. The oxide precursor of the catalyst prepared from $[Cu(en)_2]^{2+}$ showed a fewer number of the type II sites and a larger number of the type I sites compared to the distribution of sites in the sample prepared from the acetate precursor. After the reduction in CO, the number of type I sites decreased relative to the number of type II sites in the sample prepared from the ethylenediamine precursor. These results suggest that Cu^{2+} ions originally sited near isolated Al^{3+}, are converted into Cu^{1+} ions near multiple Al^{3+} ions after reduction in CO. Myamoto, *et al*, modeled the siting of Cu^{2+} ions in ZSM-5 to show that the cupric ions located near two framework Al^{3+} ions in preference to other locations (21) and these results agree with our own modeling results for $[Cu(en)_2]^{2+}$ (*vide supra*).

This suggestion is agreeable to a model where some of the $[Cu(en)_2]^{2+}$ cation ion exchanges into a site which can balance the charge; this site could be located near two framework Al^{3+} ions. Other ions site near isolated Al^{3+} ions. Under the influence of the CO and elevated temperatures, the copper ions located near isolated Al^{3+} ions migrate to another position located near multiple Al^{3+} ions. This migration of the copper ion need not be present in the catalyst synthesized from the acetate precursor in aqueous solution as the acetate species is electrically neutral, originally. The original siting of this ion may be attended by water molecules so that the divalent Cu ion may be partially charge balanced by an $[OH]^{1-}$ ion during ligand exchange of an $[OAc]^{1-}$ ion for a surface $[OH]^{1-}$. Thus, this new complex, $[Cu^{2+}(OH)^{1-}]^{1+}$ demonstrates an overall charge of 1+ even though the copper ion is divalent. With an overall charge of 1+, this complex may site to an isolated framework Al^{3+} ion. During reduction in CO, the OH leaves with a surface proton as water and the cupric ion is reduced to a cuprous ion and this ion need not migrate to another surface site to remain in charge balance. We demonstrated the importance of the OH ion, i. e., effect of solvent, in the siting of the Cu ion by impregnating the $Cu(OAc)_2$ in dry acetontrile for which $[OH]^{1-}$ concentration was presumed to be very low. This catalyst showed very little Cu ion incorporation into the zeolite and it was almost inactive for the NO reduction reaction (17).

E.   Chemical reactions

We characterized these catalysts by two probe reactions: NO decomposition and methanol conversion to higher-molecular weight hydrocarbons. The NO decomposition has been reported in the literature for Cu/ZSM-5 catalysts prepared from the acetate precursor (16, 22 and many other citations), whereas, the methanol reaction has only been reported for the undecorated H-ZSM-5 catalyst. We show here the effect of precursor upon the yields of these two reactions.

NO decomposition: The conversion to N_2 versus space-time is shown in Figure 3 for the CuZ5LH4, CuZ5En21 operating at 773 K for a feed containing 4 mol% NO and the balance He at a total pressure of 1 atm. The details of the microreactor and analytical apparatus were described previously (17). We also present in Fig. 3 a correlation of the literature results over a catalyst similar to CuZ5LH4 (22). These data show that the catalyst prepared from the acetate precursor is more active than the one described in the literature and it is 8-11 times more active than the catalyst prepared from the $[Cu(en)_2]^{2+}$ precursor. Clearly the effects of the solvent, water *vs* acetonitrile, and the effect of precursor combine to make the catalytic properties of these two samples much different.

The characterization by CO adsorbed on the Cu species suggests that the environment surrounding the Cu may play a very important role in governing the activity of the catalyst for NO decomposition. The more active catalyst (CuZ5LH4) shows a nearly equal distribution of the copper ions sited near pairs of Al^{3+}, Site II, and isolated Al^{3+} ions, Site I; whereas, the catalyst prepared from $[Cu(en)_2]^{2+}$ shows most of the copper ions sited near multiple Al^{3+}, Site II, *after reduction in CO*. The data of Li and Hall suggest that Cu^{1+} is a likely candidate as the active phase for this reaction (22), thus, the IR data of CO on a reduced catalyst is relevant to explaining the activity of these catalysts. It is speculated that the rate of reaction is influenced by the number of Al^{3+} ions residing in the neighborhood of the copper species. We speculate further that these neighboring Al^{3+} ions influence the ease with which Cu ions are reduced and reoxidized during the steps of the NO reaction. In particular, we hypothesize that Cu ions are more difficult to reduce when surrounded by more than 1 Al^{3+} ion.

Methanol oligomerization: The H-ZSM-5 catalyst is known to convert methanol to higher-molecular-weight hydrocarbons in the gasoline blending range. In this work, we focus on the yields (Table 3) of di-, tri-, and tetramethyl benzenes obtained from the following three catalysts: H-ZSM-5, CuZ5LH4, and CuZ5En21. The reaction temperature was 300°C and the feed gas was an He stream saturated with methanol at room temperature in a flow of He (30 STP cm^3/min). Rao described previously the details of the reactor and analytical system (17). The amount of catalyst was adjusted to totally convert all of the methanol at this temperature (0.71 g of CuZ5LH4; 0.75 g of CuZ5En21; and 0.75 g of ZSM-5). The effect of adding as little as 2 wt% Cu to the zeolite was to decrease the light gas yields (C_3-C_4) from 80% to 68% at constant methanol conversion and was to increase the yields of tri- and tetramethylbenzenes from 2.5% to 19.5%. The effect of the ethylenediamine precursor and acetonitrile solvent is to improve the yields of di-, tri-, and tetramethylbenzenes (13.6% to 23%) over that obtained from the catalyst prepared from $Cu(OAc)_2$. The FTIR study suggests that the CuZ5En21 sample contained more copper species located in the site II configuration (i. e., near multiple Al^{3+} ions) than the CuZ5LH4 sample *after reduction in CO*. Thus, we speculate that this siting of the Cu ions near multiple Al^{3+} ions may be responsible for the increased yields of di-, tri-, and tetramethyl benzenes.

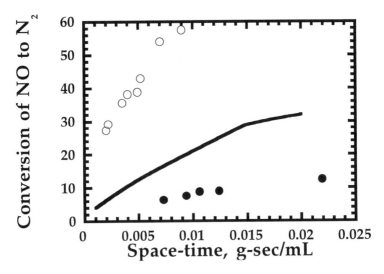

Figure 3 NO Conversion to N_2 over Cu/ZSM-5 Catalysts at 500°C
Data from Li and Hall - line; CuZ5LH4 - open circles; CuZ5En21 -- filled circles.

Table 3 Yields Obtained over H-ZSM-5, CuZ5LH4, and CuZ5En21

Species	Undecorated H-ZSM-5	Sources of Copper Precursor	
		$Cu(OAc)_2$	$Cu(en)_2(ClO_4)_2$
CH_3OH	0%	7.9%	0%
$C_3\text{-}C_4$	80.3	62.5	68.0
Total C_8	4.5	2.1	3.5
Total C_9	2.5	4.8	8.6
Total C_{10}	0	6.7	10.9
H_2O	12.7	16.0	17.5
Totals	100	100	99.8

One may explain these results of gas yield and yields of methyl benzenes as a function of Al^{3+} neighbors by a multiple site model for which *alkylation* reactions to aromatics occur at Al^{3+} sites and that alcohol conversions to alkyl intermediates occur at Cu^{1+} sites. Thus, the siting of copper ions near multiple Al^{3+} sites produced more methyl benzenes than if the copper ions were sited near multiple Al^{3+} ions. Without any copper sites, as in H-ZSM-5, we speculate that few polyalkylbenzenes will be formed and that yields of lighter gases will be favored.

Conclusions -- Metal Complexes in ZSM-5

Metal complexes may be introduced into Na-ZSM-5 from aprotic solutions to produce novel, ion exchanged metal oxide in zeolite catalysts. The use of the metal complex affords a degree of control on the siting of the metal ion in the zeolite which can be exploited to change and control the catalytic properties. In the case of NO decomposition, the siting of the copper species did not produce high activity catalysts. However, the same catalyst showed very favorable catalytic properties for the methanol conversion to higher-molecular-weight hydrocarbons. The use of the copper metal complex resulted in a catalyst which produced larger amounts of tri- and tetra-methyl benzene compared to a catalyst prepared from aqueous iion exchange of the Cu salt [$Cu(OAc)_2$].

References

1. M. G. White, Catal. Today, **vol 18**, No. 1., pp. 73-109, (1993).
2. P. Van Der Voort, M. B. Mitchell, E. F. Vansant and M. G. White, Interface Science, **5**, 179 (1997).
3. P. Van Der Voort, M. Baltes, E. F. Vansant and M. G. White, Interface Science, **5**, 209 (1997).
4. A. Zippert and M. G. White, ACS Preprints --Division of Petroleum Chemistry, Chicago, IL. Meeting of the ACS, August, 1993, pp 872-875.
5. J. A. Bertrand, D. A. Bruce, and M. G. White, A. I. Ch. E. Journal, **vol 39**, No. 12, 1966, (1993).
6. H. N. Choksi, J. A. Bertrand and M. G. White, J. Catalysis, **164**, 484-489 (1996).
7. L. D. Rollmann and E. W. Valyocsik, "Zeolite Molecular Sieves," in Inorganic Synthesis, **vol 22**, p. 67, John Wiley and Sons, (1981).
8. T. Masuda, A. Sato, H. Hara, M. Kouno and K. Hasimoto, Appl. Catal. A: Gen. **111**, 143 (1994).
9. D. W. Breck, *Zeolite Molecular Sieves: Structure, Chemistry and Use* (John Wiley and Sons, New York 1974); A. Dyer, *An Introduction to Zeolite Molecular Sieves* (John Wiley and Sons, New York 1988); R. Szostak, *Molecular Sieves: Principles of Synthesis and Identification* (Van Nostrand Reinhold, New York 1989).
10. N. Herron, Inorg. Chem., **25**, 4741 (1986).
11. C. Bowers and P. K. Dutta, J. Catal. **122**, 271 (1990).
12. D. A. Bruce, J. A. Bertrand, P. S. E. Dai, M. L. Occelli, R. Petty and M. G. White, in *Catalysis of Organic Reactions,* ed. by M. Scaros (Marcell Dekker, Inc., New York, 1994), pp. 545-52.

13.   D. A. Bruce, A. P. Wilkinson, M. G. White and J. A. Bertrand, J.
      Chem. Comm., Royal Society of Chemistry Journal, 2059-2060,
      (1995).
14.   D. A. Bruce, A. P. Wilkinson, M. G. White and J. A. Bertrand, J. of
      Solid State Chemistry, **125**, 228-233 (1996).
15.   J. A. Bertrand, D. A. Bruce, A. P. Wilkinson, P. S. E. Dai, R. H. Petty
      and M. G. White, in *Catalysis of Organic Reactions*, Ed. by Russel E.
      Malz, Jr. (Marcel Dekker, Inc. 1996), p 435.
16.   M. Iwamoto, H. Yahiron, K. Tanda, N. Mizuno, Y. Mine and S.
      Kagawa, J. Phys. Chem., **95**, 9, 3727, (1991).
17.   S. N. R. Rao, Ph. D. thesis, Georgia Institute of Technology (1996).
18.   W. Geilmann and A. Voight, Z. Anorg. Chem. **193**, 311 (1930).
19.   A. W. Aylor, S. C. Larsen, J. A. Reimer and A. T. Bell, J. Catal. **157**,
      592 (1995).
20.   C. Marquez-Alvarez, G. S. McDougall, A. Guerrero-Ruiz and I.
      Rodrigues-Ramos, Appl. Surf. Scie, **78**, 477 (1994).
21.   A. Miyamoto, H. Himei, Y. Oka, E. Maruya, M. Katagiri, R. Vetrivel
      and M. Kubo, Catal. Today, **22**, 87 (1994).
22.   Y. Li and W. K. Hall, J. Catal. **129**, 202 (1991).

Enantioselective Hydrogenation of Trifluoroacetophenone Over Pt/alumina

Michael Bodmer, Tamas Mallat and Alfons Baiker

Laboratory of Technical Chemistry, Swiss Federal Institute of Technology, ETH Zentrum, CH - 8092 Zürich, Switzerland

Abstract

The enantioselective hydrogenation of 2,2,2-trifluoroacetophenone (TFAP) over chirally modified Pt/Al$_2$O$_3$ afforded up to 61 % ee to (R)-1-phenyl-2,2,2-trifluoroethanol. Enantiodifferentiation is favored by applying weakly polar solvents and low surface hydrogen concentration (low hydrogen pressure, reactor operated in the transport limited regime). It is proposed that both polar solvents and hydrogen can efficiently compete with TFAP for the Pt sites and diminish ee. Among the modifiers tested only cinchonidine and (R)-2-(1-pyrrolidinyl)-1-(1-naphthyl)ethanol were efficient. Molecular modelling studies provided a feasible structure for the diastereomeric transition complex formed between TFAP and cinchonidine and an explanation for the observed enantiodifferentiation.

Introduction

The discoveries of the Ni - tartrate and the Pt - cinchona alkaloid systems represent two milestones in the development of solid enantioselective hydrogenation catalysts. The systematic effort aiming at understanding the reaction

75

mechanism and optimizing the catalyst preparation and reaction conditions resulted in high ee's in the original reactions, in the hydrogenation of β- and α-ketoesters, respectively {for details see some recent reviews (1-3)}.

Most of the studies of the cinchona-modified platinum focus on the reduction of ethyl pyruvate, and only a few attempts have been made to broaden the scope of the reactants. Up to 20 % enantiomeric excess (ee) has been achieved in the hydrogenation of some β-ketoesters, aryl-alkyl ketones, α-methoxy ketones and β-diketones (4, 5). 33-38 % ee has been reported for the hydrogenation of α-diketones (6) over Pt/SiO$_2$ modified with cinchonidine (CD). These values are far below the efficiency achieved in the hydrogenation of α-ketoesters and α-ketoacids {up to 85-95 % ee, depending on the structure of the reactant (7, 8)}.

We have recently reported the successful application of CD-modified Pt for the enantioselective hydrogenation of ketopantolactone {79 % ee (9)} and some α-ketoamides {up to 60 % ee (10)}. The hydrogenation of 2,2,2-trifluoroacetophenone (TFAP) to 1-phenyl-2,2,2-trifluoro-ethanol (Scheme 1) represents another extension of the application range of chirally modified Pt (11).

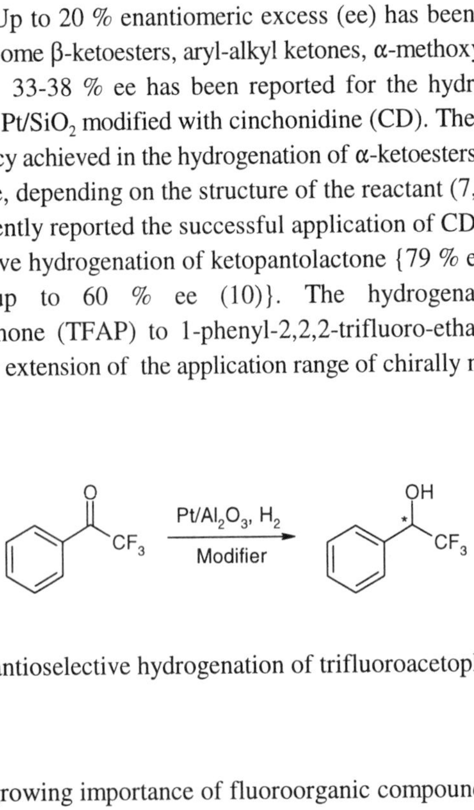

Scheme 1   Enantioselective hydrogenation of trifluoroacetophenone (TFAP)

Due to the growing importance of fluoroorganic compounds in agricultural and pharmaceutical chemistry, the asymmetric reduction of various aryl α-fluoroalkyl ketones has been investigated using chiral reagents (12-14) or homogeneous complex catalysts (15, 16). 90-95 % yield and 90 % ee have been obtained in the reduction of TFAP with organoborane reducing agents (12, 14). An early attempt for the asymmetric hydrogenation catalyzed by Co with a quadridentate mononegative ligand (15) failed to produce any ee. Recently, the Rh-catalyzed asymmetric reduction by hydride transfer from 2-propanol in the presence of a C2 symmetric chiral diamine as ligand provided 33 % ee to the (*R*)-enantiomer, in a rather slow reaction (16).

Intrigued by the obvious difficulties of this asymmetric catalytic reaction, we continued our study with chirally modified Pt/alumina, aiming at clarifying the role of some important reaction parameters and the chemical structure of the modifier.

Experimental

A 5 wt-% Pt/alumina (Engelhard 4759) was pretreated in a fixed bed reactor in flowing nitrogen at 400 °C for 30 min, followed by a reductive treatment in hydrogen for 90 min and cooling to room temperature in hydrogen. The metal dispersion after the heat treatment was 0.27, as determined from TEM images.

Hydrogenations were performed in a 100 ml stainless steel autoclave (Baskerville) with magnetic stirring (n = 1000 min^{-1}). A 50 ml glass liner with a PTFE cap and stirrer was used to keep the reaction mixture inert. Under standard conditions, 90 mg 5 wt% Pt/alumina (after reductive treatment), 1.28 g (7.35 mmol) 2,2,2-trifluoroacetophenone (TFAP, Fluka), 13.6 μmol cinchonidine (CD, Fluka) and 20 ml solvent were mixed under air for 20 min before reaction (in the glass liner of the autoclave). This procedure was found to be the most efficient pretreatment for obtaining good ee (11). The hydrogenation reaction was carried out at room temperature and 10 bar pressure. The pressure was held at a constant value by a computerized constant volume - constant pressure equipment (Büchi BPC 9901).

The enantiomeric excess (ee) and conversion were determined by an HP 5890A gas chromatograph equipped with a Chirasil DEX CB (Chrompack) capillary column. The standard deviation of the analytical method was ± 0.5 %. The enantioselectivity is expressed as ee (%) = 100 × ([|R]-[S])/([R]+[S]).

Results and Discussion

Solvent Effect

The enantioselective hydrogenation of TFAP has been carried out in a wide range of solvents using standard catalyst pretreatment and reaction conditions. Table 1 illustrates that the chemical nature of solvent had a strong influence on the average reaction rate and the final ee. For the interpreation of the solvent effect, the enantioselectivities were plotted as a function of solvent polarity characterized by the empirical solvent parameter E_T^N (17). It can be seen from Fig. 1 that apolar or weakly polar solvents favor the enantiodifferentiation, but there is no clear correlation in the range of polar and protic polar solvents.

Table 1   Hydrogenation of TFAP in various solvents listed in the order of polarity (standard conditions)

N°	Solvent	Time min	Conv. %	ee %
1	cyclohexane	105	100	47
2	n-hexane	54	93	49
3	mesitylene	n.d.*	93	50
4	toluene	122	100	39
5	*t*-butyl methyl ether	146	100	47
6	1,2,4-trichlorobenzene	103	97	46
7	chlorobenzene	94	100	41
8	1,2-dichlorobenzene	84	100	52
9	2-propyl acetate	100	99	44
10	dichloromethane	112	97	50
11	1,2-dichloroethane	101	90	49
12	N,N-dimethyl formamide	140	73	11
13	acetonitrile	70	100	20
14	2-propanol	67	91	42
15	acetic acid	100	43	6
16	ethanol	n.d.*	30	24

* - not determined

A feasible interpretation is that not (only) the solvent polarity, but the interaction between Pt and solvent is important. Assuming that TFAP adsorbs weakly on Pt, those solvents which interact strongly with the metal surface can compete with the reactant for the active sites and hinder the proper interaction between reactant, modifier and Pt. For example, the destructive adsorption of ethanol on Pt produces strongly adsorbed CO and hydrocarbon fragments, while this degradation could not be detected for 2-propanol (18-20). The presence or absence of surface impurities can rationalize the remarkably different ee's obtained in these two alcoholic solvents.

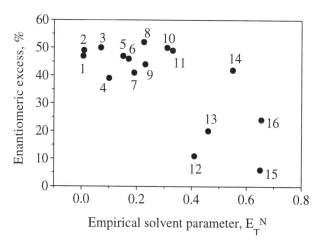

Figure 1   Influence of solvent polarity on the enantiomeric excess (standard conditions, numbers according to Table 1).

Influence of Hydrogen Pressure

The hydrogen pressure, or more precisely the surface hydrogen concentration, is another important parameter of the reaction. The measure of enantiodifferentiation in toluene between 1 and 80 bar is shown in Fig. 2. Obviously, high hydrogen concentration diminished the enantioselectivity. A similar correlation was observed in other solvents, such as 1,2-dichlorobenzene, as illustrated in Fig. 3. The optimum hydrogen concentration depended on other parameters, such as the reactant concentration (Fig. 3). When repeating the experiment in dichlorobenzene at 2 bar (shown in Fig. 3), but varying the mixing frequency between 1000 (standard conditions) and 100, a small but significant change in ee was observed (Fig. 4). Low mixing efficiency, which provided lower surface hydrogen concentration under otherwise identical conditions, was advantageous for the enantiodifferentiation. Note that the opposite effect of hydrogen concentration was observed in the enantioselective hydrogenation of α-ketoesters, α-ketoamides and ketopantolactone over cinchona-modified Pt (3, 9, 10).

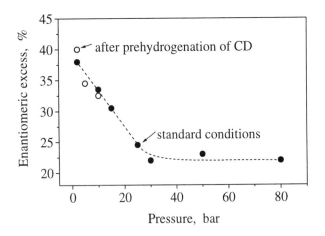

Figure 2   Influence of hydrogen pressure on the ee in toluene under standard conditions (●), and when using CD prehydrogenated in toluene at 40 bar (O)

A possible explanation is that the quinoline ring system of CD is (partially) saturated at high hydrogen pressure, resulting in weaker anchoring of the modifier onto the Pt surface and a loss in ee (21). In control experiments CD was prehydrogenated at 40 bar for 1 h in the autoclave before TFAP was added to the slurry. The minor changes observed in these experiments (open circles in Fig. 2), compared to the reactions performed under standard conditions, indicate that the modifier basically retained its enantiodifferentiating ability.

NMR analysis of CD after prehydrogenation at 40 bar indicated that the quinoline rings were unattacked, but 75 % of the vinyl groups were reduced to ethyl groups to form 10,11-dihydrocinchonidine.

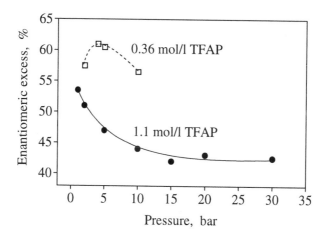

Figure 3 Influence of hydrogen pressure and reactant concentration on the ee in 1,2-dichlorobenzene; standard conditions, TFAP concentration: 0.36 mol/l (standard conditions) or 1.1 mol/l

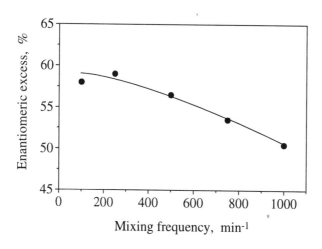

Figure 4 Influence of mixing frequency on the enantioselectivity (1,2-dichlorobenzene, 2 bar, standard conditions)

Influence of Reaction Temperature

The hydrogenation of TFAP has been carried out in the temperature range -12 to 70 °C, in mesitylene. The reaction time, conversion and enantioselectivity data are collected in Table 2. The ee decreased with ascending temperature above room temperature. Similar observations have been reported for the enantioselective hydrogenation of other activated carbonyl compounds over CD-modified Pt (2, 3, 9).

The unusual decrease of ee below room temperature is likely due to kinetic effects. It has been shown above that mass transport limitations in the slurry reactor favor the enantiodifferentiation, due to the lower hydrogen concentration on the Pt surface under steady state conditions. At room temperature or above the reactor operated in the mass transport limited regime, but at low temperature the transport limitation gradually decreased and the reaction became kinetically controlled. The resulting high surface hydrogen concentration diminished the ee.

Table 2   Influence of reaction temperature on the rate and enantioselectivity of TFAP hydrogenation (solvent: mesitylene, standard conditions)

Temp. °C	Time min	Conv. %	ee %
-12	431	10	32
4	423	17	39
20	n.d.*	93	50
42	120	66	40
70	120	83	28

* - not determined

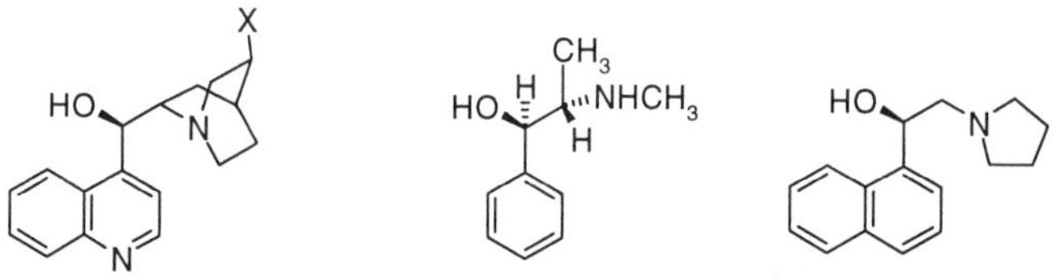

Cinchonidine (X = C$_2$H$_3$)
Dihydrocinchonidine (X = C$_2$H$_5$)

(-)-Ephedrine

1
(*R*)-2-(1-Pyrrolidinyl)-
-1-(1-naphthyl)ethanol

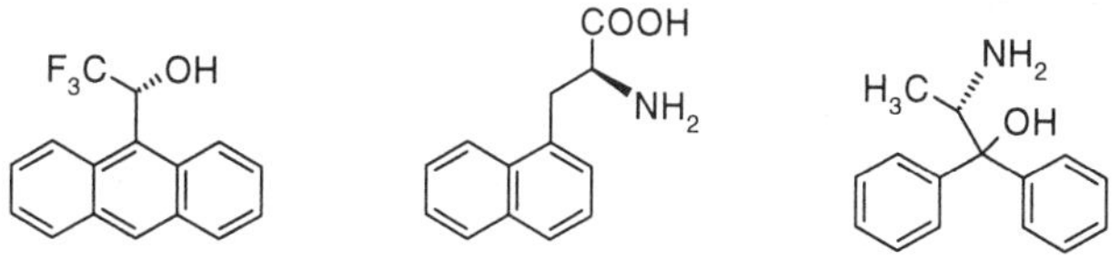

2
(*R*)-1-(9-Anthracenyl)-
2,2,2-trifluoroethanol

3
3-(1-Naphthyl)-
-L-alanine

4
(*S*)-2-Amino-1,1-diphenyl-
-1-propanol

Scheme 2

Table 3   Enantioselective hydrogenation of TFAP in the presence of various chiral modifiers (13.6 µmol modifier, standard conditions).

Modifier	Time min	Conversion %	Ee %
Cinchonidine	104	95	44 (*R*)
Cinchonidine hydrochloride	166	100	47 (*R*)
Dihydrocinchonidine	125	94	44 (*R*)
Ephedrine	152	66	3 (*S*)
1	135	100	35 (*R*)
2	110	66	4 (*S*)
3	130	63	3 (*R*)
4	135	69	2 (*R*)

Change of Modifier Structure

It has been attempted to substitute CD with other alkaloids, chiral 1,2-aminoalcohols, amines or alcohols. Some examples are collected in Table 3 and Scheme 2. Most of the modifiers were barely efficient, except some simple derivatives of CD and pyrrolidinyl-naphthylethanol (**1**). In this respect the Pt-catalyzed enantioselective hydrogenation of TFAP is rather similar to that of other activated carbonyl compounds: only cinchona alkaloids and some derivatives (8), pyrrolidinyl-naphthylethanol (22), pyrrolidinyl-anthracenylethanol (23), naphthyl ethylamine and derivatives (21), and dihydrovinpocetine (24,25) provided high to medium enantioselectivity.

Molecular Modelling

A molecular modelling study has been performed to rationalize the nature of enantiodifferentiation in this reaction. The mechanistic model is based on earlier observations, according to which CD adsorbs on the Pt surface via the quinoline moiety, and the quinuclidine N-atom is responsible for interaction with the carbonyl group of α-ketoester (26). The so-called "open 3"conformation of CD has been found to be energetically most favorable (9, 27, 28).

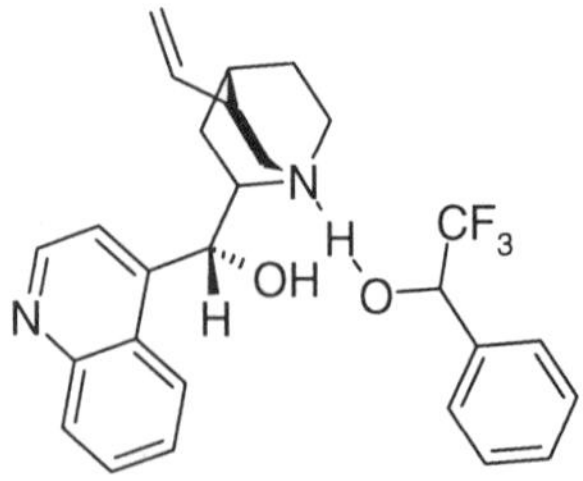

Scheme 3   Top view of the CD-TFAP adduct leading to (*R*)-product. Platinum surface (not shown) lies parallel to the drawing plane.

Molecular mechanics (MM+, Hyperchem) was used for calculating the structures and energies of complexes formed by a 1:1 interaction of CD and TFAP. The atomic charges were calculated at the semiempirical level (AM1). Scheme 3 presents the calculated minimum energy conformation of the activated complex, leading upon hydrogenation to the formation of (R)-1-phenyl-2,2,2-trifluoroethanol (assuming *cis* addition of two hydrogen atoms adsorbed on the Pt surface). The carbonyl O-atom of TFAP and the quinuclidine N-atom of CD are bound via H-bonding. The activated complex resembles the half-hydrogenated state of the carbonyl group {late transition state (29)}. The complex leading to the formation of (S)-product is energetically less favorable (by 9 kJ.mol^{-1}), in line with the experimentally observed enantiodifferentiation.

The interaction of Pt with the reactant and modifier could not be involved in the calculations. Scheme 3 shows the top view of the adduct positioned over a flat (Pt) surface, assuming that the aromatic rings of CD (30) and TFAP are adsorbed parallel to the surface.

Conclusions

The enantioselective hydrogenation of 2,2,2-trifluoroacetophenone (TFAP) over chirally modified Pt is the first successful example of the asymmetric catalytic hydrogenation of this compound. The 61 % ee to (R)-1-phenyl-2,2,2-trifluoroethanol, achieved in the presence of cinchonidine, is not an optimized value. It is assumed that optimization of the hydrogen concentration on the Pt surface, applying apolar solvents and appropriate modifier and reactant concentrations (11) at low temperature, will provide significantly higher ee.

Still, we do not expect that the outstanding ee's reported for the enantioselective hydrogenation of α-ketoesters over cinchona-modified Pt can be achieved in this reaction. In the latter case the rate acceleration observed in the presence of CD (as compared to the rate of unmodified reaction) was in the range of 10 - 20 or sometimes even higher (3). On the contrary, the rate acceleration is rather low, only 2 - 2.5 in the enantioselective hydrogenation of TFAP (11).

Acknowledgement

Financial support of this work by the Swiss National Science Foundation (Program Chiral 2) is gratefully acknowledged. Thanks are also due to Olivier Schwalm for molecular modelling calculations.

References

1. T. Osawa, T. Harada and A. Tai, *Catal. Today* **37**, 465 (1997).
2. A. Baiker, *J. Mol. Catal.*, **115**, 473 (1997).
3. A. Baiker and H. U. Blaser, *Handbook of Heterogeneous Catalysis*, G. Ertl, H. Knözinger and J. Weitkamp, Eds., VCH, 1997, Vol. 5, p. 2422.
4. H. U. Blaser, H. P. Jalett, D. M. Monti, J. F. Reber and J. T. Wehrli, *Stud. Surf. Sci. Catal.*, **41**, 153 (1988).
5. S. Bhaduri, V. S. Darshane, K. Sharma and D. Mukesh, *J. Chem. Soc. Chem. Comm.*, 1738 (1992).
6. W. A. H. Vermeer, A. Fulford, P. Johnston and P. B. Wells, *J. Chem. Soc., Chem. Comm.*, 1053 (1993).
7. H. U. Blaser and H. P. Jalett, *Stud. Surf. Sci. Catal.*, **78**, 139 (1993).
8. H. U. Blaser, H. P. Jalett and J. Wiehl, *J. Mol. Catal.*, **68**, 215 (1991).
9. M. Schürch, O. Schwalm, T. Mallat, J. Weber and A. Baiker, *J. Catal.*, **169**, 275 (1997).
10. G.-Z. Wang, T. Mallat and A. Baiker, *Tetrahedron: Asym.*, **8**, 2133 (1997).
11. T. Mallat, M. Bodmer and A. Baiker, *Catal. Lett.*, **44**, 95 (1997).
12. E. J. Corey and R. K. Bakshi, *Tetrahedron Lett.*, **31**, 611 (1990).
13. P. V. Ramachandran, A. V. Teodorovic and H. C. Brown, *Tetrahedron*, **49**, 1725 (1993).
14. P. V. Ramachandran, A. V. Teodorovic, B. Gong and H. C. Brown, *Tetrahedron: Asym.*, **5**, 1075 (1994).
15. R. W. Waldron and J. H. Weber, *Inorg. Chem.*, **16**, 1220 (1977).
16. P. Gamez, F. Fache and M. Lemaire, *Tetrahedron: Asym.*, **6**, 705 (1995).
17. C. Reichardt, *Solvents and Solvent Effects in Organic Chemistry*, VCH, Weinheim, 1988.

18. L-W. Leung and M. J. Weaver, *Langmuir*, **6**, 323 (1990).
19. T. Iwasita and E. Pastor, *Electrochim. Acta*, **39**, 531 (1994).
20. T. Mallat, Z. Bodnar, B. Minder, K. Borszeky and A. Baiker, *J. Catal.*, 168, 183 (1997).
21. B. Minder, M. Schürch, T. Mallat, A. Baiker, T. Heinz and A. Pfaltz, *J. Catal.*, **160**, 261 (1996).
22. K. E. Simons, G-Z. Wang, T. Heinz, T. Giger, T. Mallat, A. Pfaltz and A. Baiker, *Tetrahedron: Asym.*, **6**, 505 (1995).
23. M. Schürch, T. Heinz, R. Aeschimann, T. Mallat, A. Pfaltz and A. Baiker, *J. Catal.*, in press.
24. A. Tungler, T. Màthé, K. Fodor, R. A. Sheldon and P. Gallezot, *J. Mol. Catal. A: Chem.*, **108**, 145 (1996).
25. A. Tungler, T. Tarnai, T. Màthé, G. Vidra, J. Petrò and R. A. Sheldon, in M. G. Scaros and M. L. Prunier (Eds.), *Catalysis of Organic Reactions*, M. Dekker, New York, 1995, p. 201.
26. H. U. Blaser, H. P. Jalett, D. M. Monti, A. Baiker and J. T. Wehrli, *Stud. Surf. Sci. Catal.*, **67**, 147 (1991).
27. G. D. H. Dijkstra, R. M. Kellogg and H. Wynberg, *J. Org. Chem.*, **55**, 6121 (1990).
28. O. Schwalm, J. Weber, B. Minder and A. Baiker, *Int. J. Quantum Chem.*, **52**, 191 (1994).
29. G. Webb and P. B. Wells, *Catal. Today*, **12**, 319 (1992).
30. G. Bond and P. B. Wells, *J. Catal.*, **150**, 329 (1994).

Heterogeneous Enantioselective Hydrogenation Catalysed by Platinum Modified by Some Morphine Alkaloids

S.P. Griffiths and P.B. Wells
Department of Chemistry, University of Hull, Hull, HU6 7RX, UK

K.G. Griffin and P. Johnston
Johnson Matthey PMD-Chemicals, Orchard Road, Royston, SG8 5HE, UK

Abstract

Methyl pyruvate hydrogenation at 10 bar pressure and 298 K has been studied over 6.3% Pt/silica (EUROPT-1) modified by various morphine alkaloids. Two categories of reaction are evident (i) modification by codeine, 7,8-dihydrocodeine, or thebaine induces enantioselectivity in favour of S-lactate (ee = 2 to 5%) which is not influenced by quaternisation of the alkaloid, and (ii) modification by oxycodone or naloxone provides enantioselective reaction in favour of R-lactate (ee = 4 to 10%), the rate and selectivity being destroyed by alkaloid quaternisation. Catalyst modified by pholcodine gives only racemic product. Category (ii) reactions have a formal similarity to those modified by cinchona alkaloids, enantioselectivity being associated with the presence of the =N-C-C-O-H sequence in the modifier structure. Butan-2,3-dione hydrogenation was racemic with codeine as modifier but enantioselective with oxycodone as modifier (ee = 13%(R)). Adsorption of these alkaloids appears to require the participation of edge or step sites at the Pt surface. Mechanisms that account for these observations are proposed.

Introduction

In recent years the use of conventional supported platinum-group metal catalysts to achieve heterogeneous enantioselective hydrogenation has received considerable attention (1-5). Enantioselective reaction may occur following adsorption of a chiral modifier at the surface of a supported metal. In that environment the pro-chiral reactant may undergo selective enantioface adsorption (a thermodynamic effect) and the adsorbed enantiofacial forms of the reactant may be converted to product at the same or at different specific rates (a kinetic effect). The interplay of these thermodynamic and kinetic factors determines the sense and magnitude of the observed enantiomeric excess (5-9).

Cinchonidine (5, Fig.1), cinchonine, alkaloids of related structure, and synthetic variants (1-3, 10,11) have been the studied as modifiers of supported Pt, and hydrogenations that occur enantioselectively at these surfaces include those of α-ketoesters and α-ketoacids (1-3,12), adjaccnt dikctoncs (13), trifluoro-acetophenone (14) and pyruvic acid oxime (15). Cinchonidine-modified Pt catalyses pyruvate hydrogenation to lactate; an enantiomeric excess of 70%(R) can be readily achieved (16) and 96%(R) has been obtained under optimised conditions (17). Cinchonine directs reaction towards selective S-lactate formation. Such enantioselective reaction is rate-enhanced by a factor of 20 to 50 (18). Considerable effort has been expended to achieve an understanding of the mechanisms of these reactions and of the molecular interactions involved (19-24). The varied properties of the platinum group metals as enantioselective hydrogenation catalysts has recently been reviewed (5); suffice it to say that (i) of these metals, Pt has been the most widely studied and (ii) substantially different conditions are achieved in pyruvate hydrogenation at each platinum group metal surface when cinchonidine is used as modifier.

The search for alternative modifiers has proved a difficult task (11). Most approaches have centred on the use of molecules which, like the cinchonas, contain an aromatic function (to facilitate adsorption) and an appropriate stereogenic environment incorporating a tertiary N-atom (to provide an enhanced rate at the enantioselective site via H-bonding between the adsorbed modifier and the reacting molecule). Consequently, quaternisation of this N-atom leads to a dramatic loss of both enantioselectivity and rate (3). Vinca alkaloids are effective modifiers (25) and a preliminary report has appeared describing the use of strychnos and morphine alkaloids (26). In this paper we explore the use of codeine and alkaloids of related structure as modifiers for the enantioselective hydrogenation of methyl pyruvate and of butan-2,3-dione over Pt/silica. These alkaloids show two remarkable properties (i) in some cases quaternisation does not affect rate and enantioselectivity and (ii) a small structural change induces a switch in the sense of the enantioselectivity. These characteristics are unique and were judged sufficiently important to justify a detailed study notwithstanding the low values of the enantiomeric excess obtained.

Experimental

Materials. The catalyst used in this investigation was the standard reference material EUROPT-1, a 6.3% Pt/silica, which has been well characterised (27-30). Pt particles range in size from 0.9 to 3.5 nm with a maximum in the distribution at 1.8 nm (dispersion ca. 60%). As received the platinum is in an oxidised state; reduction was achieved in an atmosphere of hydrogen at 393 K over 0.5 h. Reduced under these conditions the Pt particles may be raft-like with preferential (111) orientation (31). Pure samples of the alkaloids codeine, dihydrocodeine, thebaine, pholcodine, and oxycodone were kindly donated by Macfarlan Smith; naloxone was obtained from Sigma. The structures of these alkaloids are shown in Fig. 1. Pure samples of some N-methyl quaternary salts were obtained by reaction of the parent alkaloid with iodomethane under reflux; purity was determined by TLC and NMR. Methanol and ethanol (BDH), methyl pyruvate and butan-2,3-dione (Fluka) were used as received or after distillation as appropriate.

Catalyst modification. The process whereby alkaloid is adsorbed onto the catalyst surface is termed *modification*. Solutions of modifier (see Tables for concentrations) were injected onto samples of reduced catalyst (0.1 g) under hydrogen. The resulting slurry was transferred to a small beaker and stirred in air at 293 K for 1 h after which it was centrifuged, the supernatant liquid decanted, a know volume of fresh solvent introduced, and the slurry transferred cleanly to the reactor. The solvent used for modification was the same as that used subsequently for reaction.

Reactors and analytical procedures. Reactions were conducted in stirred high pressure reactors (a Fischer-Porter reactor in glass and a Baskerville reactor in stainless steel fitted with a glass liner). Reaction solutions consisted of 10 ml methyl pyruvate (113 mmol) or butan-2,3-dione and 20 ml solvent. Hydrogen pressure was 10 bar, the value being maintained throughout by computer control. The reactors were stirred at 850 and 1200 rpm respectively. Hydrogen uptake increased with time and initial (maximum) rates were obtained over the first 10% or so of reaction (accelerating rates, as are common with cinchona-modified catalysts (16) were not observed). After the desired conversion, catalyst was separated by filtration, and separation of solvent, reactant and product enantiomers was achieved by chiral glc (column: 50 m Cydex B

(SGE)). Results obtained by chiral glc were checked, from time to time, by polarimetry.

Results

Methyl pyruvate hydrogenation at 10 bar pressure and 298 K over unmodified EUROPT-1 proceeded slowly in alcoholic solution (Table 1, entries 1 and 2). When the catalyst was modified by adsorption of codeine a rate enhancement was observed, the extent of which diminished as the concentration of the solution used in catalyst modification was increased (entries 3 - 5). Products formed over modified catalyst showed an enantiomeric excess of 3% in favour of S-lactate. Mass spectrometric and NMR analysis showed that codeine was hydrogenated to 7,8-dihydrocodeine during reaction. Modification with an authentic sample of 7,8-dihydrocodeine provided a similarly enhanced rate and enantiomeric excess (entry 6). Pyruvate hydrogenation in methanol catalysed by codeine quaternised by methyl iodide behaved in a closely similar fashion to the catalyst modified by the parent alkaloid (entries 7, 8). To test the integrity of the quaternary salt under hydrogenation conditions it was dissolved in methanol and exposed to 10 bar hydrogen at 298 K in the presence of catalyst. Thin layer chromatography showed that the N-atom remained quaternised, and mass spectrometry revealed that the double bond at C7-C8 was hydrogenated. Codeine-N-oxide was a similarly effective modifier (entry 9).

Thebaine adsorbed onto EUROPT-1 was an effective modifier, S-lactate again being formed in a small excess. The enhanced rate passed through a maximum as the concentration of thebaine used in the modification step was increased (entries 10 - 13).

Pholcodine-modified EUROPT-1 showed no enantioselectivity in pyruvate hydrogenation and reaction rate was not enhanced (entries 14,15). The alkaloid was effectively adsorbed on the catalyst surface as judged by the appearance of dihydropholcodine in solution.

Adsorption of oxycodone onto EUROPT-1 induced enantioselectivity in favour of R-lactate, the enantiomeric excess ranging from 7 to 10%. In ethanol there was no rate enhancement and no dependence of rate or of enantiomeric excess on the concentration of the alkaloid solution used in the modification step (entries 16 - 19). Remarkably, rate was enhanced in methanolic solution (entry 20) but modification with the quaternary salt greatly reduced activity and provided racemic product (entry 21). Naloxone followed oxycodone in

Table 1. Methyl pyruvate hydrogenation catalysed by alkaloid-modified 6.3% Pt/silica (EUROPT-1) at 10 bar pressure and 298 K

Modification solution		Reaction[a]	
Alkaloid	Solvent / Vol	Initial rate /mmol $h^{-1}g(cat)^{-1}$	Enantiomeric excess / %
none	EtOH / 40 ml	50	0
none	MeOH / 20 ml	30	0
codeine 100 mg	EtOH / 70 ml	310	3(S)
codeine 200 mg	EtOH / 40 ml	260	3(S)
codeine 200 mg	EtOH / 20 ml	140	3(S)
dihydrocodeine 200 mg	EtOH / 20 ml	180	3(S)
codeine 50 mg	MeOH / 20 ml	120	3(S)
Me(codeine)I 50 mg	MeOH / 20 ml	110	3(S)
codeine-N-oxide 100 mg	EtOH / 40 ml	90	4(S)
thebaine 10 mg	EtOH / 40 ml	140	2(S)
thebaine 20 mg	EtOH / 40 ml	180	3(S)
thebaine 100 mg	EtOH / 40 ml	150	3(S)
thebaine 200 mg	EtOH / 40 ml	130	4(S)
pholcodine 20 mg	EtOH / 40 ml	60	0
pholcodine 100 mg	EtOH /100 ml	40	0
oxycodone 20 mg	EtOH / 20 ml	50	8(R)
oxycodone 100 mg	EtOH / 40 ml	50	10(R)
oxycodone 100 mg	EtOH / 20 ml	40	7(R)
oxycodone 200 mg	EtOH / 20 ml	60	8(R)
oxycodone 50 mg	MeOH / 20 ml	180	8(R)
Me(oxycodone)I 50 mg	MeOH / 20 ml	v. slow	0
naloxone 20 mg	EtOH / 40 ml	250	5(R)
naloxone 50 mg	EtOH / 40 ml	320	4(R)
naloxone 50 mg	MeOH / 20 ml	90	4(R)
Me(naloxone)I 50 mg	MeOH / 20 ml	v. slow	0

[a] modification and reaction conducted in the same solvent

Table 2.   Butan-2,3-dione hydrogenation catalysed by alkaloid-modified 6.3%
          Pt/silica (EUROPT-1) in ethanol at 10 bar pressure and 298 K

Modification solution		Reaction	
Alkaloid	Solvent / Vol	Initial rate /mmol $h^{-1}g(cat)^{-1}$	Enantiomeric excess / %
none		EtOH / 20 ml	560	0
codeine	200 mg	EtOH / 40 ml	5000	0
oxycodone	100 mg	EtOH / 20 ml	1030	13(R)

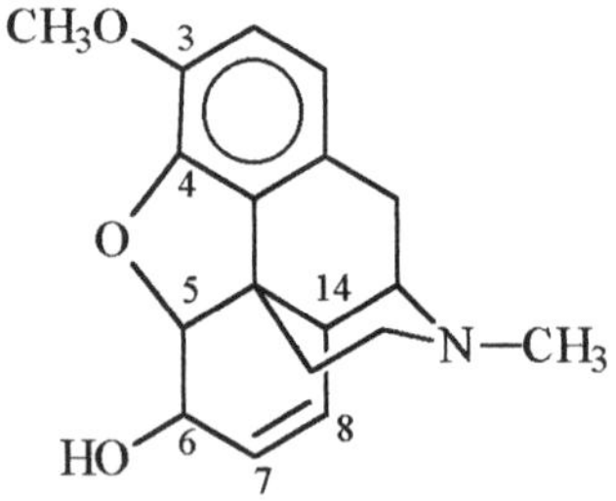

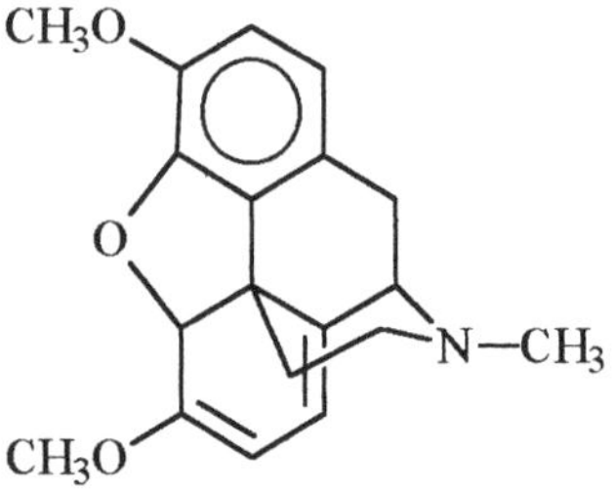

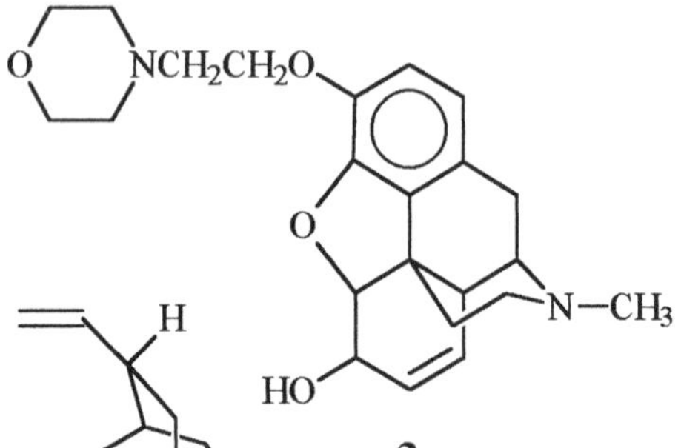

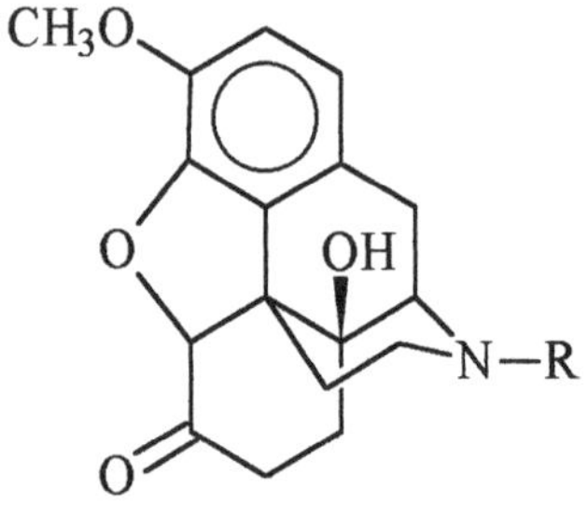

Fig. 1. Structures of alkaloids: **1**, codeine; **2**,
thebaine; **3**, pholcodine; **4**, oxycodone (R =
–CH₃), naloxone (R = –CH₂CH=CH₂); **5**,
cinchonidine.

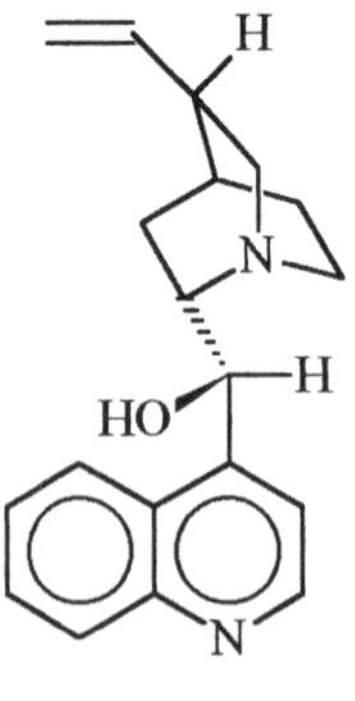

promoting enantioselectivity towards R-lactate in both methanolic and ethanolic solutions and at enhanced rates; again, quaternisation of the alkaloid greatly reduced activity and destroyed enantioselectivity (entries 22 - 25).

Codeine and oxycodone were examined as catalyst modifiers for butan-2,3-dione hydrogenation in ethanol at 10 bar pressure and 298 K (Table 2). Both induced a rate enhancement but only oxycodone-modified catalysts were enantioselective, the favoured product being R-lactate.

Discussion

The main polycyclic component of the structures of alkaloids **1 - 4** is T-shaped. Consequently, if strong π-chemisorption involving the aromatic ring is to be achieved between a codeine molecule and a (111) plane of Pt atoms, such adsorption seems to require edge sites as shown in Fig. 2. Such adsorption may be achieved by *either* face of the aromatic ring (Figs. 2a and 2b).

A. Reactions over Catalysts Modified by Codeine, Thebaine, and Pholcodine

Methyl codeinium iodide was as effective a modifier as codeine for pyruvate hydrogenation, both in terms of enantioselectivity and the extent of the enhanced rate. For the adsorption model shown in Fig. 2 the N-atom is directed away from the Pt surface and hence quaternisation should not influence any enantioselective site on the metal surface. Furthermore, the mechanism of rate enhancement cannot involve H-bonding at the N-atom (as is the case when the modifier is a cinchona alkaloid [3]). Codeine-N-oxide was an effective modifier, comparable to the quaternary salt, and the same considerations apply.

By contrast, the bulky substituent at carbon atom 3 in pholcodine eliminates both enantioselectivity and enhanced rate. It is inferred that this substituent blocked the potentially enantioselective site and that such reaction as occurred took place at unmodified sites on the Pt surface giving racemic product with no enhancement of rate. *From these observations the enantioselective site in the case of adsorbed codeine is identified as one or more of the surface platinum atoms located adjacent to the oxygen atom bridging carbon atoms 4 and 5 in adsorbed codeine.*

The values of the enantiomeric excess and enhanced rate observed for dihydrocodeine and thebaine as modifiers were closely similar to those observed for codeine, from which it is inferred that the slightly greater

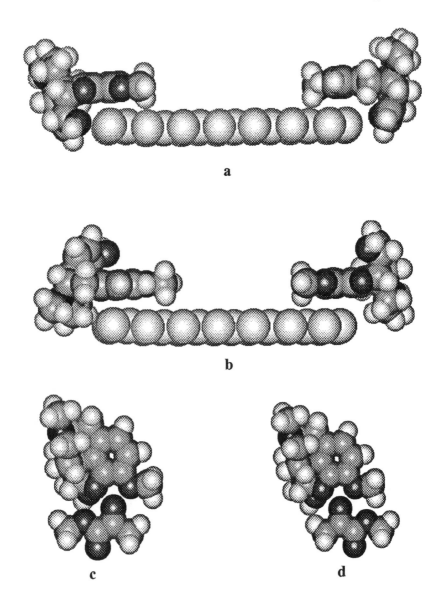

FIG. 2. (a) (b) Representations of codeine π-adsorbed at the edges of a (111)
plane of Pt atoms by each face of the aromatic ring.
(c) (d) Approach of the two enantiofaces of methyl pyruvate to codeine
(see text).

molecular complexity provided by each of these molecules does not significantly modify the molecular geometry adjacent to the enantioselective site. This is consistent with the location of the enantioselective site as proposed above.

In the presence of codeine, butan-2,3-dione hydrogenation showed a rate enhancement by a factor of 9 but no enantioselectivity (Table 2). Standard molecular modelling procedures (19) have been used to define the energy surface for the approach of a butan-2,3-dione molecule to codeine in the neighbourhood of the O-atom bridging C4 and C5. Approach of the dione by each enantioface gave rise to the same energy profile. If this situation applies also to the molecules in the adsorbed state, then selective enantioface adsorption will not occur and an enantioselective outcome to hydrogenation is not expected.

In the presence of codeine, methyl pyruvate hydrogenation showed a rate enhanced by a factor of 3 to 6 and an enantiomeric excess of 3% (Table 1). This enantioselectivity can be regarded as a consequence of the substitution of $-CH_3$ for $-OCH_3$ in the dione. In principle, this enantiomeric excess could originate in two distinct ways: (i) codeine may adsorb *either* as in Fig. 2a *or* as in Fig. 2b, the adjacent enantioselective site providing only weak enantioselectivity, or (ii) codeine may adsorbed *both* as Fig. 2a *and* as Fig. 2b and adjacent enantioselective sites provide higher values of the enantiomeric excess but in opposing senses thus giving a low value as a net result. Analogous calculations of energy surfaces for the mutual approach of codeine and methyl pyruvate in its two enantiomeric configurations again showed no significant differences. (The approach is visualised in Figs. 2c and 2d in such a way that, were these molecules adsorbed on a (111)-plane of Pt atoms, the aromatic ring of the alkaloid and the carbonyl groups of the ester would be located over Pt atoms). The calculations are not sufficiently accurate to identify with certainty the small energy difference required to provide an enantiomeric excess of 3%. Thus the result of the calculation is consistent with but does not prove alternative (i) above, and is probably not consistent with alternative (ii). Independent information concerning any preferred mode of adsorption of codeine is required in order to progress the interpretation further.

The origin of the enhanced rates in these reactions is not clear, but the mechanism may be analogous to that proposed for the cinchona-based systems except that, in these cases, the crucial H-bond is formed to an oxygen atom of the codeine structure, possibly that situated between carbon atoms 4 and 5.

B. Reactions over Catalysts Modified by Oxycodone and Naloxone

Enantioselectivity in oxycodone- and naloxone-modified reactions favoured R-product formation in *both* pyruvate hydrogenation and butan-2,3-dione hydrogenation (Tables 1 and 2). However, the structures of these alkaloids are so close to that of codeine that it must be assumed that their modes of adsorption follow that shown in Fig. 2, i.e. adsorption occurs at edges or at steps in the surface of the small Pt particles. Furthermore, selective enantioface adsorption should be possible at the site proposed in Section A leading to product rich in the S-enantiomer. Thus, overall reaction favouring R-product formation signals the simultaneous operation of a second enantiodirecting pathway. These alkaloids are distinguished by their possession of an OH function at carbon atom 14; moreover, the introduction of this function provides an =N-C-C-OH sequence in the structure which is also present in the quinuclidine moiety of the cinchona alkaloids and is the locus of the enantiodirecting influence when those alkaloids are adsorbed on Pt. By analogy with the proposal of Baiker and co-workers for pyruvate hydrogenation over cinchona-modified Pt in acetic acid or ethanol solution (20,21), we propose the N-atom in oxycodone and naloxone to be protonated in the active state of the catalyst. The reactants may then form two H-bonds to the alkaloid as shown in structure (I) of Fig. 3 (for pyruvate ester R = OMe, for butan-2,3-dione R = Me). Transfer of the H-atom to the reactant then occurs as the reactant adsorbs as a half-hydrogenated state (structure (II)), and addition of a second H-atom completes product formation, the R-configuration being favoured (structure (III)).

Quaternisation of oxycodone and of naloxone eliminated enantioselectivity and greatly reduced the rate in pyruvate hydrogenation (dione hydrogenation was not studied); the mechanism in Fig. 3 is consistent with this observation.

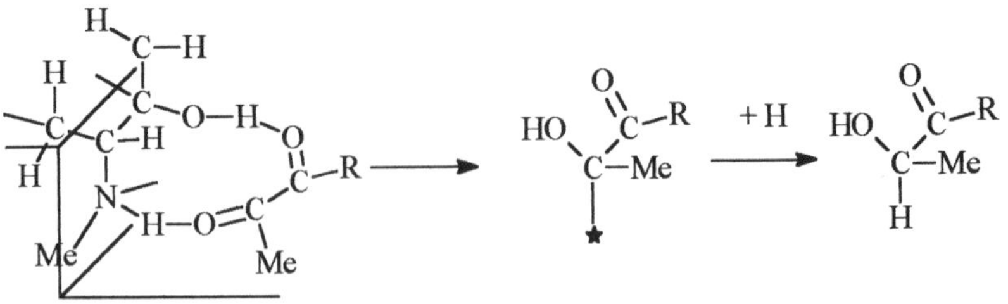

Fig. 3.   The preferential formation of R-product at a step site occupied by adsorbed oxycodone or naloxone.

Conclusion

These modifiers as a family show the remarkable property, not previously encountered, of creating two types of chiral environment at a metal surface such that pyruvate ester hydrogenation can be directed towards either S- or R-lactate. Studies of morphine alkaloid adsorption at stepped Pt surfaces would test our hypothesis that such adsorption occurs preferentially at edge and step sites.

Acknowledgments

PBW thanks Macfarlan Smith for samples of morphine alkaloids, EPSRC for the award of a studentship to SPG, and Johnson Matthey for sustained financial support of his investigations into heterogeneous enantioselective catalysis. The authors thank Dr. K.E. Simons for valuable discussions.

References

1. H-U. Blaser, *Tetrahredron: Asymm.*, **2**, 843 (1991).
2. H-U. Blaser and M. Muller, *Stud. Surf. Sci. Catal.*, **59** (Heterog. Cat. Fine Chem. II), 73 (1991).
3. P.B. Wells and G. Webb, *Catal. Today*, **12**, 319 (1992).
4. A. Baiker and H.U. Blaser in *Handbook of Heterogeneous Catalysis*, G. Ertl, H. Knozinger, J. Weitkamp Eds.; VCH, Weinheim, **5**, 2422 (1997).
5. P.B. Wells and A.G. Wilkinson, *Topics in Catal.*, in press.
6. C.R. Landis and J. Halpern, *J. Am. Chem. Soc.*, **109**, 1746 (1987).
7. M. Boudart and G. Djega-Mariadassou, *Catal. Lett.*, **29**, 7 (1994).
8. Y. Sun, R.N. Landau, J. Wang, C. LeBlond and D.G. Blackmond, *J. Am. Chem. Soc*, **118**, 1438 (1996).
9. J. Wang, C. LeBlond, C.F. Orella, Y. Sun, J.S. Bradley and D.G. Blackmond, *Stud. Surf. Sci. Catal.*, **108** (Heterog. Catal. Fine Chem. IV), 183 (1997).
10. B. Minder, M. Schurch, T. Mallat, A. Baiker, T. Heinz and A. Pfaltz, *J. Catal.*, **160**, 261 (1996).
11. A. Pfaltz, *Topics in Catal.* in press.
12. H-U. Blaser and H.P. Jallet, *Stud. Surf. Sci. Catal.*, **78** (Heterog. Cat. Fine Chem. III), 139 (1993).

13. W.A.H. Vermeer, A. Fulford, P. Johnston and P.B. Wells, *J. Chem. Soc. Chem. Commun.*, 1053 (1993).

14. T. Mallat, M. Bodmer and A. Baiker, *Catal. Lett.*, **44**, 95 (1997).

15. K. Borszeky, T. Mallat, R. Aeschiman and A. Baiker, *J. Catal.*, **161**, 451 (1996).

16. I.M. Sutherland, A. Ibbotson, R.B. Moyes and P.B. Wells, *J. Catal.*, **125**, 77 (1990).

17. H-U. Blaser, H.P. Jallet and J. Wiehl, *J. Mol. Catal.*, **68**, 215 (1991).

18. P.A. Meheux, A. Ibbotson and P.B. Wells, *J. Catal.*, **128**, 387 (1991).

19. K.E. Simons, P.A. Meheux, S.P. Griffiths, I.M. Sutherland, P. Johnston, P.B. Wells, A.F. Carley, M.K. Rajumon, M.W. Roberts, and A. Ibbotson, *Recl. Trav. Chim. Pays-Bas*, **113**, 465 (1994).

20. O. Schwalm, B. Minder, J. Weber and A. Baiker, *Catal. Lett.*, **23**, 271 (1994).

21. O. Schwalm, J. Weber, B. Minder and A. Baiker, *J. Mol. Struct. (Theochem)*, **330**, 353 (1995).

22. R.L. Augustine, S.K. Tanielyan, amd L.K. Doyle, *Tetrahedron: Asymm.*, **4**, 1803 (1993).

23. J.L. Margitfalvi and M. Hegedus, *J. Mol. Catal: Chemical*, **107**, 281 (1996).

24. J. Wang, Y. Sun, C. Le Blond, R.N. Landau and D.G. Blackmond, *J. Catal.*, **161**, 752 (1996).

25. A. Tungler, T. Tarnai, T. Mathe, G. Vidra, J. Petro and R.A. Sheldon, *15th Conference of Organic Reactions Catalysis Society*, Phoenix, Marcel Dekker, 201 (1994).

26. S.P. Griffiths, P. Johnston, W.A.H. Vermeer and P.B. Wells, *J. Chem.Soc. Chem. Commun.*, 2431 (1994).

27. G.C. Bond and P.B. Wells, *Applied Catal.*, **18**, 225 (1985).

28. J.W. Geus and P.B. Wells, *Applied Catal.*, **18**, 231 (1985).

29. A. Frennet and P.B. Wells, *Applied Catal.*, **18**, 243 (1985).

30. P.B. Wells, *Applied Catal.*, **18**, 259 (1985).

31. S.D. Jackson, M.B.T. Keegan, G.D. McLellan, P.A. Meheux, R.B. Moyes, G. Webb, P.B. Wells, R. Whyman and J. Willis, *Stud. Surf. Sci. Catal.*, **63** (Prep. of Catalysts V), 135 (1991).

New Enantioselective Heterogeneous Catalysts

Setrak K. Tanielyan and Robert L. Augustine
Center for Applied Catalysis, Seton Hall University, South Orange, NJ 07079
USA

Abstract

Interest in enantioselective synthesis has been increasing steadily over the past several years particularly because of the need for developing chiral drugs and agrochemicals. Almost all of the currently effective enantioselective catalysts presently used in organic synthesis are chiral, metal complexes which are soluble in the reaction medium. Over the past twenty-five years or so a number of attempts have been made to 'heterogenize' homogeneous catalysts. These supported catalysts, though, are not widely used since their activity is generally lower than that of the corresponding homogeneous analog. Further, it has been reported that activity is frequently lost on attempted re-use. Some success has been reported in preparing polymer supported chiral complexes, but the selectivity observed with the use of such 'heterogenized' species has generally been lower than that obtained using the homogeneous catalyst itself. In addition, the preparation of these supported chiral complexes usually requires special treatment to fix the chiral ligand to the support. The ability to attach a preformed homogeneous catalyst to a support would have a distinct advantage.

We have accomplished this by developing a method for "anchoring" preformed homogeneous catalysts to a variety of support materials. In many instances the activity and selectivity observed with the supported species were greater than those found using the corresponding homogeneous catalysts. On re-use, the supported species usually exhibited an increase in activity and selectivity which was maintained on re-use in up to fifteen cycles. Catalyst leaching was not detected. The catalysts used were a number of chiral metal complexes. The reactions studied included the hydrogenation of prochiral substrates, hydroformylations and allylations.

Introduction

As evidenced by the large number of publications concerning enantioselective synthesis, interest in this area has been increasing steadily over the past several years particularly because of the need to develop viable syntheses for chiral drugs and agrochemicals. One of the most effective means of attaining such syntheses is to use an enantioselective catalyst and, thus, transfer the chirality of the catalyst to many thousands of product molecules. Almost all of the currently effective enantioselective catalysts presently used in organic synthesis are chiral, metal complexes which are soluble in the reaction medium. Generally, high product enantiomeric excesses (ee's) have been attained using these catalysts. While these catalysts have been shown to promote many different kinds of reactions and to be quite selective, from a practical standpoint, one still has to be concerned with the removal of the complex from the reaction mixture. Further, since most of the chiral ligands used to prepare these complexes are quite expensive, ligand recovery and re-use can be a significant problem.

The difficulties associated with the separation of a homogeneous catalyst from the reaction mixture and its potential re-use could be alleviated by the use of heterogeneous catalytic species. However, there are, to date, only two viable heterogeneous catalyst systems capable of promoting an enantioselective reaction and both of these are substrate specific. One, nickel modified with tartaric acid and sodium bromide, is used specifically for the enantioselective hydrogenation of β-ketoesters or β diketones giving product alcohols having enantiomeric excesses (ee's) of 95% or greater (1,2). The catalyst in this system is believed to be the nickel tartrate formed by the corrosive interaction between the tartaric acid and the nickel. The sodium bromide is thought to selectively poison the active-non-chiral metal sites and thus minimize the racemic hydrogenation that takes place on such sites. It has also been shown that when an bulky carboxylic acid such as pivalic acid is added to this catalyst system, the resulting mixture can be used for the hydrogenation of methyl ketones with high enantioselectivity (3,4).

The second heterogeneous system in present use is a platinum catalyst modified by the addition of a cinchona alkaloid, particularly cinchonidine. This catalyst system is specific for the enantioselective hydrogenation of α-ketoesters and acids. A large amount of research has been done on this catalyst system over the past several years in an attempt to develop a sufficiently detailed understanding of the processes taking place on the catalyst surface so that this information can be used to produce other catalyst systems for other reactions (2,5,6). From this body of work some specific features of the catalyst system have come to light. First, platinum is the only viable metal for the hydrogenation of α-ketoesters and acids. No other metal will work in this reaction. However, a palladium/cinchonidine system has been used for the hydrogenation of α-phenylcinnamic acid with reasonably high enantioselectivity (7). Second, it appears that the modifier must have at least a binuclear aromatic ring system attached to a chiral β-oxo tertiary amine for optimum selectivity (8,9). Since the modified catalysts promote the hydrogenation of α-ketoesters at a rate significantly higher than the non-modified hydrogenation, some interaction between the modifier and the substrate must be taking place which results in the activation of the carbonyl group toward hydrogenation.

It is primarily this later feature of the reaction which provides the high enantioselectivity observed with these catalysts since it is reasonable to assume that there will be some active sites on the platinum surface which will not have a modifier near them and, thus, reaction at these sites will lead to the formation of the racemic product. With this in mind, it may be a significant problem to develop a similar type of system for the enantioselective hydrogenation of other prochiral substrates.

What, then, can be done to produce other enantioselective heterogeneous catalysts capable of promoting the many different types of reactions that can be run using homogeneous catalysts? Over the past twenty-five years, many attempts have been made to "heterogenize" the more versatile homogeneous catalysts, the primary aim being to maintain the reaction activity and selectivity of the homogeneous species while at the same time significantly increasing the ease of separation from the reaction medium. One such approach to achieve "heterogenization" involved reacting a metal complex or salt with a solid support such as a polymer or a metal oxide which had been previously modified by the addition of phosphine or amine ligands to the surface of the support (10-17). From a practical approach, these catalysts are not widely used since their activities are frequently lower than those of the corresponding homogeneous analogs. In addition, problems associated with polymer swelling and attendant mass transport difficulties can be encountered, as well as the finding that activity is frequently lost on attempted re-use. Some success has been reported in preparing polymer supported chiral complexes, but the selectivity observed with the use of such "heterogenized" species has generally been lower than that obtained using the homogeneous catalyst itself (18).

In rare instances, the oxide support does not have to be modified before the application of a metal complex. The interaction of $Rh(OH)(CO)(PPh_3)_2$ with an alumina surface to give a supported catalyst for the hydrogenation of alkenes and benzene has also been described (19). This report also states that the presence of the Rh-OH entity is necessary for interaction with the surface of the alumina and that other complexes could not be attached to the oxide surface.

Another problem associated with catalysts made from metal complexes which are attached to either a modified polymer or metal oxide surface is that the techniques used for their preparation are rather specific and are driven by the nature of the ligand to be attached. Hence, modification of the catalyst to introduce another, more selective ligand is usually an arduous and complex task, if it is one that can be accomplished at all. This circumstance has particular importance where the preparation of enantioselective catalysts is concerned since optimal enantiomeric excess is usually obtained using a specific ligand or class of ligands for a given reaction or substrate.

The preparation of chiral, aqueous-phase catalysts and their use in the preparation of naproxen has been reported. These heterogeneous catalysts have the same enantioselectivity as the homogeneous counterpart, but are 2 to 2.5 times less active (20).

With all of these problems associated with the present methods used to prepare supported homogeneous catalysts, it would appear that a new approach to the problem was needed. The goal would be to develop a method for preparing such supported homogenous catalysts which utilizes pre-formed complexes. Further, the resulting catalyst should retain the activity and selectivity exhibited by the homogeneous species even on repeated re-use. Leaching of the metal and/or the ligands must also be essentially non-existent.

We provide here a preliminary report on the development of a procedure by which pre-formed homogeneous catalysts have been anchored to a variety of different supports. These catalysts have been used to promote a number of different chiral and achiral reactions which will be presented in the following discussion.

TABLE 1 Reaction rate and product ee data obtained on hydrogenation of 1 over Rh(DIPAMP) catalysts on different modified supports.

Support	Use #	Rate[a]	ee
Montmorillonite K	1	0.18	76%
	3	0.56	91%
Montmorillonite K[b]	1	0.18	92%
	3	0.54	94%
Carbon	1	0.07	83%
	3	0.40	90%
Alumina	1	0.32	90%
	3	1.67	92%
Lanthana	1	0.38	91%
	3	0.44	92%

[a] Moles H_2/mole Rh/min.
[b] Catalyst stored for eight months in air.

Experimental

Hydrogenation of methyl 2-acetamidoacrylate (1). To a 25 mL reaction flask was added an amount of catalyst containing 20 micromoles of the active metal complex, Rh(COD)(DIPAMP)BF$_4$ on a modified support. A solution of 1.26 millimoles of 2-acetamidoacrylic acid methyl ester in 10 mL of methanol was then added and the flask was evacuated and filled with hydrogen three times and the reaction initiated at atmospheric pressure and room temperature with computer monitoring of the hydrogen uptake. After hydrogen absorption ceased the product was analyzed by gas chromatography using a β-cyclodextrin Chiraldex column. As listed in Table 1 with a Montmorillonite K support the hydrogen uptake occurred at a rate of 0.18 moles H$_2$/mole Rh/min with the product having an enantiomeric excess (ee) of 76%. After multiple re-use the catalyst had an activity of 0.56 moles H$_2$/mole Rh/min and the product had an ee of 91%. The activity and selectivity of this catalyst was retained after storage for at least eight months at room temperature in a screw capped vial in air as illustrated by the data given in Table 1. Rate and enantioselectivity data obtained using Rh(DIPAMP) catalysts on other support meterials are also given in Table 1 while the data found using rhodium complexes containing other chiral ligands supported on alumina are listed in Table 2.

TABLE 2 Reaction rate and product ee data obtained on hydrogenation of 1 over different rhodium complex catalysts supported on modified gamma alumina.

Catalyst	Use	Supported[a]		Soluble[b]	
		Rate[c]	ee	Rate[c]	ee
Rh(DIPAMP)	1	0.32	90%	0.25	76%
	3	1.67	92%	na	na
Rh(PROPHOS)	1	2.0	68%	0.26	66%
	3	2.6	63%	na	na
Rh(Me-DUPHOS)	1	1.8	83%	3.3	96%
	3	4.4	95%	na	na

[a] Modified alumina supported complexes.
[b] Unsupported complex in solution.
[c] moles H$_2$/mole Rh/min

TABLE 3 Data obtained on allylation of sodium dimethylmalonate by 2-butenyl acetate (3) using Pd(R,R-BINAP) on a modified carbon catalyst.

Catalyst	Rxn Time	% Conv.	% 4	% 5	% 6	ee%	4/5	4+5/6
Soluble[a]	24 hr	96	56	6	38	26	9.3	1.6
Supported[b] (1st Use)	12 hr	100	76	1.5	22	27	50	3.5
2nd Use	12 hr	100	78	1.2	21	31	65	3.8
4th Use	12 hr	100	78	1	21	30	78	3.9

[a] Unsupported complex in solution.
[b] Modified carbon supported complex.

Hydrogenation of dimethyl itaconate (2). To a 25 mL reaction flask containing the catalyst prepared from 200 mg of modified Montmorillonite K clay and 10 micromoles of the Ru(SS-BINAP)(NEt$_3$)Cl$_2$ complex was added a solution of 1.42 millimoles of dimethylitaconate in 8 mL of methanol and the flask was evacuated and filled with hydrogen three times. The reaction was initiated at atmospheric pressure and room temperature with computer monitoring of the hydrogen uptake. After hydrogen absorption ceased the product was analyzed by gas chromatography using a β-cyclodextrin Chiraldex column. The hydrogen uptake occurred at a rate of 0.06 moles H$_2$/mole Ru/min with the product having an enantiomeric excess (ee) of 81%. After multiple re-use the catalyst had an activity of 0.09 moles H$_2$/mole Ru/min and the product had an ee of 91%.

Allylation of sodium dimethylmalonate by 2-butenyl acetate (3). To a catalyst containing 180 mg of modified Norite carbon and 25 micromoles of Pd(R,R-BINAP)Cl$_2$ was added a solution of 160 mg of sodium dimethylmalonate in 5 mL of tetrahydrofuran and the suspension stirred for 10 minutes. 2-Butenyl acetate (3) (60 microliters) was then added by syringe and the reaction mixture stirred at room temperature for 12 hours. The reaction liquid was then removed and analyzed by gas chromatography. The reaction data are listed in Table 3 along with data obtained on running the reaction using a soluble Pd(R,R-BINAP)Cl$_2$ catalyst.

TABLE 4 Data obtained on hydroformylation of 1-hexene (7) over Rh(P(C₆H₅)₃)₃Cl on a modified Montmorillonite K clay.

Catalyst	% Conv.	Hydroformylation				
		%8	%9	%10	%11	9/10
Soluble[a]	90	-	48	38	9	1.3
Supported[b] (1st use)	78	15	49	29	-	1.7
(2nd Use)	65	12	40	24	-	1.6
(3rd Use)	70	10	44	25	-	1.8

[a] Unsupported complex in solution.
[b] Modified Montmorillonite K supported complex.

Hydroformylation of 1-hexene (7). To a 200 mg of catalyst containing 10 micromoles of $Rh(P(C_6H_5)_3)_3Cl$ supported on modified Montmorillonite K in a stainless steel autoclave was added 30 mL of toluene and 6 mL of 1-hexene. The autoclave was flushed three times with a 1:1 mixture of CO and H_2, pressurized to 1000 psig, the temperature was raised to 80°C and stirring initiated at 800 rpm. The reaction was run for 24 hours after which time the liquid was removed and a second portion of 1-hexene and toluene were injected into the autoclave and the hydroformylation run again. This procedure was repeated three times with the products analyzed by GC-MS using a 30m x 0.25mm HP-1 capillary column. The reaction data is listed in Table 4 along with corresponding data for the reaction run using a soluble $Rh(P(C_6H_5)_3)_3Cl$ catalyst.

Results and Discussion

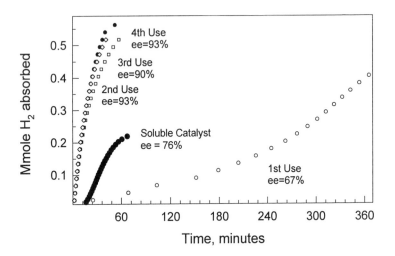

The hydrogen uptake and ee data for the hydrogenation of methyl 2-acetamidoacrylate (**1**) (Eqn. 1) over a supported Rh(DIPAMP) catalyst are shown in Fig. 1 along with the corresponding data for the reaction catalyzed by the soluble Rh(DIPAMP). Interestingly, in the first use of the supported catalyst, the rate of hydrogen uptake and the product ee were lower than those observed with the soluble catalyst. However, on reuse of the supported catalyst the rate and product ee increased significantly. Storing this catalyst in air for extended periods of time did not alter its activity or selectivity as shown by the data in Fig. 2. Fig 3 shows the hydrogen uptake data and product ee data obtained on re-use of this catalyst for repeated the hydrogenation of **1**. There was no evidence of any catalyst leaching during these reactions.

Fig. 1. Hydrogen uptake curves and observed ee's observed for the hydrogenation of **1** over Rh(DiPAMP) supported on modified Montmorillonite K. The catalyst was freshly prepared and re-used four times.

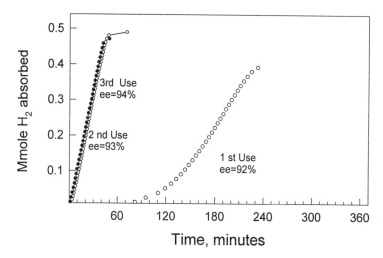

Fig. 2. Hydrogen uptake curves and observed ee's observed for the hydrogenation of **1** over Rh(DiPAMP) supported on modified Montmorillonite K. The catalyst was stored in air for eight months before use and re-used three times.

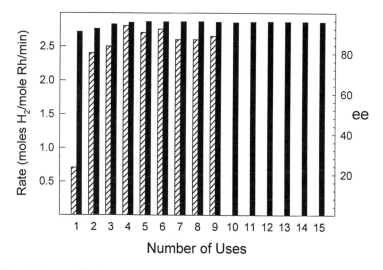

Fig. 3. Rates of hydrogen uptake (Hatched bars) and observed ee's (Solid bars) observed for multiple hydrogenations of **1** over a single portion of Rh(DiPAMP) supported on modified Montmorillonite K.

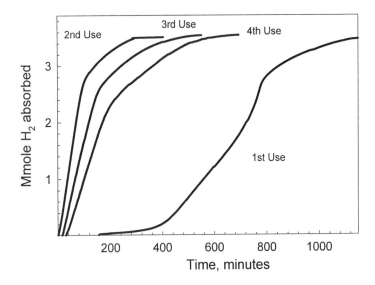

Fig. 4. Hydrogen uptake curves observed for the hydrogenation of
1-heptene over $Rh(P(C_6H_5)_3)_3Cl$ supported on modified silica.
The catalyst was freshly prepared and re-used four times.

A further example of the repeated use of the supported homogeneous
catalysts was the multiple hydrogenations of 1-heptene over a modified silica
supported $Rh(P(C_6H_5)_3)_3Cl$. Hydrogen uptake data for these reactions are
shown in Fig. 4. Here too there was no loss of activity on re-use and no metal
leaching was observed.

An important aspect of this procedure is the fact that it can be used to
anchor complexes to a variety of support materials. In Table 1 are listed the
hydrogenation rates and product ee's for the hydrogenation of **1** over
Rh(DIPAMP) supported on several different supports. For comparison
purposes, the homogeneous hydrogenation had a rate of 0.25 moles of
hydrogen/mole of rhodium/minute and gave a product with an ee of 76%. In
almost every instance, the activity and selectivity of the supported catalysts
increased on re-use. The exception here is the basic support, lanthana. We are
unsure at this time as to why this is the case.

We have also found that the nature of the ligand had little influence on
the ability to support the complex and to use the supported species several times
with no loss of activity or selectivity. Table 2 lists the hydrogen rate data and
product ee's for the hydrogenation of **1** over some of the modified alumina
supported rhodium complexes used in this work. Here, too, an increase in rate

and selectivity was observed on reuse of the supported catalysts. Also listed are the corresponding rates and product ee's obtained using the homogeneous catalysts in solution.

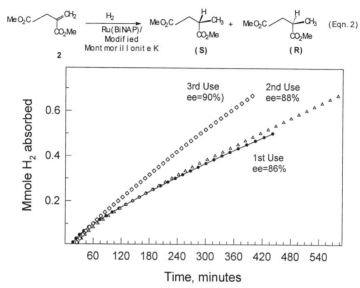

Fig. 5. Hydrogen uptake curves and observed ee's observed for the hydrogenation of **2** over Ru(BINAP) supported on modified Montmorillonite K. The catalyst was freshly prepared and re-used three times.

Fig. 5 shows the hydrogen uptake data for the hydrogenation of dimethyl itaconate (**2**) (Eqn. 2) over a supported Ru(BiNAP) catalyst. Here, too, the activity increases on re-use. The enantioselectivity, though, increases only slightly.

To illustrate the versatility of this approach to supporting homogeneous catalysts, a supported Pd(BiNAP) catalyst was used for the allylation of dimethyl malonate (Eqn. 3) with the reaction data shown in Table 3 along with the corresponding results obtained using the soluble catalyst. Table 4 lists the reaction data obtained on using a soluble and supported $Rh(P(C_6H_5)_3)_3Cl$ catalyst for the hydroformylation of 1-hexene (**7**) (Eqn. 4).

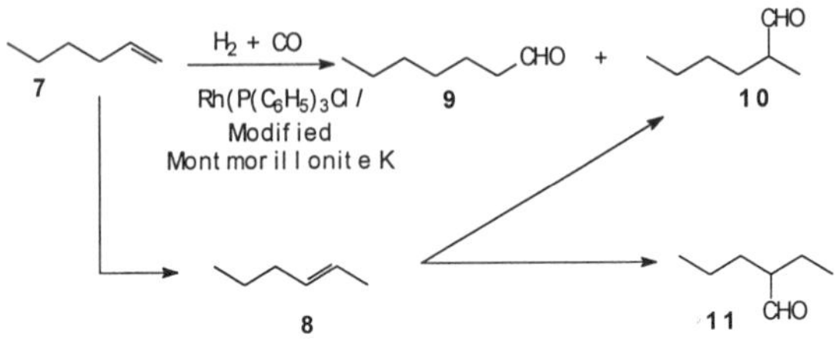

Conclusions

The results presented here show that this new method of anchoring preformed homogeneous catalysts onto a support material has considerable potential. The supported catalysts are stable in extensive re-use and leaching does not take place under the conditions we employed. Work is presently underway to determining the nature of the complex/support interaction and a possible reason for the apparent general increase in activity frequently observed on re-use of these catalysts.

Acknowledgement

This research was supported by the National Science Foundation under Grants Nos. CTS-9312533 and CTS-9708227.

References

1. Y. Izumi, *Adv. Catal.*, **32**, 215 (1983).

2. G. Webb and P.B. Wells, *Catal. Today*, **12**, 319 (1992).

3. T. Osawa, T. Harada and A. Tai, *J. Mol. Catal.*, **87**, 333 (1994).

4. T Osawa, T. Harada and A. Tai, *Catal. Today*, **37**, 465 (1997).

5. R. L. Augustine and S. K. Tanielyan, *J. Mol. Catal.*, **118**, 79 (9\1997).

6. H.-U. Blaser, H.-P. Jalett, M. Muller and M. Studer, *Catal. Today*, **37**, 441 (1997).

7. Y. Nitta and K. Kobiro, *Chem. Lett.*, 165 (1995).

8. B. Minder, T. Mallat, A. Baiker, G. Wang, T. Heinz and A. Pfalz, *J. Catal.*, **154**, 371 (1995).

9. K.E. Simmons, G. Wang, T. Heinz, T. Giger, T. Mallat, A. Pflatz and A. Baiker, *Tetrahedron: Asymmetry*, **6**, 505 (1995).

10. J.C. Bailar, *Catal. Revs.*, **10**, 17 (1974).

11. A. Akelah and D.C. Sherrington, *Chem. Rev.*, **81**, 557 (1981).

12. T. Shido, T. Okazaki and M. Ichikawa, *J. Catal.*, **157**, 436 (1995).

13. F. Minutolo, D. Pini and P. Salvadori, *Tetrahedron Letters*, **37**, 3375 (1996).

14. K. Kaneda and T. Mizugaki, *Organometallics*, **15**, 3247 (1996).

15. J. John, M.K. Dalal, D.R. Patel and R.N. Ram, *J. Macromol. Sci., Pure Appl. Chem..*, **A34** 489 (1997).

16. E. Lindner, A. Jarger, M. Kemmler, F. Auer, P. Wegner, H.A. Mayer, A. Benez and E. Plies, *Inorg. Chem.*, **36**, 862 (1997).

17. H. Brunner, P. Bublak and M. Helget, *Chem. Ber.*, **130**, 55 (1997).

18. B. Pugin and M. Muller, *Stud. Surf. Sci. Catal.*, **78** (Heterog. Catal. Fine Chem.), 107 (1993).

19. A.M. Trzeciak, J.J. Ziotkowski, Z. Jaworske-Galas, M. Mista and J. Wrzyszcz, *J. Mol. Catal.*, **88**, 13 (1994).

20. K.T. Wan and M. E. Davis, *J. Catal.*, **152**, 25 (1995).

Effect of Poisoning and Metal Adsorption on the Enantioselective Heterogeneous Catalytic Hydrogenation of Isophorone and Ethyl Pyruvate

K. Fodor, A. Tungler, T. Máthé*, S. Szabó**, R. A. Sheldon***
Department of Chemical Technology, Technical University of Budapest, 1521 Hungary
* Res. Group for Org. Chem. Technology of Hung. Academy of Sci.
** Central Research Inst. for Chemistry of Hung. Academy of Sci.
***Laboratory for Organic Chemistry and Catalysis, Delft University of Technology

Abstract
Observations were made in the Pd mediated hydrogenation of isophorone and in the Pt mediated hydrogenation of ethyl pyruvate with regard to how the catalyst modification with catalyst poisons and by catalytically inactive (underpotentially deposited) metals affects the processes of enantioselection. The poisons influence only the reaction rate, and the adsorbed metals affect both the reaction rate and enantioselectivity.

Introduction

The most successful approach in chiral heterogeneous catalysis is the modification of metal catalysts by chiral compounds. Raney-nickel modified with tartaric acid was the first heterogeneous chiral catalyst to give high enantioselectivities in the hydrogenation of b-keto esters [1]. The second example, first reported by Orito et al [2], was the highly enantioselective hydrogenation of a-keto esters over platinum catalysts modified with cinchona alkaloids. Since then a variety of the catalyst-chiral modifier-reactant systems has been tested resulting in a wide range of optical yields [3]. Although research groups proposed different rationalisations for the observed enantioselectivities and rate enhancement in pyruvate hydrogenation, they agree on the common requirements the catalyst, the reactant, the chiral modifier, the solvent, etc. have to meet in these reactions. The so called 'ligand acceleration" model [4] is valid in the hydrogenation of pyruvate where rate enhancement occurs upon modification. Whereas in the Pd mediated enantioselective hydrogenation of isophorone (modifier dihydroapovincaminic acid ethyl ester, ee 55%), of phenyl cinnamic acid (modifier cinchonidine, ee 72%) and other α,β-unsaturated carboxylic acids the reaction rate was smaller than that in the unmodified reaction. Therefore the ligand acceleration model cannot be applied as a matter of course.

The earlier results reported on the enantioselective hydrogenation over modified catalysts clearly show that the behaviour of the chiral modifier-reactant-metal systems strongly depends on the catalyst properties. Several studies on the performance of the catalyst reveal that the dispersion (the mean metal particle size), the nature and texture of the support, the preparation and reduction method (morphology, contaminants, degree of reduction of the metal) and heat-treatment of the catalyst have a significant influence on enantio-differentiation [3]. Furthermore, if two different competing alkaloid modifiers are present in the reaction mixture, their effect on enantioselectivity is also dependent on the catalyst used [5]. The differences in stereoselectivity, obtained with catalysts on different supports and different preparation methods, cannot be interpreted only by the model proposed by Margitfalvi et al. [6,8], in which the chiral modifier-reactant interaction taking place in the solution plays the key role in the reaction rate and enantioselectivity determining process. On the other hand, the existence of the chiral modifier-reactant complex in the solution was detected by CD and NMR measurements [7,8]. Nevertheless the formation of the chiral modifier-reactant complex in the liquid phase is a necessary but not a sufficient condition for inducing enantio-differentiation. We believe that the proper adsorption of this reactant-chiral modifier complex on the catalyst surface is an essential prerequisite for enantioselection. The strong adsorption of the alkaloid molecule on the metal surface (shown by adsorption measurements), the absence of corrosive chemisorption (in contrast to the tartrate modified Raney-Ni catalyst), and the fact that similar results were given by impregnation or in situ modification of the catalyst in the modifier solution also support this idea.

In this paper we report our observations with regard to how the catalyst modification affects the processes of enantioselection. To this end the effect of catalyst poisons was studied and experiments were performed with catalysts whose surface was modified by catalytically inactive (underpotentially deposited) metals.

Experimental

Materials

The catalysts used were partly commercial products: 5% Pt/C Aldrich and 5% Pt/Al$_2$O$_3$ Engelhard. Pd black catalyst was prepared according to the following procedure: 18 mmol (6.0 g) K$_2$PdCl$_4$ was dissolved in 50 ml water and reduced at boiling point with 74 mmol (5.0 g) Na(HCOO) dissolved in 20 ml water. When the reduction was completed, the pH of the suspension was basic (pH=9). The catalyst was filtered and washed several times with distilled water.

Platinum on alumina catalysts were heat-treated at 400 °C for three hours in hydrogen flow under atmospheric pressure in a glass reactor, afterwards they were cooled down to room temperature and flushed with nitrogen.

Ethyl pyruvate and cinchonidine were supplied by Merck. (-)-Dihydroapovincaminic acid ethyl ester was prepared in our laboratory by catalytic hydrogenation of apovincaminic acid ethyl ester supplied by Richter Gedeon Co., followed by separation of the epimers [7].

Catalytic hydrogenation

The hydrogenation of ethyl pyruvate and isophorone was carried out at 25 °C and under 10-50 bar hydrogen pressure in a conventional apparatus or in a Büchi Bep 280 autoclave equipped with a magnetically driven turbine stirrer and a gas-flow controlling and measuring unit. Before the hydrogenation the reaction mixtures were stirred under nitrogen for 15 minutes in the reaction vessel.

The reaction mixtures were analysed with a gas chromatograph equipped with a b-cyclodextrin capillary column (analysis temperatures: ethyl lactate at 90 °C, dihydroisophorone at 110 °C) and FID. The chromatograms were recorded and peak areas were calculated with Chromatography Station for Windows V1.6 (DataApex Ltd., Prague). Enantiomeric excess was defined as:

$$ee \ (\%) = ([R]-[S])/([R]+[S])*100$$

Results and discussion

The thoroughly studied hydrogenation of the carbonyl group of ethyl pyruvate over Pt catalysts and the hydrogenation of the carbon-carbon double bond of isophorone over Pd catalysts served as our model reactions (Fig. 1). Cinchona type alkaloids (dihydrocinchonidine (DHCND) and dihydrocinchonine (DHCNN)) and (-)-dihydroapovincaminic acid ethyl ester ((-)-DHVIN) were used as chiral modifier (Fig. 2). We applied Pt on alumina, Pt on carbon and Pd black catalysts.

In a series of hydrogenations we poisoned the catalyst surface with two common, sulphur containing poisons, namely carbon disulphide and thiophene,

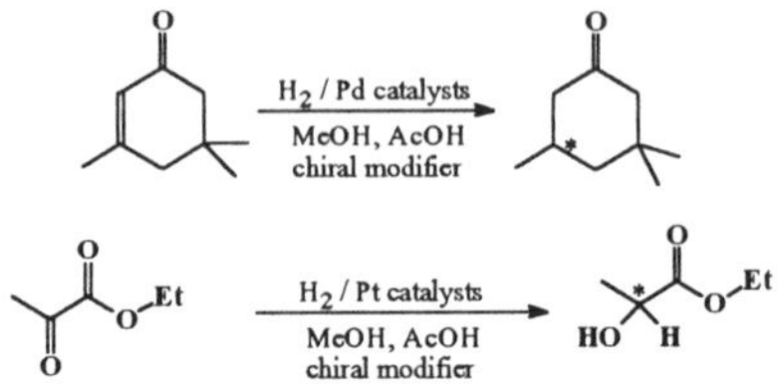

Figure 1 The model reactions

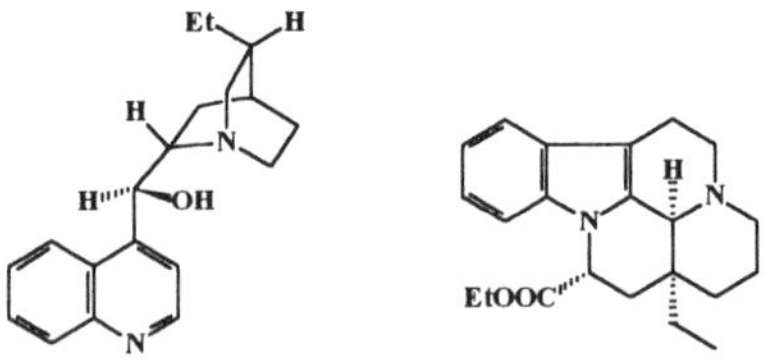

Figure 2 Structure of dihydrocinchonidine and (-)-dihydroapovincaminic acid ethyl ester

and with Captopril, which is a chiral sulphur containing proline derivative (Fig. 3). The enantioselective reaction was allowed to proceed to 20-40% conversion, and then different amounts of the poison were added to the reaction mixture. The ee values were calculated for the period after poisoning, and the reaction rate was determined from the conversion in the first 60 minutes.

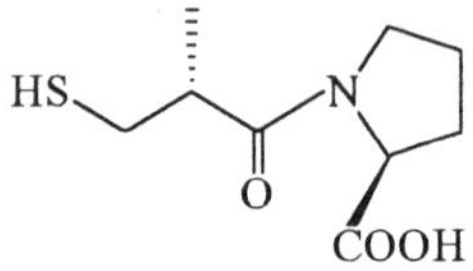

Figure 3 Structure of Captopril

Table 1, 2, 3 present data on activity and selectivity of Pt and Pd catalysts in the presence of the catalysts poisons. Generally, as the amount of the poison (either an achiral or a chiral one) was increased, enantioselectivity did not change much, and the reaction rate, not surprisingly, decreased gradually as the total number of the surface active sites decreased due to the irreversible adsorption of the poison molecules as shown in Fig. 4.

These results indicate that the poison molecules partially cover the active sites of the catalyst surface, and as a result, the reactant-modifier complex can adsorb only on smaller surface active area, but presumably in the same way as without poisoning.

Table 1 Hydrogenation of ethyl pyruvate in the presence of dihydroapovincaminic acid ethyl ester and thiophene as poison

amount of thiophene (mmol)	conversion/time (%/min)	ee (%)	poison /metal molar ratio $(x10^3)$
0	0.61	23	0
0.19	0.47	25	15
0.94	0.43	24	73
2.25	0.19	24	176
4.50	0.17	23	352
6.76	0.2	21	528
9.0	0.08	22	703
18.0	0.04	22	1406

Conditions: Pt/C 0.05 g, ethyl pyruvate 6 g, (-)-DHVIN 0.05 g, AcOH 0.1 g, MeOH 50 cm^3, 10 bar, 25 °C.

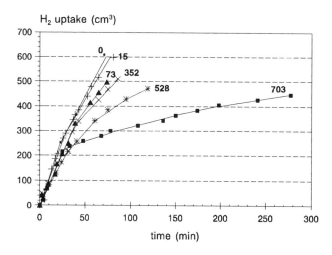

Figure 4 Hydrogenation curves in the presence of different amounts of thiophene
Conditions: see Table 1.

Table 2 Hydrogenation of ethyl pyruvate in the presence of dihydrocinchonidine and dihydrocinchonine

poison	amount of poison (mmol)	conversion/time (%/min)	ee (%)	poison /metal molar ratio $(x10^3)$
none	0	0.81	25	0
thiophene	0.19	0.79	31	15
"	6.76	0.24	21	528
"	9.0	0.2	22	703
"	18.0	0.08	21	1406
CS_2	0.49	0.22	25	38
"	0.67	0.09	27	52
"	0.97	0.03	25	76
"	1.61	0.0002	38	126
Captopril	0.54	0.31	30	42
"	1.0	0.07	25	78
"	2.0	0.05	22	156
"	6.7	0.02	35	523
none [a]	0	2.57	68	0
Captopril [a]	4.6	2.2	73	180
none [b]	0	1.5	41 (S)	0
Captopril [b]	4.6	1.12	46 (S)	180

Conditions: Pt/C 0.05 g, ethyl pyruvate 6 g, DHCND 0.05 g, AcOH 0.1 g, MeOH 50 cm^3, 10 bar, 25 °C;
[a]: Pt/Al$_2$O$_3$ (E) 0.1 g, ethyl pyruvate 6 g, DHCND 0.1 g, [b]: DHCNN 0.1 g, AcOH 0.5 g, Toluene 50 cm^3, 50 bar, 25 °C.

To make further changes to the catalyst surface, we performed experiments, where the surface of Pt and Pd catalysts were modified with adsorbed catalytically inactive metals such as cadmium, tin and lead. Several metals can be deposited on the surface of metals, such as Pt and Pd, which can adsorb hydrogen well, via ionization of the preadsorbed hydrogen [9-14]. This means that the hydrogen preadsorbed on the Pt or Pd is exchanged with the metal ions resulting in adsorbed atoms of this metal on the catalyst surface. In order to get maximum interaction between the two metals, it is important to avoid bulk deposition of the inactive metal, i.e. metal adatoms should be deposited onto the catalyst surface. Since Cd, Sn and Pb are non-noble metals, their standard electrode potential is smaller than zero, consequently, no bulk deposition can be expected under reaction conditions. As the adsorption of a second metal alters the adsorptive and catalytic properties of Pt and Pd, it is expected to affect both the activity and selectivity of the catalyst.

Table 3 Hydrogenation of isophorone in the presence of dihydroapovincaminic acid ethyl ester

poison	amount of poison (mmol)	conversion/time (%/min)	ee (%)	poison /metal molar ratio (x10^3)
none	0	0.18	49	0
thiophene	5.0	0.07	49	1
CS$_2$	1.67	0.12	54	0.4
"	3.33	0.13	51	0.7
"	6.66	0.17	53	1.4
"	13.3	0.06	50	2.8
Captopril	4.63	0.12	51	1
"	13.9	0.02	48	3

Conditions: Pd black 0.5 g, isophorone 7 g, (-)-DHVIN 0.02 g, AcOH 0.2 g, MeOH 50 cm^3, 50 bar, 25 °C.

The results of testing these modified catalysts in the hydrogenation of ethyl pyruvate and isophorone are presented in Table 4-9. In the first series of experiments (Table 4-6), the inactive metal was first deposited onto the catalyst surface by stirring the catalyst and the solution of ions of that metal in hydrogen. The enantioselective catalytic reaction initiated when the reactant and the chiral modifier were added to the mixture.

Both activity and enantioselectivity decreased in every reaction. The greater the amount of the inactive metal, the greater the enantioselectivity decreased. In some cases enantioselectivity was lost completely.

Table 4 Hydrogenation of ethyl pyruvate over Pt/Al$_2$O$_3$ catalyst

	metal/Pt atomic ratio	relative activity	ee (%)	conversion (%)
	-	1	85	100
Cd	0.5	0.022	36	71
	1	0.014	36	47
	5	0	-	0
Sn	2.5	0.014	0	46
Pb	5	0.005	9	15

Conditions: Pt/Al$_2$O$_3$ 0.05 g, ethyl pyruvate 3 g, DHCND 0.05 g, AcOH 30 cm^3, 50 bar, 25 °C, second metal deposition prior to the hydrogenation.

Table 5 Hydrogenation of ethyl pyruvate over Pt/C catalyst

	metal/Pt atomic ratio	relative activity	ee (%)	conversion (%)
	-	1	28	100
Cd	0.5	0.15	9	50
Sn	0.5	0.68	10	85
	0.8	0.24	1	32
	1	0.27	2	48

Conditions: Pt/C 0.05 g, ethyl pyruvate 6 g, (-)-DHVIN 0.05 g, AcOH 0.1 g, MeOH 30 cm^3, 50 bar, 25 °C, second metal deposition prior to the hydrogenation.

In the following series of experiments (Table 7-9), the second metal was added to the reaction mixture during the enantioselective hydrogenation reaction, at about 15-30% conversion. This means that not only the hydrogen but also the modifier, the reactant and product molecules were present in the solution and on the catalyst surface during the adsorption and reduction of the second metal. Under such conditions complexation of the second metal with both the modifier and the ethyl lactate product can take place.

Table 6 Hydrogenation of isophorone over Pd black catalyst

	metal/Pd atomic ratio	relative activity	ee (%)	conversion (%)
	-	1	54	85
Cd	1/100000	0.43	36	30
	1/10000	0.45	40	52
	1/2000	0.37	32	15
	1/500	0.31	27	22
Sn	1/10000	0.61	42	45
	1/500	0.24	34	17
Pb	1/250	0.27	23	19

Conditions: Pd black 0.3 g, isophorone 7 g, (-)-DHVIN 0.02 g, AcOH 0.2 g, MeOH 50 cm^3, 50 bar, 25 °C, second metal deposition prior to the hydrogenation.

Table 7 Hydrogenation of isophorone over Pd black catalyst

	metal/Pd atomic ratio	relative activity	ee (%)	conversion (%)
	-	1	54	100
Cd	1/10000	1.1	52	74
	1/500	1.06	50	74
	1/250	1.1	46	82
Sn	1/1000	0.95	50	47
	1/500	0.97	47	66
	1/250	1.02	47	57
Pb	1/250	0.96	44	61

Conditions: Pd black 0.3 g, isophorone 7 g, (-)-DHVIN 0.02 g, AcOH 0.2 g, MeOH 50 cm^3, 50 bar, 25 °C, second metal deposition after 15-30% conversion.

In the hydrogenation of isophorone, the modification of Pd black catalyst with Cd, Sn or Pb in this way had no significant influence on activity and enantioselectivity. On the contrary, in the hydrogenation of ethyl pyruvate over either Pt/Al$_2$O$_3$ (DHCND as chiral modifier) or Pt/C catalyst ((-)-DHVIN as chiral modifier), the decrease in the rate observed after deposition of the inactive metal was almost as much as when the inactive metal was deposited prior to the hydrogenation. Enantioselectivity also decreased, in some cases by a factor of 2, except for Cd-Pt/Al$_2$O$_3$, where it remained roughly the same as without Cd.

Table 8 Hydrogenation of ethyl pyruvate over Pt/Al$_2$O$_3$ catalyst

	metal/Pt atomic ratio	relative activity	ee (%)	conversion (%)
	-	1	85	100
Cd	1	0.033	80	47
	5	0.015	80	26
Sn	2.5	0.036	37	96
	5	0.029	40	60
Pb	5	0.045	51	70

Conditions: Pt/Al$_2$O$_3$ 0.05 g, ethyl pyruvate 3 g, DHCND 0.05 g, AcOH 30 cm^3, 50 bar, 25 °C, second metal deposition after 15-30% conversion.

Table 9 Hydrogenation of ethyl pyruvate over Pt/C catalyst

	metal/Pt atomic ratio	relative activity	ee (%)	conversion (%)
	-	1	25	100
Cd	0.5	0.20	14	73
	1	0.10	13	58
Sn	0.8	0.80	14	100
	1	0.52	11	100
	2	0.11	12	57

Conditions: Pt/C 0.05 g, ethyl pyr. 6 g, (-)-DHVIN 0.05 g, AcOH 0.1 g, MeOH 30 cm^3, 50 bar, 25 °C, second metal deposition after 15-30% conversion.

Conclusions

Blaser et al. proposed that the pyruvate is hydrogenated fast on the modified surface sites to an optically active product, while it is hydrogenated slowly on the unmodified sites to a racemic mixture of the product (ligand acceleration model) [4]. The chiral active sites are assumed to be formed by reversible adsorption of alkaloid molecules on the metal surface. If the catalyst poison is added to the reaction mixture at a certain conversion, the poison molecules will be adsorbed irreversibly on the catalyst surface replacing some of the alkaloid molecules. According to this model, enantioselectivity could remain unchanged only if the poison molecules were adsorbed to an equal degree on the modified and unmodified surface sites, which is unlikely because surface sites of a heterogeneous catalyst are not homogeneous in their activities. On the contrary, poisoning diminishes only the activity of the catalyst, and does not affect enantioselectivity. We concluded from this result that after poisoning the enantioselective sites are reformed by the interaction of the catalyst, the chiral modifier and the reactant but only on a smaller surface area. Modifying the metal surface with an inactive metal evidently influences catalytic activity since it effects the adsorption characteristics of hydrogen as well as that of the reactant, the product, chiral modifier and solvent molecules. This results in a change in the amount and in the bond strength of these adsorbed species. According to our expectations the presence of the adsorbed catalytically inactive metal led to lower activity due to the much lower number of surface sites capable of adsorbing reactant and chiral modifier molecules. Enantioselectivity decreased considerably in every reaction indicating that the modification of the catalyst with a second metal, resulting in not only quantitative but also structural and chemical changes, reduces the ability of the catalyst surface to accommodate properly the reactant-modifier complex, which is needed to induce enantioselectivity. The effect of modifying the catalyst with an inactive metal was also found to be dependent on the catalyst.

In conclusion, the common features of these enantioselective hydrogenations are the structural, reactivity and affinity characteristics of the effective chiral modifiers, the interactive functional groups of the prochiral reactants and the appropriate catalysts. Though these enantioselective systems differ in their efficiency, their mode of action must be similar. The known models: the template model [2], the ligand-acceleration model [4] and the modifier-reactant interaction in the solution [6,8] give only partial explanation of the stereoselectivity differences between the catalysts used in enantioselective hydrogenation reactions. We believe that the formation of the chiral modifier-reactant-catalyst complex on the catalyst surface is the most important factor in these enantioselective reactions. The differences in the performance of the different catalysts do not only depend on the adsorption strength of the chiral modifiers on them, but also on their ability to accommodate the reactant-chiral modifier complex in an appropriate conformation. On the surface of f.c.c. metals, chiral kinked places exist or can be formed, which can be characterised by Miller indexes of different and at least two odd numbers, for example (532, 731). We suggest that the adsorbed chiral modifier-reactant complex is able to rearrange the metal atoms of the catalyst surface to some extent, and as a result to form a chiral site on the surface. This assumption is in close relation with that of Augustine [15], he supposed that adatoms form the center of chiral sites. There is a dynamic equilibrium of the chiral modifier-reactant complex between its adsorbed and liquid phase state. After the complex has left the surface, the surface metal atoms are rearranged to a thermodynamically stable structure. This hypothesis could explain the high reactant and catalyst specificity of enantioselective heterogeneous hydrogenations. Catalysts are considered to be different in their restructuring and chiral site-forming ability. This is in accordance with the different behaviour of the different catalysts in asymmetric hydrogenations. The reactants containing two carbonyl bonds or a carbonyl and a C=C double bond are only suitable for these reactions, presumably because only this part of the reactant could induce the rearrangement of the catalyst surface. The well-known requirements that a chiral modifier for enantioselective hydrogenation has to meet are: a condensed aromatic part to facilitate adsorption on the metal and a secondary or a tertiary nitrogen atom in a rather rigid chiral environment to interact with the carbonyl group of the reactant. We suggest that the chiral modifier-reactant complex is in an equilibrium between the liquid phase and the catalyst surface as well as the "free" reactant and modifier molecules.

Acknowledgements
The authors gratefully acknowledge the financial support of the Commission of European Communities, COST PECO 12382 and the support of the Hungarian OTKA Foundation under No. T 015674, they are grateful to Gedeon Richter Co. for supplying vinpocetin.

References

1   Y. Izumi, M. Imaida, H. Fukawa and S. Akabori, *Bull. Chem. Soc. Jpn.*, 36 (1963) 21.
2   Y. Orito, S. Imai and S. Niwa, *J. Chem. Soc. Jpn.*, 8 (1979) 1118.
3   H.U. Blaser, H.P. Jalett, M. Müller and M. Studer, *Catal. Today*, in press.
4   H.U. Blaser, M. Garland and H.P. Jalett, *J. Catal.*, 144 (1993) 569.
5   A. Tungler, K. Fodor, T. Máthé, Roger A. Sheldon, Heterogeneous

Catalysis and Fine Chemicals IV. Basel, (1996) *Studies in Surface Science and Catalysis*, Vol. 108. (1997) 157.

6   J.L. Margitfalvi, *J. Catal.*, 156 (1995) 175.
7   A. Tungler, T. Máthé, K. Fodor, J. Kajtár, I. Kolossváry, B. Herényi and R.A. Sheldon, *Tetrahedron: Asymmetry* 6 (1995) 2395.
8   J.L. Margitfalvi, M. Hegedüs and E. Tfirst, *Tetrahedron: Asymmetry* 7 (1996) 571.
9   T. Mallát, S. Szabó, M. Sürch, U.W. Göbel, and A. Baiker: *Reaction Kinetics and Catalysis Letters*, in press.
10 S. Szabó, *Int. Rev. Phys. Chem.*, 10 (1991) 207.
11 E. Lamy-Pitara, L. El Ouazzani-Benhima, J. Barbier, M. Cahoreau and J. Cassio, *J. Electroanal. Chem.*, 372 (1994) 233.
12 S.A.S. Machado, A.A. Tanaka and E.R. Gonzales, *Electrochim. Acta*, 39 (1994) 2591.
13 G. Kokkinidis and A. Papoutsis, *J. Electroanal. Chem.*, 271 (1989) 233.
14 W.A.H Vermeer, A. Fulford, P. Johnston and P.B.Wells, *J. Chem. Soc., Chem. Commun.* (1993) 1053.
15 R. L. Augustine, S. K. Tanielyan and L. K. Doyle, *Tetrahedron Asymmetry.*, 4, 8, (1993) 1803.

Ultrasonics in Chiral Metal Catalysis. Effects of Presonication on the Asymmetric Hydrogenation of Ethyl Pyruvate over Platinum Catalysts

Béla Török[1]*, Károly Felföldi[2], Gerda Szakonyi[3], Árpád Molnár[2] and Mihály Bartók[2]

[1]Organic Catalysis Research Group of the Hungarian Academy of Sciences and [2]Department of Organic Chemistry, József Attila University, H-6720 Szeged, Dóm tér 8, Hungary, e-mail: torok@chem.u-szeged.hu, fax:36-62-322-668 [3]Department of Biochemistry, Albert Szent-Györgyi University, Medical School, Szeged, Hungary

ABSTRACT

Chiral sonochemical hydrogenation of ethyl pyruvate to the corresponding ethyl lactate was carried out over various platinum catalysts in different solvents under moderate hydrogen pressure (atmospheric and 10 atm). The reaction rates and the enantiomeric excesses were determined over Pt/C, Pt/Al$_2$O$_3$, Pt/SiO$_2$ and Pt/K-10 catalysts in various solvents both under conventional and sonochemical conditions. The effect of ultrasounds on the catalytic activity and enantioselectivity was tested applying sonochemical pretreatment before the reaction. The ultrasonic irradiation was found to be highly advantageous in these hydrogenations. After the presonication of the catalysts, the enantioselectivity was highly improved over Pt/SiO$_2$ and Pt/K-10 catalysts. In addition, reactions took place in quantitative yield with complete chemoselectivity and the hydrogenation rates increased by one order of magnitude despite the very mild (atmospheric hydrogen pressure, room temperature) experimental conditions. To clarify the macroscopic effect of ultrasounds on the catalysts transmission electron microscopic measurements were carried out. The sonochemical pretreatment decreases the metal particle size of the catalysts producing in some cases more selective surface environment for the enantioselective hydrogenation.

1. INTRODUCTION

The catalytic hydrogenation of the carbonyl group is one of the most important procedures in organic chemistry requiring, however, usually relatively high temperature and hydrogen pressure (1). Asymmetric hydrogenation of prochiral ketones is in the forefront of active research areas in heterogeneous enantioselective catalysis (2). In this field the asymmetric reduction of α-ketoesters has received especially significant attention (3, 4). The chiral hydrogenation of ethyl pyruvate (Scheme 1) provides excellent enantioselectivity (~ 90% ee) using cinchona alkaloids and supported platinum catalysts under mild hydrogen pressure (5-10 atm) (5). The reaction also takes place in nonpressurized systems, though with lower selectivity (50% ee) (4). Taking into account the importance of developing effective, relatively cheap heterogeneous catalytic systems for enantioselective hydrogenations extensive efforts have been made and reviewed (6) to find optimized conditions. The major experimental variables studied are the catalysts (metal, support, particle size etc.) (6, 7), the modifier and its concentration (6, 8, 9), solvents (10) mass transfer and hydrogen pressure (11).

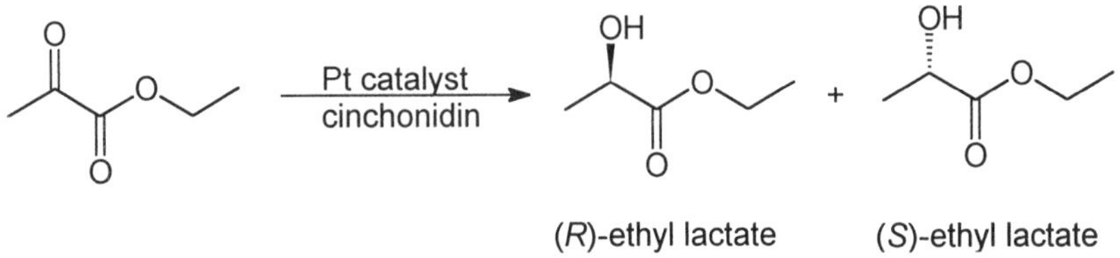

(R)-ethyl lactate   (S)-ethyl lactate

Scheme 1

The increasing number of important chiral pharmaceuticals and agrochemicals still initiates growing efforts to explore new, more convenient and economical synthetic routes for the preparation of these compounds. Due to the nature of large scale industrial processes and environmental considerations there is a strong driving force in recent years to find catalysts and experimental conditions where the hydrogenation reactions take place at high rate and with good selectivity under mild experimental conditions. One of the most promising techniques introduced recently is the application of ultrasonic irradiation in hydrogenations (12). The method was found to be highly efficient in the increase of the yield and the hydrogenation rate of olefins (13). Platinum catalysts, mainly Pt/C, were usually applied in sonochemical hydrosilylation and hydrogenation of alkenes (14,15), while palladium and nickel catalysts were also tested in these reactions (16). It was stated that the sonochemical treatment provided highly dispersed metal loaded on the surface of inert supports (17). Taking into account the importance of these reactions, it is surprising, however, that the effect of ultrasonic irradiation on the enantioselective pyruvate hydrogenation has never been tested. The only report with respect to the application of ultrasounds in asymmetric hydrogenations has been published by Tai et al. (18). They found that the presonication reduced the reaction time and slightly increased the enantioselectivity in the hydrogenation of β-diketones and ketoesters catalyzed by modified Raney-Ni.

Continuing our recent studies on the application of ultrasonics in enantioselective catalytic hydrogenations (19), the heterogeneous chiral hydrogenation of ethyl pyruvate was chosen for testing the influence of sonochemical pretreatment on the C=O hydrogenation reaction of high practical importance. The emphasis is placed on the improvement of the reaction rates and the enantiodifferentiation during the catalytic reaction, however, the actual transformations of the catalysts are also discussed.

2. EXPERIMENTAL

2.1. Materials Ethyl pyruvate was of analytical grade and purchased from Fluka, while the solvents with minimum purity of 99.5% were Reanal and Fluka products. The cinchonidin used as modifier (minimum purity >98%) was purchased from Fluka. The organic compounds including the reactant and the solvents were freshly distilled before each run. The catalysts used in the hydrogenations were 5% Pt/C (Engelhard, S_{BET}=950 m^2g^{-1}), 5% Pt/Al$_2$O$_3$ (Aldrich, S_{BET}=250 m^2g^{-1}), 3% Pt/SiO$_2$ (S_{BET}=240 m^2g^{-1}) (20) and 5% Pt/K-10 (S_{BET}=270 m^2g^{-1}) (21).

2.2. Methods

2.2.1. Hydrogenations under atmospheric H$_2$ pressure The hydrogenation reactions were carried out at room temperature (25 °C) in a conventional atmospheric batch reactor system equipped with a rubber septum to introduce the reactants. A Realsonic 40SF ultrasonic bath (20 kHz, 30 W) was used for the presonication of the catalysts or the reaction mixtures. High purity, oxygen-free hydrogen was prepared by a Whatman Model 75-34 hydrogen generator directly connected to the system. In a typical run 50 mg of the catalyst and 5 mg of cinchonidin was suspended in 5 ml solvent and the reactor was flushed with hydrogen for 15 min. Then the catalyst was activated in a closed system either by stirring for 1 h or sonication for 30 min under atmospheric hydrogen pressure. After the activation period 0.25 ml of ethyl pyruvate was injected into the reactor vigorously stirred and the hydrogen uptake was followed by a gas burette.

2.2.2. Hydrogenations in pressurized systems The hydrogenation under 10 atm hydrogen pressure was carried out in a Berghof Bar 45 autoclave. The catalytic system including the catalyst, solvent and modifier was activated under 10 atm hydrogen pressure and the reactant was introduced. During the reaction samples were withdrawn to determine the actual yield and selectivity. The sonochemically activated catalysts were prepared as described in *2.2.1.* and transferred into the autoclave in hydrogen atmosphere and then the reactant introduced. After that the autoclave was flushed with 10 atm hydrogen several times and filled to 10 atm and strirred (1300 rpm) for the required reaction time.

2.2.3. Product analysis The ethyl lactate isomers were identified on the basis of their mass spectra using a HP 5890 GC/HP 5970 MS system equipped with a 50 m long HP-1 capillary column, while the quantitative analysis including enantiomeric separation was performed with a HP 5890 GC-FID gas chromatograph using a 30 m long Cyclodex-B (J&W Scientific) capillary column. The optical yield is expressed as the enantiomeric excess (ee%) of the (R)-(+)-ethyl lactate and calculated according to the literature (ee%= ([R]-[S]) x 100 / ([R]+[S])).

2.2.4. Transmission electron microscopy (TEM) Measurements were performed with a Philips CM10 electron microscope at 90 kV at a magnification of 300 000. Each catalyst was tested before and after sonication (30 min). Samples were dispersed in toluene and mounted and air-dried on a plastic film supported by a Formvar grid. The metal particle size distribution was determined and the mean particle diameters were calculated as average of individual diameters ($\Sigma n_i d_i/\Sigma n_i$, $n=10^3$) from the magnified TEM images.

3. RESULTS AND DISCUSSION

3.1. Effect of Ultrasounds on the Hydrogenation Rate

Atmospheric systems The rate of the hydrogenation reactions was followed on the basis of hydrogen uptake of the reaction mixture during the reaction. The results were determined in ethanol and toluene both in conventional systems and with presonicated catalysts. As a representative example, the results of the Pt/SiO_2-catalyzed hydrogenation of ethyl pyruvate are plotted in Fig. 1.

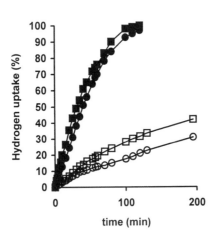

The differences in the hydrogenation rates were found to be significant when Pt/SiO_2 (Fig. 1) and Pt/K-10 (not shown) were used. The reaction rates increased in these cases at least by one order of magnitude as a result of the presonication before the hydrogenation. In addition, in contrast with the conventional activation by stirring, 100 % yield was usually achieved with presonicated catalysts. Without sonication very long reaction

Figure 1   Hydrogen uptake *vs.* time functions of the enantioselective hydrogenation reaction of ethyl pyruvate over 3% Pt/SiO_2 catalyst (□-EtOH, silent, ■-EtOH, presonicated, O-toluene, silent, ●-toluene, presonicated)

times were needed for obtaining medium yield in Pt/SiO$_2$ or Pt/K-10 catalyzed reactions. In contrast, in the case of Pt/C and Pt/Al$_2$O$_3$ catalysts (not shown) only a slight difference was observed between the reaction rates. The presonication of these catalysts also increased the reaction rate, however, only by a factor of 1.1-1.2. The change of the reaction medium does not affect the reaction rates significantly. It is interesting to note that the ultrasonic irradiation during both the activation period and the hydrogenation decreased the rate of hydrogen uptake in the Pt/C catalyzed systems.

3.2. Effect of Ultrasounds on the Enantioselectivity

3.2.1. Atmospheric systems When the hydrogenation of a prochiral carbonyl group is carried out the enantioselectivity is the most important feature as it is usual in chiral hydrogenations. The enantioselectivity data expressed in ee% are tabulated in Table 1.

Table 1 Enantioselective hydrogenation of ethyl pyruvate over various platinum catalysts under conventional conditions and ultrasonic irradiation in atmospheric systems (25 °C, 1 atm hydrogen pressure)

Catalyst	Solvent	Yield[a] / % (time) ⌡	Yield[a] / % (time))))	ee / % ⌡	ee / %)))
5% Pt/C	EtOH	95 (73 min)	95 (58 min)	20.2	23.2
	toluene	100 (65 min)	100 (45 min)	30.5	31.7
	AcOH	100 (27 min)	100 (40 min)	39.0	55.0
3% Pt/SiO$_2$	EtOH	95 (600 min)	100 (52 min)	20.2	34.2
	toluene	74 (16h)	100 (110 min)	24.8	50.5
	AcOH	100 (45 min)	100 (50 min)	61.0	72.0
5% Pt/K-10	EtOH	80 (480 min)	93 (210 min)	23.0	40.4
	toluene	46 (720 min)	100 (320 min)	25.3	44.1
	AcOH	88 (10 h)	77 (9 h)	37.0	21.0
5% Pt/Al$_2$O$_3$	EtOH	100 (95 min)	100 (45 min)	31.4	29.8
	toluene	100 (130 min)	100 (130 min)	49.6	34.2
	AcOH	100 (35 min)	100 (51 min)	67.8	67.2

EtOH - ethanol, AcOH - acetic acid
[a]Yields are given in mol%

As the data indicate the variation of solvents in the conventional hydrogenations significantly modifies the enantioselectivity: higher ee values were obtained in acetic acid. As expected, the order of solvents producing increasing ee values is acetic acid > toluene > ethanol.

The effect of presonication on the enantioselectivity shows similar increase as observed in the case of the reaction rates. As a general rule, the ultrasonic irradiation increases the enantioselectivity. Whereas the increase in the Pt/C catalyzed reactions is moderate, the ee values are highly improved in the case of Pt/SiO_2 and Pt/K-10 catalysts. The ee increase was found to be between 15-26%, which is a significant improvement indicating important and intriguing changes in the catalysts during sonication. The best enantioselectivity (ee 72%), which is by far the best up to now in atmospheric systems, was observed in the Pt/SiO_2 catalyzed system in acetic acid. However, beside this effect, the enantioselectivities obtained after presonication in acetic acid are similar to those observed in conventional systems over Pt/Al_2O_3 and Pt/K-10. This observation suggests that the nature of the influence strongly depends on the catalyst used.

3.2.2. Pressurized systems Since at low hydrogen pressures a moderate mass transfer limitation was observed the hydrogenation reactions were also carried out in an autoclave under 10 atm to study the effect of higher hydrogen pressure and ultrasounds on the reaction. The enantioselectivity data obtained are collected in Table 2.

Table 2 Enantioselective hydrogenation of ethyl pyruvate over various platinum catalysts with and without ultrasonic pretreatment in pressurized systems (25 °C, 10 atm hydrogen pressure, 3 h)

Catalyst	Solvent	Yield[a] / % ⤵	Yield[a] / %)))	ee / % ⤵	ee / %)))
5% Pt/C	EtOH	100	100	20.2	46.4
	toluene	100	100	61.2	66.3
	AcOH	100	100	50.6	56.0
3% Pt/SiO_2	EtOH	100	100	50.2	53.2
	toluene	100	100	71.0	77.4
	AcOH	100	100	70.2	75.4
5% Pt/K-10	EtOH	100	100	66.4	43.2
	toluene	100	100	67.8	49.2
	AcOH	100	100	45.4	39.0
5% Pt/Al_2O_3	EtOH	100	100	39.6	50.2
	toluene	100	100	68.6	64.2
	AcOH	100	100	78.2	80.2

[a]Yields are given in mol%

As the table shows the hydrogenation took place quantitatively in each case. Moreover, comparing the ee values in Tables 1 and 2 significantly higher enantioselectivities can be achieved in pressurized systems than those obtained under atmospheric conditions. Similarly to the atmospheric systems a solvent dependence of the enantioselectivity was observed. However, in this case toluene was found to be the best solvent except for the alumina supported catalyst. The presonication, as above, resulted in ee% increase, the only exception is the K-10 supported catalyst, where the presonication resulted in decreasing selectivity. It is important to note that the selectivity is just slightly, however, reproducibly higher than in conventional systems. The spectacular increase in enantioselectivity found in some cases in above mentioned atmospheric system was not observed here.

3.3. Effect of Ultrasounds on the Catalysts

Taking into account literature preliminaries with respect to the effect of ultrasonic irradiation on solid particles regardless of their material (22) it was an obvious necessity to study our catalysts as well. In order to determine the influence of the presonication on the metal particle size of the catalysts, transmission electron microscopy (TEM) was applied. As representative examples the TEM images of the Pt/SiO$_2$ and Pt/Al$_2$O$_3$ catalysts before and after ultrasonic irradiation are displayed in Fig. 2. On the basis of electron micrographs, the metal particle size distribution of both catalysts determined are illustrated in Fig. 3.

In addition, the mean metal particle diameters of each catalysts as received and after 30 min ultrasonic irradiation used in the present study are also collected (Table 3).

Table 3 The mean metal particle diameters of the supported platinum catalysts before and after 30 min presonication

Catalyst	Particle diameter (nm)	
	as received	)))
5 % Pt/C	3.0	2.5
3 % Pt/SiO$_2$	6.7	4.2
5 % Pt/K-10	3.7	3.8
5 % Pt/Al$_2$O$_3$	23.1	12.6

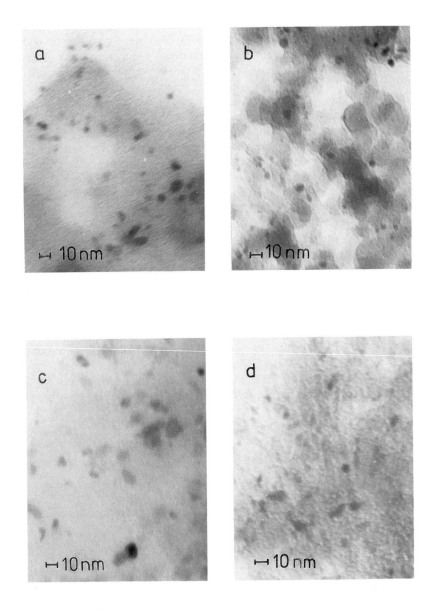

Figure 2 TEM micrographs of supported platinum catalysts: (a) 3% Pt/SiO$_2$, as received, (b) 3% Pt/SiO$_2$ after 30 min presonication, (c) 5% Pt/Al$_2$O$_3$ as received, (d) 5% Pt/Al$_2$O$_3$ after 30 min presonication.

(a)

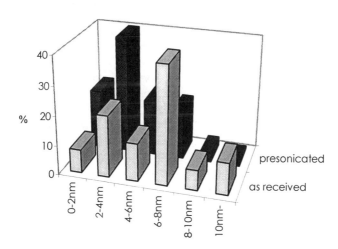

(b)

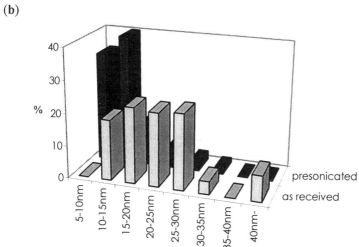

Figure 3 Metal particle size distributions of supported platinum catalysts as received and after 30 min ultrasonic irradiation (**a**) 3% Pt/SiO$_2$, (**b**) 5% Pt/Al$_2$O$_3$

As it is shown, the size of the metal particles decreased significantly in almost each case, as a result of the irradiation. As an additional effect, a change in particle size morphology was also observed. As the images (Fig. 2) and Fig. 3 indicate the particle size of the platinum crystallites in the pristine samples was found to be rather inhomogeneous, ranging between 5-60 nm. In contrast, the predominance of smaller particles are seen in both Figs. 2 and 3. Additionally, although the particle size range is nearly the same, the distribution is altered to a large extent. The small particles became predominant and the distribution is close to homogeneous in the range of 3-5 and 10-15 nm. Summarizing the effect of ultrasounds on the catalysts, it can be stated that the sonochemical pretreatment produces smaller particles and more uniform size distribution.

Earlier literature observations state, that ultrasonics decrease the particle size in systems containing solid particles (23) independently from the material applied. This phenomenon has been described recently (17) with respect to metal particles of supported metal catalyst as well. A characteristic decrease in particle size especially at low metal loading has been found. It is well known that the decreasing metal particle size i.e. more efficiently dispersed metal over the support brings about higher efficiency in heterogeneous metal-catalyzed hydrogenations. In the light of earlier reports concerning C=C double bond hydrogenation (12, 15, 16) this effect is responsible for the rate increase, which is in accordance with our observations. As an additional contribution of the ultrasounds to the activity enhancement, the removal of the impurities from the metal surface and the more effective activation during the prehydrogenation should also be considered. The effect of the increasing dispersion has already been studied in conventional systems on the enantioselectivity of ethyl pyruvate hydrogenation (24). Blaser and coworkers prepared supported platinum catalysts of various dispersions and used them under conventional experimental conditions. It was found that increasing metal dispersion results in increasing enantioselectivity (24) up to ~25% dispersion. Although the effect of ultrasonics on the enantioselectivity is not clearly understood yet, the above mentioned phenomenon may be responsible for the ee changes. Since the particle size of the catalysts used is generally high (e.g. metal dispersion except Pt/SiO$_2$ is under 15%) the sonication can produce more optimal catalyst dispersion through particle size decrease for enantiodifferentiation. Beside, modifier-metal interaction can produce more effective active centers as a special case of the template effect, since the presence of the modifier is crucial during presonication: the ee increase is lower if cinchonidine is added after the sonochemical treatment (25). However, the effect is very specific, highly depends on the individual catalysts and solvents as shown by the different solvent and support dependencies.

4. CONCLUSIONS

The experimental results described above unambiguously prove that ultrasonic irradiation of the catalyst—modifier system before reaction is an efficient method to enhance the enantioselectivity in the hydrogenation of ethyl pyruvate. In addition, the presonication was found to be effective in the acceleration of the hydrogenation of C=O group in ethyl pyruvate, and to increase the product yield. The mechanistic background of the phenomenon is not well understood yet, however, at the present stage of our study the emphasis is placed on the metal particle size modification brought about by ultrasonic pretreatment. Parallel this phenomenon, the catalyst-modifier interaction during presonication is considered to be an effect of crucial role. Eventually, the sonochemical pretreatment is a very intriguing and promising method for metal-catalyzed enantioselective reactions.

ACKNOWLEDGMENT

Financial support by the Hungarian National Science Foundation (OTKA Grants T016109 and F023674) and the Hungarian Academy of Sciences is highly appreciated.

REFERENCES

1. M. Bartók and K. Felföldi,. in *Stereochemistry of Heterogeneous Metal Catalysis* (Ed.: Bartók, M.), Wiley, Chichester (1985), Chapter VII, p 335.
2. M. Bartók and Gy. Wittman, in *Stereochemistry of Heterogeneous Metal Catalysis* (Ed.: Bartók, M.), Wiley, Chichester (1985), Chapter XI, p 511; H.-U. Blaser, *Chem. Rev.* **92**, 935 (1992).
3. H.-U. Blaser and M. Müller, *Stud. Surf. Sci. Catal.* **59**, 73 (1991); *Chiral Reactions in Heterogeneous Catalysis* (Eds.: Jannes, G. and Dubois, V.), Plenum Press, New York, London (1995).
4. R. L. Augustine, S. K. Tanielyan and L. K. Doyle, *Tetrahedron: Asymmetry* **4**, 1803 (1993).
5. H. U. Blaser, H.-P. Jalett and F. Spindler, *J. Mol. Catal. A*, **107**, 85 (1996).
6. A. Baiker, *J. Mol. Catal. A*, **115**, 473 (1997).
7. J. T. Wehrli, A. Baiker, D. M. Monti and H. U. Blaser, *J. Mol. Catal.* **61**, 207 (1990).
8. H. U. Blaser, H. P. Jalett, D. M. Monti, A. Baiker and J. T. Wehrli, *Stud. Sci. Surf. Catal.* **67**, 147 (1991).
9. G. Bond, P. A. Meheux, A. Ibbotson and P. B. Wells, *Catal. Today* **10**, 371 (1991).
10. B. Minder, T. Mallat, P. Skrabal and A. Baiker, *Catal. Lett.* **29**, 115 (1994).
11. Y. Sun, R. N. Landau, J. Wang, C. Leblond and D. G. Blackmond, *J. Am. Chem. Soc.* **118**, 1348 (1996); J. Wang, Y. Sun, C. LeBlond, R. N. Landau and D. G. Blackmond, *J. Catal.* **161**, 752 (1996); Y. Sun, J. Wang, C. LeBlond, R. N. Landau and D. G. Blackmond, *J. Catal.* **161**, 759 (1996).
12. T. J. Mason and J. P. Lorimer, *Sonochemistry*, Ellis Horwood, Chichester, (1988).

13. P. Boudjouk, in *Ultrasounds. Its Chemical, Physical, and Biological Effects* (Ed.: Suslick, K. S.), VCH, New York, Weinheim (1988), p.165

14. B. H. Han and P. Boudjouk, *Organometallics*, **2**, 770 (1983).

15. P. Boudjouk and B. H. Han, *J. Catal.* **79**, 489 (1983).

16. P. W. Cains, L. J. McCausland, D. M. Bates and T. J. Mason, *Ultrasonics Sonochem.* **1**, S45 (1994).

17. C. L. Bianchi, R. Carli, S. Lanzani, D. Lorenzetti, G. Vergani and V. Ragaini, *Ultrasonics Sonochem.* **1**, S47 (1994).

18. A. Tai, T. Kikukawa, T. Sugimura, Y. Inoue, T. Osawa and S. Fujii, *J. Chem. Soc. Chem. Commun.* 795 (1991).

19. B. Török, K. Felföldi, G. Szakonyi and M. Bartók, *Ultrasonics Sonochem.* (in press).

20. F. Notheisz and M. Bartók, *J. Catal.* **71**, 331 (1981); G. V. Smith, M. Bartók, D. Ostgard and F. Notheisz, *J. Catal.* **101**, 212 (1986).

21. 5 % Pt/K-10 was prepared by impregnation of K-10 montmorillonite with $H_2[PtCl_6]$ followed by a reduction with refluxing EtOH similarly as described by M. Komiyama and H. Hirai, *Bull. Chem. Soc. Jpn.* 2833 (1983).

22. B. Török and Á. Molnár, *Organic Reactions by Microwave and Sonochemical Activation* (in Hungarian) in *New Results of Chemistry*, Vol. 84, Akadémiai Kiadó, Budapest, 1997.

23. J. Lindley, in *Chemistry with Ultrasound* (Critical Reports on Applied Chemistry) Vol. 28, SCI&Elsevier, London, New York (1990), P. 27.

24. H.-U. Blaser, H. P. Jalett, D. M. Monti, A. Baiker and J. T. Wehrli, in *Structure-Activity and Selectivity Relationships in Heterogeneous Catalysis* (Eds.: Grasselli, R. K. and Sleight A. W.), Elsevier, Amsterdam (1991), p. 147.

25. B. Török, K. Felföldi, G. Szakonyi, K. Balázsik and M. Bartók, *Catal. Lett.* (submitted for publication)

Racemization of (R)-(-)-10-Methyl-Δ-1(9)-Octalin: Double Bond Migration During Hydrogenation over Pt Catalysts

G.V. Smith, Y. Wang, R. Song, and J. Tyrrell

Department of Chemistry and Biochemistry
Southern Illinois University at Carbondale
Carbondale IL 62901-4409

Abstract

(**R**)-(-)-10-methyl-$\Delta^{1(9)}$-octalin was synthesized and hydrogenated over three different Pt catalysts. Hydrogenation and double bond migration did not occur over PtO_2, slow hydrogenation but not double bond migration occurred over 5% Pt/C, and slow hydrogenation and slow double bond migration occurred over 5% Pt/Al_2O_3. The stereochemistry of (**R**)-(-)-10-methyl-$\Delta^{1(9)}$-octalin suggests that the formation of π-allyl surface species requires an allylic hydrogen virtually perpendicular to the planes of the double bond and the surface site.

Introduction

Apparent *trans*-addition of hydrogen to carbon-carbon double bonds has been the subject of numerous investigations. As a result, several mechanisms have been proposed[1], such as a dissociatively adsorbed alkene[2], **A**; a topside addition of hydrogen[3], **B**; a 1,3 hydrogen shift over the topside of the adsorbed alkene[4], **C**; and double bond migration followed by desorption, flipping, and readsorption[5], **D**.

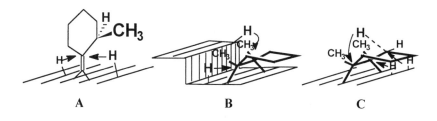

A   **B**   **C**

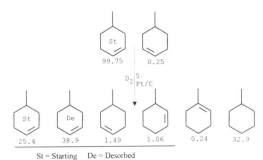

1,2-dimethyl cyclohexene   *cis*   *trans*

	Pt	80	20
(isomerize)	Pd	25	75

2,3-dimethyl cyclohexene   (flip)

(readsorb upside down)

D

This latter mechanism is the most widely accepted[1]; however, it lacks full accounting for the formation of 20% *trans* product from the hydrogenation of 1,2-dimethylcyclohexene over Pt[6]. The arguments for this are linked to the extent of double bond migration in different alkenes.

Double bond migration occurs to different extents on different catalysts. On Pd it usually occurs readily, but on Pt it is slow. The argument that *trans*-1,2-dimethylcyclohexane is formed from 1,2-dimethylcyclohexene by first double bond migration, then desorption, next readsorption on the opposite side of the ring, and finally, *cis*-addition of hydrogen (mechanism **D** above) requires 70% incursion of this path to produce 20% *trans* product (2,3-dimethylcyclohexene itself yields only 30% *trans*)[3]. Such a large amount of double bond migration from the more to the less stable isomer in the cyclohexene ring over Pt is difficult to rationalize by available data. For example, deuteriumation of (*R*)-4-methylcyclohexene on Pt reveals only 2.55% double bond migration at 33% addition[7] (Scheme 1). Double bond migration in the cyclohexene ring is

St = Starting   De = Desorbed

Scheme 1: Deuteriumation of (*R*)-4-Methylyclohexene

slower than for a double bond migration in a linear chain[8] (Scheme 2), and

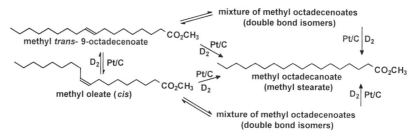

Scheme 2: Deuteriumations of methyloctadecenoates

slower still than in the strained ring system of apopinene[9] (Scheme 3).

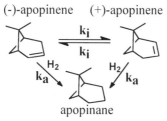

Scheme 3: Hydrogenation of Apopinenes

Table 1 compares the relevant data.

Table 1. *Relative Rates of Double Bond Migration on Pt.*

Structural type	rel rate migr	migr/add
cyclohexene	1.0	0.078
straight chain hydrocarbon (oleic acid, *cis*-)	3.7	0.28
apopinene	7.9	0.61

Looking at just this evidence in which double bond migration occurs very slowly in cyclohexene, it seems unlikely that double bond migration can fully account for the formation of *trans*-1,2-dimethylcyclohexane from the hydrogenation of 1,2-dimethylcyclohexene over Pt. In addition, a similar situation occurs during the hydrogenation of $\Delta^{9(10)}$-octalin over Pt, in which *cis*- and *trans*-decalin are formed in approximately equal amounts (Scheme 4).

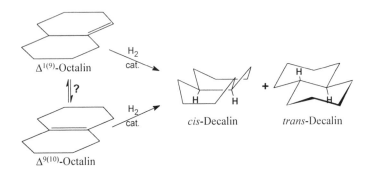

Scheme 4: Hydrogenation of $\Delta^{1(9)}$- and $\Delta^{9(10)}$-Octalins

The isomeric $\Delta^{1(9)}$-octalin is not observed to isomerize to the more stable $\Delta^{9(10)}$ isomer[10]. Therefore, in this case, as in the case of 1,2-dimethylcyclohexene, the argument follows the lines that the tetrasubstituted alkenes are weakly adsorbed and cannot compete with their trisubstituted isomers for surface sites, and in the pores of the catalysts these isomers interconvert but are not observed in the liquid phase. Support for this argument comes from the facts that the quantities of *trans* isomers increase with the double bond migration ability of various catalysts[11], which suggest that double bond migration sites (or the sites nearby) and pore diffusion are indeed involved in apparent *trans*-addition.

However, if pore diffusion were allowing hidden double bond migration followed by readsorption and hydrogenation in the cases of 1,2-dimethylcyclohexene and $\Delta^{9(10)}$-octalin, then it also should allow it in the case of (*R*)-4-methylcyclohexene. But it doesn't. On the other hand, rampant isomerization is observed on both Pt and Pd during hydrogenation of the apopinenes[9], which raises the question of substrate stereochemical influence on double bond migration during hydrogenation.

To further understand the influence of substrate structure on double bond migration (and possibly *trans*-addition), we have now synthesized (*R*)-(-)-10-methyl-$\Delta^{1(9)}$-octalin and measured its racemization (double bond migration) over Pt.

Experimental

The catalysts, 5% Pt/C, 5% Pt/Al$_2$O$_3$, and PtO$_2$ were purchased from Engelhard Industries. Hydrogen and helium were purchased from GENEX company.

Hydrogenations were run at ambient temperature (23-25°C) and 2.0 psig hydrogen in an apparatus previously described[9]. Essentially, it consists of a small glass slurry reactor (15 mm by 100 mm) containing double helical Vigreux indentations, with a neck at the top (for connection to the hydrogen handling system) adjacent to an injection port capped by a nylon bushing and a silicon septum. After appropriate introduction of catalyst, evacuations, flushings with hydrogen, and introduction of solvent and substrate, the reactor is shaken in a vortex manner at 2500 revolutions per minute. Because hydrogenations of these compounds are so slow, there is little danger of mass transfer problems (with the improbable exception of pore diffusion) influencing results.

(*R*)-(-)-10-methyl-$\Delta^{1(9)}$-octalin was synthesized in a mixture with *cis*- and *trans*-(*R*)-(-)-10-methyl-$\Delta^{1(2)}$-octalin in a five step sequence (Scheme 5). The first four steps follow the procedure of Pfau[12]. The last step is a new example of the selective reduction of the carbonyl group of an aliphatic α,β-unsaturated ketone. NMR spectra agreed with those published [13]. The $\Delta^{1(2)}$-octalins were

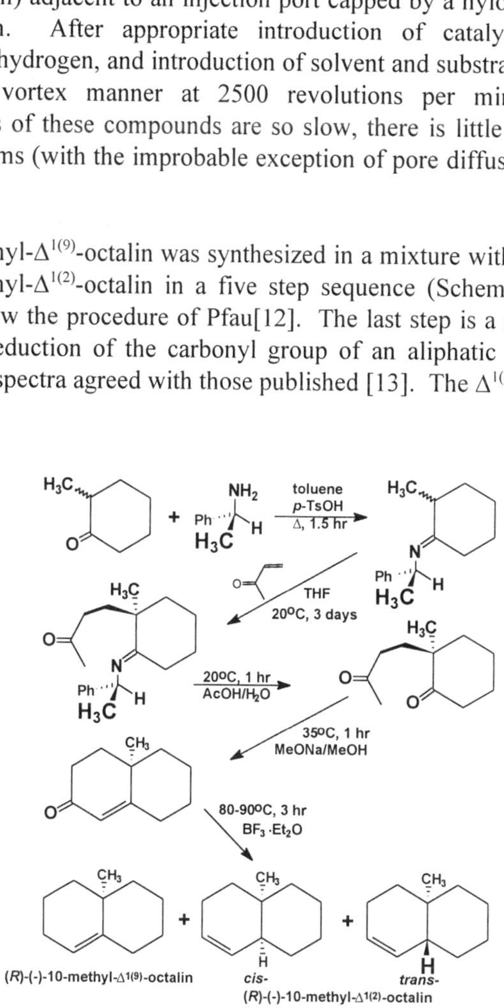

Scheme 5: Preparation of (*R*)-(-)-10-Methyl-$\Delta^{1(9)}$-Octalin

removed by hydrogenation (*n*-hexane solvent) over 5%Pt/C, and the resulting mixtures containing $\Delta^{1(9)}$-octalin, now diluted with *cis*- and *trans*-9-methyldecalin, were further hydrogenated to different extents (Scheme 6) over the different Pt catalysts, and their optical rotations measured.

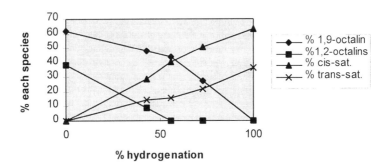

cis and trans   cis- and trans-9-methyldecalin

Scheme 6: Partial Hydrogenation of Synthetic Mixture of 10-Methyloctalins

Results

In a mixture with its isomer, the $\Delta^{1(9)}$-octalin hydrogenates over 5%Pt/C more slowly than its $\Delta^{1(2)}$-isomer, so after approximately 50% hydrogenation, none of the $\Delta^{1(2)}$ remains, and any change in optical rotation of the $\Delta^{1(9)}$ can be followed during further hydrogenation (Figure 1).

Figure 1. Hydrogenation of 10-methyl-$\Delta^{1(9)}$- octalin and cis- and trans-10-methyl-$\Delta^{1(2)}$- octalin over 5% Pt/C

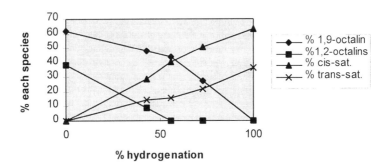

However, when the optical rotation (D line of Na at 25°C) of the reaction mixture was plotted against the concentration of 10-methyl-$\Delta^{1(9)}$-octalin (Figure 2), it was found that no loss in specific rotation occurs; that is, no racemization and, therefore, no double bond migration occurs. The specific rotation of the remaining $\Delta^{1(9)}$-octalin was calculated in the following manner. As Fig. 2 shows, the optical rotation is not zero at 100% hydrogenation. Rather it is -1.12°, which

Figure 2. Optical rotation of reaction mixture during hydrogenation of 10-methyl-Δ $^{1(9)}$-octalin after hydrogenation of all 10-methyl-Δ $^{1(2)}$-octalin

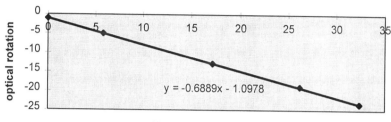

% 10-methyl-Δ $^{1(9)}$-octalin

is confirmed by extrapolation of the optical rotation of the reaction mixture to zero percent 10-methyl-Δ$^{1(9)}$-octalin. The intercept occurs at -1.10°, which indicates an unhydrogenatable impurity in the reaction mixture. Based on this small correction, the calculated specific rotation of the 10-methyl-Δ$^{1(9)}$-octalin remains virtually unchanged during hydrogenation with an average value of -68.7±0.3°. Any change is buried in experimental error.

Hydrogenation on 5% Pt/Al$_2$O$_3$ showed small racemization outside experimental error. Over this catalyst the specific rotation decreased approximately 11% during 22% hydrogenation indicating a 5.5% double bond migration, which is similar to that of oleic acid (Scheme 2 and Table 1).

On the other hand, to our surprise, we could not hydrogenate 10-methyl-Δ$^{1(9)}$-octalin over PtO$_2$. We don't know if this results from traces of Na[14], but this, as well as the slow hydrogenation over the other catalysts, suggested severe congestion around the double bond and caused us to more carefully examine the structure of the more stable molecular conformations of the molecule.

Another surprising result was the predominant production of the *cis*-saturate rather than the expected *trans*. Correcting for the ratio of *cis/trans* (3.3) produced from the Δ$^{1(2)}$ isomers, the ratio produced by the Δ$^{1(9)}$-octalin is 1.6 (62% *cis*, 38% *trans*). We expected hydrogen to add to the side opposite to the 10-methyl group and produce predominantly *trans*-decalin. However, in all cases, the *cis*-decalin dominates. This result along with the unexpectedly slow hydrogenations and the lack of hydrogenation over PtO$_2$ suggested unique stereochemical factors existing between the octalin and the catalyst. Factors which not only inhibit hydrogenation, but also inhibit double bond migration.

Discussion

In his review of stereochemistry of hydrogenation of unsaturated hydrocarbons, Siegel drew attention to critical steric factors influencing product stereochemistry[15]. A major focus was on the stereochemistry of the half-hydrogenated state and on steric factors within the hydrogenating molecules. Careful analysis of these factors have predicted many stereochemical results. At that time, some attention was given to steric interactions between substrate and surface; however, the current understanding of surfaces developed by surface science was just emerging, so that could not form a foundation for mechanistic speculation. Rather, emphasis was placed on a single metal atom as the active site after the model of newly discovered organometallic complex homogeneous catalysts. Later, Siegel drew analogies between the single metal atom in homogeneous catalysts and the corner atoms of metal crystallites and invented the nomenclature, ^1M, ^2M, ^3M, to describe the degree of coordinative unsaturation of exposed metal atoms[16]. These have come to be known respectively as plane, face, or terrace sites; edge or ledge sites; and corner, kink or adatom sites. Work by Augustine[17] and us[9] has drawn attention to the mild structure sensitivity of hydrogenation and double bond migration and pointed to the probable involvement of edges and corners as active sites for hydrogenation and double bond migration. Thus we have come to think of active sites on metal surfaces as having definite shapes on which organic molecules should fit.

Thinking of active sites as having definite stereochemistry is not new. In fact Balandin's multiplet theory was an early attempt to carefully describe the structure of active sites[18]. Indeed, Balandin's ideas might well describe what we now call corner and edge sites.

Assuming the substrate should fit the active site, we can ask the question: why do the apopinenes hydrogenate so readily accompanied by rapid double bond migration, while the 10-methyl-$\Delta^{1(9)}$-octalins hydrogenate so slowly accompanied by virtually no double bond migration? The apopinenes possess a bridging methylene below the plane of the double bond, so we ascribe this structure to fit nicely on an edge or corner site.

On the other hand, molecular modeling of 10-methyl-$\Delta^{1(9)}$-octalin reveals it also has a somewhat curved structure such that neither side of the double bond may fit flat on the surface (Figure 3). The 10-methyl group projects above the plane on one side of the double bond and one portion of the saturated ring projects on the other side. By the same reasoning as with the apopinenes, 10-methyl-$\Delta^{1(9)}$-

octalin should fit nicely on an edge or corner site. So fit is not enough to explain surface-substrate reactivity. Whatever is the structure of the active site, the 10-methyl-$\Delta^{1(9)}$-octalin may fit on it but clearly does not readily react on it.

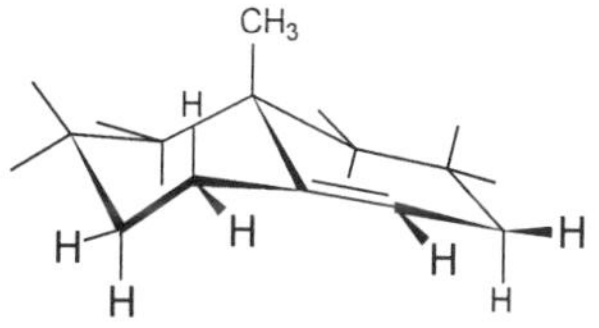

Figure 3. (*R*)-(-)-10-Methyl-$\Delta^{1(9)}$-Ocatlin

Further examination of the structures of apopinene and 10-methyl-$\Delta^{1(9)}$-octalin reveals another important stereochemical difference. In the structure of apopinene, the allylic hydrogen opposite the *gem*-dimethyl group is involved in double bond migration. When this allylic hydrogen is substituted with deuterium, it is seen to migrate so rapidly that it does not mix well with the surface hydrogen[19]. Such a favorably situated allylic hydrogen does not exist

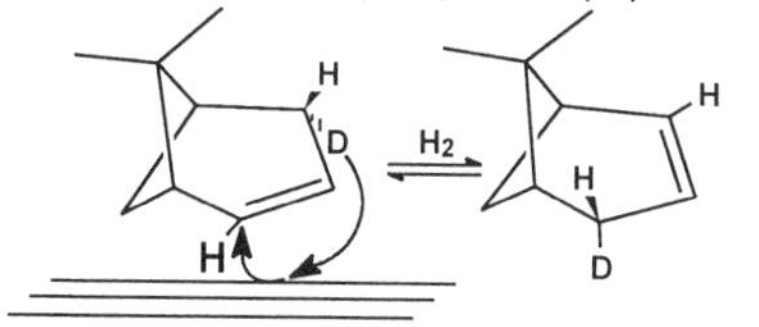

(-)-4D-apopinene   (+)-4D-apopinene

Scheme 7: Isomerization of Deuterated Apopinene during Hydrogenation

in 10-methyl-$\Delta^{1(9)}$-octalin. Rather, the appropriate allylic hydrogen at the 8 position, which is on the opposite side of the rings as the methyl group, is parallel to the plane of the double bond and not in position to be easily abstracted by the surface. Nevertheless, the other allylic hydrogen at the 8 position possesses almost the stereochemistry to be abstracted if the molecule adsorbs with the methyl group towards the surface, and since the major product is *cis*-9-methyldecalin, (*R*)-(-)-10-Methyl-$\Delta^{1(9)}$-octalin must be adsorbed on a site with its methyl towards the surface. That this other allylic hydrogen is not abstracted, signals a critical steric requirement for formation of a π-allyl surface species. Not only must a π-allyl structure be possible in the substrate, but also the allylic hydrogen must be abstractable. Apparently, this stereochemistry is not as important for addition as it is for double bond migration.

We can conclude, therefore, that congestion around the double bond inhibits addition, but that stereochemistry of the allylic hydrogens and their ability to move into a position nearly perpendicular to the plane of the double bond and the surface is critical.

References

1. see Chapter III by Á. Molnár in "Stereochemistry of Heterogeneous Metal Catalysis", (M. Bartók) John Wiley & Sons, Chichester, 1984, pp 53-210
2. J.F. Sauvage, R. Baker, and A.S. Hussey, *J. Am. Chem. Soc.,* **82** (1960) 6090; R.L Burwell, Jr, B.K.C. Chim,., and H.C. Rowlinson, *J. Am. Chem. Soc.,* **79** (1957) 5142.
3. F.G. Gault, J.J. Rooney, and C. Kemball, *J. Catal.,* **1** (1962) 255 ; G.C. Bond, *Fourth International Congress on Catalysis, Moscow, 1986,* (Ed. J. Hightower), Rice University Printing & Reproduction Dept., Houston, Texas, (1969), vol. 3, pp 1217-1234l.
4. G.V. Smith and J.R. Swoap, *J. Org. Chem.,* **31** (1966) 3904.
5. S. Siegel, P.A. Thomas, and J.T. Holt, *J. Catal.,* **4** (1965) 73.
6. S. Siegel, and G.V. Smith, *J. Am. Chem. Soc.,* **82** (1960) 6082.
7. G.V. Smith, J.A. Roth, D.S. Desai, and J.L. Kosco, *J. Catal.,* **30** (1973) 79.
8. E. Selke, W.K. Rohwedder, C.R. Scholfield, and H.J. Dutton, *Petrol. Prepr.,* 11(4), A-63 (1966); H.J. Dutton, C.R. Scholfield, E. Selke, and W.K. Rohwedder, *J. Catal.,* **10** (1968) 316.
9. G.V. Smith, F. Notheisz, Á.G. Zsigmond, D. Ostgard, T. Nishizawa, and M. Bartók, *Proc. Ninth Internat. Cong. Catal.,* **3**, *Calgary, June 26-July 1,* The Chemical Institute of Canada, Ottawa, 1988, pp 1066-1073.
10. G.V. Smith, and R.L. Burwell, Jr., *J. Am. Chem. Soc.,* **84** (1962) 925; A.W. Weitkamp, *J. Catal.,* **6** (1966) 431.
11. S. Nishimura, H. Sakamoto, and T. Ozawa, *Chem. Lett.,* (1973) 855.
12. M. Pfau, G. Revial, A. Guingant, and J. d'Angelo, *J. Amer. Chem. Soc.,* **84** (1985) 237.
13. D.V. Avila, A.G. Davies, and I.G.E. Davison, *J. Chem. Soc. Perkin Trans,* II (1988) 1847; T.L. Underiner and H.L. Goering, *J. Org. Chem.* **52**(1987) 897
14. P.N. Rylander, "Catalytic Hydrogenation over Platinum Metals", Academic Press, New York, 1967, p 23.
15. S. Siegel, *Advances in Catalysis,* **16** (1966) 123-177.
16. S. Siegel, J. Outlaw, Jr., and N. Garti, *J. Catal.,* **52** (1978) 102-115.
17. R.L. Augustine, "Heterogeneous Catalysis for the Synthetic Chemist", Marcel Dekker, N.Y., 1996, pp 34-46.
18. A.A. Balandin, *Advances in Catalysis,* **19** (1969) 1-210.
19. G.V. Smith, B. Rihter, Á. Zsigmond, F. Notheisz, and M. Bartók, "Studies in Surface Science and Catalysis 101, 11th International Congress on Catalysis - 40th Anniversary", (Eds. J.W. Hightower, W.N. Delgass, E. Iglesia, and A.T. Bell), Elsevier, Amsterdam, 1996, pp 251-255.

1998 Murry Raney Plenary Lecture: Traditional and Novel Modifications of Raney® Catalysts: Advances in Theory and Technology

Anatoly B. Fasman
Consultant, San Francisco, California

Abstract

The findings obtained by the author with his associates over a long period of time regarding the mechanism of the Raney®catalysts genesis during selective leaching of binary and multicomponent parent alloys, their morphology, bulk and surface layers chemical and phase composition, peculiar features of the solid state interactions between zerovalent and ionic forms of catalytically active, leachable, and alloying metals as well as their adsorption properties, activity, selectivity, and stability in a variety of liquid- and gas-phase organic reactions are summarized and discussed. A range of metallic systems potentially favorable for the preparation of effective Raney catalysts is substantiated. The classification of modifiers is given. Several novel types of Raney catalysts such as nonpyrophoric modification that can be considered as a promising alternative to the fire-hazardous and explosive conventional one, derived from energy-saturated metastable precursors, and supported on the inert carriers are developed. The approaches to the target-oriented selection and optimization of Raney catalysts based on the comprehensive study of their physicochemical characteristics are suggested.

Introduction

This paper represents the results of systematic and extensive studies in the field of sponge metal catalysts theory and technology that have been started more than 40 years ago, after my graduation from the department of Catalysis and Technical Chemistry at Alma-Ata State University, Kazakhstan. Experiments were then performed at the same University and later at the Institute of Organic Catalysis and Electrochemistry established in 1969 by my unforgettable adviser late Professor D.V. Sokolsky who has drawn my attention to this field

Raney catalysts are active and selective in a variety of fine organic syntheses and industrial processes (1). Since 1954, they have found wide use in the different types of low- and medium-temperature fuel cells (2). Yet, both the fundamental principles of their optimization and the technology of their preparation and manufacturing do not meet the up-to-date requirements of resources saving, environment protection, operating safety, etc. Therefore, it has to be a radical change in the basic aspects of these crucial problem.

The method of selective leaching of non-noble components (Mg, Zn, Sn, Cu, Fe, etc.) from their binary alloys with platinum group metals by strong electrolyte solutions to obtain high-porous metallic residues, the so-called " Explosive powders", has been well-known since the beginning of the last century (3). This subtractive technique is a routine and widely used in hydrometallurgy. The extraction of leachable components leads to formation of metallic sponges with highly-developed surface. Without any doubt, such highly-explosive and costly platinum-based powders could not find their way into industry. M. Raney became the world's first who using this approach created a new class of important commercial catalysts (4-7), and he deserves the indisputable credit for this invention.

An overview of thousands of publications and patents devoted to Raney catalysts and their application in organic syntheses has been presented in several books (1, 2, 8, 9) and in a number of articles. The results of a wide range of later works have not been summarized and discussed with the exception of few relatively short reviews focused on particular aspects of this problem although over the last 15 years their structure and physicochemical properties have been profoundly studied by the abundance of currently available sophisticated methods.

The catalysts of the greatest commercial importance are Raney nickel, copper, and cobalt. They are widely used in such reactions as hydrogenation, hydrogenolysis, dehydrogenation, cyclization, isomerization, condensation, different rearrangements as well as in asymmetric syntheses, biologically active substances preparation, monomers purification, etc. (1, 8). Much efforts have been directed towards replacing conventional Raney catalysts by multicomponent ones, for example (1, 2, 8, 9, 10-17). Unfortunately, although doped Raney catalysts have been used for a long period, the empirical approach to selection of modifying additives in many cases, as can be seen from reviews (1, 2, 8, 9), was a basic one.

The vital issue and the major goal of our research were to lay the foundations and to create the scientific bases for manufacturing of new industrially important powerful Raney catalysts with predetermined operating characteristics steming from the comprehensive studies of their genesis, structure, and physicochemical properties. In order to accomplish this and to gain a complete understanding of interrelations between the most important parameters of these catalysts, a wide complex of modern physical methods was used.

Experimental Methods

Details of the initial alloys and Raney catalysts preparation as well as peculiarities of methods and types of equipment used for their characterization have been reported in the past (see the list of references) and are only briefly outlined here.

Starting alloys were prepared from the reagent-grade metals in high-frequency or arc furnaces and subjected to homogenazing annealing. Throughout this paper their compositions are given in wt %. They were crushed to powders with subsequent screening into narrow fractions in a gravitational air separator. Partially leached fixed-bed catalysts as well as thin films of leached parent alloys deposited on a NaCl monocrystal have been also prepared and investigated. Initial alloys were leached by NaOH or KOH solutions and then thoroughly washed off. In the case of precious metal-based catalysts, when needed diluted mineral acids or even electrochemical pickling (18-20) were used.

To obtain the information about the contribution of every individual phase to the overall properties of parent alloys and Raney catalysts, the residues after leaching of specially fabricated binary $NiAl_3$ and Ni_2Al_3 (1,2,21-31), $TiAl_3$, $ZrAl_3$, $NbAl_3$, $TaAl_3$ (32), and ternary $NiTiAl_2$ (32), $NiMo_5Al_{10}$ (24, 25, 32, 33), $NiFeAl_9$ and $NiFe_3Al_{10}$ (34), $CaNi_2Al_8$, $SrNi_2Al_9$, $BaNi_2Al_9$ (35) nickel and aluminium based intermetallides as well as platinum bismuthides, antimonides, and indides (18) and ruthenium gallides (36) were studied.

To examine their bulk and surface chemical composition, structure and properties, a range of characterization methods was used : X-ray and electron diffraction, small-angle X-ray scattering, X-ray fluorescence spectroscopy, transmission and scanning electron microscopy, electron probe microanalysis, IR spectroscopy, Mössbauer spectroscopy, thermal analysis (TG, TPR, DSC), Auger spectroscopy, XPS, and SIMS.

Particle size distribution was evaluated by TA-4 Coulter counter, their surface area and pore structure were determined from nitrogen and argon adsorption-desorption isoterms using a standard BET-apparatus and pore volume analyzer. A number of catalysts were studied by a mercury penetration technique giving close results. To determine their adsorption properties and the state of adsorbed and absorbed hydrogen, neutron diffraction, TPD, and potentiodynamic methods were used. Their catalytic and electrocatalytic properties in a wide range of temperatures and pressures were evaluated using perfect-mixing and flow reactors as well as low-temperature fuel cells. Liquid- and gas-phase hydrogenation of organic compounds with different types of unsaturated bonds and functional groups, Fisher-Tropsh synthesis, and anodic oxidation of hydrogen have been studied.

To follow the reactant concentrations on the active surface in situ, we measured the electrochemical potential of catalyst particles during hydrogenation. In the last few years, this method due to Prof. D. V. Sokolsky was successfully employed by authors (17) for the same purpose.

Raney Catalysts Genesis and Structure

The terms "Raney catalyst"and "Skeleton catalyst" are deeply embedded in the chemical literature. It is generally believed that their meaning is restricted only to sponge residues formed by dissolution of the non-noble component⁻ or components from intermetallides or solid solutions with catalytically active metals. However, in a more comprehensive sense, they can be reasonably expanded to a variety of metal blacks, formed by decomposition of oxides, metal complexes, and organometallic compounds (37). Porous body of such catalysts also forms by a removal of the closest environment atoms and ions (oxygen, chalcogens, halogens, organic radicals, etc.) from their chemical compounds with catalytically active metals with subsequent formation of metal-metal bonds, microcrystals, and eventually a new phase having a defective crystal lattice and a great length of grain boundaries. From this approach inevitably follows the conclusion that mechanisms of all these catalysts genesis as well as the ways of their optimization are bound to be in principle rather similar. This idea is open to discussion but I will try to validate it below by comparison of classical Raney catalysts with different types of noble metal blacks.

The range of potential binary alloys feasible for the Raney catalysts preparation has been pointed out in (37). It should be emphasized that a number of foretold metallic systems, for example gallium-based (36) were successfully used within a few years to prepare a series of new Raney catalysts.

The nature of leachable components, a fraction of which always occur in Raney catalysts, may has a profound impact on their properties. For example, Raney Pt from Pt_2Bi_3 has an extremely high isomerization ability in the course of 1-hexen hydrogenation (18, 37).

The scheme [see (37)], stated that leaching starts throughout a diffusive recrystallization mechanism and keeps pace with a consecutive aggregation of primary structural elements formed from the braking dendrite branches of parent alloys, without a doubt, takes into consideration only the most common features of this process. It has been found that etching of aluminium from Ni-Al alloys starts into intercrystalline gaps followed at the later stages by the dendrites desintegration and rearrangement to yield high-porous particles retaining the dimensions of the original phase grains (38, 39). Formation of a skeletal structure can be described by the mechanism with rather similar coefficients of surface and volume diffusion (40).

Intermetallides rich in nickel resist leaching because their anodic behavior in alkaline solutions is determined by passivated nickel constituent (21). Formation of a nickel phase goes through the stages of origination and subsequent decomposition of unstable hydrides NiH_x (12,27, 41). The bulk of a catalyst from $NiAl_3$ is comprised of the closely interwoven chains of nickel globules whereas Ni_2Al_3 transforms into structure with outer layers of sponge nickel followed in the depth by the layers of spatially conjugated lower nickel aluminides (27, 38). Results of other authors are closely related to ours (42-44). A variety of amorphous hydroxides comprised of nickel, aluminium, and alkaline metal ions are formed. The secondary processes that lead to catalysts stabilization involve partial oxidation of highly dispersed nickel, crystallization of amorphous hydroxides and formation of spinels and other multicomponent oxide phases. Raney copper and platinum form in a similar way (27, 41, 45).

Both the rate and the extent to which aluminides decomposition proceeds depend on their crystal lattice structure (2, 45). Phases with a loose crystalline lattice considerably accelerate corrosion of others. Oxides and hydroxides hinder recrystallization even under very mild conditions at ordinary temperatures (46). In service, a rise in the degree of oxide phases crystallinity renders Raney catalysts more stable to deactivation (12, 41, 47, 48). In Raney catalysts from higher intermetallides ($NiAl_3$, $PtAl_4$, etc.) cylindrical mesopores are dominant, whereas ones prepared from Ni_2Al_3 and similar phases less abundant in leachable components contain a high proportion of spherical pores with narrow inlets and micropores as well (22, 49). The bottle-like shape of mesopores is mainly due to the location of oxides and hydroxides in the above-mentioned ones. The degree of sponge metal catalysts recrystallization under operating conditions correlates with the quota of micropores less than $15 \cdot 10^{-10}$m in radius and it is important to keep in mind that predominantly mesoporous Raney catalysts are far more stable than microporous blacks precipitated from solutions (50).

Catalysts activity, selectivity, and stability depend on the state of adsorbed and absorbed hydrogen, that is why the latter has been evaluated by several different methods. Neutron scattering (51) and Mössbauer spectroscopy (52) data provide strong evidence for the absence of nickel hydrides in the interior of Raney nickel particles, whereas in the thin surface layers they were identified by electron microdiffraction (27, 41) and SIMS

(29). It is well-established now that the major portion of Raney nickel's hydrogen is adsorbed on the surface. From TPD spectra it follows that catalysts prepared by leaching NiAl$_3$ and other higher aluminides contain much more weakly-bounded hydrogen (25, 53). Its removal has no noticeable effect. Drastic changes occur only after desorption of strongly- bounded hydrogen (41).

In the case of Raney nickel, intragranular diffusional limitations during liquid-phase hydrogenation under atmospheric pressure can be ignored only when the size of catalyst particles is less than 10^{-6}m (37). But under real service conditions, when their dimensions are equal to several tens of microns and much more, the degree of pore surface utilization (the efficiency factor) is very small (53, 54). Its greatest magnitude is peculiar to the catalysts with wide cylindrical pores, whereas the least one to the ink-pot type that are not easily-accessible for reactants. A fall of hydrogen content on the surface into the deep regions of pores causes the symbate dependence between catalyst particle dimensions and their electrochemical potential shift during hydrogenation. The similar statement is true for selectivity of hydrogenation that provides a possibility of increasing some highly-reactive intermediates yield, using nothing but variation of pores shape and distribution as well as catalyst particle size. For example, during nitrobenzene hydrogenation over Raney nickel only by this expedient an unusually high phenylhydroxylamine yield was obtained.

The same conclusion has been drawn regarding Raney platinum from Pt-Al alloys of variable composition (37, 45). It was found its activity is determined mainly by structure rather than by the surface chemical composition.

Multicomponent Catalysts

As can be seen from Table, starting in the mid-1930s, all metals with exception of one lanthanide and radioactive actinides have been the subject of extensive research with the aim of finding highly efficient promoters for Raney nickel. The literature on this subject is quite voluminous. Many hundreds articles and patents on this matter have been published. It has been found that only a few of them, primarily Ti, Cr, and Mo, can do the job and offer the greatest promise from the industrial standpoint.

In the typically metallic Pt-based Raney catalysts segregation of components and their distribution between bulk and surface layers have a profound impact on operating characteristics. For example, the intensity of double bond migration during hydrogenation of α-olefins over catalysts from

PtAl$_3$- (Pd, Rh, Ru, or Ir) alloys goes down in a sequence Pd >> Rh > Ir > Ru (55), being in excellent agreement with experimental results for the same doping metals in the pure state (56) as well as with our own data for a variety of Pt-, Pd-, Rh-, Ir-, and Ru-blacks (57). The enrichment of Raney Pt-Rh catalysts surface layers with rhodium has been proved in (58).

Table
Raney nickel modifiers

Period, years	Investigated modifiers
1931-1940	Mg, Zn, Cu, Cr, Co
1941-1950	Mo, W, Fe
1951-1960	B, Sn, Pb, Ag, Ti, Zr, V, Ta, Mn
1961-1970	Ga, In, Tl, Sb, Bi, Ce, Nb, Re, Ru, Rh, Pd, Pt
1971-1980	Cd, La, Pr, Nd, Sm, Eu, Gd, Tb, Dy, Ho, Er, Tu, Yb, Lu, Ge, Os, Ir
1981-1990	Ca, Sr, Ba, Hf, U

But it cannot be too highly stressed that during leaching the majority of alloying metals undergo partial or even full oxidation (2, 59) and redistribution accompanied by the formation of new compounds. For example, a series of tri-, tetra-, penta-, and hexavalent molybdenum oxides was identified (24, 26, 28, 32, 33). Iron in such catalysts is present in the form of Fe$_3$O$_4$, γ-Fe$_2$O$_3$·H$_2$O and NiFe$_2$O$_4$ (24, 34), zirconium transforms into ZrO$_2$ and K$_2$Zr$_2$O$_5$ (60, 61), chromium into α- and β-Cr$_2$O$_3$·3H$_2$O (12, 31), manganese into MnO,Mn$_2$O$_3$, KMnO$_2$, and NiMn$_2$O$_4$ (60),etc. The same conclusions regarding Co-, Cu-, and Pt-Raney catalysts from ternary alloys were made in (41, 58, 62). An important point is that structure and permeability of the superficial oxide layers can vary over a broad range: from loose and " island-like " to tightly blocking the most part of active metal surface. The literature on this subject is quite extensive (1, 2, 15, 63, etc.).

It was shown (14) that the direction of migration and the location of components during and after leaching are governed by the tendency of establishing thermodynamic equilibrium. They can be predicted on the basis of phase diagrams of metallic, oxide, and hydroxide systems comprising all catalyst constituents. For example, the antibate distribution of aluminium and zirconium compounds on the catalyst surface is due to a lack of interaction in the Al_2O_3-ZrO_2 system. Therefore, multicomponent Raney catalysts are heterogeneous and involve a variety of zones differing from each other in chemical and phase composition.

If Raney catalysts are abundant in oxides sandwiched between metallic grains, they are not too different from those on supports, where active metallic phase is distributed on the highly-developed surface of oxide carriers.

With all these results in mind, we can turn to a classification of modifiers initially suggested in 1971 (37) and to a phenomenological model of multicomponent Raney catalysts proposed in (13, 26).

Preparation of chemically similar high-porous hybrid metal-oxide blacks is rendered possible by the reduction of ions from solutions, decomposition of melts, etc. (41, 57, 64). From the above-stated reasons, it can be assumed that the most part of Raney catalysts and doped metal blacks is not fundamentally different from supported catalysts. All three types fall into the class of metall-oxide systems peculiar features of which are intensive spillover as well as SMSI phenomena. These oxides protect metallic constituents against recrystallization.

Formation of binary and multicomponent oxides and hydroxides into Raney nickel body and on the surface leads to increase in the bond energy of adsorbed hydrogen associated with changes in their activity and selectivity. As a rule, dispersing of Raney catalyst particles as a result of modification strongly
diminishes the role of intradiffusion hindrance (37). Therefore, great care must be exercised regarding a large body of publications and patents that carry information on an abrupt jump of powdery Raney catalysts activity under the influence of alloying modifires because it may originates not only from the well-known phenomenon of truly chemical promotion, but more often, as it has been shown in (37), from the simple dispersing of catalyst particles into finer ones.

By contrast to activity, modifires due to a change of the Raney catalysts adsorption properties beyond any doubt can favor one reaction pathway over another and thus have a beneficial or negative affect on their selectivity (1). For example, the presence of oxides and strongly-bounded hydrogen increases the selectivity of doped Raney nickel catalysts in hydrogenation of dienes and hinders double bonds migration along the carbon chain (37, 65). A good case in point is also the hydrogenation of substituted anhraquinones for H_2O_2 manufacturing because   reduction of their side aromatic rings inhibits   the oxidation stage of the process. It was shown that for selective reduction of quinone groups, catalysts containing only strongly-bounded hydrogen are required (10, 11, 66). Raney Ni-Ti catalysts have higher selectivity that undoped samples towards lower olefins production in the Fisher-Tropsh synthesis (67).

Developed in Kazakhstan multicomponent Raney catalysts have passed successful pilot and the industrial scale tests at quite a number of chemical, petrochemical, and pharmaceutical plants in the processes of hydrogenation and hydrogenolysis of fats, sugars, furfurol, nitrocompounds, removal of impurities from monomers, etc. More than a hundred of former USSR Author's Certificates and foreign patents, for example (68), have been obtained.

As already noted, the development of multicomponent Raney catalysts as well as investigation of their structure and physicochemical characteristics have been the subject of so many publications. To quote even the smallest fraction of them in a short article is extremely difficult if not absolutely impossible. However, we will cite a few recent papers supporting some of our long-obtained results, for instance (17). These authors found that doping Raney nickel by metals oxidizing during leaching tends to decrease the electrochemical potential shift in the course of liquid-phase hydrogenation. The explanation offered by these researches stems from the fact that metal oxides can be viewed as Lewis acids capable to adsorb the electron-rich functional groups (carbonyl, nitrile, etc.) of unsaturated compounds and thereby freeing metallic surface for the hydrogen chemisorption and activation. This interpretation being basically correct tends to overlook the important fact that all of such modifires increase the strongly-bounded hydrogen content, which in a lesser extent can be removed from the surface by molecules of unsaturated compounds. In addition, it is well to bear in mind that dispersing of porous Raney catalysts under the action of oxidizable alloying additives reduces the degree of intradiffusion hindrance and as a consequence increases hydrogen concentration in the interior of catalyst particles. Both of these factors may be attributable to the noted reduction of potential shift.

Nonpyrophoric Raney Catalysts

Many advantages of Raney nickel catalysts are to a great extent limited by their pyrophority, especially at the stages of discharging from reactors and separation from reaction products. To remove these limitations, a new technology has been developed. It is based on the thermal treatment of fresh catalysts with subsequent passivation by air (15, 25, 69-72). These catalysts are intended for hydrogenation and hydrogenolysis of various organic compounds as well as for electrochemical generators instead of conventional fire-hazardous Raney nickel.

The thermal treatment results in sintering of Raney catalysts which involves decreasing their specific surface and number of lattice defects. Varying operating conditions and gas medium composition, it is possible to increase the efficiency factor by producing mainly mesoporous structure with a cylindrical shape of pores. Mild oxidation leads to formation of surface layers consisting of non-stoichiometric oxygen-containing phases with low reduction energy barrier due to which such catalysts can be easily reactivated by hydrogen under 60-80°C in the liquid- or 150-200°C in the gas phases respectively, achieving the activity of a freshly- prepared Raney nickel.

The obtained results have been interpreted in terms of a phenomenological model describing transformation of these catalysts at the stages of thermal treatment, passivation, and reduction (15, 72). It was found that coarse particles of Raney nickel after reactivation functioning as a support for a new structural component that can be called " surface nickel black ". The amount of this active form is proportional to the content of non-stoichiometric oxides on the surface of passivated catalyst.

All the stages of preparation proceed in the same reactor of special design and are totally automated under computer control. The worked out process makes it possible to minimize the amount of effluent and exhaust gases. These catalysts can be easily transported. Their stability for storage in the air without a significant change in activity is not less than three months.

The elaborated catalysts have been successfully tested on the pilot and industrial scales in the processes of the folic acid semiproducts [2,4-diamino-5-isonitroso-6- hydroxypyrimidine and (p-nitrobenzoyl)-L (+)-glutamic acid (73)]as well as 1-phenyl-3-methyl-pyrazoline-5 hydrogenation, dehydrogenation of 16β-bromo-17α-oxyprogosteron, reducing amination in the N-methylglucamine manufacturing and in a diversity of other reactions. In all cases the yields of desired reaction products were comparable with those obtained through the use of a freshly prepared Raney nickel. After separation from reaction products, the nonpyrophoric catalyst, unlike the commercial one, does not burn. High selectivity of nonpyrophoric Raney nickel in hydrogenation of compounds with double and triple carbon-carbon bonds with preservation of a benzene ring can be provided by modifying (15). Successful tests of these catalysts in H_2-O_2 low-temperature fuel cells have been also carried out.

Mechanochemical Methods of Raney Nickel
Catalysts Preparation and Regeneration

Initial Ni-Al and Ag-Al alloys were prepared in an energy-saturated atrittor (74) and in a cooled planetarium-type ball mill (75-77). In the first case, the synthesis is incepted by a latent period during which melted aluminium occurs as the result of a local heating followed by wetting of nickel particles and initiation of a rapid self-propagation high-temperature reaction. In the second case, the mechanism of alloy formation is close to a diffusion type. Phase composition of mechanochemical alloys is far from an equilibrium one. Both devices allow for the production of alloy powders within one stage.

The unique feature of these alloys is liability to leaching over a much wider range of nickel concentrations caused by their significantly greater lattice distortion and nonequilibrium structure. By this expedient it became possible to prepare Raney nickel and silver from equiatomic and even more aluminium-deficient alloys (NiAl, Ag_2Al). This opens up new vistas for reducing losses of expendable alloy components, especially aluminium.

Structurally, Raney nickel from mechanochemical alloys can be viewed either as grains of an unleached NiAl covered by a highly-dispersed FCC- Ni (76, 77) or as a metastable BCC-Ni (78). The issue remains open to discussion. It should be only emphasized that formation of the unusual intermediate phases during leaching of quenched Ni_2Al_3 has been long-observed in (43). Such catalysts are more active that conventional ones.

In addition to a number of advantages over the traditional method, mechanochemical alloying can provide the basis for a new technology of Raney nickel regeneration. Mechanochemical alloying of spent and deactivated Raney nickel with aluminium powder yields catalysts that are superior in activity to commercial ones.

Fixed-bed and Superficial Raney Catalysts

By applying the Raney technique, it is possible to develop outer surface layers of lumpy alloys, remaining their cores unleached, and prepare on this basis a variety of stationary catalysts for flow-type reactors. Thin layers of precursor alloys on any irregularly shaped smooth metal surfaces can be formed by thermal diffusion, ionic implantation, and other methods of solid state chemistry and physics (79, 80). This method appears to be particularly promising for membrane catalysis, where overall reaction rate is often limited by a small gas- or liquid-metal interface.

Raney Catalysts on Supports

A high proportion of active metals in traditional Raney catalysts is localized in their bulk and virtually cannot be involved in catalysis. As a result, expensive metals, especially belonging to the platinum group, do not play any serious role in the process. This generates a need for the development of a new type of Raney catalysts that combines their unique advantages with those for the supported ones. Such catalysts can be prepared via the impregnation of supports by salts of noble and non-noble (Sn, Pb, Cu, Zn, etc.) metals with their subsequent reduction and treatment with acidic or alkaline solutions (37). It has been proved that reduction of the above-mentioned mixtures gives rise to intermetallides or solid solutions functioning as Raney catalyst precursors. This method can be particularly important for the preparation of oil refining and petrochemical catalysts.

The formation mechanism and physicochemical characteristics of Pt /Al_2O_3 Raney catalysts prepared by leaching Pt-Cu/Al_2O_3 precursors were thoroughly studied in (20,81-84). It involves formation of Cu[Pt(H_2O)$_x$Cl$_{6-x}$] complex, the reduction of which gives Pt-Cu solid solution. After the leaching, residual copper ions bound with support act as centers retarding migration and aggregation of small platinum clusters. These catalysts due to their structure are several times more active than conventional ones that can be explained by a sharp rise in platinum dispersity.

Conclusion

A wide range of problems significant for the Raney catalysts theory and technology received the bulk of attention in this paper. It must be underscored that that Raney's invention, made 70 years ago, has provided a sound basis for the development of a new family of effective catalysts for a variety of processes- from almost entirely metallic platinum- and silver-based electrocatalysts for fuel cells to closely approximated typical supported catalysts for fine organic syntheses and petrochemistry, key features of which are SMSI and spillover phenomena.

A comprehensive study of Raney catalysts structure and properties using a complex of integral and local physical methods points to their heterogeneity. They are formed by aggregation of noble component crystallites via leaching of parent alloys, while fully or partially oxidized less-noble elements migrate towards alloy-alkali interface transforming initially metallic precursors into metal-oxide compositions. Such mixtures involve particles and epitaxial layers of zerovalent and ionic forms of catalytically active, leachable, and alloying metals entering into a great number of solid-state reactions with final distribution determined by the tendency of establishing thermodynamic equilibrium. Oxides and hydroxides, acting as a structure forming carriers as well as a support for metal crystallites, not only stabilize Raney catalysts, but also form their active surface providing spillover and altering energetic characteristics of adsorbed molecules.

Extensive investigation of catalysts from monophases constituting multicomponent parent alloys was used as a starting point for classification of modifiers and the interpretation of obtained results in terms of phenomenological model describing their main features.

Several innovative modifications of Raney catalysts have been developed. Among them are Raney nickel and multicomponent nickel-containing catalysts, non-inflammable in air after drying, Raney catalysts on supports as well as mechanochemical technology of their preparation. It should be emphasized that our nonpyrophoric catalysts can be stored for a long time and then easily reactivated under mild conditions. The mechanochemical technology can be utilized for regeneration of the spent Raney catalysts deactivated in a variety of industrial processes. Commercialization of these non-standard catalysts and processes could add a new chapter to the Raney catalysts epopee.

Acknowledgments

I greatfully and cordially recognize the enthusiasm and efforts of many Ph.D. students, postdoctoral fellows, and professional researches who collaborated with the author in these long-standing studies. Their names can be seen from the list of references.

References

1.   E. I. Gildebrand and A. B. Fasman, *Raney Catalysts In Organic Chemistry*, Nauka, Alma- Ata, 1982 (in Russian).

2.   A. B. Fasman and D. V. Sokolsky, *Structure and Physicochemical Properties of Raney Catalysts*, Nauka, Alma-Ata, 1968 (in Russian).

3.   E. Cohen and T. Strengers, *Z. phys. Chem.*, **61**, 698 (1908).

4.   R. B. Seymour, *Chem. Eng. News*, **25**, 2628 (1947).

5.   J. Tröger and G. Vollheim, *Chem. Ztg.*, **99**, 446 (1975).

6.   D. S. Tarbell and A. T. Tarbell, *J. Chem. Education*, **54**, 26 (1977).

7.   R. B. Seymour and S. R. Montgomery, in *Heterogeneous Catalysis* (*ACS Symp. Series*, **222**), 491 (1983).

8. B. M. Bogoslovskii and Z. S. Kazakova, *Raney Catalysts, Their Properties and Utilization in Organic Chemistry*, Goskhimizdat, Moscow, 1957 (in Russian).

9. T. Kubomatsu and S. Komatsu, *Raney Catalysts*, Koritsu, Osaka,1971.

10. A.B. Fasman, S. D. Mikhailenko, N. A. Maksimova, Zh. A. Ikhsanov, V. Ya. Kitaigorodskaya and L.V. Pavlyukevich, *Applied Catal.*, **6**, 1 (1983).

11. S. D. Mikhailenko, A. B. Fasman, N. A. Maksimova, E. V. Leongard, E.S. Shpiro and G.V.Antoshin, *Applied Catal.*, **12**, 141 (1984).

12. A. B. Fasman, E. A. Vishnevetskii, N. A. Maksimova, V. N. Ermolaev, and A. I. Savelov, in *Catalytic Reactions in Liquid Phase*, Alma-Ata, Nauka, 192 (1985).

13. E. A. Vishnevetskii, S. D. Mikhailenko, N. A. Maksimova and A.B. Fasman, *React. Kinet. Catal. Lett.*, **31**, 445 (1986).

14. N. A. Maksimova, E. A. Vishnevetskii, V. Sh. Ivanov and A. B. Fasman, *Applied. Catal.*, **35**, 59 (1987).

15. A. B. Fasman, T. A. Khodareva, S. D. Mikhailenko, E. V. Leongard and A. I. Lyashenko, in *Razvitie Rabot v Oblasti Kataliza v Kazakhstane*, Alma-Ata, Nauka, Part 1, 149 (1990).

16. S. Hamar-Thibault, J. Masson, P. Fouilloux and J. Court, *Applied Catal.* (A), **99**, 131 (1993).

17. J. Pardillos-Guindet, S. Metais, S. Vidal, J. Court and P. Fouilloux, *Applied Catal.* (A), **132**, 61 (1995).

18. A. B. Fasman, R. Kh. Ibrasheva, T. A. Sergeeva, M. M. Kalina, V. P. Polyakova, L. I. Voronova, V. B. Chernogorenko, S. V. Muchnik, K. A. Lynchak, M. N. Abdusalyamova and N. A. Verbitskaya, *Kinet. Katal.* , **21**, 287 (1980).

19. A. B. Fasman and T. V. Kuzora, *Kinet. Katal.*, **23**, 144 (1982).

20. G. M. Khutoretskaya, T. V. Kuzora and A. B. Fasman, *Kinet. Katal.*, **28**, 467 (1987).

21. S. M. Aleikina, I. K Marshakov, A. B. Fasman and I. V. Vavresyuk, *Elektrokhimia*, **6**, 1648 (1970).

22. A. B. Fasman, V. F. Timofeeva, V. N. Rechkin, Yu. F. Klyuchnikov and I. A. Sapukov, *Kinet. Katal.*, **13**, 1513 (1972).

23. A. B. Fasman, B. K. Almashev and V. N. Rechkin, *Zh. Prikl. Khim.*, **46**, 282 (1973).

24. G. Sh. Talipov, T. N. Nalibaev, A. B. Fasman and A. S. Sultanov, *Kinet. Katal.*, **15**, 744 (1974).

25. A. B. Fasman, V. A. Zavorin and G. A. Pushkareva, *Kinet. Katal.*, **15**, 994 (1974).
26. E.A. Vishnevetskii and A.B. Fasman, *Zh. Fiz. Khim.*, **55**, 2084 (1981).
27. V.N. Ermolaev, A.B. Fasman, S.A. Semiletov and V.I. Khitrova, *Izv. Akad. Nauk SSSR, Ser. Fiz.*, **47**, 1218 (1983); *Kristallografiya*, **28**, 611 (1983)
28. G.N. Baeva, A. B. Fasman, E. S. Shpiro, G. V. Antoshin, L. V. Pavlyukevich, Zh. A. Ikhsanov and Kh. M. Minachev, *Kinet. Katal.*, **26**, 1220 (1985)
29. E. S. Shpiro, G. N. Baeva, A. B. Fasman and Kh. M. Minachev, *React. Kinet. Catal. Lett.*, **32**, 499 (1986).
30. V. N. Ermolaev, G. A. Pushkareva and A. B. Fasman, *Kinet. Katal.*, **29**, 431 (1988).
31. E. A. Moroz, S. D. Mikhailenko, A. K. Dzhunusov and A. B. Fasman, *Rect. Kinet. Catal. Lett.*, **43**, 63 (1991).
32. A. B. Fasman, G. A. Pushkareva, B. K. Almashev, V. N. Rechkin, Yu. F. Klyuchnikov and I. A. Sapukov, *Kinet. Katal.*, **12**, 1271 (1971).
33. A. B. Fasman, G. E. Bedelbaev, N. A. Maksimova, V. N. Ermolaev and A. Sh. Kuanyshev, *Kinet. Katal.*, **29**, 437 (1988).
34. A. B. Fasman, G. E. Bedelbaev, G. K. Alekseeva, V. N. Ermolaev and A. Sh. Kuanyshev, *React. Kinet. Catal. Lett.*, **39**, 81 (1989).
35. N. A. Maksimova, G. A. Kil'dibekova, G. E. Bedelbaev, A. Sh. Kuanyshev, E. I. Gladyshevskii, N. B. Manyakov and A. B. Fasman, *Kinet. Katal.*, **29**, 662 (1988).
36. N. A. Maksimova, G. L. Padyukova, A. Sh. Kuanyshev, V. M. Zolotarev, S. P. Yatsenko and A. B. Fasman, *Kinet. Katal.*, **28**, 954 (1987).
37. A. B. Fasman, in *Trudy Inst. Org. Katal. Electrokhim. Akad. Nauk Kaz. SSR*, **1**, 24 (1971); *Mater. IV Vsesoyz. Konfer. Katal. Reakts. Zhidk. Faze*, Alma- Ata, Part 1, 15 (1974); *III Soviet- Japanese Seminar on Catalysis*, Alma- Ata, Preprint No. 30 (1975);*Nanesennye Metall. Katalizatory Prevrascheniya Uglevodorodov*, Novosibirsk, Nauka, 71 (1978); *Catalytic Reactions in Liquid Phase*, Alma- Ata, Nauka, 260 (1980).
38. A. A. Presnyakov, K. T. Chernousova, T. Kabiev, A. B. Fasman and T. T. Bocharova, *Zh. Prikl. Khim.*, **40**, 958 (1967).
39. A. B. Fasman, S. D. Mikhailenko and E. A. Vishnevetskii, *Zh. Prikl. Khim.*, **56**, 1269 (1983).

40.   S. D. Mikhailenko, G. N. Baeva, B. F. Petrov and A. B. Fasman,
       Poroshkov. Metallurg., No.7, 8 (1986);   *Reactiv. Solids*, **2**, 373
       (1987).
41.   A. B. Fasman, V. N. Ermolaev and G. L. Padyukova, in *Razvitie
       Rabot   v Oblasti Kataliza v Kazakhstane*, Alma- Ata, Nauka, Part 1,
       (1990).
42.   F. Delannay, *Reactiv. Solids*, **2**, 235 (1986).
43.   J. Gros, S. Hamar- Thibault and J. C. Joud, *Surf. Interface Anal.*, **11**,
       611 (1988).
44.   S. Hamar- Thibault, J. Thibault and J. C. Joud, *Z. Metallkunde*, **83**,
       258 (1992).
45.   G. L. Padyukova, V. G. Shalyukhin, E. N. Belozerova and A. B.
       Fasman, *Zh. Fiz. Khim.*, **61**, 1832 (1987).
46.   T. V. Kuzora and A. B. Fasman, *Zh. Prikl. Khim.*, **57**, 522 (1984).
47.   A. I. Savelov and A. B. Fasman, *Zh. Prikl. Khim.*, **55**,76 (1982); *Zh.
       Fiz. Khim.*, **56**, 1837, 2459 (1982), **59**, 1027 (1985).
48.   A. I. Savelov, A. B. Fasman, A. I. Lyashenko and O. I. Yuskevich, *Zh.
       Fiz. Khim.*, **63**, 1918 (1989).
49.   A. B. Fasman, D. V. Sokolsky, V. F. Timofeeva. D. K. Bazhakov and
       G. A. Pushkareva, *Proceed. Internat. Symp. RILEM / IUPAC, Final
       Report*, Part 1, Academia, Prague, B85 (1974).
50.   G.G. Kutyukov, E.V.Kolodii and A.B. Fasman, *Elektrokhimia*, **13**,
       400 (1977).
51.   S. Sh. Shilstein, E. A. Vishnevetskii, V. A. Somenkov and A. B.
       Fasman, *Izv Akad. Nauk SSSR, Neorg. Mater.*, **16**, 2144 (1980).
52.   A. I. Klystov and A. B. Fasman, *Zh. Fiz. Khim.*, **53**, 2459 (1984).
53.   A. B. Fasman, V. M. Safronov, G. L. Padyukova, G. A. Pushkareva,
       V. N. Ermolaev, M. M. Kalina and B. Ya. Pel'menshtein, *Kinet.
       Katal.*, **24**, 695 (1983).
54.   V. M. Safronov, V. I. Vorob'eva and A. B. Fasman, *Zh. Prikl. Khim.*,
       57, 1335 (1984).
55.   A. B. Fasman, D. K. Bazhakov and A. D. Dembitskii, *React. Kinet.
       Catal. Lett.*, **1**,389 (1974).
56.   P. N. Rylander, *Catalytic Hydrogenation over Platinum Metals*, Acad.
       Press, New York, London, 1967.
57.   V. Ya. Kitaigorodskaya, A. B. Fasman and A. P. Gorokhov, *Zh.
       Organ. Khim.*, **14**, 1165 (1978).
58.   V. G. Shalyukhin, G. L. Padyukova, A. Sh. Kuanyshev and A. B.
       Fasman, *Elektrokhimia*, **24**,541 (1988).

59. S.M. Aleikina, I. K. Marshakov, A.B. Fasman and I.V.Vavresyuk, *Zaschita Metallov*, **6**, 691 (1970).

60. A.B. Fasman, K.A. Nurumbetov, B.K. Almashev and T.N.Nalibaev, *React. Kinet. Catal. Lett.*, **2**, 89 (1975).

61. V. G. Shalyukhin, E. V. Leongard, G. L. Padyukova, A. B. Fasman and M. V. Ulitin, *Zh. Fiz. Khim.*, **64**, 3003 (1990).

62. A. B. Fasman, V. A. Lifanova, V. N. Ermolaev and M. M. Kalina, *Zh. Prikl. Khim.*, **57**, 2049 (1984).

63. S. Hamar- Thibault, J. Gros, J. Masson, J. P. Damon and J. M. Bonnier, *Stud. Surf. Sci. Catal.* (*Preparation of Catalysts V*), **63**, 435 (1991).

64. G. G. Kutyukov, A. B. Fasman, E. V. Kolodii and V.S. Tyrtyshnaya, in *Splavy Blagorod. Metallov*, Nauka, Moscow, 233 (1977).

65. A. B. Fasman, G. A. Pushkareva and V. F. Timofeeva, *Zh. Fiz. Khim.*, **52**, 358 (1978).

66. Zh. A.Ikhsanov, N. A. Maksimova, V. Ya. Kitaigorodskaya, A. B. Fasman and L. V. Pavlyukevich, *Zh. Organ. Khim.*, **17**,1533 (1981).

67. S. D. Mikhailenko, G. N. Baeva, E. S. Shpiro and A. B. Fasman, *Kinet. Katal.*, **32**, 928 (1991).

68. A. B. Fasman, D. V. Sokolsky, G. I. Rutman, J. I. Michurov, V. A. Zavorin, Z. S. Shalimova, V. F. Timofeeva, T. G. Dautov, D. K. Bazhakov and J. M. Sivakov, *US Patent* 4,107,090 (1978).

69. A. B. Fasman, E. V. Leongard, E. A. Vishnevetskii, A. I. Lyashenko and S. D. Mikhailenko, *Zh. Fiz. Khim.*, **57**, 1401 (1983).

70. A. B. Fasman, E. V. Leongard, A. I. Lyashenko and S. D. Mikhailenko, *Zh. Prikl. Khim.*, **59**, 2280 (1986).

71. A. B. Fasman, S. D. Mikhailenko, E. V. Leongard, T. A. Khodareva and A. I. Lyashenko, *React. Kinet. Catal. Lett.*, **38**, 181 (1989).

72. S. D. Mikhailenko, T. A. Khodareva, E. V. Leongardt, A. I. Lyashenko and A. B. Fasman, *J. Catal.*, **141**, 688 (1993).

73. A. I. Lyashenko, O. I. Yuskevich, T. A. Khodareva, E. V. Leongard, A. B. Fasman, B. D. Statkevich, G. E. Chernysh and M. E. Butovskii, *Khim.- Farm. Zh.*, **21**, 279 (1987).

74. S. D. Mikhailenko, B. F. Petrov, O. T. Kalinina and A. B. Fasman, *Poroskov. Metallurg.*, No.10, 44 (1989).

75. E. Ivanov, T. Grigorieva, G. Golubkova, V. Boldyrev, A. B. Fasman, S. D. Mikhailenko and O. T. Kalinina, *Mater. Lett.*, 7,51, 55 (1988).

76. A. B. Fasman, S. D. Mikhailenko, O. T. Kalinina, E. Yu. Ivanov and G. V. Golubkova, *Stud. Surf. Sci. Catal.* (*Preparation of Catalysts V*), **63**, 591 (1991).

77. S. D. Mikhailenko, O. T. Kalinina, A. K. Dzhunusov, A. B. Fasman, Ye. Yu. Ivanov and G. V. Golubkova, *Sibirsk. Khim. Zh.*, No.5, 93 (1991).

78. E. Ivanov, K. Suzuki, K. Sumiyama, S. E. Makhlouf and H. Yamauchi, *Solid State Ionics*, **60**, 229 (1993).

79. A. P. Gorokhov, A. B. Fasman, V. V. Pudikov, S. I. Vladimirtsev and M. M. Kalina, *USSR Inventor's Certificate* 601,042 (1977).

80. A. B. Fasman, P. I. Zabotin, N. A. Maksimova, S. V. Druz, A. M. Zagor'ev and S. D. Mikhailenko, *Zh. Prikl. Khim.*, **56**, 2569 (1983).

81. G. M. Khutoretskaya, V. N. Ermolaev and A. B. Fasman, *Kinet. Katal.*, **29**, 180 (1988).

82. G. M. Khutoretskaya, V. F. Vozdvizhenskii, A. A. Il'ina, A. Sh. Kuanyshev and A. B. Fasman, *Kinet. Katal.*, **29**, 185 (1988).

83. G. M. Khutoretskaya, V. N. Vorob'ev and A. B. Fasman, *Kinet. Katal.*, **29**, 668 (1988).

84. G. M. Khutoretskaya, A. B. Fasman and T. B. Molodozhenyuk, *Neftekhimia*, **29**, 319 (1989).

Diastereoselective Hydrogenation of Cyclic β-Ketoesters over Modified Ru/Zeolite Catalysts

S. Coman, C. Bendic, M. Hillebrand, E. Angelescu, V. I. Pârvulescu
University of Bucharest, Faculty of Chemistry, B-dul Republicii 13, Bucharest, Romania
A. Petride
Institute of Organic Chemistry, Splaiul Independentei 208, Bucharest, Romania
M. Banciu
Polytechnic University of Bucharest, Department of Organic Chemistry, Str. Polizu 1, Bucharest, Romania

Abstract

Diastereoselective hydrogenation of a cyclic-β-ketoester over Ru supported catalysts was investigated. The catalysts were prepared by deposition of Ru on different molecular sieves (L, ZSM-5 or APO-54), Al_2O_3 and carbon. These catalysts were characterized by elemental analysis, adsorption of N_2 at 77 K, H_2 chemisorption, XRD, TPR and XPS. Cinchonidine (CD) and L(+)-tartaric acid (TA) were used as chiral modifiers. Pivalic acid (PA) was added as chiral modifier of TA in the reaction mixture. The nature of the interactions in the substrate-modifier complex and substrate-catalysts complex was discussed on the basis of the reaction rate compared with results obtained by using molecular modeling of the possible forms of these complex. The results showed a very complex activity-selectivity-catalyst relationship.

1. Introduction

The use of catalysis in fine chemistry synthesis is one of the topics that has attracted a large interest over the time (1). More and more complicated organic reactions have been investigated in the presence of the catalysts. Two are the main reasons for this extraordinary interest: the ability to increase the reaction rates and the selectivities of these processes.

The chiral reactions are among these. Up to now, most of these reactions have been carried out under homogeneous conditions (2), but recent studies have shown that in the presence of heterogeneous catalysts these systems could also be very effective (3, 4, 5, 6).

Because of its practical applications, the enantioselective hydrogenation of β-ketoesters and α-ketoesters has received a special interest. In this respect, two catalytic systems have mainly been investigated: (a) nickel catalysts, usually as Raney Ni, modified by α-hydroxy or α-amino acids (3, 4), and (b) platinum catalysts modified with the cinchona alkaloid derivatives (5, 6). Only a few studies on other platinic metals such as Pd, Ru and Rh, have been reported. The first results have indicated that these catalysts are non-selective and only slightly active. Contrary to these reports, our studies on the hydrogenation of prostaglandin intermediates on Ru supported on different molecular sieves have shown that these catalysts are active and selective, yielding prostaglandin intermediates with natural-like configuration (7).

To date, however, the information on using molecular sieves as supports in such reactions are scarce and, to the best of our knowledge, the diastereoselective hydrogenation of cyclic-ketoesters under heterogeneous catalysis has not yet been investigated.

Several objectives have guided the present study. The first one has been to investigate the diastereoselective hydrogenation of β-ketoesters. In this respect, the hydrogenation of 7-carboxymethyl-5H-6,7-dihydro-dibenzo [a,c] cyclopheptan-6-one (noted as substrate **1**) has been used as a model. Another objective has been to determine whether the hydrogenation of this substrate is sensitive to the catalyst features and to the nature of the modifiers. To answer these questions,the ruthenium metal supported on different molecular sieves has been used as catalyst. For comparison, ruthenium supported on large porous alumina and carbon has also been investigated. As chiral modifiers, L(+)-tartaric acid (L(+)-TA) and (-)-cinchonidine (CD) have been used.

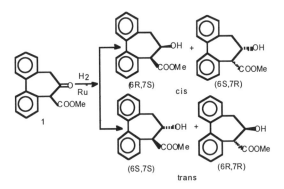

Scheme 1. Reaction scheme for diastereoselective hydrogenation of the substrate **1**

2. Experimental

2. 1. Materials and synthesis of the substrate: The synthesis of the ketoester **1**, (7-carboxymethyl -5H-6,7-dihydro-dibenzo [a, c] cyclopheptan-6-one), was performed following the procedure described previously in Ref. 8, starting with diphenyl-o,o,-diacetic acid. The ketoester **1**, m. p. 103-105 °C, gave IR- and H-NMR spectral data identical to that previously reported (8).

2. 2. Catalysts preparation and characterization: Ru supported catalysts were obtained by deposition of Ru on molecular sieves (L, APO-54, ZSM-5) in the potassium form from a solution containing 0.4 M $RuCl_3$ $3H_2O$ (Fluka) following a procedure reported elsewhere (9). For comparison, Ru/Al_2O_3 (Engelhard 4872) and Ru/carbon (Engelhard 4857) were also used. All catalysts were reduced in hydrogen (30 ml/min) at 500°C for 6 hours. The heating rate was of 1°C/min.

The samples were characterized by elemental analysis, adsorption of N_2 at 77 K, XPS, and temperature programmed reduction (TPR). Elemental analysis of Ru, Si, Al and P was performed by atomic emission spectroscopy with inductively coupled plasma atomization (ICP-AES). The adsorption and desorption curves of N_2 at 77 K were obtained with a Micromeritics ASAP 2010 apparatus after degassing the samples at 200°C under vacuum. XPS spectra were recorded using a SSI X probe FISONS spectrometer (SSX-100 /206) with monochromatic AlKα radiation. The energy scale was calibrated by taking the Au $4f_{7/2}$ binding energy at 93.98 eV. The binding energy (B. E.) scale was referred to Si $_{2p}$ =103 eV for K-L and Na-ZSM-5 and to Al $_{2p}$ =73.7 eV for Al_2O_3 and APO-54 -supported samples,respectively. Due to overlapping of the Ru_{3d} and C_{1s} lines, the Ru_{3p} was used.

2. 3. Modification process: The chiral modifiers were added to the investigated catalysts following two procedures: i) cinchonidine (CD) was introduced in the reaction mixture in an amount corresponding to a molecular ratio alkaloid-to-substrate of 1.0 and ii) the catalysts were introduced in an autoclave containing 200 ml solution of L(+)-tartaric acid (0.07 g, pH=3.2). The suspension was vigorously stirred for 1h at 100°C. After decanting the solution, the catalysts were successively washed with water, methanol and tetrahydrofuran (THF).

2. 4. Hydrogenation and analytical method: The hydrogenation of the substrate **1** was carried out in a stainless steel stirred autoclave under 10 atm. and at a temperature range of 60-80°C. Standard experiments used 10 mg substrate dissolved in 10 ml ethanol or THF (in the case of the modified catalyst with L(+)-TA). In order to improve the selectivity, some experiments using L(+)-TA modified catalysts were carried out in the presence of pivalic acid (PA) as co-modifier. The molar ratio L(+)-TA-to-PA was in the range of 0.13-0.23.

The conversions and the selectivities were determined from GC-MS data. The reaction products were analyzed using a Carlo-Erba HRGC 5300 equipped with two capillary columns, one with permethyl-β-cyclo-dextrin and another with dimethyl-penthyl-β-cyclo-dextrin. The diastereoisomeric excess (d.e.) of the *cis*-(6S,7R) and (6R,7S) products were calculated as: d.e. % = |[6S, 7R] - [6R,7S] | / | [6S,7R] +[6R,7S] | x 100 or: d.e. % = | [6R,7R

] - [6S,7S] | / | [6R,7R] + [6S,7S] |x 100 for the *trans*-(6R,7R) and (6S,7S) products.

In the experiments where the pivalic acid was added to the reaction mixture, the reaction product was neutralized with K_2CO_3, extracted in 100 ml anhydrous ethyl ether and dried on Na_2SO_4.

2. 5. Molecular Modeling: The semiempirical AM1-method (MOPAC-version 7) and the molecular mechanics method, MMX force field (MM2 molecular force field including π-electron calculations) (PCWIN-program) was used. It was found that there are no esential differences in the results obtained by the two techniques. Therefore, all the calculations concerning the interactions with the modifier were performed by molecular mechanics technique.

3. Results

3.1 Characterization of catalysts: The chemical composition and the textural properties of the investigated catalysts are presented in Table 1. Comparative Langmuir and t-plot surface areas indicate that even for high loadings of supported metal, as those used in the present study, the active surface accesible to large molecules as substrate or cinchonidine is still high.

Table1. Catalysts used in diastereoselective hydrogenation of substrate **1**

Cat.	Support	Ru loading wt %	Support surface area (m^2 / g)	Catalysts Langmuir surface area (m^2 / g)	Catalysts t-plot surface area (m^2 / g)	Micropore volume (cm^3 / g)
Ru-1	L	12.02	220	95	39	0.016
Ru-2	L	5.84	220	104	37	0.015
Ru-3	ZSM-5	3.11	430	396	271	0.109
Ru-4	APO-54	3.4	80	17	5	0.002
Ru-5	Al_2O_3	5.0	-	220	0	
Ru-6	C	5.0	-	380	149	0.083

The XPS data regarding the above catalysts are given in Table 2. In order to give a measure of Ru distribution inside the catalysts, the XPS atomic ratio of Ru to another dominant element was determined. In the case of Ru-1, Ru-2, and Ru-3 catalysts, Si was chosen as reference element because it exhibits a very well positioned band associated with the Si_{2p} state. One can see that the binding energies corresponding to the Ru_{3p} state vary within a rather large scale. Both the reduced Ru(0) and the oxidized Ru(IV) coexist, irrespective of the support nature (10). These data show that in the case of Ru-1, Ru-2, and Ru-3 the population of the oxidized species is increased comparative to the Ru-4, Ru-5 and Ru-6 catalysts. On Ru-APO-54, the Ru-to-reference element ratio from XPS data is higher than that determined from ICP-AES measurements. This indicates that the majority of the ruthenium is present on the outer surface of the support.

TPR data gave supplementary evidence in this sense. The reducibility of the metal on different supports can be characterized by the degree of the reduction. The data presented in Table 3 are in perfect agreement with the XPS measurements. The hydrogen chemisoption data are also given in the same Table 3. Columns 6 and 7 show the diameter of the particles (nm) calculated from hydrogen chemisorption (column 6) and from XRD data considering Ru (101) line (column7).

Table 2. XPS results

Catalyst	Ru_{3p} (eV)	O_{1s} (eV)	Ru/ reference element (XPS)	Ru/ reference element (chemical composition)
Ru-1	462.5	532.2	0.23 (Ru/ Si)	0.13 (Ru/ Si)
Ru-2	462.3	532.2	0.14 (Ru/ Si)	0.06 (Ru/ Si)
Ru-3	462.6	532.2	0.16 (Ru/ Si)	0.02 (Ru/ Si)
Ru-4	461.3	532.3	0.06 (Ru/ Al)	0.05 (Ru/ Al)
Ru-5	461.2	532.3	0.15 (Ru/ Al)	0.03 (Ru/ Al)
Ru-6	461.1	532.3	0.20 (Ru/ C)	0.01 (Ru/ C)

Table 3. TPR data, H_2 chemisorption and dispersion

Catalyst	Extend of reduction [a] (%)	H_2 uptake (cm^3 / g)	D (%)	ΔV	d (nm) chem.	d (nm) XRD
Ru-1	47.06	0.61	4.86	1.88	27.57	43.2
Ru-2	33.36	0.30	6.76	1.25	19.82	39.1
Ru-3	45.22	0.10	1.99	1.81	67.33	73.0
Ru-4	66.22	0.11	1.49	2.65	89.93	-
Ru-5	95.32	1.93	18.25	3.81	7.34	21.07
Ru-6	96.06	0.04	0.4	3.93	33.5	2.46*

*-this method does not reveal the metal particles inside the pores
[a] - calculated from the hydrogen uptake

The hydrogen consumption during TPR can be represented as the change in valence of Ru: ΔV= (2x moles H_2 consumed) / moles Ru (column 5). For 161 μmole of Ru, the calculated amount of H_2 necessary to reduce it to Ru(0) is 322 μmole (H_2 / Ru =2 ; ΔV= 4).

3. 2. Catalytic tests

3.2.1. Effect of cinchonidine: Figure 1 shows the change of the reaction rate and of the selectivity to the hydroxyester on the investigated catalysts in the presence of CD as chiral modifier and of ethanol as solvent. If one expresses the reaction rate as the number of moles of substrate **1** transformed per number of moles of reduced Ru (as those determined from TPR) and time (TOF), the following reactivity order can be observed: Ru-6 > Ru-2 >Ru-5 > Ru-1. Data given in Figure 1 show that Ru-2 and Ru-6 are about two times more active than Ru-5 or Ru-1. One should note that the increase of the Ru content from Ru-2 to Ru-1 does not contribute to the increase of the activity, but on the contrary. The Ru-3 and Ru-4 catalysts exhibited very low catalytic activities under the same conditions.

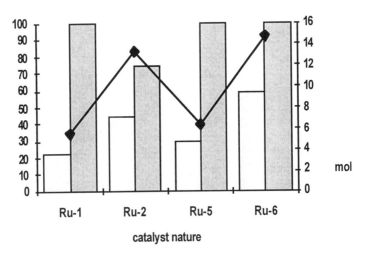

Figure 1.   Evolution of the reaction rate   (–◆–), of the chemoselectivity to hydroxyester (▫) and of the regioselectivity to cis-hydroxyester (▪), 80°C, 10 atm., 4h.

The hydrogenation of the substrate **1** could proceed at different sites and this behaviour could be expressed by: i) chemoselectivity, when we can distinguish between the hydrogenation of ketonic C=O bond, keto-esteric C=O bond, and of the aromatic ring, ii) regioselectivity, when we can distinguish between *cis*- and *trans*-isomers, and iii) diastereoselectivity, when we can discern between the R and S configurations.

Concerning the chemoselectivity, Ru-6 exhibits the best properties and, in this case, the hydroxyester represents 58% of the reaction products. Except for Ru-2, all catalysts are regioselective, having the *cis*-isomers the only reaction products. On Ru-2, which exhibited the higher TOF, the *trans*-hydroyesters are formed in about 26%. There is no direct correlation between the chemo-, regio- and diastereoselectivity. A high chemoselectivity to hydroxyester could be accomplished by a total regioselectivity in the *cis*-configuration. However, the d.e. in the (6R, 7S)-configuration was very low in this case (< 9%). On Ru-6, which exhibited the highest regioselectivity to *cis* isomers, d.e. was zero. For Ru-2, on which *trans*-hydroyesters are also produced, d.e. in (6R, 7R) - configuration was 67.6%.

The increase in the reaction time for the Ru-2 catalyst enhanced both the chemoselectivity to hydroxyesters and the regioselectivity to *cis*-hydroxyester. After 8 h of reaction, the chemoselectivity to hydroxyesters was 49.4% and the regioselectivity to *cis*-hydroxyester was 81.16%. D.e. in *cis*-(6R,7S)-configuration increased by only 1% versus that recorded after 4h.

Table 4 gives the influence of the CD on the reaction rate and on the selectivity for the modified and the non-modified Ru-2 catalyst. The presence of the chiral modifier decreases both the reaction rate and the chemoselectivity to the hydroxyester, but does not influence *cis-trans* ratio. The important role of the modifier resides in the fact that its presence induces diasteroselectivity both for the *cis* (even though d.e. is small) and *trans*-isomers. In the absence of CD, only the racemic was obtained, irrespective to the catalyst used and the reaction conditions.

Table 4. Influence of the CD about the reaction parameters

Cat.	Reaction rate (mole/moles Ru x s) x 10^5	S (%)	Regio-selectivity in the *cis*-hydroxy-ester (%)	Regio-selectivity in the *trans*-hydroxy-ester (%)	*Cis* (6R,7S) d.e. (%)	*Trans* (6R,7R) d.e. (%)
Ru*-2	1.33	44.06	73.9	26.10	8.9	67.6
Ru-2	2.24	55.93	73.14	26.86	racemic	racemic

Ru*-2 -catalyst in the presence of the CD.
S (%) - hydroxyester chemoselectivity

3.2.2. Effect of tartaric acid: The experiments carried out using L(+)-TA modified catalysts showed higher activities than those recorded in the presence of CD. As it was presented above, the Ru-3 catalyst exhibited almost no activity in the presence of CD as chiral modifier. However, the modification of this catalyst with L(+)-TA was found to induce catalytic activity. The chemoselectivities to hydroxyester are lower than those obtained in the similar conditions on the other catalysts investigated. The regioselectivities to *cis*-isomers are rather high, about 88-90%.

The hydrogenation to the *cis*-isomer on the same catalyst occurs with a 10% d.e. toward the (6R,7S)-configuration, while the hydrogenation to the *trans*-isomer yields 16% d.e. in the (6R,7R)-configuration (Figure 3). More important seems to be the effect of PA upon the diastereoselectivity. Thus, for the *cis*-isomer diastereoselective hydrogenation, the decrease of TA-to-PA ratio leads to a progressive diminution in the d.e. to (6R, 7S) configuration. An inversion of the optical selectivity occurs for a TA-to-PA ratio of 0.13. In these conditions, the d.e. in the (6S, 7R) configuration was measured, as shown in Fig. 3.

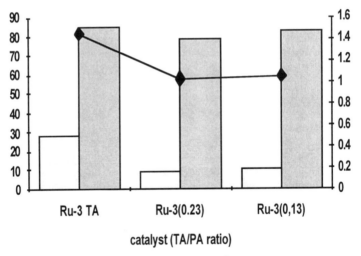

Figure 2. Evolution of the reaction rate (◆) , of the chemoselectivity to hydroxyester (▫) and of the regioselectivity to *cis*-hydroxyester (▨) as a function of TA/PA ratio, on Ru-3 catalyst. 80° C, 10 atm.

The same behavior occurs also in the evolution of the *trans*-isomer. Thus, the increase of PA content leads to a diminution of d.e. in (6R, 7R)-configuration. One should note that for the same TA-to-PA ratio of 0.13 the reaction products exhibit only a d.e. in (6S, 7S)-configuration.

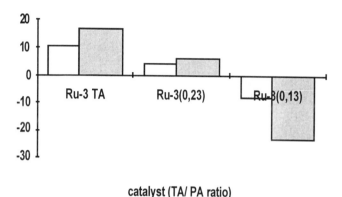

catalyst (TA/ PA ratio)

Figure 3. D.e. for *cis*-hydroxyester (☐) and *trans*-hydroxyester (■) as a function of TA/PA ratio, on Ru-3 catalyst. 80° C, 10 at., 4 hr.

3.3 Molecular modeling: The net charges on the oxygen atoms in both C=O groups of the substrate (-0.276 for keto-carbonyl and -0.341 for ester-carbonyl) point out that these can be involved in hydrogen bond formation with the modifier. Therefore, a starting CD-substrate complex was chosen for the molecular mechanics model in which both carbonyl groups of the β-ketoester can be involved in hydrogen bonds. Using a minimization-docking procedure, several possible complexes stabilized by two hydrogen bonds were found. Among these, two exhibit a minimum energy: a) one between the protonated quinuclidine nitrogen and the keto-carbonyl group, and b) the complex formed by the C9-OH group and the oxygen of the ester-carbonyl. Since the protonated CD can adopt two minimum-energy open forms, for each β-ketoester isomer (*cis* or *trans*) two complexes were obtained. In these complexes it was found that the quinoline ring and phenyl rings of the substrate adopt a non-parallel position. Therefore, it is impossible to assume that both the alkaloid and the ketoester are adsorbed on the Ru surface without the distruction of the weak interaction of the complex.

These results point in the same direction with the other literature data (11), and suggest that the role of the modifier is to provide a specific diastereoisomeric complex which is formed in the liquid phase. This complex can be further adsorbed on the Ru surface only through the unshielded site of the keto-carbonyl group. The optimal interaction energy was obtained for the modifier-substrate structure leading to the (6R, 7R) -configuration (Figure 4). This result also explains the high d.e. obtained for the *trans*-hydroxyester and the low or zero d.e. in the case of the *cis*-hydroxyester.

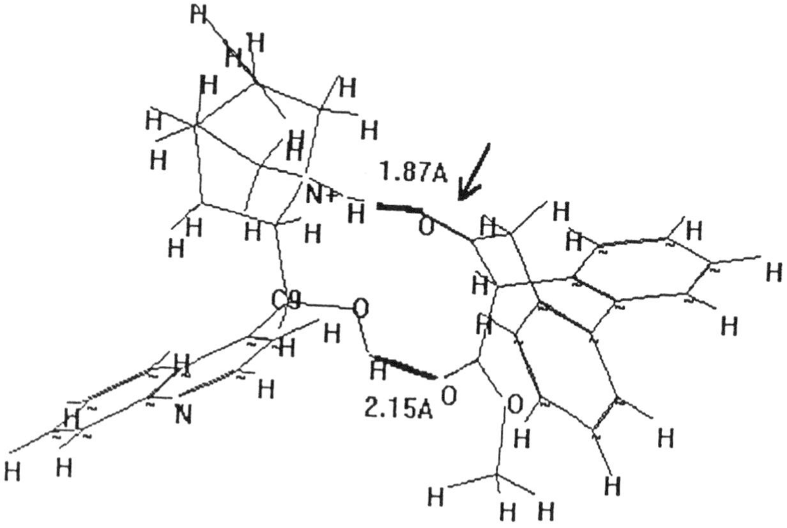

Figure 4. The complex formed between the protonated CD and *trans*-substrate leading to (6R, 7R) - configuration

4. Discussion

The catalyst characterization indicates that the Ru deposition on different molecular sieves occurs mainly on the external surface. A similar behavior takes place for Al_2O_3 or carbon supports. Because of the interaction with the support, the reducibility of ruthenium is different and therefore both reduced and oxidized (Ru(IV)) species coexist.

Catalysts prepared by deposition of the Ru on more acidic supports, such as L or ZSM-5, contain more ruthenium in oxidized states, while those prepared by Ru deposition on less acidic support contain more ruthenium in metallic state. However, under the conditions of such high metal loading as that used in the present study,large metal crystallites are formed irrespective of the nature of the support. This behavior was clearly evidenced by hydrogen chemisorption and XRD measurements.

Under such conditions, the hydrogenation of the substrate **1** occurs with different reaction rates as a function of the support nature and the metal loading. Only the ruthenium particle sites at the external surface of the zeolites can participate in the diastereoselective hydrogenation because both the reactant molecules and the molecules used as chiral modifiers (especially cinchonidine alkaloid) are too large to penetrate inside the zeolitic pore systems. However, the catalyst-activity relationship is not a simple one. Less acidic supports contained more reduced Ru species and it was expected that these catalysts would exhibit higher reaction rates. If we consider the TOF values, one can see that the higher ones were obtained on catalysts with a reduced degree of reduction and, in addition, with a lower surface area and accessibility. Therefore, the dispersion of the metal particles seems to be extremely important for this reaction and lower values such as 5 or 7 could exhibit an effective contribution. In the case of the Ru-3 catalyst, this value is even lower (only 1.99), indicating the existence of large reduced particles. Simply an external surface area of 80 $m^2.g^{-1}$ could generally play an important contribution in the catalytic reactions. This catalyst was found to be almost inactive both in the presence and in the absence of cinchonidine. On the contrary, Ru-2 exhibits smaller Ru particles and a lower reduction extent and it was found to be catalytically active. This allows us to speculate that there is also an interaction between the reduced and the unreduced Ru species, and this interaction has a contribution in such reactions.

L(+)-TA exhibits a particular effect upon the activity of the catalysts. The modification induced by L(+)-TA upon the catalysts makes it a component of the active surface. A good example for this behavior is the Ru-3 catalyst which, after modification with tartaric acid, exhibits a certain activity in this reaction. These data are in agreement with the accepted model (3) of the interaction between TA and the active metals.

But the most important property of these catalysts is the selectivity. In this respect, we found that the reaction follows three levels of selectivity: chemo-, regio-, and diastereoselectivity, respectively. Therefore, we will try to analyse separately the factors that control each kind of selectivity:

a) factors controlling the chemoselectivity. Data presented in Figures 1 and 2 show the following order of chemoselectivity: Ru-6>Ru-2>Ru-5>Ru-1>Ru-3. This trend is similar to the change in conversion, and one should note that, in this case, high selectivities correspond to high conversions. A careful analysis of the experimental data led us to speculate that the chemoselectivity is controlled kinetically. The hydrogenation of the ketonic C=O requires active sites with a smaller number of atoms than those required by the aromatic rings, and also less energetic than those required by keto-esteric C=O bond. How these factors contribute to the high reaction rates and to the high chemoselectivities Ru-2 and Ru-6 the catalysts, we are not able to explain yet. The decreased values of the chemoselectivity for Ru-3 should be correlated with the very low reactivity of this catalyst. The addition of L(+)-TA substantially increased the reaction rate. However, the increase of PA content in the reaction mixture determined both a decrease of the reaction rate and of the chemoselectivity.

The increase of the reaction time was found to have a positive effect upon the chemoselectivity. It is very difficult to explain what it is the real behaviour of the catalyst in such conditions but assuming a kinetically controlled reaction it is very likely that the structures found by the molecular modeling analysis require more time to be formed and, in consequence, good selectivities are obtained only on the catalysts on which the reaction rate is high.

b) factors controlling the regioselectivity: Except for Ru-2 and to a small extent for Ru-3, the *cis*-hydroxyester was the only geometric isomer found on all the investigated catalysts. This behavior should be correlated with the Ru topography on these catalysts. Molecular modeling data showed that the formation of *cis*-isomers is facilitated by the Ru(101) face. In the case of Ru-1, a high Ru concentration allows a very nice growing of the(101) face, as it was observed from XRD patterns. A similar behavior occurs on large porous supports (APO-54, Al_2O3 or C). On the Ru-2 and Ru-3 catalysts, both the low metal loading and the microporous structure do not allow a good growing of this face.

c) factors controlling the diastereoselectivity: Diastereoselectivity was found to be controlled by the chiral modifiers. In the absence of modifiers, only the racemic was obtained, irrespective of the reaction conditions. The addition of cinchonidine oriented the reaction towards the (6R, 7S)-configuration for the *cis*-isomer, and towards the (6R, 7R)-configuration for the *trans*-isomer. The lower and even zero d.e. obtained for the *cis*-isomer is related to the hindered adsorption of the modifier-substrate complex formed in the liquid phase. The catalysts modified with L(+)-TA oriented the diastereoselectivity in the same direction as CD. The addition of PA determines an inversion in the d.e. Thus, for the *cis*-isomer, the d.e. associated to the (6R, 7S)-configuration decreased progressively. When the ratio of TA-to-PA reached 0.13, the d.e. corresponded to the (6S, 7R)-configuration. A similar behaviour was observed in the case of the *trans*-isomer. A similar effect of PA was previously reported by us in the case of prostaglandin hydrogenation (7).

5. Conclusions

Hydrogenation of 7-carboxymethyl-5H-6,7-dihydro-dibenzo [a,c] cycloheptan-6-one on Ru-supported catalysts indicated a very complex relationship between catalyst activity and selectivity.

The activity of the catalysts seems to be controlled both by the dispersion of the metal and by its degree of reduction. Both factors are in direct correlation with the nature of the support. The selectivity of this reaction is a quite complicated problem because the hydrogenation of 7-carboxymethyl-5H-

6,7-dihydro-dibenzo [a,c] cycloheptan -6-one that could occur simultaneously on different sites, leading to chemo-, regio- and distereoselectivity. Each type of selectivity seems to be controlled by another factor. Thus, the chemoselectivity seems to be kinetically controlled, the regioselectivity by the topography of Ru and the diastereoselectivity by the chiral modifiers.

In conclusion, by modifying the support nature and metal loading one can achieve a fairly good chemoselectivity in this reaction with a total regioselectivity in the *cis*-isomer. In these conditions d.e. have values lower than 10% for the *cis*-isomer, and maximum 67.7% for the *trans*-isomer. These values indicate that both CD and L(+)-TA are not the best modifiers for such molecules.

References:

1. L. Ghosez, in *Chiral Reactions in Heterogeneous Catalysis* , G. Jannes and V. Dubois Eds.; Plenum Press: New York and London, 1995
2. M. Beller, *Stud. Surf. Sci. Catal.*, **108**, 1 (1997)
3. T. Osawa, T. Harada, A. Tai, *J. Mol. Catal.*, **87**, 333 (1994)
4. H. U. Blaser, *Tetrahedron: Asymmetry*, **2**, 843 (1991)
5. B. Minder, M. Schurch, T. Mallat and A. Baiker, *Cat. Lett.*, **31**, 143 (1995)
6. R. L. Augustine, S. K. Tanielyan, *J. Molec. Cat. A: Chemical*, **118**, 79(1997)
7. F. Cocu, S. Coman, C. Tånase, D. Macovei, V. I. Pârvulescu, *Stud. Surf. Sci. Catal.*, **108**, 207 (1997)
8. E. Ciorånescu, M. Banciu, M. Elian, A. Bucur and C. D. Nenitzescu, *Liebigs Annal. Chem.*, **739**, 121 (1970)
9. V. I. Pârvulescu, V. Pârvulescu, S. Coman, D. Macovei and Em. Angelescu, *Stud. Surf. Sci. Catal.*, **91**, 561 (1995)
10. G. Filloti, S. Coman, E. Angelescu and V. I. Pârvulescu, *Appl. Stud. Surf. Sci.,* in press
11. J. L. Margitfalvi, M. Hegedus and E. Tfirst, *Tetrahedron: Asymmetry*, **7**, 571 (1996)

Homogeneous Catalytic Hydrogenation and Reductive Hydrolysis of Nitriles

Richard P. Beatty
Dupont Co., PO Box 80203, Wilmington, DE 19880-0302

Abstract

This paper describes hydrogenation and reductive hydrolysis of nitriles to amines or alcohols, and selective hydrogenation of dinitriles to aminonitriles, using homogeneous, organophosphine ruthenium hydride catalysts. It also describes simple methods for preparation of the catalysts.

Introduction

Nitrile hydrogenation is practiced industrially on a huge scale, and is of particular interest to Dupont. Adiponitrile (ADN), prepared by hydrocyanation of butadiene, is hydrogenated to hexamethylenediamine (HMD), which is reacted with adipic acid to make Nylon 6,6. Another commercial specialty diamine, Dytek A (2-methyl-pentamethylenediamine) is obtained by hydrogenating 2-methylglutaronitrile (MGN), the major coproduct from the ADN process. 1,12-Dodecanediamine, from hydrogenation of dodecanedinitrile, has been used for Nylon 12,12.

Selective hydrogenation of ADN to aminocapronitrile (ACN) is also of great interest because ACN can be converted to Nylon 6. Since ADN can supply both the diamine and diacid component of Nylon 6,6, this offers the potential for a hydrocyanation-based, single-intermediate route to both Nylon 6 and 6,6.

Current commercial ADN hydrogenation processes all use heterogeneous catalysts, covered extensively in the open literature and patents (1). There are two dominant types of catalyst. Fixed-bed iron oxide catalysts are very inexpensive and afford high yield of diamine, but require severe conditions (5000 psi, 180-250 °C) and use large amounts of ammonia to suppress formation of secondary amines. Raney catalysts operate under milder conditions ($\leq$100 °C, 500 psi), but are significantly more expensive.

The art on homogeneous nitrile hydrogenation catalysts is much less extensive, and reported catalysts almost always suffer from very low rates or poor selectivity to primary amines (2). Perhaps the best of the many reported ruthenium catalysts is an incompletely characterized complex formulated as $K[Ru_2H_6(PPh_3)_6]$, prepared by reaction of $[RuHCl(PPh_3)_2]_2$ with KH, which reportedly hydrogenated 53 ADN/Ru in 16 hours (95 °C, 90 psi, all substrate/catalyst ratios in this paper are expressed on a molar basis) (3).

Rhodium hydride complexes with bulky phosphine ligands, such as HRh(P-i-Pr$_3$)$_3$, or Rh$_2$H$_2$(μ-N$_2$)(PCy$_3$)$_4$, were reported to hydrogenate nitriles to primary amines even at 25 °C and 1 atm (4).

The iron oxide and Raney nickel catalysts used commercially to hydrogenate ADN to HMD generally offer only statistical selectivity to ACN. Statistical selectivity, that expected assuming equal and independent reactivity of the nitrile ends of ADN and ACN, varies with ADN conversion. At low conversion, ADN is hydrogenated to ACN with high selectivity. As conversion increases, and the concentration of ACN builds up, ACN is hydrogenated more rapidly to HMD, decreasing ACN selectivity. Because of this, it is essential to consider conversion when comparing selectivity of different catalysts. One convenient way to do this is to compare experimental selectivities to the statistical selectivity-conversion curve.

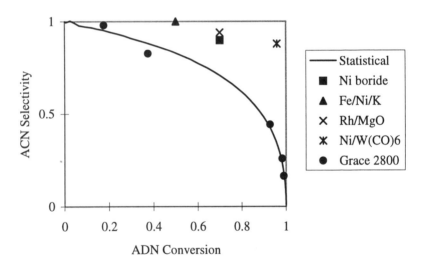

Figure 1. ACN selectivity vs. ADN conversion for selected heterogeneous catalysts, as reported in the literature

The calculated statistical ACN selectivity curve is shown in Figure 1, along with points taken from the literature for catalysts claimed to offer high ACN selectivity (5). For comparison, unpromoted Raney Ni (e.g. W.R. Grace "2800") closely follows the statistical curve. Although some of these catalysts offer high selectivity when operated under the reported conditions, closer review of the literature and lab testing usually reveals technical questions, complications in scaleup, or economic issues that make commercial feasibility questionable.

One obvious approach to ACN, use of common Raney Ni at low conversion, where high selectivity can be achieved, is not as simple as it appears. Economics dictates that unconverted ADN must be recycled. Unfortunately, Raney catalysts typically require small amounts of base (e.g. aqueous NaOH) to maintain activity, which causes unacceptable byproduct formation from ADN when recycle is attempted by distillation.

There are very few reports of selective homogeneous hydrogenation of dinitriles to aminonitriles. Ziegler-Sloan-Lapporte catalysts have been reported which appear to give ≈75% ACN selectivity at 80% ADN conversion (6). A homogeneous rhodium system has also been reported to give high selectivity, but use of strong base in the catalyst system results in unacceptable base-catalyzed byproduct formation (7).

Reductive hydrolysis of nitriles to alcohols has received much less attention than nitrile hydrogenation. Alcohols have long been found as byproducts of nitrile hydrogenation under aqueous conditions, but selectivity and yield are generally low (8). The initial addition of H_2 to a nitrile forms an imine, which undergoes competitive hydrogenation to amine or hydrolysis to aldehyde followed by hydrogenation to alcohol. Alcohol selectivity can be improved by

$$RC\equiv N \xrightarrow{\text{ } H_2 \text{ }} RHC=NH \xrightarrow{\text{ } H_2 \text{ }} RH_2C-NH_2$$

$$NH_3 \Updownarrow H_2O \tag{1}$$

$$RHC=O \xrightarrow{\text{ } H_2 \text{ }} RH_2C-OH$$

limiting hydrogen and conducting the hydrogenation under acidic conditions with Raney Ni, but reaction rate is low and the lifetime of Raney catalysts under acidic conditions is not expected to be long (9). In fact, the redox chemistry between base metal catalysts and acids has been used to generate hydrogen in situ for hydrogenations. Improved selectivity toward alcohols has also been achieved in a two stage process involving a first step of hydrogenating a nitrile using Pd/C at low pressure and temperature, under acidic conditions, with limited hydrogen, trapping the intermediate imine as a hydrogenation-resistant iminium salt. The second step proceeded by hydrolyzing the iminium salt and completing the hydrogenation of aldehyde to alcohol with Raney Ni (10). Nitriles have also been converted to alcohols in a vapor phase hydrogen transfer reaction over a hydrated zirconia catalyst, but large amounts of isopropanol hydrogen donor were required and temperature control was critical (11). A vapor phase process obviously has limited utility for less volatile nitriles.

The catalysts described herein are more active and more stable than other reported homogeneous nitrile hydrogenation catalysts but, at their current state of development, have not been demonstrated to be economically competitive with heterogeneous catalysts for large scale hydrogenations or reductive hydrolyses such as ADN to HMD or hexanediol. Economic viability will be easier to achieve for reactions involving higher value materials, where mild conditions or selectivity are critical, such as in the agricultural and pharmaceutical fields. For example, enantioselective catalytic hydrocyanation can convert alkenes to specific optical isomers of nitriles in high enantiomeric excess (12). Though not yet fully explored, the catalysts discussed herein offer an alternative to other known catalytic and stoichiometric transformations of such nitriles to amines or alcohols, under mild conditions, without racemization of adjacent chirality.

Experimental

Pressure reactions below 110 psig were performed in 50 mL quartz Fisher-Porter tubes. A 50 cc Hastalloy C autoclave from Autoclave Engineers was used for pressure reactions up to about 1000 psig. Both reactors were equipped with dip legs to allow periodic sampling over the course of a reaction. All reactions discussed herein were batch reactions under constant hydrogen pressure.

Reaction samples were quantitatively analyzed by gc, unsplit, on a 30 m DB-5 megabore column from J&W Scientific, with the column effluent divided to allow simultaneous FID and NPD detection. Intermediates and byproducts were identified by gc-ms and comparison to authentic materials. Cyclododecane or t-butylbenzene were used as internal standards, to allow calculation of a mass balance, which we found essential, even in scouting experiments. Small amounts of HMD can sometimes get "lost" on the gc column, resulting in falsely high ACN selectivities. This can be minimized by initially conditioning the gc column with HMD injections, heavy use of the column, and frequent analysis of appropriate standards.

All manipulations of air sensitive compounds were performed inside a Vacuum Atmospheres glovebox under continuous nitrogen purge. The Fisher-Porter and Autoclave equipment was designed to be brought into the box for loading and unloading of air-sensitive catalysts.

(COD)RuCl$_2$ (13), K[Ru$_2$H$_6$(PPh$_3$)$_6$] (3), and RuHCl(CO)(P-i-Pr$_3$)$_2$ (14) were prepared according to published methods. RuHCl(CO)(PCy$_3$)$_2$ was prepared analogously to RuHCl(CO)(P-i-Pr$_3$)$_2$ (15). Other catalysts were prepared as described below.

Synthesis of Ruthenium Phosphine Hydride Catalysts

In the course of this work, two new, general methods were discovered for preparation of RuH$_2$(L)$_x$(PR$_3$)$_{4-x}$, L = H$_2$ or N$_2$, x = 0, 1, or 2. Although a wide variety of ruthenium phosphine hydride complexes are already known, their preparation involves many different starting materials and reactions. Descriptive catalytic chemistry is very heavily weighted toward triphenylphosphine complexes because of availability, relatively high stability, ease of preparation, and lack of general synthetic methods for other phosphine complexes (16). The methods used to prepare PPh$_3$ complexes generally cannot be used to prepare complexes of trialkylphosphines such as PCy$_3$.

One of our methods involves hydrogenation of (COD)Ru(2-methylallyl)$_2$ in the presence of phosphine ligands (17). This method was initially devised as a rapid way to generate catalysts in situ, to facilitate scouting of ligand steric and electronic effects on catalyst activity. Indeed, with small to medium sized ligands, such as P(n-Bu)$_3$ or PPh$_3$, hydrogenation proceeds cleanly to give RuH$_2$P$_4$ or RuH$_2$(H$_2$)P$_3$ complexes. With bulky phosphine ligands, inconsistencies were noted. Although known RuH$_2$(H$_2$)$_2$(PCy$_3$)$_2$, prepared by the literature method (18), was a very active nitrile hydrogenation catalyst, hydrogenation of toluene solutions of (COD)Ru(2-methylallyl)$_2$ in the presence of PCy$_3$, which might be expected to form the same hydride, did not generate an active catalyst. NMR spectra revealed formation of (toluene)RuH$_2$(PCy$_3$) instead of RuH$_2$(H$_2$)$_2$(PCy$_3$)$_2$. The desired RuH$_2$(H$_2$)$_2$(PR$_3$)$_2$ can be obtained by conducting the reaction in a non-aromatic solvent such as THF.

A second method, which we have found useful for preparation of large quantities of catalyst, involves a one-pot reaction of (COD)RuCl$_2$ with hydrogen and phosphine ligand, in biphasic organic (e.g. toluene)/aqueous media, with aqueous base (e.g. NaOH) and a phase transfer catalyst (e.g. benzyltriethyl-ammonium chloride) (19). (COD)RuCl$_2$ can be easily prepared on a large scale from RuCl$_3$(H$_2$O)$_x$, a primary ruthenium starting material, and can be handled briefly in air. This reaction is thought to involve intermediate formation of ruthenium phosphine hydridochloride complexes, which require base for deprotonation and complete hydrogenation to halide-free complexes (20). The products of this reaction depend on steric bulk of the phosphine ligand and the ligand/Ru mole ratio. Small ligands (e.g. PBu$_3$) and chelating ligands form RuH$_2$P$_4$ complexes. Bulky ligands (e.g. PCy$_3$) form RuH$_2$(H$_2$)$_2$P$_2$ complexes. Intermediate size ligands (e.g. PPh$_3$) tend to form mixtures of P$_2$, P$_3$, and P$_4$ complexes, with the exact composition controllable to some extent by varying the ligand/Ru ratio. Formation of arene complexes did not interfere with synthesis of RuH$_2$(H$_2$)$_2$(PCy$_3$)$_2$ using toluene as the organic phase, making this a much more convenient preparation than the original synthesis, which requires Ru(cyclooctadiene)(cyclooctatriene).

The products from either of these two methods can be identified easily from their distinctive and characteristic ^{31}P and ^{1}H NMR spectra. H$_2$ ligands in RuH$_2$(H$_2$)P$_3$ and RuH$_2$(H$_2$)$_2$P$_2$ complexes exchange readily with N$_2$, so that clean spectra of the hydrides can only be obtained on samples sealed under H$_2$. Clean spectra of N$_2$ complexes can be obtained by sparging solutions with N$_2$. Casual handling of complexes in the glovebox affords mixtures of H$_2$ and N$_2$ complexes.

Large crystals of $RuH_2(N_2)_2(PCy_3)_2$ were grown by evaporation of a hexane solution of the species in the drybox. X-ray diffraction revealed trans PCy_3 ligands, cis hydrides, and N_2 ligands trans to the hydrides. Other workers have prepared this complex by a different method and reported it to be very unstable (21), but as prepared by our method it is easily isolable and stable at room temperature under N_2.

Nitrile Hydrogenation

The reaction profile for a typical ADN hydrogenation using $RhH(P-i-Pr_3)_3$ is shown in Figure 2. This is a remarkably active catalyst in comparison to other homogeneous nitrile hydrogenation catalysts, with an initial reaction rate of greater than 2 nitrile group hydrogenations/minute under very mild conditions. This reaction proceeded to completion, showing no signs of catalyst deactivation. The ultimate yield of HMD was very high, and ACN selectivity throughout the reaction was much higher than statistical (Figure 3). When the pressure was increased to 1000 psi, rate increased but selectivity was identical to that at lower pressure. Despite the strong performance of this catalyst, due to the expense of Rh, we immediately focused on catalysts with less expensive metals such as Ru.

An ADN hydrogenation using $K[Ru_2H_6(PPh_3)_6]$ was considerably slower, even under much more severe conditions (Figure 4). Although not obvious from Figure 4, catalyst deactivation was occurring before complete ADN conversion. Formation of byproducts lowered ACN selectivity, especially early in the hydrogenation (Figure 3). The main byproduct, 2-cyano-cyclopentylimine, is known to form from ADN by base-catalyzed (presumably $K[Ru_2H_6(PPh_3)_6]$) cyclization. At higher conversions, selectivity was greater than statistical.

Finally, an ADN hydrogenation using $RuH_2(H_2)_2(PCy_3)_2$ is shown in Figure 5. This hydrogenation was run under conditions similar to those of Figure 4, showing that the activity and stability of this catalyst is significantly higher. ACN selectivity was greater than statistical throughout the reaction (Figure 3)

Dihydrogen is a remarkably good ligand in $RuH_2(H_2)_2(PCy_3)_2$. NMR experiments showed that the primary form of this catalyst under hydrogenation conditions in the presence of nitriles was still $RuH_2(H_2)_2(PCy_3)_2$. Pressure NMR experiments over the range 50-600 psig H_2 showed $K_{eq} \approx 0.005$ (all concentrations in moles/liter) for the following equilibrium at 20 °C.

$$RuH_2(H_2)_2(PCy_3)_2 + \text{Valeronitrile (VN)} \rightleftharpoons RuH_2(H_2)(VN)(PCy_3)_2 + H_2 \quad (2)$$

When hydrogen pressure was decreased, the concentration of the nitrile complex increased, but after hydrogenation was complete the Ru returned to the

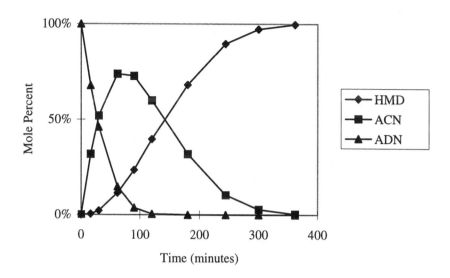

Figure 2. ADN hydrogenation with RhH(P-i-Pr₃)₃, 133 ADN/Rh, 110 psig H₂, 50°C, toluene solvent.

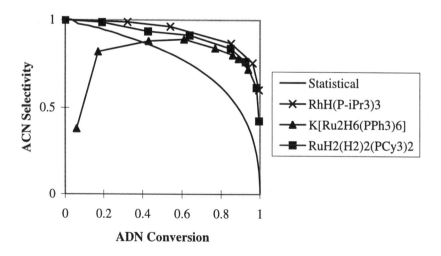

Figure 3. ACN selectivity vs. ADN conversion for selected homogeneous catalysts.

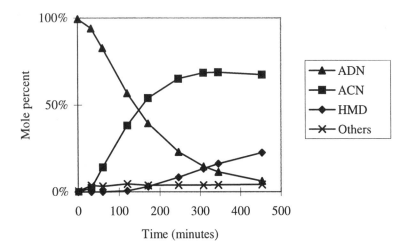

Figure 4. ADN hydrogenation with K[Ru$_2$H$_6$(PPh$_3$)$_6$], 51 ADN/Ru, 80°C, 1000 psig H$_2$, toluene solvent.

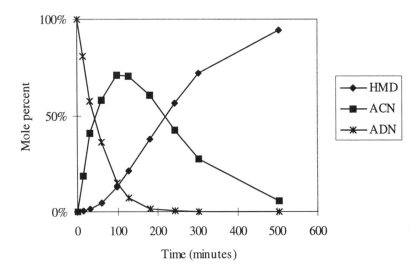

Figure 5 ADN hydrogenation with RuH$_2$(H$_2$)$_2$(PCy$_3$)$_2$, 46 ADN/Ru, 80 °C, 1000 psig H$_2$, toluene solvent.

form $RuH_2(H_2)_2(PCy_3)_2$. Nitrile complexes prepared by this exchange reaction can be used as hydrogenation catalyst precursors.

A wide variety of ruthenium-phosphine-hydride complexes was scouted for ADN hydrogenation activity and selectivity, covering a range of ligand steric and electronic properties. As noted earlier, these complexes were of the type RuH_2P_4, $RuH_2(H_2)P_3$, or $RuH_2(H_2)_2P_2$, mainly depending on steric bulk of the phosphine. Our goal was to maximize both catalyst hydrogenation activity and stability, ultimately striving to increase catalyst utility, or total nitrile turnovers/catalyst consumed.

At constant temperature and pressure, hydrogenation rate increased in the order $P_4 < P_3 < P_2$ complexes, with catalyst stability following the reverse trend. It was even possible to prepare apparently homogeneous catalyst solutions by hydrogenation of $(COD)Ru(2\text{-methylallyl})_2$ in the presence of only 1 P/Ru, and these had the highest activity of all.

Unfortunately, activity and stability were a close tradeoff. Increasing temperature increased reaction rate but also decreased catalyst stability so that the total number of moles of nitrile hydrogenated per mole Ru remained essentially unchanged.

Attempts to increase catalyst utility of $RuH_2(H_2)_2(PCy_3)_2$ by increasing the ADN/Ru ratio from 53, as in Figure 5, to about 250 resulted in catalyst deactivation at about 75% ADN conversion when the reaction solvent was toluene. NMR analysis showed that this was due to formation of $(\text{toluene})RuH_2(PCy_3)$, which is not an active hydrogenation catalyst. A toluene solution of $RuH_2(H_2)_2(PCy_3)_2$ stored at room temperature slowly (months) converted to $(\text{toluene})RuH_2(PCy_3)$, even under 1 atm H_2. Use of amine or ether solvents eliminated this deactivation and enabled us to achieve complete hydrogenation with ADN/Ru > 250, but as we increased ADN/Ru even more, catalyst deactivation again occurred. Because of the low concentration of catalyst (≤ 0.003 M) we were unable to determine the cause of deactivation.

Another approach to increasing catalyst utility is to recycle the catalyst. Since $RuH_2(H_2)_2(PCy_3)_2$ is water stable but not water soluble, and HMD is water soluble, it is possible to water-extract HMD from an organic solution containing the catalyst. Extraction avoids exposing catalyst to more extreme conditions such as might be necessary for distillation.

Out initial efforts to use aqueous extraction for catalyst recycle were very encouraging. We next attempted to carry out the hydrogenation in biphasic aqueous-organic media, hoping to continuously feed ADN and water, separate the phases after hydrogenation, and recycle the organic catalyst phase directly back to the hydrogenation reactor. The presence of a separate water phase during hydrogenation might also improve ACN selectivity by extracting ACN, as it forms, into the water phase, away from the catalyst. Our first attempt to demonstrate this concept revealed a typical disappearance curve for the ADN, showing that the catalyst had normal activity in the biphasic media. The reaction rate was as expected and GC analysis was very clean. Surprisingly, instead of ACN, the major intermediate was 6-hydroxy-capronitrile and, instead of HMD, the final product was 1,6-hexanediol.

Nitrile Reductive Hydrolysis

Although $RuH_2(H_2)_2(PCy_3)_2$ does catalyze reductive hydrolysis, $RuHCl(CO)P_2$ (P = P-i-Pr$_3$ or PCy$_3$) are preferred. A reductive hydrolysis of MGN to 2-methyl-1,5-pentanediol (MPDO) using $RuHCl(CO)(PCy_3)_2$ is shown in Figure 6. Since MGN is not symmetical, two intermediate hydroxynitriles (HOCN1 and 2) are observed. GC-MS suggests that the one formed in greater abundance is the product of reductive hydrolysis of the less hindered end of MGN.

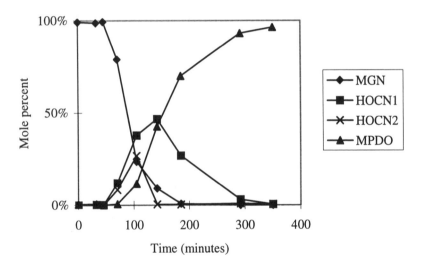

Figure 6. MGN reductive hydrolysis using RuHCl(CO)(PCy$_3$)$_2$, 72 H$_2$O/CN, 61 MGN/Ru, 1000 psi H$_2$, 80 °C, THF solvent.

The induction period apparent in Figure 6 is believed to be due to slow formation of $RuH_2(H_2)(CO)(PCy_3)_2$. This species can be prepared by hydrogenation of $RuHCl(CO)(PCy_3)_2$ in the presence of aqueous base and phase-transfer catalyst, and is active for reductive hydrolysis without an induction period (22). Interestingly, neither $RuHCl(CO)(PCy_3)_2$ nor $RuH_2(H_2)(CO)(PCy_3)_2$ are active ADN hydrogenation catalysts; in the absence of water or amines, no reaction occurs. Amines promote nitrile hydrogenation with $RuH_2(H_2)(CO)(PCy_3)_2$, but not with $RuHCl(CO)(PCy_3)_2$.

As indicated in eq. 1, hydrogenation and reductive hydrolysis are competing reactions. In the case of benzonitrile, benzaldehyde was actually observed. Which reaction dominates depends on the nitrile and solvent. With ADN or MGN, reductive hydrolysis predominated even in toluene-water media. With the more hydrophobic dodecanedinitrile, under the same conditions, only hydrogenation occurred. Dodecanedinitrile, and other hydrophobic nitriles, can be reductively hydrolyzed by using a homogenizing solvent such as THF instead of toluene. However, some surprising phase behavior must be considered under pressure: THF and water are not miscible at temperatures between about 72 and 136 °C, and ADN and water become miscible above about 102 °C.

References

1. K. Weissermel and H. –J. Arpe, Industrial Organic Chemistry, VCH, Weinheim, 1993.

2. A.M. Joshi, K.S. MacFarlane, B.R. James, and P. Frediani, *Chem. Ind. (Dekker)*, **53** (Catal. Org. React.), 497 (1994) and *Prog. in Catal.*, **73**, 143 (1992); I.S. Thorburg, S.J. Rettig, and B.R. James, *J. Organomet. Chem.*, **296**, 103 (1985); C.S. Chin and B. Lee, *Catal. Lettters*, **14**, 135 (1992); Z. He, D. Neibecker, N. Lugan, and R. Mathieu, *Organometallics*, **11**, 817 (1992); D.K. Mukherjee, B.K. Palit, and C.R. Saha, *J. Mol. Catal*, **88**, 57 (1994); A. Bose and C.R. Saha, *J. Mol. Catal.*, **49**, 271 (1989); T. Suarez, and B. Fontal, *J. Mol. Catal.*, **45**, 335 (1988); B. Fell and G. Gurke, *Chem. Zeit.*, **115**, 83 (1991); V. Balladur, P. Fouilloux, and C. de Bellefon, *Appl. Catal.*, *A*, **133**, 367 (1995). E. Band, W.R. Pretzer, M.G. Thomas, and E.L. Muetterties, *J. Am. Chem. Soc.*, **99**, 7380 (1977); K.C. Dewhirst, US 3,454,644 (1969) to Shell; D.R. Levering, US 3,152,184 (1964) to Hercules.

3. R. A. Grey and G. P. Pez, US 4,254,059 (1981) and 4,268,454 (1981) to Allied. Also see: R.A. Grey, G.P. Pez, and A. Wallo, *J. Am. Chem. Soc.*, **103**, 7536 (1981); J. Halpern, *Pure Appl. Chem.*, **59(2)**, 173 (1987); D.E. Linn and J. Halpern, *J. Am. Chem. Soc.*, **109**, 2969 (1987).

4. T. Yoshida, T. Okano, and S. Otsuka, *J. Chem. Soc. Chem. Commun.*, 870 (1979).

5. Ni boride: L. K. Friedlin, and T. A. Sladkova, *Izv. Akad. Nauk SSSR*, 151 (1961); Fe/Ni/K: F. Medina, P. Salagre, J. E. Sueiras, and J. L. G. Fierro, *Appl. Catal. A: Gen.*, **92**, 131 (1992); Rh/MgO: F. Mares, J. E. Galle, S. E.

Diamond, and F. J. Regina, *J. Catal.*, **112**, 145 (1988), US 4,389,348 (1983), and US 4,601,859 (1986) to Allied; Raney Ni/W(CO)₆: K. Sanchez, WO 9,316,034 (1993) to Dupont.

6. B. Fell and G. Gurke, *Chem. Zeit.*, **115**, 83 (1991).

7. S. E. Diamond, F. Mares, and A. Szalkiewicz, US 4,362,671 (1982) to Allied.

8. H. Rupe and F. Becherer, *Helv. Chim. Acta*, **6**, 865 (1923) N. Takenaka, C. Shimakawa, and K. Tachiyama, Japanese Kokai: Hei 4-36250 (1992) to Mitsui Toatsu Chemicals, Inc.

9. E. Moeltgen and P. Tinapp, *Liebigs Ann. Chem.*, 1952 (1979).

10. A. Kleeman, K. Deller, K. Drauz, and B. Lehmann, DE 3,232,749 (1984) to Degussa.

11. H. Matsushita, M. Shibagaki, and K. Takahashi, EP 271092 (1988) to Japan Tobacco.

12. A. L. Casalnuovo, T. V. Rajanbabu, L. W. Gosser, R. J. McKinney, and W. A. Nugent, Jr., U.S. 5,312,957 (1994) to Dupont.

13. M. O. Albers, T. V. Ashworth, H. E. Oosthuizen, and E. Singleton, *Inorg. Synth.*, **26**, 68 (1989).

14. M. A. Estreulas and H. Werner, *J. Organomet. Chem.*, **303**, 221 (1986).

15. F. G. Moers and J. P. Langhout, *Recueil*, **91**, 591 (1972).

16. W. H. Knoth, *Inorg. Synth.*, **15**, 31 (1974).

17. R. P. Beatty and R. A. Paciello, U.S. 5,559,262 (1996) to Dupont.

18 B. Chaudret and R. Poilblanc, *Organometallics*, **4**, 1722 (1985).

19. R. P. Beatty and R. A. Paciello, U.S. 5,599,962 (1997) to Dupont.

20. V. V. Grushin, *Acct. Chem. Res.*, **26**, 279 (1993); C. Grunwald, O. Gevert, J. Wolf, P. Gonzalez-Herrero, and H. Werner, *Organometallics*, **15**, 1960 (1996).

21. S. Sabo-Etienne, M. Hernandez, G. Chung, B. Chaudret, and A. Castel, *New J. Chem.*, **18**, 175 (1994).

22. Identified spectroscopically by analogy to the known P-iPr₃ complex: D. G. Gusev, A. B. Vymenits, and V. I. Bakhmutov, *Inorg. Chem.*, **31**, 1 (1992).

Selective Hydrogenation of Butyronitrile Over Promoted Raney® Nickel Catalysts

S. N. Thomas-Pryor, T. A. Manz, Z. Liu, T. A. Koch*, S. K. Sengupta*, and W. N. Delgass

School of Chemical Engineering, Purdue University, West Lafayette, IN 47906; *DuPont Nylon, Wilmington, DE 19880

Abstract

In this study, several reaction and characterization methods have been combined to better understand the hydrogenation of nitriles over Raney® nickel catalysts. Hydrogenation of butyronitrile was carried out at pressures of 1 to 14 bar in liquid-phase batch reactors. The addition of sodium hydroxide was found to increase the selectivity to butylamine. Catalysts promoted with iron or iron and chromium were found to have improved activity. Competitive adsorption studies showed that butylamine competes with hydrogen for adsorption sites and hydroxide ions for a second type of adsorption site. XPS measurements showed that the active surface is composed of zero valent nickel and aluminum atoms.

Introduction

Catalytic hydrogenation of nitriles is important in the manufacture of the nylon monomer hexamethylenediamine (1,6-hexanediamine) from adiponitrile (ADN). ADN is first partially hydrogenated to 6-aminohexanenitrile and then to 1,6-hexanediamine [1]. The industrial preparation of these primary amines is usually accomplished in the liquid phase at elevated hydrogen pressures (270-600 bar) and at 30 bar in some recent processes [2], where the catalyst represents a key factor in determining the activity and selectivity with respect to primary amines. A potential catalyst for this process is Raney® nickel promoted with metals which improve their selectivity, activity, and stability. Our objective here is to examine the relation between the selective hydrogenation and surface composition and structure. In this paper these issues are addressed using butyronitrile (BN) as a model reactant because its single nitrile group affords a simpler reaction system compared to the more complex ADN reaction.

Butyronitrile is hydrogenated to form butylimine (BI), an intermediate which is further hydrogenated to form butylamine (BA), the desired reaction product. The unwanted by-products are formed by nucleophilic addition of butylamine to butylimine. This produces an intermediate 1-amino-dibutylamine which subsequently loses ammonia to form N-butylidene-butylamine (X), which is subsequently hydrogenated to form dibutylamine (DBA) [3, 4]. The reaction network is shown in Figure 1 below.

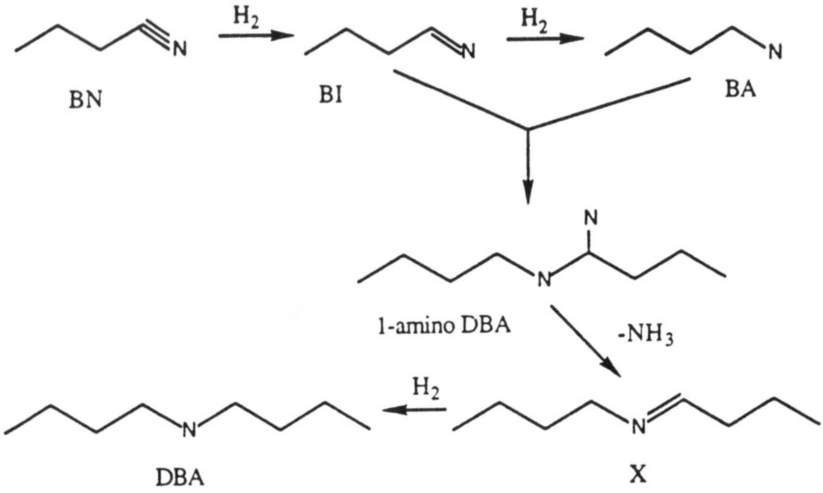

Figure 1: Reaction schematic for butyronitrile hydrogenation.

Neither butylimine nor 1-amino-dibutylamine was detected in the reaction liquid, indicating that both of these species are highly reactive and that little, if any, desorbs from the surface before hydrogenation or ammonia elimination, respectively.

Experimental Methods

Activation of Raney® Nickel Alloys

The catalysts used in this study were 1) unpromoted, 2) iron promoted, and 3) iron and chromium promoted Raney® nickel catalysts. The starting alloys had ca. 1-4 wt % promoters and were provided by GRACE Davison.

The alloy powders (-140 mesh) were activated in 30 wt. % sodium hydroxide solution at 90°C for 1 hr. Seventy percent excess NaOH over that needed to react with all the aluminum in the alloy was used. The powder was slowly added to the leaching solution in order to prevent excess foaming due to hydrogen evolution:

Leaching reaction: $Al + OH^- + H_2O \rightarrow AlO_2^- + 3/2\ H_2$

The catalysts were rinsed several times in deionized distilled water and stored in sealed bottles containing 3 wt. % aqueous solution of NaOH to prevent surface aluminum hydrolysis.

Surface Area Measurements

Total surface areas were measured from BET analysis of nitrogen adsorption isotherms at 77 K using a Micromeritics ASAP 2010 adsorption unit after outgassing the sample at 50°C overnight. A value of 0.164 nm^2 was used for the cross section of the nitrogen molecule.

CO adsorption was performed on the same adsorption unit to determine nickel surface area. An adsorption temperature of 35°C was chosen as a conveniently controllable value and was shown to give the same quantitative results as the more common 0°C value. Prior to data collection, the sample was degassed *in situ* at 120°C for 8-12 hr. Linear adsorption of CO with nickel was assumed [5].

For both BET and metallic surface area measurements, the samples were introduced wet into the apparatus, in order to prevent exposure to air and subsequent oxidation.

X-Ray Photoelectron Spectroscopy

The chemical states of both Ni and Al in the activated catalysts were examined on a Perkin Elmer PHI 5300 ESCA system with a Mg anode operated at 15 KV and 300 W. Typically, a sample slurry was washed with distilled and deionized water for five times, then loaded into the XPS sample holder. In order to protect the sample from air exposure, it was always covered with a layer of water. The wet sample was degassed overnight at room temperature in the XPS introduction chamber, prior to data collection. Ni and Al foils were used to determine binding energies corresponding to the metallic and oxidized states.

Low Pressure Reactions

The low pressure reaction setup consisted of a custom built 100 ml stirred glass reactor with a cooling jacket, sample port, and gas inlet and outlet ports. The reactor could be fed with either hydrogen or nitrogen gas at one atmosphere. An electrically driven impeller (Caframo Ltd., Model 1: RZR 1), rotating at 2000 rpm, was used to provide intense mixing. The reactor was equipped with baffles to improve agitation.

An appropriate amount of catalyst (1-3 g) was rinsed with DI water and weighed according to the following procedure: 1) a 10 ml volumetric flask was filled with DI water and weighed, 2) the catalyst sample was placed in the flask and the liquid level readjusted to 10 ml, 3) the flask was weighed again, and 4) the catalyst weight under water was given by the difference in the two weights. The dry catalyst weight is given by 1.15 times the weight under water (to account for the displacement of water by the catalyst).

The catalyst and water were placed in the reactor together with 50 ml of methanol solvent. For reactions performed with base, 0.5 g of NaOH was added at this time. The reactor was purged with hydrogen for several minutes during which time the catalyst adsorbed hydrogen.

To initiate the reaction, 5 ml of butyronitrile was added with a syringe through the rubber septum sampling port. Samples were withdrawn through the port using a filtered syringe to separate catalyst from the liquid. The samples were analyzed on a gas chromatograph equipped with a 5 ft. Chromosorb 103 packed column.

High Pressure Reactions

High pressure reactions were performed in a 300-ml autoclave reactor (Autoclave Engineers, Inc.). Experiments were carried out at 75°C, 14 bar for two hours. A gauge and thermocouple were inserted into the autoclave to measure the hydrogen pressure and temperature of the reaction. A pressurized addition cylinder was attached to the reactor to introduce the butyronitrile into the reactor to start the reaction.

The butyronitrile solution was prepared by weighing 10.35 grams of butyronitrile and 0.04 grams of cyclododecane, which was used as an internal standard. Methanol (109 grams), the solvent, was added to the reactor. The catalyst was rinsed three times in distilled water, decanted, and then rinsed three times in excess methanol. Next, 0.6 grams of catalyst was added to the reactor. The reactor was purged in nitrogen, then hydrogen. After purging, the reactor was agitated at 1500 rpm and the solution was heated to the desired temperature and allowed to equilibrate. Butyronitrile was added to initiate the reaction.

All samples collected were identified with respect to the elapsed reaction time. After purging the sample line, samples were taken in 1.0 ml aliquots, placed in a GC vial, and diluted with a measured amount of methanol. The samples were analyzed using a HP-5 (Crosslinked 5% PH ME Siloxane) 25 m x 0.32 mm (0.25 μm film thickness) capillary column.

Butylamine Adsorption Experiments

Adsorption isotherms were measured in the glass reactor setup under flowing hydrogen at 1 bar and 20°C with stirring at 2000 rpm. The first step was to prepare a BA stock solution by adding 5.0 ml of BA to 250.0 ml of methanol. A background curve was prepared by running the adsorption measurement with no catalyst. A second adsorption measurement was made using 2-5 g catalyst.

A calibration curve was prepared by chromatographic analysis of samples after sequential addition of 5 ml aliquots of the butylamine (BA) stock solution to 50 ml methanol and 10 ml of water. The adsorption was done according to the procedure below:
1) A catalyst sample was placed in a beaker and rinsed with water several times to remove traces of sodium hydroxide. An amount of catalyst which gave approximately 1 mmole of total BA adsorption sites was used.

2) The catalyst was weighed under water, as discussed above, and placed in the reactor together with a total of 10 ml water and 50 ml methanol. The reactor was purged with hydrogen.

3) A filtered syringe was used to withdraw a liquid sample. The sample was analyzed chromatographically to verify that the solvent was free of impurities.

4) 5.0 ml of BA stock solution was added to the reactor. After the hydrogen flow was turned back on, about 3 minutes were needed for equilibrium to be established. A sample was taken and analyzed in the GC. A second sample was taken and analyzed after 5 minutes. If the two samples did not agree within 3 mM, more samples were taken to establish the correct BA concentration in the reactor. The syringe was cleaned and the filter replaced.

5) Additional BA was added to the reactor, according to the procedure in step 4, until the total concentration in the reactor reached 25-45 mM. After each addition, samples were taken and analyzed according to the procedure in step 4.

The uptake was calculated as (mmoles BA added - mmoles BA in solution)/grams of catalyst. The mmoles of BA in solution was calculated by converting chromatographic areas to concentration and multiplying by the volume of liquid in the reactor.

Results and Discussion

Surface Area Measurements

The bulk compositions of the activated catalysts were measured by atomic absorption spectroscopy and are presented in Table 1 below together with the total and metallic surface areas measured by di-nitrogen and CO adsorption as previously described.

The activated catalysts contained significant amounts of aluminum and promoters after leaching. Compositions of the catalysts suggest that iron and chromium are not lost during activation. These promoters cause an increase in the total BET surface areas, but have no effect on the CO adsorption surface areas.

Low Pressure Reactions

Reactions were carried out at 20°C using 5 ml of BN, 50 ml methanol, 10 ml water, and 1-3 grams of catalyst. A typical reaction profile is shown in Figure 2 below. The concentration of N-butylidene-butylamine (X) goes through a maximum as it is formed and then declines as it is hydrogenated to form DBA.

Table 1. Summary of compositions, surface areas, and reactivities of catalysts.

Catalyst	Composition wt %				Surface Area m²/g		$k_{1/2}$ *
	Ni	Al	Fe	Cr	BET	CO	
Unpromoted	83.7	16.3	-	-	70	39	0.010
Low Fe	84.8	13.0	2.1	-	103	33	0.024
High Fe	79.5	13.0	7.5	-	98	39	0.033
Fe & Cr	73.4	22.9	1.5	2.2	127	50	0.042

* half order rate constant, (mole/L)$^{1/2}$/(min. g catalyst)

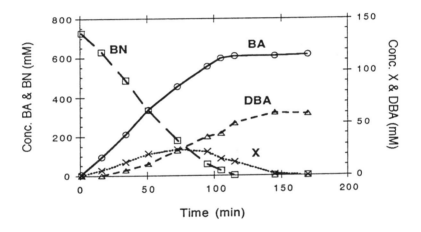

Figure 2:   Concentration versus time plot for unpromoted Raney® Ni.

1.) Effect of Promoters

The reaction was found to fit the rate equation $r = k_{half} *C_{BN}^{1/2}$ from the linearity of $C_{BN}^{1/2}$ versus time plots. A first order model did not fit at conversions above 90 percent. The non-linearity of the first order plots became severe as the reaction approached completion, as shown in Figure 3.

The last column in Table 1 shows that the iron and chromium doubly-promoted catalyst had the highest activity. The iron-only promoted catalysts had intermediate activity, while the unpromoted catalyst had the lowest activity.

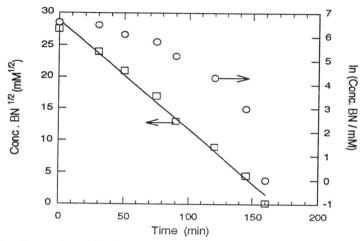

Figure 3: Comparison of half order and first order plots for the unpromoted catalyst.

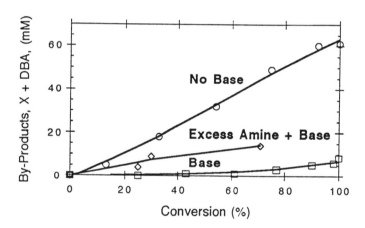

Figure 4: Effects of base and butylamine addition on by-product formation, unpromoted catalyst. Conditions, 5 ml BN added to: 1)10 ml water + 50 ml methanol, 2) 10 ml water + 50 ml methanol + 0.5 g NaOH, 3) 10   ml water + 25 ml methanol + 25 ml butylamine + 0.5 g NaOH.

2.) Effect of Base Addition

As shown in Figure 4, the addition of base reduced the amount of by-product formation dramatically. Base could also control the amount of by-products formed even when the solvent was made 40 vol. % in BA. A similar effect was observed for the doubly-promoted catalyst. For the unpromoted catalyst, the final yield of by-products formed was 1) 16.8% , 2) 2.1 %, and 3) 6 %. For the doubly-promoted catalyst, the final yield of by-products was 1) 21.0 %, 2) 2.1 %, and 3) 6 %. The conditions for 1), 2) and 3) appear in the caption of Figure 4.

The addition of excess BA had a detrimental effect on the reaction rate. For the unpromoted catalyst, the rate in the BA-rich solvent was only 25 % of that in a solvent without excess BA. For the doubly-promoted catalyst, the rate decreased by a smaller amount, to 63 % of the rate without BA.

High Pressure Reactions

The data obtained at 14 bar and the higher temperature of 75°C fits both first order and half order kinetics at conversions up to 90%, while at higher conversions, the rate deviates from half order. We take this data to indicate that the order approaches 1 at higher temperatures. The change from half order at low temperature to first order at higher temperature is consistent with a simple Langmuir-Hinshelwood analysis assuming that BN adsorbs on the surface. Finite coverage of the surface at low temperature decreases the order because $r = kKC_{BN}/(1 + KC_{BN})$ can be approximated as $r = k(KC_{BN})^{1/n}$. Since the adsorption is exothermic, K decreases with temperature and r approaches kKC_{BN} when KC_{BN} is sufficiently less than 1. A detailed kinetic analysis is in progress.

Table 2. Summary of Kinetic data for promoted Raney® nickel catalysts. Experiments performed at 14 bar, 75°C.

Catalyst	Conversion %	Selectivity %	k_1 min^{-1}	initial rate, r (mmol g^{-1} min^{-1})
Unpromoted	26.5	80.7	0.0025	0.6
Low Fe	69.9	86.6	0.0096	2.4
High Fe	57.9	86.1	0.0068	1.7
Fe & Cr	96.0	76.4	0.0287	7.0
Fe & Cr,w/base	99.9	98.5	0.0605	15.0

* Catalysts leached in 25 wt% NaOH at 90 °C for 3 hr.

Table 2 gives a summary of the observed first order rate constant, k_1 and rate data for experiments carried out for two hours at 75°C and 14 bar with an initial butyronitrile concentration of 1.1 mole/L using 0.6 grams of catalyst. The reaction products which were monitored included butylamine, dibutylamine, and N-butylidene-butylamine (identified from GC-mass spectrometry data). The selectivity was defined as follows: moles BA produced / (initial moles BN - final moles of BN) and was found to be constant after the first 15 minutes of reaction.

No significant difference in selectivity between the singly and doubly promoted catalysts was observed. For the catalysts examined, the selectivity was approximately 80+/-5%. However, in the case of the doubly promoted catalyst with base, the selectivity increased to nearly 100%, showing that base has a dramatic effect on selectivity.

Table 2 also shows that single promotion of Raney nickel with iron produces a maximum in the rate at the low level of Fe addition. The doubly promoted catalyst has a rate 3.5 times higher than the optimum Fe promoted catalyst, suggesting a beneficial effect of chromium. The turnover frequencies (rate normalized to # of Ni sites measured by CO adsorption) also show the doubly promoted catalyst (TOF = 0.2 sec^{-1}) to be the more reactive than the unpromoted catalyst (TOF = 0.01 sec^{-1}).

X-Ray Photoelectron Spectroscopy

The chemical states of Ni and Al in the activated Raney Ni catalysts are reflected in their core electron binding energies. The XPS results from the Ni and Al foils showed that metallic Ni and Al have 2p binding energies of 852.5 and 73 eV respectively, while for oxidized Ni (2+) and Al (3+) the corresponding binding energies shift to higher values, 855.2 and 75.7 eV. These values are in good agreement with the literature [6]. Our results show that the Ni 2p binding energy in the active Raney Ni catalysts is 852.6 eV and Al 2p 72.5. Therefore, it can be concluded that both Ni and Al remain in metallic form. This finding is in agreement with a previous study [7]. It should be noted, however, that these two elements can exist in both metallic and oxidized forms if an activated Raney Ni is stored in water rather than dilute sodium hydroxide solution, as shown in Figure 5. We also note that the absence of an Na (1s) peak indicated that the NaOH was removed by the washing procedure.

Butylamine Adsorption Experiments

As shown in Table 3, the addition of sodium hydroxide decreases the uptake of BA on the catalyst surface. However, the addition of more base did not completely block BA uptake. In the presence of excess base, the uptake was 0.2 mmol/g. This suggests that BA may adsorb on two different sites, one which is blocked by base and one which is not.

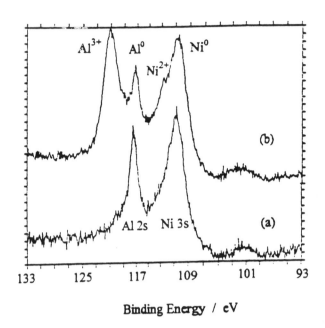

Binding Energy / eV

Figure 5. XPS of unpromoted Raney® Ni, (a) stored in 3% NaOH aqueous solution; (b) stored in distilled and deionized water.

Table 3. Effect of NaOH addition on butylamine adsorption on unpromoted catalyst.

NaOH / M	0	0.08	0.24	0.53
BA Uptake (+/- 0.05) mmol/g catalyst	0.34	0.23	0.17	0.20

Adsorption experiments were also done to determine the effects of hydrogen on BA uptake in the presence of sodium hydroxide (0.24 M). Hydrogen was flowed over the catalyst for approximately 5 minutes and the BA uptake was measured to be 0.17. The hydrogen flow was shut off and the reactor was purged with nitrogen for approximately 5 minutes. Then it was found that the BA uptake increased to 0.31 mmol/g. The purging gas was switched back to hydrogen and the uptake decreased to 0.18 mmol/g. These results suggest that hydrogen may compete with BA for adsorption sites on the surface.

Carbon monoxide was used to further examine the nature of the BA adsorption sites.

The reactor was loaded with catalyst and solvent, agitated at 2000 rpm, at 20°C then exposed to a flow of CO of about 0.5 mol/sec for 5 minutes. The

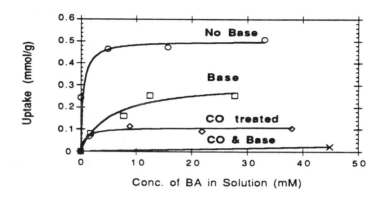

Figure 6: Comparison of BA uptake isotherms in the presence and absence of CO and OH-.

hydrogen flow was then re-established and a BA adsorption isotherm obtained, as shown in Figure 6. CO blocked 80% of the BA uptake. When this catalyst was flushed with methanol and then used for BN hydrogenation, the rate was 25% of the value for a fresh catalyst. This behavior is consistent with irreversible adsorption of CO on nickel. As shown in Figure 6, the total BA uptake on the unpromoted catalyst was affected by both base and CO addition. Addition of base decreased the uptake of BA from 0.5 to 0.28 mmol/g. Addition of CO decreased the uptake from 0.5 to 0.1 mmol/g. In the presence of both CO and hydroxide the uptake was nearly eliminated.

The improvement in catalyst selectivity by base addition suggests that the base adsorbing sites play an important role in the condensation reaction. If BA adsorbed on these sites condenses with BI, a decrease in BA coverage in the presence of base could then account for the decrease in condensation.

Table 4: BA Adsorption sites on unpromoted and doubly-promoted catalysts.

Catalyst	% of BA sites blocked by OH-	Total BA ads sites, mmol/g	Ni sites* mmol/g
Fe/Cr Promoted	56 +/- 15 %	0.09	1.33
Unpromoted+	34 +/- 5%	0.37	1.04

* measured by CO adsorption; + average of 3 experiments.

Table 4 shows the total number of BA adsorption sites on unpromoted and doubly promoted catalysts as well as the percentage of BA adsorption sites blocked by base. The doubly-promoted catalyst had a significantly smaller BA uptake than the unpromoted catalyst, consistent with its greater resistance to BA poisoning mentioned earlier. The saturation uptake, extrapolated from the isotherms, was 0.09 mmol/g for the doubly-promoted catalyst compared to 0.37 mmol/g for the unpromoted catalyst. Approximately 40 percent of the total BA adsorption sites were blocked by hydroxyl ions.

Conclusions

The addition of iron or iron and chromium as promoters enhanced the catalyst activity. The addition of promoters had only a small effect on the selectivity and CO uptake, but subtantially increased BET areas.

XPS studies showed that storage of catalysts in dilute sodium hydroxide solutions protects them from hydrolysis or oxidation. Surprisingly, the primary Al state in the active surface is Al°.

The addition of sodium hydroxide to the reaction mixture was found to have the strongest effect on reaction behavior, causing an order of magnitude decrease in the amount of by-products formed. Liquid phase BA adsorption isotherms suggest that hydroxyl ions compete with BA for some adsorption sites. BA also competes with CO and hydrogen for adsorption on additional sites.

Studies at 14 bar hydrogen indicate that the reaction order varies from half to first order in BN concentration as the reaction temperature increases. When the pressure was increased, the reaction rate was enhanced, but the selectivity remained about the same.

Acknowledgments

Special thanks to DuPont Nylon for funding this research and to Steve Schmidt at GRACE Davison for providing us with alloy samples.

References

1. K. Weissermel and H.J. Arpe, *Industrial Organic Chemistry*, Third Ed., VCH Publishers, New York, 249 (1997).
2. F. Medina, P. Salagre, and J.E. Sueiras, *Journal of Molecular Catalysis*, **81**, 363-371 (1993).
3. C.D. Bellefon and P. Fouilloux, *Catal. Rev.--Sci. Eng.*, **36**, 459-506 (1994).
4. J. Volf and J. Pasek, in *Catalytic Hydrogenation* (L. Cerveny, ed.), Elsevier Science Publishers, Amsterdam, 105-144 (1986).
5. S.R. Schmidt, *Catalysis of Organic Reactions* (M.G. Scaros and M.L. Dranier, eds.), Marcel Dekker, Inc., New York, 45-59 (1995).
6. C.D. Wagner, W.M. Riggs, L.E. Davis, J.F. Moulder, and G.E. Muilenberg, *Handbook of X-Ray Photoelectron Spectroscopy*, Perkin-Elmer Corporation, Eden Prairie, Minnesota (1979).
7. F. Delanny, J.P. Damon, J. Masson, and B. Delmon, *Applied Catalysis*, **4**, 169-180 (1982).

The Selective Hydrogenation of Functionalized Nitroarenes: New Catalytic Systems

Urs Siegrist*, Peter Baumeister, Hans-Ulrich Blaser and Martin Studer

Scientific Services, Novartis Service AG, CH-4002 Basel, Switzerland

Abstract

The development of two novel heterogeneous catalyst systems, Pt-Pb-CaCO$_3$ and Pt-C modified with H$_3$PO$_2$, for the chemoselective hydrogenation of several functionalized nitro compounds to the corresponding anilines is presented. The Pt-Pb-catalyst gives the best results in aprotic polar solvents in presence of FeCl$_2$ and tetrabutylammonium chloride. The hypophosphorous acid modified Pt-charcoal catalyst works best in toluene-water in the presence of VO(acac)$_2$ as promoter. Both catalyst systems are able to hydrogenate aromatic nitro compounds containing functional groups such as iodide, C=C, C≡N and even C≡C. The accumulation of hydroxylamines can be suppressed by iron or vanadium promoters. A working hypothesis to explain the effects of the modifiers as well as promoters is presented.

Introduction and Background

The catalytic hydrogenation of aromatic nitro compounds is a well-known and established method for the large-scale industrial manufacture of aromatic amines, key intermediates in the synthesis of a wide variety of fine chemicals (1). The hydrogenation of simple nitroarenes is relatively unproblematic and commercially available heterogeneous catalysts can be used. However, the selective reduction of a nitro group in the presence of other reducible functions is very often difficult, and especially olefinic and halogen functions will be hydrogenated simultaneously with the nitro group when transition metal catalysts are applied (2). Because functionalized anilines are important intermediates for a number of herbicides (3), pharmaceuticals, polymers and dyestuffs, we have studied the modification of commercial heterogeneous catalysts in order to obtain better chemoselectivities.

An especially challenging problem was the chemoselective hydrogenation of the allyl-nitrobenzoate 1a to the corresponding aniline 2a, an intermediate for a new herbicide of Novartis (Scheme 1). A search of the literature showed that mainly stoichiometric reducing agents such as iron (3), tin (6), sodium hydrosulfite (4) or zinc in ammonium hydroxide (5) have been used successfully to reduce aromatic nitro compounds containing olefinic bonds. While the reported yields ranged from poor to excellent, most of these reductions operated either in high dilution, required a large excess of reducing agent or involved filtration of colloidal precipitates. For the *catalytic hydrogenation* we found only two publications: Braden et al. (7) reported cobalt and ruthenium sulfide catalysts for the hydrogenation of nitro groups in the presence of C=C-bonds, and Onopchenko et al. (8) described Ru/C catalysts for the hydrogenation of nitro-phenylacetylene. When we tested these catalysts they were indeed reasonably selective concerning hydrogenation of the C=C bond, but the yield was very low (Table 1). With CoS_x and RuS_x catalysts, sulfur containing byproducts were formed in the course of the hydrogenation.

The present contribution describes the development of two, economically and ecologically feasible catalyst systems for the production of aniline **2a**. In addition, we show their scope and limitations for the preparation of aromatic amino compounds substituted with functional groups such as C=C, C≡C, C-X, C=N and C≡N bonds.

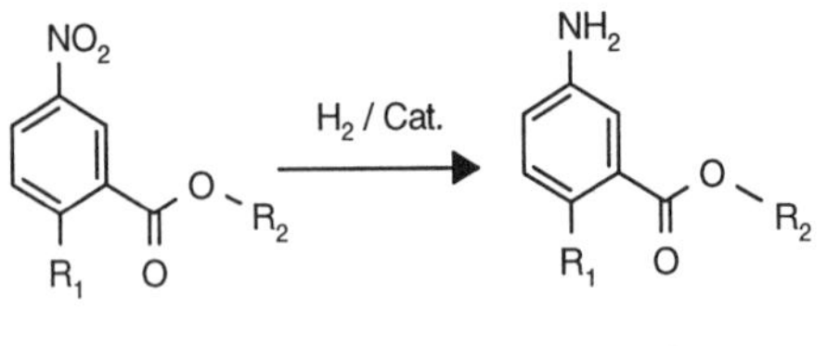

1a,2a,3a R_1: Cl

1a,2a R_2: $C(CH_3)_2COOCH_2CH=CH_2$

3a R_2: $C(CH_3)_2COOCH_2CH_2CH_3$

1b,2b,3b R_1: H

1b,2b R_2: $CH_2CH=CH_2$

3b R_2: $CH_2CH_2CH_3$

Scheme 1: Reaction scheme and structures of the starting materials **1**, products **2**, and over-hydrogenated byproducts **3**.

Experimental

Materials

All chemicals used in the hydrogenation studies were of reagent grade quality or better and used without further purification. Calibration compounds used for analysis were prepared in pure form using literature procedures. **1a** was prepared as described in (3). **1b** was synthesized from m-nitro-benzoylchloride and allyl alcohol. The supported metal catalysts were obtained from Degussa Corporation.

Typical hydrogenation procedures

Using Pt-Pb-CaCO₃-catalysts. The nitro compound and the catalyst were placed in a stainless steel autoclave equipped with a hollow shaft stirrer. After the addition of the solvent and further additives, the autoclave was sealed, flushed three times first with 10 bar nitrogen, then with hydrogen and finally charged with hydrogen to the specified reaction pressure. The hydrogenation mixture was heated to the required temperature and stirred magnetically at constant H_2-pressure until the hydrogen up-take stopped (pressure drop in a gauged H_2-reservoir). The autoclave was cooled, flushed three times with nitrogen and opened. After the filtration, the solvent was removed and the conversion was determined in the crude reaction product.

Using the Pt/C-phosphite catalyst system. The aqueous slurry of the Pt/C-catalyst was magnetically stirred in a small glass-pot with the required amount of modifier (e.g. hypophosphorous acid). After 10 min., the vanadium co-catalyst was added to the mixture and stirring was continued for an additional 10 min. The catalyst slurry, the solvent and the nitro compound were placed the in autoclave and the reaction was carried out as described above.

Analytical procedures

• *Glc*: Column: Carbowax 20M 10m, temperature program: 120°C to 200°C, ΔT: 5°C/min. Used for a quick estimation of yield and selectivity (area %).
• *Hplc:* Column: Nucleosil C 18, 5μm, 250x4mm, temperature: 35°C, flow-rate 10ml/min, injection volume: 10μl, mobile phase: A acetonitrile, B 0,1% phosphoric acid, gradient: 0-30 min: A/B 20/80; 30-45 min: A/B 90/10; 45-50 min: A/B 20/80. Used with internal standard for the quantitative determination of yield and product composition.
• *Tlc:* Plates: Kieselgel 60F254 (Merck), mobile phase: toluene-ethylacetate = 7:3. Used for the semi-quantitative estimation of the (unstable) hydrogenation intermediates and the cleavage byproducts (with reference mixtures).

Results

Preliminary Catalyst Screening

First, Co and Ru sulfide (7), Raney-nickel as well as several types of noble metal catalysts were screened for their selectivity in the hydrogenation of **1a** and of the model compound **1b** (Table 1). As expected, the C=C bond was easily hydrogenated by most catalysts. Only the Co and Ru systems (exp. 1'1 - 1'3) showed a high selectivity. However, their activity was unsatisfactory and the yields very low. The main side reactions were hydrolysis of the allylester to give m-aminobenzoic acid and allyl alcohol, and the well known metal catalyzed formation of N-allylated byproducts [20]. The results with commercial Ni (1'4), Pd (1'6-1'8) and Pt (1'9-1'12) catalysts from various catalyst suppliers were catastrophic and confirmed our expectations.

Fortunately, there were also some encouraging signals. First, the addition of Bi to a Pt-C catalyst (1'12) and also to the Lindlar catalyst (1'7, 1'8) improved the selectivity. Similarly, the modification with dicyandiamide (DCDA) increased the selectivity especially of Raney-nickel (1'5) but also of Pt-C (1'10). However, the yields were again not satisfactory, especially for the DCDA / Raney-nickel system. In addition, an extraordinarily high hydroxylamine accumulation and a Ni-leaching of nearly 50% made this catalyst very unattractive.

Nevertheless, these results convinced us to further pursue two approaches:
    i) the modification of Pt catalysts with lead,
    ii) the addition of phosphorous process modifiers to Pt catalysts

Lead modified platinum catalysts

When we started our studies with lead modified Pt-catalysts, we realized from the

Table 1. Catalytic hydrogenation of **1a** and **1b** in THF, catalyst screening

Entry	Catalyst/Modifier	Sub.	Temp. [°C]	pH$_2$ [bar]	H$_2$ up take [%]	Yield [2]+[3] [%]	Sel. [2]/([2]+[3]) [glc%]
1'1	CoSx[f] [7]	1a	110	65	102	58	> 99
1'2	RuSx[b] [7]	1a	80	10	99	30	97
1'3	5%Ru-C-H$_2$O[b] [8]	1b	120	10	27	n.d.	> 95
1'4	Raney-nickel[b]	1b	25	1	99	n.d.	44
1'5	1'5 + DCDA[d]	1a	120	6	112	55[a]	97[a]
1'6	5%Pd-CaCO$_3$[b]	1b	25	0.1	103	n.d.	0
1'7	5%Pd,3%Pb-CaCO$_3$[d]	1b	25	0.1	76	n.d.	3
1'8	1'7+ MnCl$_2$/Quinoline	1b	60	2	26	n.d.	32
1'9	5%Pt-C[c]	1b	25	5	86	n.d.	0
1'10	1'9 + DCDA[e]	1b	60	5	89	n.d.	91
1'11	5%Pt-C sulf[b]	1b	25	10	106	n.d.	0
1'12	3%Pt,3%Bi-C[c]	1b	70	10	66	n.d.	66

Yield and Selectivity via glc a) Hplc with internal standard. Catalyst / substrate ratio:
b) 10% w/w; c) 5% w/w; d) 20% w/w; e) 50mol%; f) 25% w/w

beginning that both *catalyst activity and selectivity* would be critical issues. The first experiments with a 5%Pt-3%Pb-CaCO₃-catalyst were quite encouraging. While the conversion to **2b** was indeed very low (< 10%), the catalyst was very selective concerning the double bond. Hydrodehalogenation as a further possible side reaction was never detected. We therefore decided to further optimize this catalyst in collaboration with Degussa AG and Table 2 summarizes relevant results. $CaCO_3$ was the best carrier material, other supports (e.g. charcoal or alumina) gave catalysts with low selectivity. The *lead content* was very important; the highest activity as well as a good selectivity were obtained with a lead content of 1% (2'2) and this combination was used for further investigations. In general, to get a reasonable catalytic activity, a *reaction temperature* ≥ 120°C was necessary.

The catalysts were prepared according to the Lindlar procedure (9,10), but it was difficult to achieve a constant performance. The selectivity was reproducible but the catalyst activity and the leaching of lead during the hydrogenation varied from sample to sample. These problems are not yet solved completely.

The following extensive screening of solvents, additives as well as an optimization of the reaction conditions were carried out with substrate **1a** using a 5%Pt-1%Pb-CaCO₃ catalyst. Selected results are listed in Tables 3 - 5. The polarity of the *solvent* had a strong influence on both catalyst activity and yield (Table 3). The best over - all results were obtained with ethylmethyl-ketone (MEK) and the solvent mixture THF/i-PrOH. The fact that MEK is not hydrogenated confirms the extraordinary selectivity profile of this catalyst. Unfortunately, the reaction of **1b** with MEK to the imine in the amount of 3-6%

Table 2. Effect of the lead content on catalyst activity and selectivity in the hydrogenation of **1b**

Entry	Catalyst	Temp.	pH_2	Time	H_2 up take	Sel. [2]/([2]+[3])
	5%Pt-...-CaCO₃	[°C]	[bar]	[h]	[%]	[glc%]
2'1	-3%Pb-	160	5	4.5	9	100
2'2	-1%Pb-	120	20	5	99	99.7
2'3	-0.75%Pb-	120	20	10	93	98.2
2'4	-0.5%Pb-	30	20	2	100	0

Reaction conditions: [**1b**] = 10% in THF; [cat] = 10%w/w of **1b**

Table 3. Catalytic hydrogenation of **1a**, solvent screening

Entry	Solvent	Time [h]	Yield of 2a[d] [%]	Sel. [2]/([2]+[3]) [glc%]
3'1	toluene[b]	19[a]	41	99.8
3'2	MeOH	1.5	64	99.6
3'3	EtOH	1	54	99.7
3'6	t-BuOH	5	74	99.5
3'7	THF	15	84	99.5
3'9	ethylacetate	31	88	99.6
3'10	MEK	3	79	99.8
3'11	N-methyl-2-pyrrolidone	1	65	99.7
3'12	N,N-dimethylformamide	1	69	99.8
3'13	THF/n-PrOH=4:1[c]	15	88	> 99.9
3'14	acetonitrile	23	n.d.	> 99.9

Reaction conditions: [**1a**] = 15% w/w of solvent; 5%Pt-1%Pb-CaCO$_3$: 2%w/w of **1a**; Temp.: 140°C; pH$_2$: 20 bar. a) conversion: 47%; b) cat.: 5%w/w; c) cat.: 1% w/w; d) hplc of the reaction mixture, estimated value

Table 4. Catalytic hydrogenation of **1a**, effect of FeCl$_2$, TMAC and water

Entry	Addition of FeCl$_2$ [mol% of 1a]	Addition of TMAC [mol% of 1a]	Addition of water [mol% of 1a]	Time [h]	Yield of 2a[a] [%]
4'1[b]	-	-	-	3	79
4'2	0.1	-	-	13	87
4'3	0.1	0.1	-	10.5	90
4'4	0.1	0.5	-	4	71
4'5	0.25	0.1	-	8	90
4'6	0.5	0.1	-	5	92
4'7	0.5	-	-	5.5	88
4'8	0.5	0.1	55	5.5	91
4'9	0.5	0.1	110	9.5	91
4'10	0.5	0.1	220	18	86

Reaction conditions: Solvent: MEK; Temp.: 140°C; pH$_2$ = 15 bar; Cat.: 5%Pt-1%Pb-CaCO$_3$, 0.5%w/w of **1a** a) estimated values b) 2% catalyst, pH$_2$ = 20 bar

could not be eliminated completely. The highest hydrogenation rate was observed in polar solvents such as alcohols or amides (3'2,3'3,3'11,3'12). While the selectivity towards the allylic bond was comparable to MEK, the yields were much lower due to the formation of hydrazo and azoxy products, a consequence of hydroxylamine accumulation.

This accumulation of hydroxylamine, azoxy and hydrazo compounds with modified catalysts was not unexpected. Sine *promoters* such as NH_4VO_3, $FeCl_2$ are known to improve the selectivity of nitroarene reduction (11,12), we tested several additives. It was found that the Pt-Pb-catalyst is very sensitive, e.g. catalytic amounts of CuCl, NH_4VO_3 or $VO(acac)_2$ led to a complete deactivation. However, the addition of small amounts of $FeCl_2$ and tetramethylammonium chloride (TMAC) had a beneficial influence on the hydrogenation rate and to a lesser extent also on the yield (Table 4). Other iron salts such as $FeSO_4$ or Fe(II)oxalate were less effective. It seems that the Fe^{2+} is responsible for the better selectivity, whereas the increased activity is mainly due to Cl⁻, a possible indication of an electrochemical reaction mechanism.

The *water content* of MEK also influenced the catalyst performance, e.g., 2% water in MEK prolonged the hydrogenation time by a factor of 3-4 and decreased the yield to 86% (4'10). This effect is not understood so far, but as a

Table 5. Scope of the new „Pt-Lindlar-catalyst".

Entry	R_1	R_2	R_3	Solv.	Temp. [°C]	Time [h]	Yield [%]
5'1[a)e)]	3-CH=CH₂	-H	-H	A	120	21	75
5'2[a)e)]	3-CONHCH₂CH=CH₂	4-Cl	-H	B	140	38	70
5'3[b)e)]	3-COOCH₂C≡CH	-H	-H	A	140	14	65[g)]
5'4[a)e)]	4-CH=NOH	-H	-H	C	120	8.5	51
5'5[d)e)]	2-NH₂	3-J	5-Cl	B	140	6	>90
5'6[d)f)]	2-CH₂CN	5-F	-H	B	140	4	75
5'7[a)e)]	2-Br	5-COCH₃	-H	C	120	37	83
5'8[a)e)]	2-Cl	3-NO₂	5-CN	A	120	19	88[h)]

Reaction conditions: $pH_2 = 20$ bar; [5%Pt-1%Pb-CaCO₃]: a) 2%w/w to nitro, b) 5%w/w, c) 20%w/w, $FeCl_2$: e) 0.5mol% to substrate, f) 1.5mol%; solvent: A=MEK, B=THF/n-PrOH 5:1, C=THF; g) propargyl/allyl 4:1 (after purification by chromatography), h) isolated 2,4-diamino-3-chloro-benzonitrile

consequence, the water content in MEK should not be higher than 0,5%. Due to the observed increase in the hydrogenation activity, the catalyst loading could be lowered to a technical feasible 0.5%w/w of 5%Pt,1%Pb-CaCO$_3$. Moreover, the catalyst could be recycled with the same selectivity and only a small loss in activity.

The preparative scope of the catalytic system is shown in Table 5. Except for the C≡C bond, all other functions are not reduced. The sometimes low yield is due to side products and intermediates. Noteworthy is the high yield of 2,4-diamino-3-chloro-benzonitrile without ring formation with the cyano group (entry 5'8).

H$_2$PO$_3$ modified Pt-catalysts

Process modifiers have been used successfully to enhance catalyst selectivity for the hydrogenation of various substituted nitro compounds (18,19). In the context of our problem we were especially attracted by some results of Kosak, who described Pt/C modified with H$_3$PO$_3$ as selective catalyst for the hydrogenation of *iodo*nitrobenzenes (13). When we tried *phosphorous based modifiers* for the hydrogenation of **1a**, the modified Pt/C catalyst showed outstanding selectivity towards the allylic double bond (Table 6). While the best results were obtained with H$_3$PO$_2$, other additives such as H$_3$PO$_3$, (PhO)$_2$P(O)H and HPPh$_2$ showed also an excellent selectivity. P(OPh)$_3$ and especially PPh$_3$ were less efficient. Kosak (13) postulated that the P(H)(O)OH-structure is crucial for the inhibition of hydrodehalogenation. Our results indicate to get a very high chemoselectivity for the hydrogenation of NO$_2$ vs. C=C the presence of a P-H bond is important. As expected, the *H$_3$PO$_2$ concentration* had a strong effect on rate and selectivity (Table 7). The critical level was at 2,5% H$_3$PO$_2$ relative to the Pt-catalyst,

Table 6. Catalytic hydrogenation of **1a** with Pt/C: Screening of P-modifiers

Entry	Modifier	Phosphorous/Platinum weight ratio	Sel. [2]/([2]+[3]) [glc%]
6'1[a]	H$_3$PO$_3$	1	98.6
6'2[a]	H$_3$PO$_2$	1	> 99.9
6'5[b]	(PhO)$_2$P(O)H	5.6	99.7
6'6[b]	P(OPh)$_3$	5.6	96
6'7[b)c]	HPPh$_2$	5.6	99.1
6'8[b)c]	PPh$_3$	5.6	48

Reaction conditions: Temp.: 100°C, pH$_2$ = 20 bar, cat.: 5%Pt/C, solvent: toluene. a) cat.: 1%w/w of **1a**, b) 0.5%w, c) with of NH$_4$VO$_3$-modified charcoal

Table 7. Catalytic hydrogenation of **1a**, effect of H_3PO_2 concentration

Entry	H_3PO_2 (% w/w of catalyst)	mol H_3PO_2/ mol Pt	Complete conversion [h]	Sel. [2]/([2]+[3]) [glc%]
7'1	2.5	1.5	3.25	96.5
7'2	5	3	7	> 99.9
7'5	10	6	7	> 99.9
7'6	213	125	6.5	> 99.9

Reaction conditions: Temp.: 100°C, pH_2 = 5 bar, cat.: 5%Pt/C, 1%w/w of **1a**, [**1a**] = 16%w/w in toluene

optimum results were obtained with 5% or higher H_3PO_2 concentrations (7'5, 7'6).

The influence of different *solvents* on catalyst activity and selectivity was not studied systematically. We found that polar and protic solvents are not favorable for Pt-C modified with H_3PO_2, in contrast to Pt-Pb-CaCO$_3$. With respect to process integration, the preferred solvent was toluene. Concerning reproducibility and practicability, best results were obtained with a two phase mixture of toluene/water 9:1.

As already reported by Kosak (13]) and Craig et al. (14), the catalyst modification very often causes the accumulation of hydroxylamine intermediates. Metastable hydroxylamines not only lead to azoxy and hydrazo byproduct formation, sluggish hydrogen consumption at the end of the reaction but, even worse, might also be responsible for runaway reactions.

According to TLC results, more than 20% hydroxylamine accumulated with the H_3PO_2 modified Pt-system. H-NMR results support these findings even though an exact determination was difficult due to the instability of the hydroxylamine. But it was very clear that the hydroxylamine problem had to be solved before a technical application of this catalytic system was feasible. Fortunately, in a study that was carried out almost simultaneously, Studer and Baumeister (12]) found vanadium compounds to be very effective for the prevention of hydroxylamine accumulations. When different vanadium salt were added to the H_3PO_2 modified catalyst system, small amounts of vanadyl acetylacetonate indeed reduced the hydroxylamine accumulation to less than 1%! Later we found that the *modification procedure* is also very important to get reproducible results. The best performance was achieved by first stirring the Pt-charcoal catalyst in an aqueous suspension with H_3PO_2 for approximately 10 minutes, then adding the VO(acac)$_2$ and stirring for additional 10 minutes. In contrast, an in situ modification led to a much lower selectivity.

Table 8. Scope of the new Pt/C-H₃PO₂ catalyst

Entry	R₁	R₂	Yield [%]
8'1	3-CH=CH₂	H	95
8'2	3-COO-C(CH₃)₂COOCH₂CH=CH₂	4-Cl	98
8'3	3-CH=CH-COOH	H	95
8'4	3-COOCH₂C≡CH	H	99
8'5	Br	COCH₃	93
8'6	4-CH=NOH	H	> 50

Reaction conditions: [5%Pt/C]: 1%w/w of substrate; [H₃PO₂]: **10%w/w of catalyst;** [VO(acac)₂]: **16%w/w of catalyst;** solvent: toluene; T: 100°C; p: 5bar H₂

This breakthrough allowed a technically feasible process (15). The major advantages of the Pt/C-H₃PO₂ system compared to the Pt-Pb-system are a higher selectivity and activity and the fact, that standard Pt/C-catalysts can be used. Also, a higher space-time yield can be obtained because of higher substrate concentrations. Under optimized reaction conditions, the reduction of **1a** with this Pt-H₃PO₂-VO(acac)₂-C-catalyst at 100°C and 5 bar H₂ afforded **2a** in 100% conversion, > 98% yield, and > 99,9% selectivity, without any trace dehalogenation. This outstanding selectivity is also visible from the typical

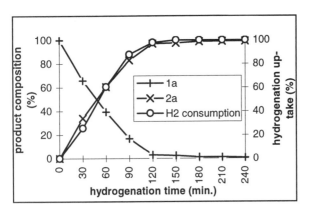

Fig. 1. Typical concentration profiles for the hydrogenation of **1a**

reaction profile (Fig.1). The hydrogenation stops completely after 100% hydrogen consumption. The fact that the formation of **1b** and the hydrogen uptake occur in parallel, confirms that no hydroxylamine accumulates.

Again, various functionalized nitro-compounds were hydrogenated with this new catalyst system to the corresponding anilines. We never detected any reduction of the second function, even a C≡C- bond remained completely unreduced! Some selected results are listed in Table 8.

Mode of Action of the Modified Catalysts

The following working hypotheses have been developed to explain the most important effects described above:

Pt-Pb-catalyst. We think that a layer of lead forms on top of the platinum (see Fig.2). This prevents the adsorption of the nitro as well as other reducible groups on Pt active sites whereas we think that the smaller hydrogen molecule can still reach the Pt and dissociate. Because the nitro group is a very strong oxidant, it can be reduced without adsorption on the surface by an electrochemical mechanism (21). The hydrogenation of the olefinic function however, can only proceed via a classical hydrogenation mechanism that requires adsorption on the active Pt site.(16). This model is in agreement with the effect of the Pt/Pb ratio; the promotion by Cl-anions, the fact that apolar solvents are not suitable and also with observed selectivity regarding C=O. C≡N and C≡C.

Pt-H₃PO₂-system. Here, a complete blocking of the Pt active sites is less probable. The observed dependence of activity and selectivity on the solvent type and the modifier concentration is more in agreement with a reversible modification of the active sites. However, low valent phosphorous compounds are known to decompose on the surface of hydrogenation catalysts (13) thereby modifying the

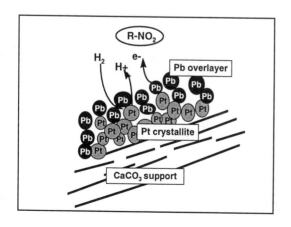

Fig. 2. Proposed mode of action of the Pt-Pb-CaCO₃-catalyst

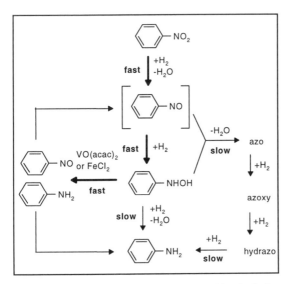

Scheme 2. Reaction scheme of nitro group reduction (17)

electronic structure of the surface and/or isolating active surface atoms. If the hydrogenation of a C=C bond requires the adsorption on an ensemble of metal atoms, then a similar mechanism as described above could be responsible for the observed effect.

While there are no details available on the mode of action of the promoters, it is interesting to speculate what their function might be. Catalytic hydrogenations of aromatic nitro compounds often have two different kinetic phases. First, the nitro compound is reduced rapidly to the corresponding hydroxylamine. Which then reacts to the corresponding aniline at a much lower rate. With vanadium promoters, the second phase is less pronounced or disappears. This suggests a mechanism which could be called 'catalytic by-pass': The hydroxylamine disproportionates rapidly (left part Scheme 2) and the nitroso intermediates re-enters the catalytic cycle. As a consequence, the amine formation is accelerated. Nitroso intermediate are not observed due a strong adsorption and fast hydrogenation reaction of the nitroso compound.

Acknowledgments

We would like to thank R. Hanreich, J. Wenger and W. Kunz for their continuous encouragement and support during the course of this project, and W. Kamke and W. Schmidt for their very careful experimental work. We also would like to acknowledge the very fruitful collaboration with Degussa AG, Hanau.

References

1　A. M. Strätz, *Catalysis of Organic Reactions*, Chem. Ind. 18: p335, Marcel Dekker, New York 1984. R.S. Downing, P.J. Kunkeler , H. van Bekkum, *Catal. Today* **37**, 121 (1997).

2　P. Rylander, *Catalytic Hydrogenation in Organic Synthesis*, p. 122 Academic Press, New York (1979).

3　M. Suchy, P. Winternitz, M. Zeller, WO Pat. 91/00278 to Ciba-Geigy (1991).

4　R.F. Kovar, F.E. Armond, U.S. Pat. 3,975,444 to the U.S. Air Force (1976).

5　A. Buraway and J.P. Critchley, *Tetrahedron*, 340 (1959).

6　J. Butera, J. Bagli, WO Pat. 91/09023 to American Home Products (1991).

7　R. Braden, H. Knupfer, S. Hartung, U.S. Pat. 4,002,673 and 4,051,177 to Bayer AG (1977).

8　A. Onopchenko, E.T. Sabourinand, C.M. Selwitz, *J. Org. Chem.*, **44**, 1233 (1979).

9　H. Lindlar, *Helv. Chim. Acta.*, **35**, 446 (1951).

10　U. Siegrist, P. Baumeister , WO Patent 95/32941 to Ciba-Geigy (1995).

11　R.L. Seagraves, U.S. Pat. 4,212,824 to Du Pont de Nemours (1980).

12　M. Studer, P. Baumeister; WO Patent 96/36597 to Ciba-Geigy AG (1996).

13　J.R. Kosak, *Catalysis in Organic Synthesis*, W.H. Jones (ed.), p. 107, Academic Press, New York (1980).
　　J.R. Kosak, U.S. Pat. 4,020,107 to Du Pont de Nemours (1977).

14　W.C. Craig, G.J. Davis, P.O. Shull, U.S. Pat. 3,474,144 to GAF Co. (1969).

15　P. Baumeister, U. Siegrist and M. Studer unpublished results

16　I. Horiuti, M. Polanyi, *Trans. Faraday Soc.*, **30**, 1164 (1934).

17　K. Haber, *Z. Electrochem.*, **4**, 506 (1898).

18　J.R. Kosak, *Ann. N. Y. Acad. Sci.*, **172**, 174 (1970).

19　P. Baumeister, H.U. Blaser, W. Scherrer, *Stud. Surf. Sci. Catal.*, **59**, 321 (1991).

20　Y. Watanabe, Y. Tsuji, H. Ige, Y. Ohsugi, *J. Org. Chem.*, **49**, 3359 (1984).

21　M. Heyrovsky, S. Vavricka, *J. Electroanal. Chem.*, **28**, 411 (1970).

Catalytic Hydrogenation of Viscous Materials in a Twin Screw Extruder with an Immobilized Palladium Catalyst

M. Ebrahimi-Moshkabad and J. M. Winterbottom
School of Chemical Engineering, The University of Birmingham, Edgbaston, Birmingham, B15 2TT, UK.

ABSTRACT

In this work the technical feasibility of using an intermeshing co-rotating twin screw extruder (ICoTSE) as a three phase reactor for highly viscous solutions was investigated. The model reaction was heterogeneous hydrogenation of dimethyl itaconate in the present of a palladium catalyst. A mixture of water and propan-2-ol was used as solvent and to increase viscosity of the solution different amount of carboxymethyl cellulose (CMC) were dissolved in the solution.

Immobilisation of palladium catalyst on the screws was also studied. To find the optimum conditions for the immobilisation, small mild steel and alumina plasma coated mild steel plates were used. Two basic coating methods, electroless and electroplating, were applied to immobilise the catalyst. It is found that the alumina plasma coated screws further coated with palladium using electroless method were significantly more active than the mild steel screws without alumina coating. The activity of palladium electroless coated catalyst was very much less than the others.

Some continuous experiments were carried out in the ICoTSE and reasonable conversions (>50%) were obtained. The behaviour of the extruder was found to be very similar to a perfect plug flow reactor by performing some residence time distribution (RTD) measurements.

1. INTRODUCTION

Using extruders as chemical reactors (REX) in polymer industries has been known for some years. I. G. Farbenindustrie patented some work in the 1940s and issued it to BASF in 19953 (1). Due to the industrial importance of REX most of findings have been patented (2). The main advantages of extruders as chemical reactors are the capability of very good mixing and handling of viscous materials, high pressure and temperature operation, multistaging capability and near plug flow reactor behaviour. By applying REX in a chemical process which deal with viscous solutions the whole process can be intensified.

REX has increasingly been studied in recent years. These investigations mainly deal with applying different types of extruders and reactions in polymer, food and pharmaceutical industries (3). Most of the reactions were in one or two phases (2). Jordi et. al. (4) introduced a twin screw extruder as a three phase reactor for hydrogenation of soybean oil in 1994. However, since soybean oil is not viscous at high temperature (150 °C), the conventional reactor (an stirred tank reactor) was found to be more economical than the extruder.

Recently we (5) carried out the hydrogenation of dimethyl itaconate in a solution of water and propan-2-ol where viscosity was adjusted by adding different amount of a type of carboxymethyl cellulose (CMC) in an intermeshing co-rotating twin screw extruder (ICoTSE) using a 5 wt% Pd/C catalyst in slurry solutions. The results showed the superiority of the ICoTSE over a stirred tank reactor (STR) when the viscosity of the solution was relatively high (> 0.5 % CMC solution, i. e. 40 cp.). The reaction rate in the STR decreased from 9.6 $mol/m^3.s$ to nearly zero when the solution changed from 0 % to 1.5 % CMC solution, while in the ICoTSE this decrease was relatively small and it approached a constant value (o.6 $mol/m^3.s$).

To make the ICoTSE more suitable for continuous operation for three phase reactions, various methods of immobilisation of palladium (as catalyst) on the surface of the screws of extruder were investigated. The flow characteristics of the ICoTSE in continuous operation were determined using residence time distribution (RTD) measurements.

2. MODEL REACTION

The hydrogenation of dimethyl itaconate (DMI) on a palladium catalyst was chosen as a model reaction since it is a fast, first order reaction (5). A type of carboxymethyl cellulose (CMC) was added to the solution to make a wide range of viscous solutions. The solvent was a mixture of water and propan-2-ol because DMI is not soluble in water and CMC is not soluble in Propan-2-ol.

3. EXPERIMENTAL

Two series of experiments were designed to investigate the performance of the ICoTSE with immobilised screws as a three phase reactor. In the first set of experiments different methods of coating were applied on small plates to find the optimum conditions of palladium immobilisation. Then the optimum method and conditions were used to immobilise the catalyst on the surface of screws and the feasibility of using the extrude with immobilised catalyst for a three phase reaction was investigated. In the last series of experiments the behaviour of the extruder for continuous operation was studied by performing some RTD measurements.

3.1 Electroplating on Mild steel

Small mild steel plates (2.8 cm x 2.2 cm) with 1 mm thickness were supplied as substrate. The plates were cleaned in a decon 90 solution, sandblasted and degreased in refluxing acetone. The plating solution used was a 5 g/l PdCl2 in ammonia solution (28%) and the procedure of Manolatose et. at. (6) was followed. The current density was adjusted between 2-3 mAm.cm^{-2} and the duration of plating varied from 1.5 to 10 minutes.

3.2. Electroless Coating on Mild Steel

The substrate was the same as electroplating method. Cleaning, sandblasting and de-greasing were performed as above method. To study the effect of pickling the surface with HCl and pre-activation with SnCl2 and PdCl2, these two steps were applied to some of the plates.

The coating solutions used were sodium tetrachloropalladate (II) (Na_2PdCl_4) and sodium chloride in water or sodium tetrachloropalladate in methanol. To optimise the conditions of the coating, the effect of concentration of palladium salt, pH and temperature of the solution was studied. Hydrochloric acid and sodium hydroxide were used to adjust the pH of the solution at the beginning of the experiments.

3.3. Electroless Coating on Alumina Pre-coated Mild steel

In these experiments , some of the plates first were coated with a thin layer of alumina using a plasma spray method. The thickness of the alumina coating should be less than 50 μm to keep enough tolerance between the screws. The powder sprayed was a supper fine α-alumina powder.

The alumina coated mild steel plates were coated using an electroless coating solution. The coating solution was the same as the electroless coating on mild steel and in these series of experiments hydrazine hydrate was used as a reducing agent. Some of the alumina coated plates were treated with aluminium nitrate and calcined at 350 °C for a few hours before palladium coating to produce γ-alumina surface.

3.4. Hydrogenation of the Model Reaction

The model reaction was carried out in two kinds of reactors, i. e. a stirred tank reactor (STR) and the ICoTSE. The activity of the immobilised catalyst was compared in these reactors and the continuous performance of the ICoTSE was studied in more details.

3.4.1. High pressure STR

The STR was a 500 cm^3 stainless steel autoclave which equipped with a changeable impeller blades (Fig. 1). The coated plates were fixed on this impeller and used as catalyst for the hydrogenation of DMI. The reactor was operated as a dead-end reactor and the reaction rates were compared as the representative for overall activity of the catalyst.

3.4.2. The ICoTSE

The extruder used in this work was an intermeshing co-rotating twin screw extruder with a 16 mm diameter and a length to diameter of 20 to 1. The barrel was made from stainless steel and its cross section was "figure 8" shape. The screws were actually a pair of shafts with a number of semi-eliptical shaped paddles on each of these shafts. The paddles were at right angle to each other on their individual shafts and also at right angles to their corresponding paddle on the other shaft (Fig. 2).

The solution was pumped into the barrel using a positive displacement pump. The pressure inside the barrel was adjusted using a variable presser relief valve that could manually be adjusted during operation. Since the screws were intermeshing, the paddles could thoroughly wipe each other and the fresh surface of immobilise catalyst continually contacted with the reactants. The experiments were carried out in batch and continuos mode.

Fig. 1: The changeable impeller blades.

Fig. 2: The intermeshing twin screws (right angle paddles).

3.4. RTD measurements

Potassium chloride (KCl) was used as a tracer and it was injected into the extruder exactly above the entrance of the barrel to study the behaviour of the ICoTSE as a continuous reactor. The most important variables studied were liquid flow rate, screw turning speed and viscosity of the solutions.

3.6. Assessments

The criteria for the best palladium coating were the amount and distribution of palladium over the whole surface, the adherence of the coating and the activity of the immobilised catalyst. The amount and consistency of palladium on the surface were assesses visually and using scanning electron microscopy (SEM) analysis. Adherence of the coating to the surface was assessed with tests such as scraping or wiping the coating and washing with solvents. The surface composition of some alumina coated plates was analysed using X-ray diffraction (XRD).

The activity of the immobilised catalysts was compared using the coated plates as blades of a changeable blade impeller in the autoclave as mentioned already. The reaction rate in the reactors was calculated using the rate of hydrogen consumption or GC analysis of the reaction products.

4. RESULTS AND DISCUSSION

4.1. Electroplating on Mild Steel

The electroplated palladium was a very thin, consistent and coherent film with metallic colour under all plating conditions. The thickness of the coating increased with increasing the current density and the duration of plating. The scanning electron micrograph of one of the plates is shown in Fig. 3. The topology of the coating was very similar to the base before coating.

4.2. Electroless Coating on Mild Steel

In the pre-treatment stage de-greasing and sandblasting of the surfaces were found to be the key processes. The results showed that pickling the surface with hot HCl did not improve the coating. Pre-activation of the surface with SnCl2 and PdCl2 improved only slightly the quality of the coating, because mild steel itself can start the substitution reaction with palladium without any pre-activation treatment.

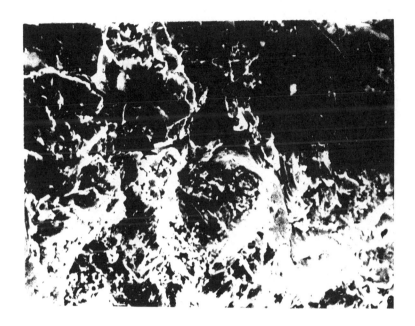

Fig. 3: SEM of palladium electroplated on mild steel (Mag. 1000)

In the coating process, it was observed that the rate of deposition of palladium on the surface in methanol solution was less (about 1/4) than that in water solution. The reasons are likely the difference of polarity of the solvents and their pH. The rate of deposition directly depended on the pH of the solution as reported by other researchers for a Pt/alumina catalyst (7) and decreased with increase of pH. The coating solution was unstable beyond pH = 6.0 and the coating was powdery black for pH $\leq$ 2.5. The optimum range of pH was found to be 3-4.

The visual appearance of successful coatings was a thin, uniform and coherent film with black or black green colour. Figure 4 shows a typical scanning electron micrograph of the coating. The clusters of palladium in this case was completely different from that in electroplating.

4.3. Electroless Coating on Alumina Pre-coated Mild Steel

It was found that de-greasing and sandblasting of the surface were essential steps for preparing an alumina coating with good adherence. X-ray diffractograms showed treating the surface with aluminium nitrate did not change the composition of the surface.

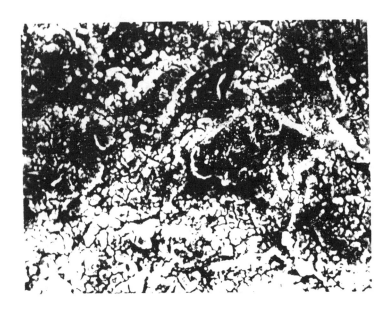

Fig. 4: SEM of palladium electroless coated on mild steel (Mag. 1000)

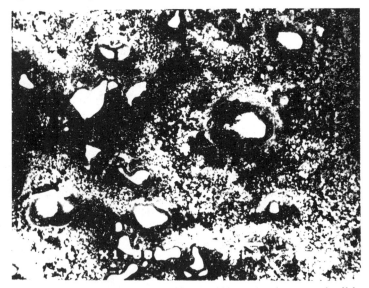

Fig. 5: SEM of palladium electroless coated on alumina coated mild steel
(Mag. 1000).

The adherence of palladium coating to the alumina was better than that for mild steel. It is believed that the structure of the support was the main cause for that. A typical scanning electron micrograph of the coating is shown in Fig. 5. The palladium clusters in this case were much smaller than those in electroless coating on mild steel.

4.4. Comparing Activities in Autoclave

The most successful coated plates from each method were used as catalyst in the hydrogenation of DMI in the autoclave. The reaction conditions and some of the results are summarised in Table 1. The reaction rate was calculated based on geometric area of the blades as catalyst surface area. The results showed that the activity of palladium electroplated was the least and no reaction rate was observed when it was used in the reaction. The reaction rate for palladium coated on alumina coated mild steel was about three times more than that for mild steel. It is believed that the difference in palladium clusters were the cause of this difference. The reaction rates were nearly the same for alumina nitrate treated and untreated plates as expected.

4.5. Reaction Rate in ICoTSE

The reaction rate of hydrogenation of DMI in the ICoTSE with immobilised screws in batch mode compiled in Table 2. It was calculated based on geometric area of the immobilised screws as catalyst surface area. The trend of the results was nearly the same as for the autoclave. Some preliminary continuous experiments were also carried out with viscous solutions. A conversion of more than 50 % was obtained even with a 2 % CMC solution (i. e. about 1100 cp. viscosity) in a short length of the barrel (ca. 260 mm).

Table 1: Initial reaction rate in STR with Immobilised Blades.

$Al(NO_3)_3$ Treating	Alumina Plasma Coating	Electroless Coating	Electroplating	R^*_A $(mol/h.m^2)$
✓	✓	✓	✗	10.48
✗	✓	✓	✗	9.23
✗	✗	✓	✗	3.1
✗	✗	✗	✓	Nil

* DMI in water - propan-2-ol, C_o = 30 g/l, T = 55 °C, P_{H2} = 8.7 atm, N_S = 1000 rpm.

Table 2: Initial reaction rate in ICoTSE with immobilised screws.

Alumina Plasma Coating	Electroless Coating	Electroplating	$R*_A$ (mol/h.m^2)
✓	✓	✗	8.38
✗	✓	✗	3.65
✗	✗	✓	Nil

* DMI in propan-2-ol, C_o = 50 g/l, T = 80 °C, P_{H2} = 7.1 atm, N_S = 250 rpm.

These experiments proved the feasibility of using the ICoTSE with immobilised screws as a continuous three phase reactor for highly viscous solutions.

4.6. RTD Measurements

Some of RTD measurement results are summarised in Table 3. It was found that screw rotation speed affected the RTD, especially when the viscosity of the solution was high. The behaviour of the extruder approached a perfect plug flow reactor (n = 15-22) when a 2% CMC solution was used. That is because of very good radial mixing of kneading paddles and lack of backmixing in the extruder. It is intended to use the dispersion number to predict the reaction conversions and compare them with experimental values.

Table 3: Mean residence time, standard deviation and number of equivalent continuous stirred tank reactors for ICoTSE.

CMC (w/w%)	V_l (cm^3/min)	N_S (rpm)	t_S (s)	σ (s)	n
0	43	0	101.0	59.4	2.9
0	43	250	119.1	49.5	5.8
0	225	250	28.4	12.6	5.1
1	43	0	158.9	55.3	8.3
1	43	250	143.5	48.2	8.9
1	225	250	25.3	9.9	6.6
2	43	0	90.6	38.5	5.5
2	43	250	124.7	26.4	22.3
2	225	250	23.6	5.7	17.1

5. CONCLUSIONS

The ICoTSE can be employed as a three phase catalytic reactor, particularly with viscous materials. For example polymers containing unsaturated (olefinic) linkages can be hydrogenated to give them high temperature stability without the use of reaction solvents (necessary in conventional stirred reactors to reduce viscosity). Furthermore, it is possible that materials such as dimer acids could be produced using immobilised acidic catalysts, in a continuous basis using the ICoTSE. The following conclusions can be drawn from the present work.

1. The activity of palladium electroplated on mild steel was much less than electroless coated palladium on mild steel and alumina coated mild steel.
2. The most active catalyst was obtained when palladium was immobilised on a surface of alumina pre-coated mild steel.
3. Treating the surface of alumina with aluminium nitrate and calcining did not produce a γ-alumina layer.
4. The technical feasibility of using an intermeshing co-rotating twin screw extruder (ICoTSE) with immobilised screws as a three phase reactor was demonstrated.
5. The behaviour of the ICoTSE approached to a perfect plug flow reactor when the viscosity of the solution was relatively high (more than 500 cp.).

ACKNOWLEDGEMENTS

The financial support of the Ministry of Culture and Higher Education of Iran is gratefully appreciated. The authors also wish to thank PRISM Consultants Ltd. for technical help and provision of a twin screw extruder.

NOMENCLATURE

C_o	Initial concentration of liquid reactant	g/l
n	Equivalent continuous stirred tank reactor	----
N_S	Screw/impeller rotational speed	rpm
P_{H2}	Partial pressure of hydrogen	atm
R_A	Initial reaction rate	mol/(h.m^2)
T	Temperature	°C
t_m	Mean residence time	s
V_l	Liquid flow rate	cm^3/min
σ	Standard deviation	s

REFERENCES

1. Farbenindustrie, German Patent 895,058 to BASF (1953).
2. Xanthos, ed., Reactive Extrusion: principle and practice, Hanser Publishers, Munich, 1992, pp1-32.

3. Orchard, Ph.D. Thesis, The University of Birmingham, 1997.
4. Jordi, S. F. Orchard and J. M. Winterbottom, *The 1994 IChemE Research Event*, **Vol. 2**, 601 (1994).
5. Ebrahimi-Moshkabad and J. M. Winterbottom, *The 1997 Jubilee Research Event*, **Vol. 2**, 1169 (1997).
6. Manolatos and M. Jerome, *Electrochemica Acta*, **41**, 354 (1996).
7. U. Olsbye, R. Wendelbo and D. Akporiaye, *Appl. Catal. A*, **152**, 127(1997).

Catalytic Hydrogenation of Aromatic Rings in Lignin

Brian R. James,[a] Yu Wang,[a] Christopher S. Alexander,[a] and Thomas Q. Hu[b]

[a]*Department of Chemistry, University of British Columbia, Vancouver, Canada V6T 1Z1*
[b]*Vancouver Laboratory, Pulp and Paper Research Institute of Canada, Vancouver, Canada V6S 2L9*

Abstract

The efficiency and the chemo- and stereoselectivity of the $RhCl_3$- and $[RhCl(diene)]_2$-catalyzed hydrogenation of 4-propylphenols (used as lignin model compounds) in aqueous/organic media are dependent on the alkyl chain-length of the phase transfer co-catalyst tetraalkylammonium salt R_4NX (R = alkyl, X = Cl, Br or HSO_4) and on the solvent; D-labeling studies show that some of the added hydrogen comes from the water. The Ziegler-Natta acetate system $Ni(OAc)_2/Zn(OAc)_2/Et_3Al$, basic solutions of $[RuCl_2(\eta^6\text{-}C_6Me_6)]_2$, and $Ru(\eta^6\text{-}C_6Me_6)(\eta^2\text{-}C_2H_4)$ and $Ru(\eta^6\text{-}C_6Me_6)(\eta^4\text{-diene})$ are effective catalyst precursors for the chemoselective hydrogenation of the phenols at 80-100°C under 50 atm H_2; stabilized M(0) colloids appear to be the catalytic species. The most effective Ru-based catalytic system is $RuCl_3/(^nC_8H_{17})_3N$ which also catalyzes the hydrogenation of a "milled wood lignin".

Introduction

Lignin, a plant constituent, is the second most abundant biopolymer on earth. The biosphere is estimated to contain 3 x 10^{11} tonnes of lignin with an annual biosynthesis rate of approximately 2 x 10^{10} tonnes (1). In chemical pulping of wood, lignin is separated from cellulose by dissolution in the pulping liquor and later used mainly as a low-grade fuel within the pulp mill. Because of the lignin removal and the accompanying degradation and loss of cellulose and hemicelluloses, the yield of chemical pulps is low, typically 45-48%. Many pulp mills worldwide have moved towards the use of mechanical pulping processes in which fibre separation is achieved through mechanical action and lignin is retained to give high-yield (95-98%) mechanical pulps. The use of mechanical pulps, however, is confined to low-quality, non-permanent papers such as newsprint and telephone directories because of the light-induced yellowing of the materials.

The yellowing is known to be due to the oxidation of the aromatic rings of lignin to chromophores such as *o*-quinones (2-4). Several chemical modification methods, such as the alkylation of lignin phenols and the reduction of lignin carbonyls, have been tested for the possible inhibition of the photo-oxidation of lignin and the yellowing of lignin-rich mechanical pulps and papers, but none of them is selective or effective (5).

Our approach for a selective and effective inhibition of such yellowing is to hydrogenate the aromatic rings of lignin (6-10). Previously, we have shown that the hydrogenation of lignin model compounds such as 2-methoxy-4-propylphenol (**1**) and 4-propylphenol (**2**) could be achieved using $RhCl_3$ (**I**), $[RhCl(\eta^4$-1,5-hexadiene)]$_2$ (**II**) or $[RhCl(\eta^4$-1,5-cyclooctadiene)]$_2$ (**III**) as the catalyst precursor and Bu_4NHSO_4 as the phase transfer co-catalyst in a 2-phase aqueous buffer (pH 7.5)/hexane medium (6,7). In this paper, we describe the effects of tetraalkylammonium salts and solvent on the activity of these Rh catalysts. We also report the chemoselective hydrogenation of **1** catalyzed by some Ru complexes and a Ziegler-Natta type Ni system. The hydrogenation of **1** and a "milled wood lignin" using $RuCl_3$ and n-trioctylamine is also described. The aim is to develop a homogeneous system to effect the hydrogenation of lignin. A recent review (11) gives access to the literature on currently available homogeneous catalysts for the hydrogenation of aromatics.

Experimental Section
The Rh and Ru complexes were prepared according to the literature methods (12-13). All catalytic hydrogenations of **1** and **2** were performed under 1-50 atm H_2 at 20-100 °C in aqueous/organic media except the reaction with the Ni system where heptane/^{i}PrOH was used. The 2-phase aqueous/hexane (1:3 V/V) Rh-based systems contained a phase transfer catalyst R_4NX (7.0 equiv/Rh) with the aqueous phase sometimes being a borate/citrate/phosphate pH 7.5 buffer, and with catalyst and substrate concentrations of 3.0 mM and 75 mM, respectively. The catalyst and substrate concentrations in the one-phase $H_2O/^i$PrOH (1:1 - 1:4 V/V) systems using Ru complexes were 1.0 - 1.3 mM and 50 - 66 mM, respectively, and in the one-phase (heptane/^{i}PrOH, 1:4 V/V) Ni(OAc)$_2$/Zn(OAc)$_2$/Et$_3$Al (2:1:9 molar ratio) system the values were 1.6 mM and 160 mM, respectively. The corresponding concentrations in the one-phase ^{i}PrOH/H_2O (1:1 V/V) $RuCl_3$/TOA (1 : 3.5 molar ratio) system were 2.0 mM and 200 - 600 mM, respectively. The hydrogenations of **1** and **2** were monitored by GC. The identification of various hydrogenated products was achieved either through GCMS or through GC co-injection of the isolated and fully characterized hydrogenated products obtained in our previous work (10). The hydrogenation of a milled wood lignin (80 mg), isolated from peroxide-bleached spruce thermomechanical pulps (14,15), was performed using $RuCl_3$ (0.04 mmol) and TOA (0.14 mmol) as the catalyst in a 3-component solvent H_2O (5-7 mL) : ^{i}PrOH (13-15 mL) : MeOCH$_2$CH$_2$OH (5 mL) at 80°C under 50 atm H_2. The characterization of the hydrogenated lignin was achieved by ^{1}H NMR spectroscopy (16).

Results and Discussion

The effect of the tetraalkylammonium salt on the activity of the Rh-catalysts **I**, **II** and **III** in the hydrogenation of the lignin model compound 2-methoxy-4-propylphenol (**1**) (eq. 1, Pr = n-propyl) is given in Table 1. As in our previous work using these catalysts with Bu_4NHSO_4 (6,7), the hexadiene precursor **II** was found to be the most effective catalyst; the longer the alkyl chain of the tetraalkylammonium salt, the higher the activity of the Rh catalyst. In the presence of $(^nC_8H_{17})_4NBr$, **II** has the highest activity (91.1% conversion), compared to the lowest 9.3% conversion when $RhCl_3$ (**I**) and Et_4NCl were used.

$$(1)$$

Table 1. Conversion (%) of lignin model **1** using R_4NX and $RhCl_3$ (**I**), $[RhCl(\eta^4\text{-}1,5\text{-hexadiene})]_2$ (**II**) or $[RhCl(\eta^4\text{-}1,5\text{-cyclooctadiene})]_2$ (**III**).[a]

R_4NX	I	II	III
Et_4NCl	9.3	43.1	8.7
Bu_4NHSO_4	10.8	52.9	36.1
$(^nC_6H_{13})_4NCl$	38.6	65.0	63.0
$(^nC_8H_{17})_4NBr$	56.8	91.1	88.2

[a] At 20 °C under 1 atm H_2 for 24 h in a 2-phase aqueous buffer (pH 7.5)/hexane solution with [**1**]/[Rh] = 25 and [R_4NX]/[Rh] = 7.

All the reactions give high chemoselectivity for 2-methoxy-4-propylcyclohexanol (**1a**) over the hydrogenolysis products 4-propylcyclohexanol (**1b**) and 4-propylcyclohexanone (**1c**), with high stereoselectivity for *cis*-2-methoxy-*cis*-4-propylcyclohexanol (*c,c*-**1a**) (for example, see Table 2). The longer alkyl chain R_4NX co-catalyst gives the higher chemoselectivity: 96.1% **1a** with $(^nC_8H_{17})_4NBr$ compared to 91.4% **1a** with Et_4NCl. The stereoselectivity for *c,c*-**1a** over *c,t*-**1a** and *t,c*-**1a** was also the highest with $(^nC_8H_{17})_4NBr$ as co-catalyst. The noticeable effect of the alkyl chain length of the R_4NX co-catalyst on the chemo- and stereoselectivity, and particularly the drastic effect on the activity (conversion of **1**) of the catalyst, are thought to reflect differences in the stabilization of the colloidal Rh(0) catalyst by the ammonium salt in the organic phase (6).

The solvent also has a profound effect on the catalytic activity of these Rh-based systems and on the chemoselectivity of the reactions. For example, optimization of **II** with $(^nC_8H_{17})_4NBr$ requires an aqueous buffer (pH = 7.5)/hexane medium (91.1% conversion at 24 h and 96.1% selectivity for **1a** vs., for example, 0.5% conversion and only 35.6% selectivity in hexane alone) (Table 3).

Table 2. Product distribution (%) in the hydrogenation of **1** using R_4NX and $[RhCl(\eta^4\text{-}1,5\text{-hexadiene})]_2$ (**II**).

R_4NX	*c,c*-**1a**	*c,t*-**1a**	*t,c*-**1a**	*c*-**1b**	*t*-**1b**	**1c**
Et_4NCl	77.4	8.7	5.3	5.7	2.1	0.8
Bu_4NHSO_4	80.2	7.3	4.9	5.2	1.8	0.4
$(^nC_6H_{13})_4NCl$	77.1	14.2	3.6	3.6	1.0	0.6
$(^nC_8H_{17})_4NBr$	85.7	5.7	4.7	2.9	1.0	0.0

Table 3. Conversion (%) and product distribution (%) in the hydrogenation of **1** using [RhCl(η^4-1,5-hexadiene)]$_2$ (**II**) and ($^nC_8H_{17}$)$_4$NBr in various solvents.

Solvent	Conversion	**1a**	**1b**	**1c**
Hexane-H$_2$O (pH 7.5)	91.1	96.1	3.9	0.0
Hexane-H$_2$O	61.0	96.7	2.9	0.4
Hexane	0.5	35.6	57.1	7.3
^{i}PrOH-H$_2$O (pH 7.5)	38.1	92.6	6.3	7.3
^{i}PrOH	0.2	27.1	48.2	32.1

In an attempt to understand further the role of H$_2$O in these reactions, deuterium labeling experiments were carried out for the hydrogenation of a simpler lignin model compound, 4-propylphenol (**2**), using the **II**/($^nC_8H_{17}$)$_4$NBr catalyst system (eq. 2). In the absence of H$_2$ or D$_2$, no hydrogenation of **2** was observed. When the reaction was carried out under 1 atm H$_2$ in hexane-H$_2$O, one of the isolated products 4-propylcyclohexanone (**1c**) has the EI mass data m/z(relative intensity): 140(38) and 141(3.8) (Fig. 1a), corresponding to **1c** (C$_9$H$_{16}$O) and C$_9$H$_{16}$OH$^+$, respectively (10). When the reaction was performed under 1 atm D$_2$ in hexane-H$_2$O, the EI mass data were 140(43), 141(25), 142(9) and 143(2) (Fig. 1b); the 141 and 142 peaks, assigned to C$_9$H$_{15}$DO and C$_9$H$_{14}$D$_2$O, respectively, show only partial incorporation of deuterium. With the use of H$_2$ and hexane-D$_2$O, the EI mass data were 141(2.8), 142(10), 143(13), 144(41), 145(43), 146(31), 147(16), 148(6) and 149(2) (Fig. 1c). The 148 peak is assigned to C$_9$H$_8$D$_8$O while the 149 peak could be assigned to either C$_9$H$_8$D$_8$OH$^+$ or C$_9$H$_7$D$_9$O. The 8 or 9 D-atoms of **1c** are likely to be all on the cyclohexanone ring, and this is possible based on the stepwise mechanism via the enol intermediate which undergoes the H-D exchange with D$_2$O (10). With the use of D$_2$ in hexane-D$_2$O, the highest mass peak for **1c** is seen at 149(7) (Fig. 1d), corresponding to the cyclohexanone ring having its maximum 9 D-atoms. The deuterium-labeling experiments clearly show the incorporation of hydrogen from water into the hydrogenated product. The key "activation role" played by water could involve a favored cleavage of a Rh-C bond by protonation (to give hydrogenated product) rather than reductive elimination of product from a hydrido-aryl species (HRh-C).

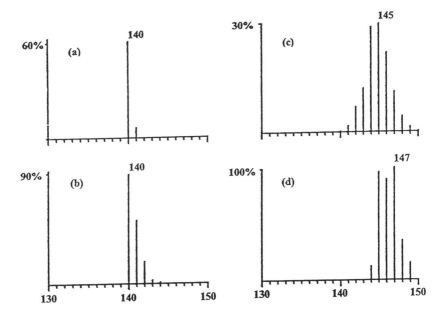

Figure 1. EI mass spectra for 4-propylcyclohexanone (1c) from the hydrogenation of 4-propylphenol (2) using **II** as the catalyst under (a) 1 atm H_2 in hexane-H_2O, (b) 1 atm D_2 in hexane-H_2O, (c) 1 atm H_2 in hexane-D_2O and (d) 1 atm D_2 in hexane-D_2O.

The colloidal Rh(0) catalyst systems described above were found to be unsuitable for the hydrogenation of lignin in mechanical pulps (6-9). We thus switched our attention to Ru-based systems, particularly as [RuCl$_2$(η^6-C$_6$Me$_6$)]$_2$/Na$_2$CO$_3$ has been reported to effect the homogeneous hydrogenation of arenes at 50 °C under 50 atm H_2 in an aqueous isopropanol medium (17).

In the presence of Na_2CO_3, $[RuCl_2(\eta^6\text{-}C_6Me_6)]_2$ (**IV**) is inactive at ~20 °C under 1 atm H_2 in $H_2O\text{-}^iPrOH$ for the hydrogenation of **1**. However, when **IV** was pretreated with aqueous Na_2CO_3 for 2 h under Ar and then used as the catalyst at 80 °C under 50 atm H_2, the hydrogenation of **1** does occur (37.0% conversion at 24 h) (Table 4); a black powder was always present at the end of the reaction and the catalytic system became inactive upon the addition of mercury, implying that the true catalyst is colloidal in nature (18). The $Ru_2Cl_2(\mu\text{-}H)_2(\mu\text{-}Cl)(\eta^6\text{-}C_6Me_6)_2$ species, generated from the reaction of **IV** and aqueous Na_2CO_3 and considered to be the catalyst in the hydrogenation of arenes (17), is probably just an intermediate en route to the active colloidal species. A recent report on closely related $[RuCl_2(\eta^6\text{-}C_6H_6)]_2$ systems for the hydrogenation of arenes in aqueous/substrate media suggests that the catalysts are polynuclear Ru-hydrides (19).

In attempts to stabilize Ru(0) as molecular catalytic species, several Ru(0) complexes and one Ru(II) complex: $Ru(\eta^6\text{-}C_6Me_6)(\eta^2\text{-}C_2H_4)_2$ (**V**), $Ru(\eta^6\text{-}C_6Me_6)(\eta^4\text{-}1,3\text{-cyclohexadiene})$ (**VI**), $Ru(\eta^6\text{-}C_6Me_6)(\eta^4\text{-}1,3\text{-cyclooctadiene})$ (**VII**), $Ru(\eta^6\text{-}C_6Me_6)(\eta^4\text{-}C_6Me_6)$ (**VIII**) and $[Ru(\eta^6\text{-}C_6Me_6)_2][PF_6]_2$ (**IX**) were prepared and used as precursors for the hydrogenation of **1** at 100 °C under 50 atm H_2. Complexes **V**-**VII** were effective for the chemoselective hydrogenation of **1** (Table 4) but black material was again present at the end of the reactions, indicating that the catalyst was probably colloidal. Complexes **VIII** and **IX** were inactive and could be recovered unchanged, suggesting that the Ru center can be stabilized against reduction to the metal by two hexamethylbenzene ligands under the above hydrogenation conditions; however, such stabilized species do not effect hydrogenation. Qualitatively it appears that these Ru-aromatic complexes could be "fine-tuned' to yield active, molecular catalysts.

The most active Ru system was generated using $RuCl_3$ (**X**) in the presence of n-trioctylamine (TOA) in $H_2O\text{-}^iPrOH$ at 50°C under 50 atm H_2 (Table 4); note that the substrate : Ru ratio was 6 times higher with **X** compared to the value for other Ru systems. The high activity of the $RuCl_3/TOA$ system is likely due to the formation and stabilization of colloidal Ru(0) species by TOA (20). Preliminary kinetic data on the hydrogenation of **1** using this catalyst in $H_2O\text{-}^iPrOH$ reveal approximate Arrhenius-type behavior at 40-80 °C with $\Delta E^* \approx 4$ kcal mol^{-1}. There is a reasonable 1st-order dependence on H_2 pressure up to 40 atm (pressure/conversion over 4 h at 80 °C: 10/26%, 30/87%, 40/97%), and increased chemoselectivity for **1a** over **1b** (less hydrogenolysis of MeO substituents) over the H_2 range of 10-40 atm (Fig. 2). The chemoselectivity for **1a** is important because it is thought to mimic the degree of breakdown of lignin polymer, and this will affect the pulp yield in the hydrogenation of mechanical pulps.

Of interest, a Ziegler-Natta type Ni catalyst (**XI**) was also effective for the hydrogenation of **1** with a very high chemoselectivity for **1a** (Table 4); however, the catalysis was again colloidal as demonstrated by the mercury test.

Table 4. Conversion (%) and selectivity (%) for **1a** in the hydrogenation of **1** using Ru complexes **IV-VII**, RuCl$_3$/TOA (**X**), and Ni(OAc)$_2$/Zn(OAc)$_2$/Et$_3$Al (**XI**) at 50-100 °C under 50 atm H$_2$ in H$_2$O (heptane for **XI**)-^{i}PrOH for 24 h with [**1**]/[Ru] = 50 and [**1**]/[Ni] = 100.

Catalyst	Temp. °C	Conversion	Sel. for **1a**
IV[a]	80	37.0	81.3
V	100	62.6	78.6
VI	100	94.7	82.1
VII	100	73.2	77.0
X[b]	50	99.7	85.6
XI[c]	90	65.2	92.2

[a] Pretreatment with Na$_2$CO$_3$ (8.0 mol equiv) for 2 h. [b] [**1**]/[Ru] = 300. [c] Pretreatment at 90°C under 1 atm N$_2$ for 2 h in heptane.

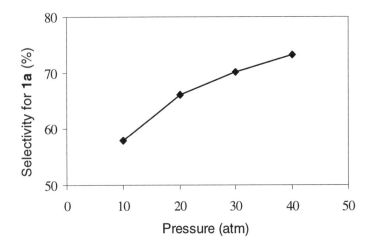

Figure 2. Dependence of chemoselectivity (%) for **1a** on H$_2$ pressure using RuCl$_3$/TOA as the catalyst at 80 °C in H$_2$O-^{i}PrOH for 4 h with [**1**]/[Ru] = 100.

The hydrogenation of an isolated lignin, the so-called "milled wood lignin" which is thought to best resemble the native lignin in wood and in mechanical pulps (21) was effected using RuCl₃/TOA as the catalyst at 80 °C under 50 atm H$_2$ in H$_2$O-^{i}PrOH-MeOCH$_2$CH$_2$OH, the methylated glycol being used to dissolve the lignin. The extent of the aromatic ring hydrogenation was established by the analysis of the ^{1}H NMR integration ratio of the aromatic protons (δ 6-8) and the protons (δ 0.8-2.1) on the hydrogenated rings that are not α-to the OH or OMe groups (16). In contrast to the hydrogenation of model compounds, reaction with lignin is extremely slow, ~38% of the aromatic rings being hydrogenated after 10 days (Fig. 3). The slow reaction is likely due to the low accessibility of the aromatic rings in the lignin polymer to the colloidal Ru(0). Increased aromatic ring hydrogenation may require cleavage of some of the interlinkages in lignin. Molecular weight distribution analysis indicates that there is a linear relationship between the extent of aromatic hydrogenation and the average molecular weight of the partially hydrogenated lignin (16).

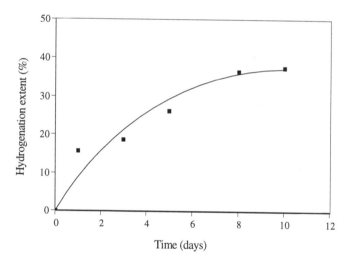

Figure 3. Extent of aromatic hydrogenation of milled wood lignin vs. time (days).

Acknowledgments
We thank: the Natural Sciences and Engineering Research Council of Canada for financial support via the Mechanical and Chemimechanical Pulps Network; Drs. R.St.J. Manley and J. Schmidt of Paprican for a sample of milled wood lignin; and Johnson Matthey Ltd. and Colonial Metals Inc. for loans of RhCl$_3$ and RuCl$_3$.

References

1. R.H. Whittaker and G.E. Likens, In *Primary Production of the Biosphere*, H. Lieth and R.H. Whitaker, Eds., Springer-Verlag, Berlin, 1975, pp. 305-328.
2. S.Y. Lin and K.P. Kringstad, *Norsk Skogindustri.* **25(9)**, 252 (1971).
3. J.K.S. Wan, M.Y. Tse and C. Heitner, *J. Wood Chem. Technol.* **13**, 327 (1993).
4. J.A. Schmidt and C. Heitner, *J. Wood Chem. Technol.* **13**, 309 (1993).
5. G.J. Leary, *J. Pulp Paper Sci.* **26(6)**, J154 (1996).
6. B.R. James, Y. Wang and T.Q. Hu, *Chem. Ind.* (Marcel Dekker), **68**, 423 (1996).
7. T.Q. Hu, B.R. James and C.-L. Lee, *J. Pulp Paper Sci.* **23(4)**, J153 (1997).
8. T.Q. Hu, B.R. James and C.-L. Lee, *J. Pulp Paper Sci.* **23(5)**, J200 (1997).
9. T.Q. Hu, B.R. James and C.-L. Lee, in *Proc. 9th Intern. Symp. on Wood and Pulping Chem.*, Montreal, Vol 1, K2-1 (1997).
10. T.Q. Hu, B.R. James, S. Rettig and C.-L. Lee, *Can. J. Chem.* **75**, 1234 (1997).
11. I.P. Rothwell, *J. Chem. Soc. Chem. Comm.* 1331 (1997).
12. M.A. Bennett, T.N. Huang, T.W. Matheson and A.K. Smith, *Inorg. Synth.* **21**, 74 (1982).
13. J. Chatt and L.M. Venanzi, *J. Chem. Soc.*, 4735 (1957).
14. H.H. Brownell, *Tappi J.* **48**, 513 (1965).
15. A. Bjorkman, *Svensk. Papperstidn.* **59**, 477 (1956).
16. T.Q. Hu, B.R. James, C.S. Alexander and Y. Wang, to be published.
17. M.A. Bennett, T.N. Huang and T.W. Turney, *J. Chem. Soc. Chem. Comm.* 312 (1979); M.A. Bennett, *Chemtech.*, 444 (1980).
18. G.M. Whitesides, M. Hackett, R.L. Brainard, J-P. P.M. Lavalleye, A.F. Sowinski, A.N. Izumi, S.S. Moore, D.W. Brown and E.M. Staudt, *Organometallics*, **4**, 1819 (1985).
19. L. Plasseraud and G. Süss-Fink, *J. Organomet. Chem.*, **539**, 163 (1997).
20. F. Fache, S. Lehuede and M. Lemaire, *Tetrahedron Lett.*, **36**, 885 (1995).
21. Y.Z. Lai and K.V. Sarkanen, In *Lignin - Occurrence, Formation, Structure and Reactions*, K.V. Sarkanen and C.H. Ludwig, Eds.; Wiley: New York, p.165 (1971).

Hydrogenation of Crotonaldehyde Over New Type of Sn-Pt/SiO$_2$ Catalysts

J. L. Margitfalvi, I. Borbáth and A. Tompos

Chemical Research Center, Institute for Chemistry, Hungarian Academy of Sciences, 1525 Budapest POB 17, Hungary.

Abstract

The hydrogenation of crotonaldehyde (CA) was studied both in gas and liquid phases over different Sn-Pt/SiO$_2$ catalysts prepared by using Controlled Surface Reactions between tin tetraethyl and hydrogen adsorbed on platinum. The change of the conditions used to anchor tin to supported platinum resulted in catalysts with high Sn/Pt ratios (Sn/Pt$_s$=3). The best performance in gas phase hydrogenation was obtained on catalysts with Sn/Pt$_s$=3.0, while in the liquid phase hydrogenation the catalyst with Sn/Pt$_s$=1.2 showed the highest selectivity towards the formation of crotylalcohol and the corresponding S$_{C=O}$ selectivity values were 70 and 60 %, respectively. In gas phase hydrogenation over Sn-Pt/SiO$_2$ catalysts a "build-up" period was observed prior to reaching the high S$_{C=O}$ selectivity region. The improvement of the S$_{C=O}$ selectivity during the time on stream period was attributed to the substrate induced catalyst activation.

Introduction

The hydrogenation of unsaturated aldehydes (UA) is a very important test reaction in the field of selective hydrogenation. In UA both the aldehyde and the olefinic double bond can be hydrogenated [1,2].

Recently it has been demonstrated that supported platinum catalysts modified either by tin (or iron) have much higher selectivities for the hydrogenation of the aldehyde group than the monometallic supported platinum catalysts [3-6]. A characteristic feature of the above bimetallic catalysts is the development of the S$_{C=O}$ selectivity during the time on stream period [5]. Results obtained in gas phase hydrogenation of crotonaldehyde (CA) indicated that a definite induction period was required to build up the active sites responsible for the formation of unsaturated alcohol [5].

With respect to the form of tin in Sn-modified Pt catalysts the role of ionic species has been proposed and the interaction of metallic platinum with ionic form of tin has been suggested [3-6]. It has also been proposed that probably the Sn^{+n} - substrate carbonyl interaction is responsible for the increased S$_{C=O}$ selectivity [5-6]. In this respect it has also been mentioned that the Sn^{+n} - substrate carbonyl interaction requires ionic Sn^{+n} species stabilized either on the metal site or at the metal-support interface.

On the other hand it has been demonstrated that supported Sn-Pt catalysts with exclusive tin - platinum interaction can be obtained upon using Controlled Surface Reactions (CSRs) between adsorbed hydrogen and tin tetraalkyls [7-10]. The surface chemistry involved in the modification of supported platinum by SnR_4 can be written by the following equations [8]:

$$MH_{ads} + SnR_4 \longrightarrow M\text{-}SnR_{(4-x)} + x \text{ RH} \qquad (1)$$
$$PSC$$

$$M\text{-}SnR_{(4-x)} + (4-x)/2 \ H_2 \longrightarrow Sn\text{-}M + (4-x) \text{ RH} \qquad (2)$$
$$SBS$$

In the formed Stabilized Bimetallic Species (SBS) (see reaction (2)), due to the control of the tin anchoring process, the tin - platinum interaction was exclusive, i. e. the amount of tin introduced onto the catalyst support was negligible. The main advantage of the use of CSRs is the exclusive formation of alloy-type bimetallic surface species [10].

Recent results obtained in the selective hydrogenation of CA over Sn-Pt/SiO_2 catalyst prepared by using conventional preparation techniques inspired us to use our Sn-Pt/SiO_2 catalysts prepared by using Controlled Surface Reactions (CSRs) mentioned above [7-10] in the selective hydrogenation of CA. It has been suggested that catalysts prepared by using CSRs, due to the intimate contact between platinum and tin in the formed supported bimetallic surface species, should result in high $S_{C=O}$ selectivity.

In this study the hydrogenation of CA has been studied over different Sn-Pt/SiO_2 catalysts prepared by using CSRs (1) and (2). In order to increase the $S_{C=O}$ selectivity of Sn-Pt/SiO_2 catalysts the amount of tin was significantly increased. In this way new aspects of the tin anchoring process had been developed. It was also very important to demonstrate that even at high Sn/Pt_s ratios the amount of Sn introduced onto the support is still negligible.

Experimental

Catalysts preparation and modification

Silicagel prepared at the Boreskov Institute of Catalysis (Novosibirsk, Russian Federation) containing 0.02 wt.% impurities was used as a support. Surface area (S= 302 m^2/g), pore volume (V_p= 0.95 cm^3/g) and mean pore diameter (d_p= 12 nm) were determined by nitrogen adsorption. The silica supported monometallic catalyst containing 3 wt. % Pt was obtained by ion-exchange technique using [Pt(NH_3)_4]Cl_2 as precursor compound. After filtration, the catalyst was washed free of chloride ion with demineralized water until pH=7, and then dried in two steps at 60 °C and 120 °C for 3 and 2 hours, respectively, (heating rate = 0.3 °C/min). The catalyst was reduced under flowing hydrogen for 4 hours at 300 °C (heating rate = 1.5 °C/min). The H/Pt ratio of the Pt/SiO_2 catalyst used was 0.52 measured by hydrogen chemisorption.

Prior to the tin anchoring step the catalyst was re-reduced in hydrogen at 300 °C for 60 minutes followed by cooling in a hydrogen atmosphere to room temperature and purging with argon for 30 minutes. The catalyst was introduced into a glass reactor and was slurred with deoxygenated benzene either in an argon or a hydrogen atmosphere. Reaction (1) was started by injection of tin tetraethyl. It is worth mentioning that in the present study the reaction time of surface reaction (1) was sufficiently increased compared to experiments described in our earlier studies [7-10].

Reaction (1) was monitored by determining the amount of ethane or ethylene. After surface reaction (1) the catalyst was washed four times with benzene at 50 °C followed by washing three times with n-hexane at the same temperature and drying in vacuum (at 5 torr) at 50 °C for one hour.

The decomposition of PSC was carried out in a hydrogen atmosphere by Thermal Programmed Reaction (TPR) technique using the following experimental parameters: heating rate = 5 °C/min, hydrogen flow rate = 30 cm^3/min, amount of catalyst = 0.4-0.6 g. The products of decomposition of PSC (C_2H_6 and C_2H_4) were analyzed by GC. After decomposition the tin content of the modified catalysts was determined by AAS and was compared with the amount of tin calculated from the overall material balance.

The material balance of tin anchoring was calculated based on the amount of C_2 hydrocarbons formed in reactions (1) and (2). In reaction (1) a correction was done to take into account the amount of hydrocarbons (ethane or ethylene) dissolved in benzene. The material balance allowed to calculate separately the amount of alkyl groups reacted in the first step of anchoring (n^I, mol/g$_{cat}$) and the amount of C_2 hydrocarbons formed in the decomposition of PSC in the TPR experiment (n^{II}, mol/g$_{cat}$). The value of n^I contains the amount of both ethane and ethylene. In this way the total amount of tin anchored could also be calculated. The amount of tin calculated in this way had a good agreement with that of determined by AAS. The material balance allowed us to calculate the value of x (see reaction (1)). In this way the stoichiometry of surface reaction (1) was determined.

Hydrogenation reaction

The hydrogenation of CA was studied both in gas and liquid phases. In gas phase the reaction was carried our under atmospheric pressure in the temperature range of 60 - 100 °C. In the liquid phase the hydrogenation was done in the pressure range of 2-6 bar, where kinetic regime could only be maintained in the temperature range of 4-40 °C. At higher temperatures the reaction became diffusion controlled.

Prior to the catalytic reaction the catalysts were pre-reduced in a hydrogen atmosphere at 300 °C. Special care was taken not to contaminate the catalysts with oxygen prior their use in the catalytic reaction.

In the gas phase hydrogenation of CA over different modified platinum catalysts fast ageing was observed [3,4]. Due to the above ageing it was difficult to get real kinetic data. To overcome the ageing induced deactivation problem in this study the conventional continuous-flow reactor was used in a periodic mode by introducing the CA-hydrogen mixture in the form of a long pulse (pulse length = 7 minutes, product analysis at the end of the pulse) followed by a long pulse of pure hydrogen (7 minutes). The use of long hydrogen pulse resulted in full or partial restoration of the catalytic activity. Upon using 15-20 consecutive pulses a constant activity (and selectivity) period was achieved allowing to carry out kinetic experiments. Analogous reactor was used to investigate the kinetic pattern of hydrocarbon reactions involved in naphtha reforming [11].

Results and discussions

Catalysts modification

Figure 1 shows the formation of ethane and ethylene during the tin anchoring step (see reaction (1)). Results given in Fig. 1 indicate that at high tin tetraethyl concentration the tin anchoring reaction has two definite steps: (i) at the very beginning of the reaction (1) the formation of ethane is fast the amount of ethylene is very small; (ii) after 40-80 minutes of reaction time ethylene appears and the form of the kinetic curve changes. The formation of ethylene was not observed when reaction (1) was carried out in a hydrogen atmosphere. In this latter case both the initial rate of reaction (1) and the amount of ethane and ethylene increased.

The TPR pattern of the decomposition of surface species formed is shown in Figure 2a and b. As emerges from figures 2a and b the decomposition pattern shows a complex character, the higher the amount of tin anchored the more the number of the TPR peaks. The summary of the TPR results are given in Table 1. From results given in Table 1 the following information can be acquired: (i) tin anchoring is a stepwise process, (ii) the amount of tin anchored can be much higher than the monolayer coverage of $Sn(C_2H_5)_{(4-x)}$.

The monolayer coverage is achieved around $Sn/Pt_s=0.4$. It means that the monolayer coverage corresponds to the TPR peaks up to 80 °C. The TPR peaks at 110, 150, and 198 °C correspond to surface organometallic moieties in the second layer. In a special experiment it was shown that tin tetraethyl adsorbed onto SiO_2 decomposes at 260 °C, consequently the TPR peak at 260 °C was attributed to tin tetraethyl adsorbed onto the support. This peak was observed only at very high concentration of tin tetraethyl (see Table 1). The above high concentrations were avoided for catalysts preparation.

Let us discuss to the formation of the second layer (Surface Complex in the Second Layer, SCSL). If the concentration of $Sn(C_2H_5)_4$ in the bulk liquid phase is high and the surface of platinum is fully covered by $-SnR_{(4-x)}$ moieties, with different extent of dealkylation, the probability to anchor SnR_4 onto $-SnR_{(4-x)}$ is high. The formation of SCSL can be written as follows:

$$\textbf{Pt-SnR}_{(4-x)} \; + \; \textbf{n SnR}_4 \quad \text{----}\text{>} \quad \textbf{Pt-\{SnR}_{(4-x)} \textbf{- (SnR}_{(4-y)}\textbf{)}_n\} \qquad (3)$$
$$\textbf{SCSL}$$

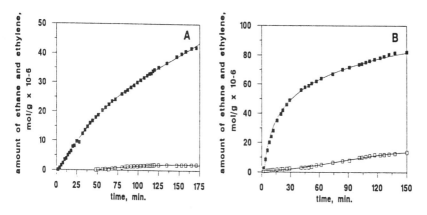

Figure 1.
Time dependence of the formation of ethane and ethylene in reaction (1). $[Sn(C_2H_5)_4]_o$, mM: A - 12.7, B - 81.0: ■ - ethane , □ - ethylene. Reaction temperature: 50 °C, solvent: benzene.

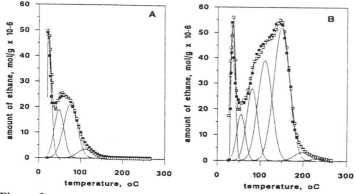

Figure 2.
Decomposition of surface complexes by TPR and deconvolution of the TPR peaks. A: $Sn/Pt_s = 0.4$ ($Sn_o/Pt_s = 1.6$), B - $Sn/Pt_s = 1.15$ ($Sn_o/Pt_s = 9.8$).

Reaction (3) describes the general equation for the formation of SCSL. In the first approximation the first layer is considered to have strongly dealkylated species, while the second layer consist of anchored SnR_4 moieties (i. e. in this case $y = 0$). We believe that this process takes place at high Sn_o/Pt_{surf} ratios in the second part of the kinetic curve. The loss of alkyl group in SCSL, i. e. $y > 1$, leads to the formation of coordinatively unsaturated sites in the second layer resulting in new anchoring sites for tin tetraethyl. In this way Multilayered Species (MLS) can be formed. The general formula of MLS is given below:

$$Pt\text{-}[\{SnR_{(4-x)} \text{ - } (SnR_{(4-y)})_n\} \text{ - } (SnR_4)_m]$$

In the formation of both SCSL and MLS the coordinative unsaturation is the driving force to anchor the next layer of SnR_4. Coordinative unsaturation (i.e. increase of x and y values in MLS) can be increased if hydrogen is used in the first step of anchoring. The influence of added hydrogen on the fate of PSC (see surface reaction (1)) has already been shown in our previous study [9]. The influence of added hydrogen on the tin anchoring process is shown in Table 2.

As it emerges from data given in Table 2 in the presence of hydrogen the following important changes were observed: (i) the rate of surface reaction (1) increased, (ii) the x value, i.e. the extent of unsaturation in PSC and SCSL strongly increased resulting in the overall increase of the amount of anchored tin. Upon using this approach the Sn/Pt_s ratio increased up to 2.

Table 2. Comparison of the tin anchoring process in the absence and presence of added hydrogen

Exp	W_o * x 10^{-6}	n^I x 10^{-6} mol/g_{cat}	C_2H_4 ** 10^{-6} mol/g_{cat}	n^{II} x 10^{-6} mol/g_{cat}	x	Sn/Pt_s at/at
Ar	3.2	95	13.0	274	1.0	1.2
H_2	8.8	354	no C_2H_4	290	2.2	2.0

$[Sn_0/Pt_s]= 9.8$; * (mol/g_{cat} x min), ** amount of ethylene formed in reaction (1).

Further increase in the Sn/Pt_s ratio was observed when small amount of oxygen was introduced into the reactor during the tin anchoring step. In this way the Sn/Pt_s ratio reached the value around 3. In none of the above two cases, i.e. in the presence of hydrogen or small amount of oxygen the TPR peak around 260 °C was not observed. Thus, the above changes in the tin anchoring process did not resulted in the introduction of SnR_4 into the support.

Hydrogenation of crotonaldehyde in gas phase reactor

Figure 3. shows the formation of products from CA over a Sn-Pt/SiO$_2$ catalyst as a function of the pulse number (time on stream). The development of $S_{C=O}$ selectivity during the time on stream is well presented. Similar phenomenon was recently described in ref. [5]. The above behaviour was attributed to the substrate induced activation of sites required to the hydrogenation of the aldehyde group. In this respect the formation of ionic forms of tin was suggested.

Results obtained at low conversion over different catalysts are summarized in Table 3. The parent Pt/SiO$_2$ catalyst had no activity towards the formation of crotonaldehyde. The $S_{C=O}$ selectivity shows a strong dependence of the amount of tin introduced onto the platinum. The highest $S_{C=O}$ selectivity measured at 5% conversion was in the range of 70 %. Even at 60 % conversion the $S_{C=O}$ selectivity was around 50 %.

It should also be emphasized that the introduction of tin in small amount resulted in a very pronounced rate increase. The reaction rate passes

1. Table. Summary of data obtained in the decomposition of surface complexes in TPR.

Exp. No	$[Sn]_o/Pt_s$	Sn_{anch}/Pt_s (at/at)	Amount of C_2H_6 form in different TPR peaks, n^{II}_{ij}, x 10^{-6} mol/g$_{cat}$ x min						
			1st peak 22-34 °C	2nd peak 48-54 °C	3rd peak 75-80 °C	4th peak 110-112 °C	5th peak 149-153 °C	6th peak 197-198 °C	7th peak 260 °C
1	0.4	0.1	9.55	7.50	5.31	0.75	0.79	-	-
2	1.6	0.4	29.75	18.32	28.50	5.13	-	-	-
3	3.0	0.5	17.54	15.94	39.20	33.06	4.22	1.41	2.03
4	5.3	0.8	23.19	19.29	36.51	74.06	10.58	3.17	2.70
5	9.8	1.2	34.79	19.05	38.65	69.98	102.88	6.53	1.98
6	23.8	1.6	40.16	16.78	38.60	65.78	106.37	105.18	12.43
7	33.5	1.6	46.46	20.36	32.23	67.62	96.60	105.36	27.57
mean values				19.29	37.04	69.36	101.95	105.27	-
				+/- 1.65	+/- 2.88	+/- 3.57	+/- 4.95	+/- 0.13	

The values given in bold were used to caluclate the mean values of hydrocarbons formed.

through a maximum, however even at Sn/Pt = 3 the rate of hydrogenation was in the same range as on the parent Pt/SiO_2 catalyst. The above rate increase indicates that in the $Sn-Pt/SiO_2$ catalysts the type of active sites is altered in comparison to the type of active sites in the Pt/SiO_2 catalyst.

Table 3. Summary of results obtained in gas phase hydrogenation of crotonaldehyde

Sn/Pt$_s$	W$_{ini.}$*	Selectivity (%) **		
	$\dfrac{\mu mol}{g \cdot s}$	SAL	SOL	UOL
0.0	2.7	100	0	**0**
0.6	6.6	47	6	**45**
1.1	5.2	41	6	**51**
2.2	4.7	35	5	**59**
2.9	2.2	20	0	**68**

Reaction temperature: 80 °C, * measured form the conversion - contact time dependencies and extrapolated to zero conversion, ** - measured at 5 % conversion. SAL - butyraldahyde, SOL - butylalcohol, UOL - crotylalcohol. The selectivity of hydrocarbons is the difference to 100 %.

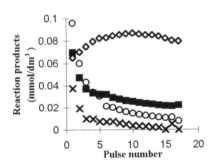

Figure 3.
Time on stream pattern of the gas phase hydrogenation of crotonaldehyde over $Sn-Pt/SiO_2$ catalysts. Reaction temperature: 80 °C, Sn/Pt$_s$ = 3.0, amount of catalyst: 100 mg, flow rate: 90 cm3/min.

◊ - crotylalcohol, **X** - butanol,
■ - butyraldehyde,
○ - hydrocarbons.

The first order rate constants were calculated both for the formation of crotylalcohol (k_1) and butyraldehyde (k_2). The dependence of these rate constants of the tin content is shown in Figure 4. Tin increases the rate constant of both reactions up to Sn/Pt$_s$ = 1. The increase of k_1 is more pronounced than that of k_2. Further increase of the Sn/Pt$_s$ ratio resulted in the drastic diminishing of k_2, while k_1 decreased only slightly. These results also indicate that upon introduction of tin new type of sites were created. Upon increasing the number of sites responsible for the hydrogenation of the aldehyde group sites involved in the hydrogenation of the double bond are strongly poisoned.

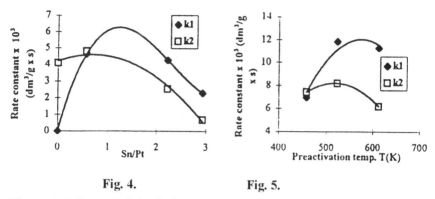

Fig. 4. Fig. 5.

Figure 4. Influence of the Sn/Pt$_s$ ratio on the first order rate constants. k_1 - hydrogenation of the aldehyde group, k_2 - hydrogenation of the olefinic double bond.

Figure 5. Influence of the pretreatment temperature in a hydrogen atmosphere on the first order rate constants. k_1 - hydrogenation of the aldehyde group, k_2 - hydrogenation of the olefinic double bond; Sn-Pt/SiO$_2$ catalyst (Sn/Pt$_s$=0.6).

The selectivity of the Sn-Pt/SiO$_2$ catalysts showed a strong dependence of the catalyst pretreatment temperature carried out in a hydrogen atmosphere prior to the hydrogenation. The related kinetic data are shown in Figure 5. It shows the dependence of the k_1 and k_2 rate constants of the temperature of pretreatment. Data indicate that sites responsible for the hydrogenation of the aldehyde group are more sensitive to the temperature increase than sites involved in the hydrogenation of the olefinic double bond.

It is also important to mention that on the monometallic catalyst the hydrogenation of butyraldehyde was also hindered (see Table 3). This fact strongly supports the general observation that in the presence of tin the reactivity of the aldehyde group is strongly enhanced. In other words it can be concluded that due to the Sn^{+n} - carbonyl interaction the aldehyde group is strongly perturbed. The above perturbation leads to the increased reactivity.

Hydrogenation of crotonaldehyde in stirred tank reactor

Results obtained in stirred tank reactor are summarized in Table 4. The performance of the Sn-Pt/SiO$_2$ catalysts in the liquid phase was different from that in the gas phase. Catalysts with Sn/Pt$_s$ > 0.8 have lower activity than the parent Pt/SiO$_2$ catalyst and the activity of catalysts with Sn/Pt$_s$ > 2 was extremely low. Similar to the findings in the gas phase the introduction of tin resulted in an overall activity increase and a high selectivity towards the formation of crotylalcohol. However, the increase of the Sn/Pt_s ratio above one did not resulted in further significant selectivity improvement.

Table 4. Summary of data obtained in the hydrogenation of crotonaldehyde over Pt/SiO_2 and $Sn-Pt/SiO_2$ catalysts in stirred tank reactor.

Sn/Pt_s	$W_{ini.}$ $\mu mol/g_{cat} \cdot s$	Selectivity (at $\alpha=3$ %)			Maximum selectivity for SOL and corresponding α	
		SAL	SOL	UOL	S	α
0	1.2	88	5	7	8	4
0.78	5.9	53	6	41	60	48
1.15	1.3	42	5	53	64	22
1.73	1.3	44	4	52	58	20
2.13	0.1	66	4	30	32	3

t=40 °C, P=3.5 bar, preactivation temperature =300 °C, solvent: cyclohexane, $[C]_o$=115 mmol/g_{cat} . total liquid volume: 20 cm^3, α = conversion.

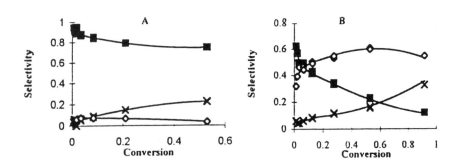

Figure 6. Hydrogenation of crotonaldehyde in liquid phase. Conversion - selectivity dependencies. A - Pt/SiO_2 catalyst, B - $Sn-Pt/SiO_2$ catalyst, Sn/Pt_s = 0.8. Experimental conditions: see table 4. ■ - SAL, X - SOL , ◊ - UOL.

Characteristic feature of this reaction is the increase of the $S_{C=O}$ selectivity with conversion (see Figure 6B). The above increase cannot be explained by kinetic reasons. The increase is attributed to the "reaction induced activation" of the catalyst. However, in the liquid phase due to the low reaction temperature and the presence of condensed phase the "reaction induced activation" cannot take place in the same way as under the condition of the gas phase reaction. We consider that the lack of "reaction induced activation" is the main reason for the low activity and relatively low selectivity of Sn-Pt/SiO$_2$ catalysts with high Sn/Pt$_s$ ratio.

Summary

Results presented in this paper indicate that Sn-Pt/SiO$_2$ catalysts prepared by CSRs are highly active and selective in the hydrogenation of crotonaldehyde into crotylalcohol. The investigation of the tin anchoring step revealed new aspects of the tin anchoring process. In addition to the formation of anchored -SnR$_{(4-x)}$ moieties with monolayer coverage the formation of multilayered surface organometallic species was evidenced. This finding opens new prospects for the preparation of new types of tin modified supported catalysts.

It was also demonstrated that the high $S_{C=O}$ selectivity of the Sn-Pt/SiO$_2$ catalyst has been developed under the condition of the hydrogenation reaction. Our results support the idea proposed recently by Ponec [5], i. e. the substrate is involved in the formation of ionic forms of tin. Ionic tin is required for the perturbation of the aldehyde group via Sn^{+n} - carbonyl interaction. The above perturbation is responsible for the increased $S_{C=O}$ selectivity. However, under condition of liquid phase hydrogenation the formation of surface species involved in the activation of the aldehyde group is strongly hindered. It was the main reason for the low activity of catalyst with high Sn/Pt$_s$ ratio.

In our Sn-Pt/SiO$_2$ catalysts due to the character of CSRs used the first layer of tin organometallic moieties (PSC) is introduced into the platinum, the subsequent layers (SCSL or MLS) are anchored to the PSC. The formed SCSL of the second layer is shown in Figure 7. As emerges from Figure 7 tin organic moieties cover both the top and the side sites of the small platinum cluster. It also means that two forms of tin should be formed after TPR: (i) tin on the top of the platinum, (ii) tin at the platinum - support interface. The higher the Sn/Pt$_s$ ratio the higher is the amount of tin introduced into the interface. The role of tin at the platinum-tin interface was also mentioned in recent studies [5,6,].

It should also be mentioned that catalysts used in this study were pre-reduced in a hydrogen atmosphere at 300 °C. The decrease of the temperature of pre-reduction strongly decreased the rate constant of the hydrogenation of the aldehyde group (k_1). These results indicate that in the formed tin - platinum alloy tin segregation can take place in a hydrogen atmosphere at high temperature. It is suggested that the formed platinum - tin alloy is probably decorated by the excess tin resulting in a random tin overlayer both at the top and the site side of the supported alloy cluster. This form of tin can easily be oxidized by the substrate onto Sn^{+n} required for the activation of the aldehyde group. The benefit of our preparation method is the fact that the required Sn^{+n} moieties are in a close contact with platinum. It is the main reason that the selectivity of catalyst prepared by CSRs exceeds the selectivity of catalysts made by using conventional techniques.

254   *Margitfalvi et al.*

Further studies are required to elucidate the character of active species in the Sn-Pt/SiO₂ catalysts with high Sn/Pt$_s$ ratio required for the selectivity in the hydrogenation of the crotonaldehyde into crotylalcohol. These investigations are in progress in our laboratories.

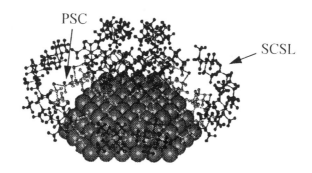

Figure 7. Computer modelling of the formation of Surface Complex in the Second Layer (SCSL) at a small platinum cluster. The first layer is $-Sn(C_2H_5)_2$.

Acknowledgment

Partial financial help provided by OTKA Grant (N° T23322). Special thanks for Dr. E. Tfirst for providing the surface model of SCSL.

References

1. P. Rylander, in *"Catalytic Hydrogenation over Platinum Metals"*, Academic Press, New York and London, 1967.
2. R.L. Augustine, Heterogeneous Catalysis for the Synthetic Chemist, Marcel, New York, N.Y. 1996.
3. T.B.L.W. Marineli, W.Nabuurs, and V. Ponec, *J.Catal.*, **151**, 431 (1995).
4. T.B.L.W. Marinelli and V. Ponec, *J. Catal.*, **156**, 51 (1995).
5. Ponec, V., *Appl. Catal.*, **149**, 27 (1997).
6. M. Englis, Vidyadhar, S. Ranade, and J. A. Lercher, *J. Mol. Catal., A: Chemical,* **121** 69 (1997).
7. Margitfalvi, J., Hegedûs, M., Gôbölös, S., Kern-Tálas, E., Szedlacsek, P., Szabó, S., and Nagy, F., in *"Proceedings, 8th International Congress on Catalysis, West-Berlin, 1984"* Vol 4, p. 903. Dechema, Frankfurt, 1984.
8. Margitfalvi, J. L., Gôbölös, S., Tálas, E., Hegedûs, M., and Ryczkowski, J. *Chem. Ind.* (Marcel Dekker), **62**, 255 (1995).
9. J. L. Margitfalvi, E. Tálas and S. Gôbölös, *Catal. Today,* **6,** 73 (1989).
10. Cs. Vértes, E. Tálas, I. Czakó-Nagy, J. Ryczkovski, S. Gôbölös, A. Vértes, J. Margitfalvi, *Appl. Catal.,* **68,** 149 (1991).
11. J. L. Margitfalvi, P. Szedlacsek, M. Hegedûs, and F. Nagy, *Appl. Catal.,* **15,** 69 (1985).

Optimization of Reaction Conditions in Single-Stage Reductive Amination of Aldehydes and Ketones

John J Birtill[+], Mark Chamberlain, John Hall, Robert Wilson & Ian Costello

ICI Amines*, Wilton, Middlesbrough TS90 8JE, England

+author for correspondence, *now Air Products (Chemicals) Teesside Ltd (same address)

Abstract

This paper is concerned with the optimization of catalyst choice and reaction conditions in several reductive amination processes. The reactions were carried out in a single process stage and the objective was to maximize yield and avoid by-products. The response of selectivity to reaction conditions was also investigated during the course of this work and the results have been interpreted in terms of the reaction mechanism. It is proposed that, under conditions whereby the aldehyde or ketone is introduced gradually, the reactions proceed via hydrogenolysis of intermediate hemi-aminals rather than hydrogenation of intermediate imines. Steric hindrance of N-adsorption or of C-OH bond breakage can affect activity or selectivity in some reactions.

Introduction

The reductive amination reaction of carbonyl compounds can be written generally as follows:

$$R_1 R_2 NH + R_3 R_4 C{=}O \longrightarrow R_3 R_4 C(OH)NR_1 R_2 \xrightarrow[\quad -H_2O \quad]{+\,H_2/\text{catalyst}} \begin{array}{c} \text{Imine or} \\ \text{Enamine} \end{array} \xrightarrow[\text{catalyst}]{+\,H_2} R_1 R_2 (R_3 R_4 CH)N$$

where R_1 - R_4 = H, alkyl or aryl . There are many examples in the literature (1-8).

The reaction may proceed via hydrogenolysis of an intermediate hemi-aminal, $R_3R_4C(OH)NR_1R_2$, or via hydrogenation of an intermediate imine or enamine (1, 2, 4, 5a).

Imine intermediates are not possible in preparation of tertiary amines and when an enamine is also not possible (eg DMBZA, see later) then the reaction mechanism must involve hydrogenolysis of the hemi-aminal. In other cases, where imine formation is possible, the imine may be isolated before hydrogenation in a 2-stage process (7b). For single stage processes some authors assume that the imine is an intermediate without direct evidence. Likewise, the reasons for choice of reaction method, reaction conditions and type of catalyst are not always very clear.

Le Bris et al studied the reductive amination of acetone with mono-isopropylamine (MIPA) over Pt/C and Ni catalysts at 29°C (7a). The reactants were premixed together. The kinetics under these conditions were consistent with the stepwise formation of the imine followed by its hydrogenation. The imine equilibration reaction is acid catalysed and is faster at higher temperature although the extent of imine formation also decreases with temperature (7b). The estimated ratio of adsorption coefficients b on Pt was 95:90:6:1 for imine:MIPA:DIPA:acetone (also $b_{imine} \gg b_{ethanol}$), and the estimated ratio of the hydrogenation rate constants $k_{imine}/k_{acetone}$ was ~23/1, thus explaining why traces of by-product isopropanol only appeared at the very end of the reaction.

The present work was concerned with the development of a single-stage process and the preferred method initially was to carry out the hydrogenation of the intermediate simultaneously with the direct injection of the carbonyl compounds into the amine/catalyst mixture at a temperature $\geq$ 100°C and in the absence of any acid catalyst. In these circumstances hydrogenolysis of the hemi-aminal may be more rapid than imine formation.

Various by-products can occur, e.g., the alcohol or alternative amines, and the combination of catalyst choice with reaction conditions should be optimized so as to avoid their formation (1-8). Avoidance of alcohol by-products is one reason why isolation of an intermediate imine is sometimes preferred before hydrogenation.

Experimental

Reactions were carried out in a stirred autoclave of 1L capacity unless otherwise stated. General details have been reported previously (5a). Target pressure was 24 bar (absolute) for process reasons. The wet slurry of precious metal catalyst, Pd/C or Pt/C, was precharged with excess amine and partially pressurized with hydrogen before being heated to reaction temperature. The aldehyde or ketone feedstock was injected slowly at reaction temperature and hydrogen was fed continuously to maintain reaction pressure. At the end of the injection period, the reaction conditions were maintained until the reaction was complete. In a variation of this procedure, some aldehyde or ketone could be precharged with the amine.

Reaction products were analysed by gas chromatography and the results quoted as peak area % values after excluding excess amine reactant. The capillary GC methods employed were developed during the course of the work:

(a) EDMPA: Chrompack WCOT fused silica, 25m x 0.32mm ID, CP-Sil 5CB, 40 - 200°C. (b) DMBZA: Chrompack WCOT fused silica, 50m x 0.32mm ID, CP-Sil 5CB, 60 - 250°C. (c) ENBA: Chrompack WCOT fused silica, 25m x 0.32mm ID, CP-Sil 19CB, 40 - 200°C.

1   The Preparation of N-Ethyl-3-methyl-2-butanamine or Ethyldimethylpropylamine (EDMPA) from Methylisopropylketone:

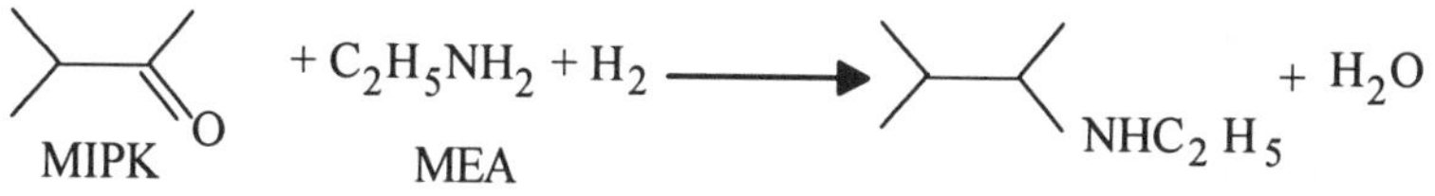

It was reported previously that 5% Pd/C catalysts were selective to EDMPA but were not very active and so careful optimisation of reaction conditions and a high charge of a catalyst with high Pd dispersion were necessary to achieve complete reaction (5a).

5% Pt/C catalysts were very active but produced too much by-product 3-methyl-2-butanol (3M2B). It had been thought that 3M2B might form in the latter stages of the injection period when [MEA] was low and [H_2O] was high, thus pushing the equilibrium away from an imine intermediate back to the ketone. However, the amount of excess MEA (5-20%) had no effect on selectivity and little or no effect on conversion with 5% Pd/C or 5% Pt/C catalyst.

The dependence of selectivity on reaction time was investigated for Pt/C catalyst by repeating the reaction four times under similar conditions. Three tests were shut down prematurely, in two cases part way through the ketone injection. ie, heater and hydrogen supply turned off and cooling water turned on. The results are shown in Table 1 and Figure 1. The hydrogen uptakes were very similar in all tests up to the time when the reaction was ended, showing that the hydrogenation reaction proceeded almost as rapidly as the ketone injection.

Table 1: Variation of EDMPA Selectivity over Pt/C with Reaction Time

Run No	Ketone Injection Time	Total Reaction Time	Analysis (excluding MEA)			Relative Selectivity	
		mins	MIPK	3M2B	EDMPA	EDMPA	3M2B
				peak area %		%	
10	75	75	2.1	0.94	96.0	99.0	1.0
9	150	150	2.8	0.94	95.5	99.0	1.0
8	185	225	0.92	1.06	96.9	98.9	1.1
7	185	300	0.15	1.02	97.8	99.0	1.0

Conditions: 1L autoclave, 23 bar, 100°C, 3.09 moles MEA (70% in aqueous solution), 2.6 moles MIPK. Catalyst: Heraeus 5% Pt/C type K-0125, 0.43g dry weight washed in with 0.83 moles water.

The instantaneous concentration ratios [MIPK]/[Hemi-aminal] or [MIPK]/[imine] should tend to increase throughout the experiment because both [MEA] and [MEA]/[H_2O] fall progressively. If the alcohol 3M2B were formed by direct hydrogenation of MIPK then the 3M2B selectivity would tend to increase throughout the experiment. In fact, the 3M2B selectivity was invariant.

Fig 1 Invariance of Alcohol Selectivity

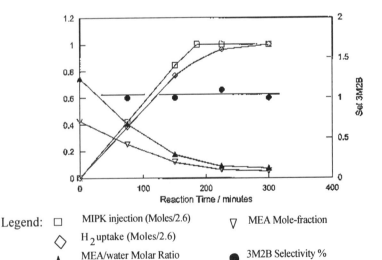

Legend: □ MIPK injection (Moles/2.6)   ▽ MEA Mole-fraction
         ◇ H_2 uptake (Moles/2.6)
         ▲ MEA/water Molar Ratio   ● 3M2B Selectivity %

In another test, the ketone injection rate was slowed down considerably. ie, injection time 323 minutes and overall reaction time 420 minutes. Except for the slower rate the reaction went to completion as normal. There was only a slight improvement in reaction selectivity.

Table 2: Effect of Water on EDMPA Selectivity

| Water added | MIPK | Relative Selectivity | |
| | Conv | EDMPA | 3M2B |
moles	%	%	%
0.8	99.5	98.9	1.09
2.5	99.5	98.8	1.16
3.6	99.1	98.6	1.41

Addition of up to 2.5 moles of water did not appear to affect conversion or selectivity. See Table 2. Addition of 3.6 moles appeared to affect conversion slightly and selectivity more so. However, the degree of phase separation in the cold reaction products increased with amount of water added and only the organic layer was analysed.

The experiment with variable reaction-time was not repeated with Pd/C catalyst. However, two runs were carried out with high and low catalyst charge and intermediate sampling before completion of the reaction. See Table 3.

Table 3: Variation of EDMPA Selectivity over Pd/C

| Run | Catalyst Charge g | MIPK Injection Time | Total Reaction Time | MIPK Conv % | Relative Selectivity EDMPA % | 3M2B |
			Minutes			
98	4.34	180	180	70	99.4	0.56
			360	99.2	99.7	0.3
100	0.87	180	360	55	99.4	0.6
			1191	98.9	99.4	0.6

The selectivity data for the intermediate sample in run 98 with a high charge of Pd/C catalyst are not consistent with those of the final sample, possibly due to sampling error, and so this experiment was inconclusive. The results from run 100 show that for a low charge of Pd/C catalyst the 3M2B selectivity is constant throughout the run, ie, similar to the behaviour of Pt/C catalyst but with less 3M2B.

The relative invariance of EDMPA and 3M2B selectivity throughout the reaction over Pt/C or Pd/C catalyst and the modest influence of water on selectivity suggest that these products are formed by parallel routes from a common intermediate with the same rate-determining step. The reaction scheme is drawn out in Figure 2. The only possible common intermediate is the hemi-aminal $(CH_3)_2CHC(CH_3)(OH)NHC_2H_5$.

The reaction rate is limited by ketone injection rate when catalyst activity is high. Step A2 is faster than step C1. When catalyst activity is limiting then the relative importance of the imine route C may increase. The absorption coefficients b for primary amines are large compared to ketones, e.g., $b_{MIPA}/b_{acetone} = 90/1$ for Pt/C (7a), and so route B, the direct hydrogenation of the ketone is suppressed in the presence of amine.

Figure 2   Reaction Scheme for EDMPA

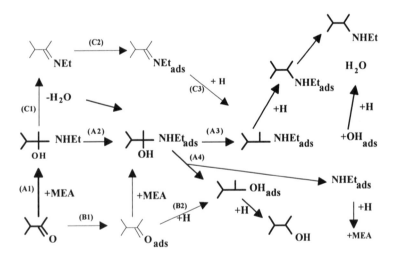

2 The Preparation of N,N-Dimethylbenzylamine (DMBZA) from Benzaldehyde and Dimethylamine (DMA)

$$PhCHO + (CH_3)_2NH + H_2 \xrightarrow[Pd/C]{} PhCH_2N(CH_3)_2 + H_2O$$

This reaction has been reported to take place over Pt catalyst (16). Although 5% Pt/C catalyst was active for this reaction, once again, an excessive amount of by-product, benzyl alcohol, was produced. A sample of sulphided 3% Pt/C was more selective. The selective properties of PtSx/C have been reported elsewhere (5b). 5% Pd/C catalyst was less active and produced a significant amount of by-product toluene but the lowest amount of benzyl alcohol, especially so when all the benzaldehyde was added at the start of the reaction (eg, run LA21).

Some typical results of non-systematic catalyst screening are presented in Table 4. The GC peak assignments were confirmed by GC-MS coupled with GC analysis of product samples spiked with individual pure components. Minor impurities included toluene (<0.2% in benzaldehyde feedstock), benzoic acid (<0.1% in benzaldehyde feedstock) and an unknown by-product D (up to 0.3 PA%) which eluted late from the GC. The hemi-aminal (H-A) $C_6H_5CH(OH)N(CH_3)_2$ was prepared by mixing DMA and benzaldehyde (an exothermic reaction) and its structure was confirmed by NMR. Its GC retention time was also checked by GC-MS. Note that neither an imine nor an enamine intermediate is possible in this reaction.

On account of the low yield of by-product alcohol the reaction conditions were investigated further for 5% Pd/C in a 1L autoclave. See Table 5.

Table 4: Catalyst Screening for the Preparation of DMBZA

Run No	Catalyst Type	Ketone Injection Time	Total Reaction Time	Temp °C	Press Bar	Analysis (excluding DMA)				
		mins				PhCHO	PhCH$_2$OH	DMBZA	H-A	Tol
							peak area %			
LA4	5%Pd/C	180	270	130	24	1.6	1.5	92.0	0	2.6
LA5	5%Pd/C	180	270	145	27	2.2	0.2	91.6	0.02	4.2
LA9	5%Pt/C	180	270	130	27	0.2	7.9	91.5	0.02	0.3
LA6	5%Pt/C	180	270	145	27	0.2	3.9	95.1	0.01	0.4
LA19	3%Pt-S/C	180	270	120	27	0.2	2.2	97.2	0	0.4
LA18	3%Pt-S/C	180	270	130	27	0.2	2.2	97.2	0	0.3
LA16	3%Pt-S/C	180	270	145	27	0.2	1.6	97.8	0	0.4
LA21	5%Pd/C	0	240	130	27	0.03	0.17	99.4	0.1	0.1
LA24	5%Pd/C	0	240	130	24	0	0.21	98.2	1.0	0.1
LA20	3%Pt-S/C	0	240	130	27	0.06	1.5	97.7	0.2	0.4

Conditions: 300 ml autoclave. Charge: PhCHO 1 mole, anhydrous DMA 1.2 moles (20% excess), water 1.1 mole. Autogenous pressure: 16 bar @ 120°C, 18 bar @ 130°C, 24 bar @ 145°C, 8 bar in runs 20, 21, 24. Catalysts from Johnson-Matthey (5% Pd/C; FR307), Heraeus (5% Pt/C; K0125), Engelhard (3% Pt-Sx/C; 7035). Catalyst dry wt 0.32g, except run LA19 0.53g.

Table 5: Optimization of Reaction Conditions for DMBZA over Pd/C

Run No	Temp °C	Initial Autog Press	Total Press Bar	PhCHO Inj Time	Total Reaction Time	Product Analysis (excluding DMA)				
				mins		PhCHO	DMBZA	PhCH$_2$OH	Tol	H-A
5	130	22	29	171	261	-	90.9	5.5	1.0	-
6	130	21	24	164	254	-	91.5	4.1	1.0	-
7[g]	130	21	24	170	270	-	93.9	4.0	2.1	-
4	80	9	24	163	253	0.03	32.5	66.6	0.5	-
2	100	-	24	157	247	0.18	37.5	59.4	2.7	-
3	120	17	24	170	260	0.01	91.2	8.2	0.6	-
6	130	21	24	164	254	-	91.5	4.1	1.0	-
14	145	26	29	175	265	0.08	87.8	5.5	6.2	-
8	130	21	24	168	1,308	0.01	92.5	5.1	2.2	-
23[f]	130	21	24	183	273	-	91.2	5.9	2.7	-
10[a]	130	17	24	176	266	0.04	91.4	5.9	2.6	-
11[b]	130	13	24	176	266	0.04	85.3	6.6	7.9	-
17[b]	145	20	24	166	256	0.03	69.1	4.2	13.7	-
18[c]	130	21	24	173	263	0.01	99.3	0.14	0.15	-
9[c]	145	28	31	180	270	0.03	98.7	0.27	0.17	-
15[c]	145	28	31	165	255	0.02	98.9	0.32	0.21	-
20[d]	130	12	24	120	249	0.02	99.5	0.16	0.13	-
21[d]	145	16	24	180	300	-	99.2	0.28	0.15	-
19[e]	130	8	24	0	203	-	99.1	0.07	0.12	0.29
22[ef]	130	8	24	0	145	-	99.4	0.07	0.15	0.13

Conditions: 1L autoclave; Catalyst 5% Pd/C, Johnson-Matthey Type FR307 (Pd metal surface area 19 M^2/g cat), dry wt 1.3g in all runs except run 7. Standard Charge: PhCHO 4.0 moles, DMA 4.8 moles (20% excess), water 1.67 moles. Variations: (a) Run 10: water 4.2 moles. (b) Runs 11, 17: 5.5 moles MeOH added instead of water. (c) Hydrogen supply isolated for first 50 minutes of ketone injection in runs 9, 15, 18; autogenous pressure only. (d) 1/3 PhCHO precharged rather than injected (e) All PhCHO precharged with other components and allowed to stand for 60 minutes at reaction temperature before hydrogen supplied. (f) Additional samples were taken for analysis during these runs, (g) catalyst wt 2.5g.

Comparing runs 6 and 7, increased catalyst charge had no effect on the level of benzyl alcohol. Reaction temperature influenced the autogeneous pressure of DMA and, due to the constraint of 24 bar total pressure, the partial pressure of hydrogen was low in reactions at high temperature until some DMA had been consumed. The level of benzyl alcohol was least at 130°C and increased significantly at lower temperatures. Extension of the post-injection reaction time in run 8 made little difference. In run 10 more water was added in order to reduce the autogenous DMA pressure. The level of benzyl alcohol increased in comparison to run 6 but the effect may not be significant given the results of runs 8 and 23. Alternatively, C-OH bond breakage may be reversible when [OH] $_{ads}$ is high or the higher $P(H_2)$ may favour direct hydrogenation of benzaldehyde over reductive amination. Addition of methanol instead of water caused the levels of by-products benzyl alcohol and, more significantly, toluene to increase.

The experimental procedure was adjusted to encourage formation of hemi-aminal before the hydrogenation stage. In runs 9, 15, 18 the hydrogen supply was isolated for the first 50 minutes of benzaldehyde injection. The reaction selectivity improved dramatically, i.e., product yield around 99%. Similar results were obtained when part of or all of the benzaldehyde was precharged with the other ingredients rather than injected. A 95% yield of DMBZA by hydrogenation of the preformed intermediate with Pd/C catalyst at 80°C was reported elsewhere after this work was completed (15).

The product profiles during the reactions were monitored by analysis of intermediate samples for the reaction methods with benzaldehyde injection in run 23 and benzaldehyde pre-addition in run 22. The results are shown in Table 6. With benzaldehyde injection, little benzyl alcohol appeared to form during the first 30 minutes but the amount increased significantly thereafter, when the DMA concentration was lower and $p(H_2)$ was higher. Toluene increased progressively throughout the reaction, presumably formed by hydrogenolysis of benzaldehyde, possibly also of DMBZA or benzyl alcohol (8). With benzaldehyde pre-addition, the level of benzyl alcohol remained low throughout the reaction and the proportion of toluene remained constant, suggesting that hydrogenolysis of the DMBZA product was not the cause.

Table 6: Progress with Reaction Time of DMBZA Preparation Over Pd/C

Run	Time minutes	Product Composition (excluding DMA) PA%				
		PhCHO	DMBZA	PhCH$_2$OH	Tol	H-A
22	0	45.1	0.7	0.07	0.13	49.8
	15	33.2	31.7	ND	0.12	34.0
	25	17.1	65.3	ND	0.12	17.1
	145	ND	99.4	0.07	0.15	ND
23	30	8.0	47.2	0.2	0.28	41.5
	60	0.8	92.7	5.1	0.51	ND
	120	0.3	91.5	6.0	1.3	ND
	270	ND	91.3	5.9	2.7	ND

Note: Run 23, Benzaldehyde all injected continuously over 180 mins; Run 22, Benzaldehyde all precharged with amine.

The mechanism proposed to account for these observations is shown in Figure 3. The selective route A to DMBZA involves preformation of the hemi-aminal intermediate which undergoes hydrogenolysis at a favourable rate at $\geq$ 130°C. The less selective route B involves adsorption on the catalyst of benzaldehyde, which, due to the its aromaticity, competes with amines for adsorption sites more effectively than other carbonyl compounds. The by-products benzyl alcohol and toluene are formed to a greater extent when some DMA has been consumed, ie, higher p(H$_2$). The formation of toluene is encouraged at high temperatures and in the absence of water (runs 11, 17) and may proceed via double hydrogenolysis of the adsorbed intermediate or possibly via a benzylidene intermediate PhCH$_{ads}$. Such an intermediate could also react with DMA to yield DMBZA. The involvement of an ethylidene intermediate CH$_3$CH$_{ads}$ has previously been proposed for the amination of ethanol (9).

Figure 3 Reaction Scheme for DMBZA

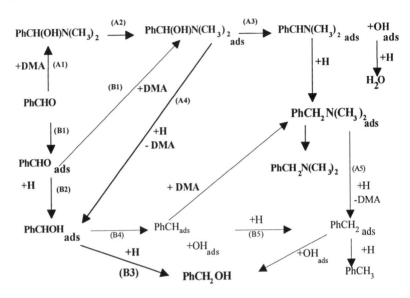

The intermediate sampling in run 22 showed that some free benzaldehyde was present throughout the run, i.e., that formation of the hemi-aminal was not complete. However, hydrogenation of the free benzaldehyde was almost negligable. It is inferred that the hemi-aminal is adsorbed preferentially to benzaldehyde itself.

3 The Preparation of N-Ethyl-n-butylamine (ENBA) from n-Butyraldehyde and Mono-ethylamine (MEA)

$$n\text{-}C_3H_7CHO + C_2H_5NH_2 + H_2 \rightarrow n\text{-}C_4H_9NHC_2H_5 + H_2O$$

The analogous preparation of N-methyl-n-butylamine (NMBA) over nickel has been described previously (10). Under the basic conditions of the reaction medium the aldehyde tends to undergo self condensation to 2-ethylhexanal, leading to the amine impurity $CH_3NHCH_2CH(C_2H_5)C_4H_9$. The condensation reaction was avoided by control of the injection rate of n-butyraldehyde so that its instantaneous concentration in the reaction mixture was low. However, nickel catalyst produced significant amounts of by-products n-butanol and the tertiary amines, dimethylbutylamine and the methyldibutylamine.

In this work, the preparation of ENBA was investigated using Pd/C and Pt/C catalysts. A 44% yield of ENBA from direct reductive amination using reduced platinum oxide catalyst had been reported previously (11). A yield of 84% ENBA was achieved quite easily using 5% Pt/C catalyst and the general reductive amination method described earlier.

The by-products to the reaction were investigated by GC-MS. Ethyldi-n-butylamine (EDNBA), formed by consecutive reaction of ENBA or a reaction intermediate with n-butyraldehyde, was a significant by-product (8.7%). By-products B-F were formed by the self-condensation reaction of n-butyraldehyde to $C_3H_7CH = C(C_2H_5)CHO$ which reacts with MEA at either the -CH=C or -CHO positions, leading ultimately to various saturated and unsaturated products with end-members $C_3H_7CH(NHC_2H_5)CH(C_2H_5)CH_2NHC_2H_5$ and $C_4H_9CH(C_2H_5)CH_2NHC_2H_5$. The tendency of conjugated unsaturated ketones to react at both C=C and C=O positions has been noted previously (5a, 12).

The yield of ENBA relative to EDNBA was improved by operation with high MEA concentration, i.e., 100% MEA excess or greater.

In line with previous work (10, 13), it was found that continuous slow injection of n-butyraldehyde was effective for avoidance of condensation products. Typical results are described in Table 7.

Table 7: ENBA Preparation over 5% Pt/C

Run	MEA Excess %	Catalyst Charge (Dry) g	Reaction Time h Injection	Reaction Time h Total	BuOH	Imine	ENBA	EDNBA	Cond B-F
49	100	0.35	5	22	0.3	-	83.6	3.4	6.2
51	100	0.70	4	23	0.3	0.2	94.1	3.7	1.2
52	160	0.70	6	23	0.3	-	94.6	2.7	1.2
53	160	1.05	4.5	5	0.3	0.3	94.6	2.8	0.4

Conditions: 1L autoclave; catalyst 5% Pt/C Heraeus type K-0125, 80°C, 24 bar; 3.3 or 4.3 moles MEA (70 wt % in water) and 15 g water precharged with catalyst, 1.65 moles n-butyraldehyde injected continuously.

Reactions were generally complete shortly after completion of n-butyraldehyde injection but the autoclave was sometimes left overnight under agitation before the mixture was cooled. Pt/C catalyst was preferred to Pd/C on account of a higher yield of ENBA. See runs 53, 56 in Table 8. A slightly improved yield of ENBA was achieved in run 58 using 5% Rh/C catalyst (Johnson Matthey) under similar conditions to run 52. See Tables 7, 8 for details.

Table 8: Choice of Catalyst for Preparation of ENBA

Run	Catalyst Type		Catalyst Charge	Reaction Time h		Product Analysis PA%				
			g	Injection	Total	BuOH	Imine	ENBA	EDNBA	Cond B-F
53	5% Pt/C	Heraeus K0125	1.05	4.5	5	0.3	0.3	94.6	2.8	0.4
56	5% Pd/C	Heraeus 91709	1.41	5	5.3	0.24	0.4	83.4	14.3	0.7
57	Ni	Degussa B113RW	4.0	5	22	1.7	-	63.6	2.4	2.9
58	5% Rh/C	Johnson-Matthey 20A	0.70	5	24	0.36	-	96.0	0.7	1.4
59	4.5% Pd/ 0.5% Rh/C	Heraeus K-0237	0.71	5	23	0.30	0.2	82.3	8.6	7.9
60	5% Ru/C	Johnson-Matthey	0.72	5	23	0.89	1.6	82.0	1.3	8.4

Conditions: 4.3 moles (160% excess) MEA (70 wt% in water); otherwise as Table 7.

It has been reported previously that rhodium catalysts are effective in reductive amination for avoiding alcohol condensation products and also that they are directive towards secondary amines as opposed to primary or tertiary amines (14).

Discussion

It has been reported previously that Pt/C catalysts have high activity for the reductive amination of cyclohexanone with MEA to form ethylcyclohexylamine (ECHA) by the continuous ketone injection method but the level of by-product cyclohexanol (0.76%) is too high (5a). Pd/C catalysts are less active but produce much less cyclohexanol. The reaction intermediate for ECHA could be an imine but no imine is possible for the reaction of cyclohexanone with dimethylamine which shows similar kinetics and by-products (17). It is proposed that these reactions, when carried out by the continuous ketone injection method, might proceed via the same mechanism which does not involve the imine.

Pd/C catalysts have low activity for the reductive amination of MIPK with MEA but well-dispersed Pd/C can be used effectively with a relatively high charge. Once again, Pt/C catalysts are more active but produce too much by-product alcohol, apparently from a common intermediate to the main reaction product, i.e., the adsorbed hemi-aminal. The reaction over Pd/C catalysts is sterically hindered for ketones with a 2-methyl branch (5a).

Pt/C catalysts are also more active than Pd/C catalysts for reductive amination of benzaldehyde with dimethylamine (DMA) but once again produce too much by-product alcohol. The formation of hemi-aminal from benzaldehyde and DMA is rapid. The main product DMBZA and the alcohol by-product may be formed from the adsorbed hemi-aminal intermediate over both Pt/C and Pd/C catalysts. Direct hydrogenation of benzaldehyde appears to be a competing reaction in the continuous ketone injection method but not when the hemi-aminal is preformed over Pd/C catalyst.

Taken altogether, the results suggest that these reductive amination reactions all proceed, not via an imine intermediate as is commonly supposed whenever possible, but via N-adsorption of the hemi-aminal and subsequent hydrogenolysis mainly of the C-OH bond to the amine product, and to some extent also of the C-N bond to the alcohol by-product, more so over Pt than over Pd. The hemi-aminals adsorb more readily via their amino-groups than the parent carbonyl compounds, e.g., $b_{amine} >> b_{ketone}$ (7a). The metal surface catalyses dehydroxylation of the hemi-aminal faster than the homogeneous, acid-catalysed dehydration to an imine. The dehydroxylated species could form an imine by breakage of a C-H bond but hydrogenolysis of the C-M bond leads directly to the reaction product.

Steric hindrance occurs for Pd catalysts in cases where there is a methyl-group at the 2-C position of the ketone (5a). The use of model-building is helpful here. See Figure 4. The adsorption of the hemi-aminal via formation of the M-N bond is followed by breakage of the C-OH bond and formation of a C-M bond which results in close interaction between the metal surface and the 2-methyl groups. Pt may be more active than Pd in such cases because of more accessible sites with less steric interaction.

Figure 4   Steric Interaction and Bond-breaking Steps

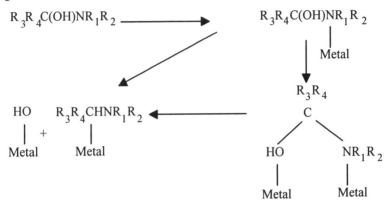

High selectivity to secondary amines when tertiary amine by-products are possible (1) can be explained either by steric hindrance of N-adsorption for the hemi-aminal or of the C-M bond after C-OH bond breakage. For example, non-observation of tri-isopropylamine in di-isopropylamine (7a) due to hindered N-adsorption, non-observation of ethyldicyclohexylamine in ECHA (5a) due to hindered C-M bonding. The tendency to EDNBA by-product in preparation of ENBA occurs because there is little steric hindrance of N-adsorption and none at all for the C-M bond.

The greater selectivity of Pd compared to Pt towards amine versus alcohol product is presumably a consequence of the relative stabilities of the various adsorbed fragments on the respective metal surfaces.

References

1   M Freifelder, in <u>Practical Catalytic Hydrogenation</u>, Wiley-Interscience, 1971, pp 333-389.

2   J March, in <u>Advanced Organic Chemistry</u>, 3rd Edition, Wiley-Interscience 1985, pp 798-800.

3   R E Malz, C-Y Lin and H Greenfield, in <u>Catalysis of Organic Reactions</u>, (W Pascoe, ed) Marcel Dekker, NY 1992, pp 369-371.

4   H Greenfield in <u>Catalysis of Organic Reactions</u>, (J Kosak and T Johnson, eds), Marcel Dekker, NY 1994, pp 265-277.

5 (a) J J Birtill pp 249-262
   (b) R E Malz, E H Jancis, M P Reynolds and S T O'Leary, pp 263-271
   (c) M G Scaros, P K Yonan, M L Prunier, S A Laneman, O J Goodmonson
        and R M Friedman, pp 457-460
        in Catalysis of Organic Reactions, (M G Scaros and M Prunier, eds),
        Marcel Dekker, NY 1995.

6 V L Mylroie, L Valente, L Fiorella and M A Osypian in Catalysis of
   Organic Reactions, (R E Malz, Jr, ed), Marcel Dekker, NY 1996, pp
   300-312.

7 (a) A Le Bris, G Lefebvre and F Coussemant, Bull. Soc. Chim. Fr 1964
        (6), 1366-74, 1347-80; 1964 (7), 1584-94, 1594-1600.
   (b) D C Norton, V E Haury, F C Davis, L J Mitchell and S Ballard, J. Org.
        Chem, 19, 1054 (1954).

8 R L Augustine in Heterogeneous Catalysis for the Synthetic Chemist,
   Marcel Dekker New York, 1996, pp 440, 530-531.

9 J R Jones, A P Sharratt, S D Jackson, L F Gladden and G Webb, Paper 133
   in Proc. 5th Int. Symp. on Synthesis and Applications of Isotopically
   Labelled compounds, (J Allan and R Voges, eds) (1994).

10 G B Patent No 1,116,610 to Ugine Kuhlmann (1968).

11 A C Cope, N A Lebel, H-H Lee, W R Moore, JACS 79, 4720 (1957).

12 GB Patent No 615,715 to Sharples Chemicals Inc, (1949).

13 J F Olin and E J Schwoegler, US Patent No 2,373,705 to Sharples
   Chemicals Inc, (1945).

14 L D Brake, US Patent No 3,597,438 to DuPont (1971).

15 H Moryama, Japanese Patent No 94-112135 to Koei Chemical Company
   (1994).

16 A Skita, F Keil, H Havemann & K P Lawrowsky., Ber., 63B, 34-50 (1930).

17 J J Birtill, unpublished work.

Acknowledgements
Mark Chamberlain and John Hall were on 1 year secondment from Newcastle
University. Additional experimental work was carried out by C L Elston, S
Patel, J M Allan, T R Glanvill. Thanks are due to S D Jackson (ICI Katalco) and
N Hindle (ICI Performance Chemicals) for helpful comments.

1997 Paul N. Rylander Plenary Lecture. Organic Reaction and Catalysis: A Marriage of Convenience

R.A. Sheldon

*DELFT UNIVERSITY OF TECHNOLOGY, Department of
Organic Chemistry and Catalysis,
Julianalaan 136, 2628 BL Delft, The Netherlands*

Abstract

The concepts of atom efficiency and E factors (kg waste/kg product) of chemical processes are explained by reference to catalytic versus stoichiometric methods in organic synthesis. This general theme is richly illustrated with examples of catalytic reduction, oxidation and carbonylation, e.g. zeolite-catalyzed Meerwein-Ponndorf-Verley reduction, olefin epoxidation over heterogeneous titanium catalysis, palladium-catalyzed biphasic carbonylations and (enantio)selective oxidations with chloroperoxidase.

Introduction

The terms organic chemistry and catalysis were both coined by the Swedish chemist, Berzelius, in 1807 and 1835, respectively. Subsequently catalysis and organic synthesis evolved along different paths. Catalysis initially developed largely as a sub-discipline of physical chemistry. Following the advent of the petrochemicals industry in the 1920s catalysis (mainly heterogeneous) was widely applied in oil refining and, to a lesser extent, in bulk chemicals manufacture. Fine chemicals manufacture, generally regarded as being synonymous with industrial organic synthesis, remained the domain of synthetic organic chemists who adhered to the time-honored stoichiometric reagents developed in the nineteenth century. A good example is oxidation chemistry where traditional stoichiometric oxidants such as potassium dichromate are still widely used.

What is wrong with classical organic syntheses employing stoichiometric inorganic reagents? The answer is simple: they produce copious amounts of inorganic salts as waste. Up until fairly recently this was not considered to be a problem. In the last decade, however, chemical producers have been subjected to increasing pressure to minimize waste at source, i.e. primary pollution prevention rather than incremental end-of-pipe solutions. Consequently, traditional concepts of process efficiency are changing from an exclusive focus on chemical yield to one that assigns economic value to waste elimination (1,2).

The E factor, atom efficiency and catalysis

The need for a drastic reduction in waste is readily apparent from a consideration of what I call the E factor - the amount of waste generated per kilogram of product - of chemical processes (Table 1) (3-5). In this context waste is defined as everything generated in the overall process with the exception of the desired product. The waste consists predominantly of inorganic salts and the higher E factors in fine chemicals and pharmaceuticals manufacture is a reflection of the more prevalent use of stoichiometric reagents and multi-step syntheses in these industry segments.

Table 1.   The E factor.

Industry segment	Product tonnage	kg waste/kg product
Oil refining	10^6-10^8	< 0.1
Bulk chemicals	$< 10^4$-10^6	$< 1 \to 5$
Fine chemicals	10^2-10^4	$5 \to > 50$
Pharmaceuticals	10-10^3	$25 \to > 100$

Fine chemicals manufacture is rampant with classical 'stoichiometric' technologies such as reductions with metal hydrides and dissolving metals, oxidations with permanganate and hexavalent chromium compounds, halogenations, nitrations, Grignard reactions and a wide variety of reactions employing stoichiometric amounts of mineral or Lewis acids, e.g. Friedel-Crafts acylations.

Another way of looking at this problem is from the viewpoint of *atom efficiency* or *atom utilization*, introduced independently by us (6) and Trost (7). Atom efficiency is defined as the ratio of the molecular weight of the desired product to the sum total of the molecular weights of all the materials produced in the stoichiometric equation. It is a useful tool for quickly evaluating the amount of waste generated in alternative routes to a particular product. For example, the oxidation of α-methylbenzyl alcohol to acetophenone using stoichiometric chromium(VI), as shown in Figure 1, has an atom efficiency of 42%. Catalytic oxidation with O_2, in contrast, has an atom efficiency of 87% and the byproduct is water.

Stoichiometric oxidation :

$$3 \text{ PhCH(OH)CH}_3 + 2 \text{ CrO}_3 + 3 \text{ H}_2\text{SO}_4$$

$$\longrightarrow 3 \text{ PhCOCH}_3 + \text{Cr}_2(\text{SO}_4)_3 + 6 \text{ H}_2\text{O}$$

$$\text{atom efficiency} = \frac{360}{860} = 42\%$$

Catalytic oxidation :

$$\text{PhCH(OH)CH}_3 + \tfrac{1}{2}\text{O}_2 \xrightarrow{\text{catalyst}} \text{PhCOCH}_3 + \text{H}_2\text{O}$$

$$\text{atom efficiency} = \frac{120}{138} = 87\%$$

Fig. 1. Atom efficiencies of stoichiometric vs catalytic oxidation of an alcohol

In practice, of course, much more waste is produced than one would expect from the atom efficiency since in practice an excess of chromium trioxide and/or sulfuric acid is used and the product yield is less than 100%. Moreover, for calculation of E factors other waste, that does not appear in the stoichiometric equation, e.g. solvent losses, drying agents, etc., have to be taken into account. A meaningful comparison should also take into account the energy consumption of alternative routes, which also can be translated into waste generation.

The key role of catalysis in waste minimization is evident: by employing catalysts chemical conversions with 'clean' reagents are rendered possible, e.g. hydrogenation (H_2), carbonylation (CO) and hydroformylation (CO/H_2) or more selective, e.g. oxidations with O_2. An additional benefit of catalytic conversions is that they often provide short-cut routes as alternatives to classical multi-step syntheses.

Comparing alternative processes solely on the basis of the *amount* of waste is obviously a gross oversimplification. Consequently, we introduced the *'environmental quotient'* (EQ) which is obtained by multiplying the E factor by an arbitrarily assigned unfriendliness quotient, Q (4). Thus, depending on the value of Q, even catalytic amounts of certain materials, e.g. heavy metals, may be undesirable components of process effluents. This raises the question of catalyst recycling which is important from both an economic and an environmental point of view. I shall return to this point later. In the ensuing discussion I shall illustrate the foregoing general theme with examples of catalytic reactions of long standing interest in my group, i.e. catalytic reduction (8), oxidation (9) and carbonylation (10).

Zeolite-catalyzed MPV reductions

The Meerwein-Ponndorf-Verley (MPV) reduction of aldehydes and ketones (Reaction 1) was discovered in 1925 (11). It involves the reaction of the substrate with a hydrogen donor, usually isopropanol, in the presence of an aluminium alkoxide, e.g. Al(OPr^i)$_3$. Although the latter is, in principle, a catalyst MPV reductions often employ stoichiometric quantities owing to the low rate of exchange of the alkoxy group in aluminium alkoxides.

$$\text{R}_2\text{CO} + (\text{CH}_3)_2\text{CHOH} \xrightleftharpoons{\text{Al(OPr}^{\text{i}}\text{)}_3} \text{R}_2\text{CHOH} + (\text{CH}_3)_2\text{CO} \qquad (1)$$

Recently, the zeolite beta-catalyzed MPV reduction of cyclohexanones has been reported by van Bekkum and coworkers (12). A major advantage of this method is that it is truly catalytic and the catalyst can be readily separated, by simple filtration, and recycled. An additional benefit is that reduction of 4-tert-butylcyclohexanone affords the thermodynamically less stable *cis*-alcohol in high selectivity (> 95%). In contrast, classical MPV reduction of this ketone affords mainly the *trans*-isomer. Preferential formation of the *cis*-isomer was attributed to transition state selectivity imposed by confinement of the substrate in the zeolite pores. The reaction is of commercial interest as the *cis*-isomer is a fragrance chemical intermediate.

Heterogeneous catalysts for liquid phase epoxidations

The catalytic epoxidation of olefins by alkyl hydroperoxides (Reaction 2) was independently discovered by Halcon and Atlantic Richfield workers in the early sixties (13). Soluble compounds of d^o metal ions, e.g. Mo(VI), W(VI), V(V) and Ti(IV) catalyze Reaction 2, molybdenum(VI) being the most effective.

$$\text{R}^1\diagdown + \text{R}^2\text{O}_2\text{H} \xrightarrow{\text{catalyst}} \text{R}^1\diagup\overset{\text{O}}{\diagdown} + \text{R}^2\text{OH} \qquad (2)$$

At Shell we developed a heterogeneous Ti(IV)/SiO$_2$ catalyst for Reaction 2 exhibiting selectivities comparable to homogeneous molybdenum (14). The superior activity of Ti(IV)/SiO$_2$, compared to soluble titanium(IV) compounds which are mediocre catalysts, was attributed (14) to site isolation of discrete Ti(IV) centres on the silica surface preventing oligomerization to less reactive μ-oxo species, which readily occurs in solution. Moreover, electron withdrawal by silanoxy ligands increases the Lewis acidity of Ti(IV) and, hence, its catalytic activity.

We proposed a mechanism involving oxygen transfer from an electrophilic alkylperoxometal complex to the olefin (14, 15). The primary function of the metal ion is to increase the electrophilic character of the peroxidic oxygens by withdrawing electrons from the O-O bond. Hence, superior catalysts are strong Lewis acids in their highest oxidation state and relative activities follow the order Mo^{VI} >> W^{VI} > Ti^{IV}, V^V. The metal ion should also be a weak oxidant to minimize competing one-electron redox processes (9), which explains why chromium(VI) is a poor catalyst for Reaction 2.

Both the Ti(IV)/SiO$_2$ and homogeneous titanium catalysts are seriously inhibited by water which means that they are unsuitable for epoxidations with H$_2$O$_2$. Hence, one can imagine our amazement at the reports of Enichem workers (16) describing the remarkable activity of titanium(IV) silicalite (TS-1) as a catalyst for olefin epoxidations with 30% aqueous hydrogen peroxide. Two materials, Ti(IV)/SiO$_2$ and TS-1, with the same elemental composition, i.e. 2% TiO$_2$ / 98% SiO$_2$ exhibited totally different catalytic properties. TS-1 contains titanium isomorphously substituted for silicon in the framework of silicalite-1, a hydrophobic, medium pore molecular sieve possessing a three dimensional pore system with pore diameters of 5.3 x 5.5 Å and 5.1 x 5.5 Å. Its remarkable activity can be attributed to site isolation of Ti(IV) centres in the hydrophobic micropores of silicalite. The hydrophobic environment of the active site favors the adsorption of the hydrophobic olefin, thus circumventing inhibition by adsorbed water molecules observed with the hydrophilic Ti(IV)/SiO$_2$. The unprecedented activity of TS-1 as an (ep)oxidation catalyst is reflected in the facile epoxidation of relatively unreactive olefins, such as propylene and even allyl chloride, at temperatures close to ambient in methanol as solvent (Table 2) (17).

Table 2. TS-1 catalyzed epoxidations with 60% aq. H_2O_2[a,b].

Olefin	T (°C)	Time (min)	H_2O_2 conv. (%)	Epoxide sel. (%)
Propylene	40	72	90	94
1-Hexene	25	70	88	90
Cyclohexene	25	90	10	n.d.
Allyl chloride	45	30	98	92

[a] Olefin/H_2O_2 molar ratio = 5; MeOH solvent.
[b] Adapted from ref. 17.

The success of TS-1 led to a flourish of activity in the synthesis of metal-substituted molecular sieves. This was based on the assumption that TS-1 was the progenitor of a broad family of novel catalysts for liquid phase oxidations, for which we coined the term redox molecular sieves (18).

For example, substitution of titanium in larger pore molecular sieves could circumvent the size restriction imposed by the micropore diameter (5.3 x 5.5 Å) of TS-1. Thus, 1-hexene is readily epoxidized with aq. H_2O_2, in the presence of TS-1, while cyclohexene is essentially unconverted (see Table 2). Corma and coworkers (19) synthesized titanium-substituted zeolite beta, which has pore diameters of 7.6 Å, and showed that it catalyzes the H_2O_2 oxidation of 1-hexene and cyclohexene at roughly the same rate. However, in contrast to with TS-1, the main product was the glycol monomethyl ether, formed by acid-catalyzed ring opening of the epoxide by the methanol solvent (Table 3).

We subsequently showed (20) that the epoxide could be obtained in high selectivity by simply neutralizing the Brønsted acid aluminium sites by reaction with an alkali metal acetate and recalcination (Table 3). Similarly, 1-octene afforded the epoxide in high selectivity when treated with TBHP in the presence of alkali metal exchanged titanium-beta (21). More recently, aluminium-free titanium-beta was synthesized and shown to catalyze the epoxidation of 1-octene with H_2O_2, albeit with ring opening in methanol as solvent (22). In this case the ring opening could be suppressed by performing the reaction in bulkier alcohols, e.g. isopropanol and tert-butanol (22).

Table 3. Ti-Beta catalyzed epoxidations with H_2O_2 in MeOH.

Olefin	Catalyst	H_2O_2 conv. (%)	Product sel. (%)		Ref
			epoxide	glycol ether	
1-Hexene	TS-1	98	96	4	19
	Ti-Beta	80	12	80	
Cyclohexene	TS-1	< 5	100	--	
	Ti-Beta	80	0	100	
1-Octene	TS-1	95	76	24	20
	Li-TS-1	85	98	0	
	Ti-Beta	48	0	97	
	Li-Ni-Beta	31	87	5	
	Na-Ti-Beta	22	84	6	

In some cases the presence of both redox and acid sites in a catalyst could be an asset. Such a bifunctional catalyst could be used to catalyze a one-pot conversion involving an (ep)oxidation step and an acid-catalyzed step. For example, Corma described (23) the one-pot conversion of linalool to a mixture of a furan and pyran via titanium-beta catalyzed epoxidation and intramolecular ring-opening (Reaction 3). Similarly, we have employed titanium-beta for the one-pot conversion of olefins to aldehydes or ketones via epoxidation with H_2O_2 followed by acid-catalyzed rearrangement, e.g. the conversion of 2,3-dimethyl-2-butene to pinacolone (Reaction 4) (24).

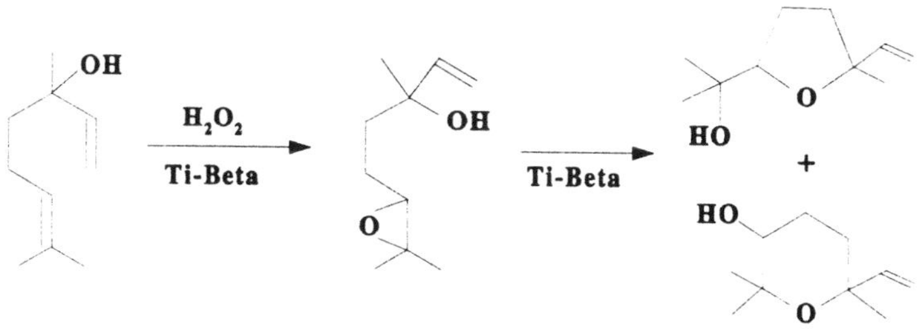

$$(3)$$

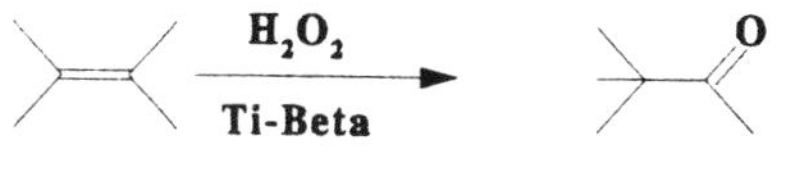

$$(4)$$

The synthesis of titanium-substituted molecular sieves (18, 25) and related materials, such as titanium aerogels (26) and xerogels (27) continues to attract attention because of their enormous potential as heterogeneous catalysts for liquid oxidations (18).

Palladium-catalyzed carbonylations in aqueous media

As noted earlier, catalytic carbonylation is another example of an atom efficient, low-salt technology which can provide alternate, clean routes to a variety of fine chemicals. A good example is provided by the Hoechst Celanese process for the manufacture of the anti-inflammatory drug, ibuprofen (28). A key step involves a 100% atom efficient carbonylation of a benzylic alcohol catalyzed by a palladium-triphenylphosphine complex in an organic solvent (Reaction 5).

$$\text{IBPE} \xrightarrow[\text{Pd}^{II}/\text{Ph}_3\text{P}]{\text{CO}} \text{ibuprofen} \tag{5}$$

In contrast to oxidations and reductions, practical examples of carbonylations and related hydroformylations are known only with homogeneous transition metal complexes. Although heterogeneous equivalents have been synthesized, by covalent attachment of metal complexes to organic or inorganic polymers, for various reasons they were not applied successfully as catalysts. Since recycling of expensive homogeneous catalysts can be an economic and/or environmental problem alternative approaches have been sought (29). One attractive method involves 'immobilization' of the catalyst in a separate liquid phase. The use of water as the second phase has the added advantage that it obviates the use of toxic organic solvents, such as chlorinated hydrocarbons which is prohibitive. Another problem associated with the use of soluble metal complexes is contamination of the product by trace amounts of the metal, in particular when the product is a pharmaceutical as is the case with ibuprofen. The application of biphasic catalysis in organic/aqueous media can also result in a substantial reduction or elimination of such contamination.

Hence, there is currently considerable interest in the application of biphasic organometallic catalysis (30). We became interested in catalytic carbonylations in water in connection with catalytic conversions of carbohydrates as renewable raw materials. As a simple model for carbohydrates we chose to study the carbonylation of hydroxymethylfurfural (HMF), itself a renewable raw material available from acid-catalyzed dehydration of carbohydrates, such as fructose

(31). We found that the water soluble palladium complex, Pd(tppts)$_3$ (tppts = trisulfonated triphenylphosphine) catalyzes the selective carbonylation of HMF to 5-formyl-furan-2-acetic acid (FFA) in an acidic aqueous medium at 70 °C and 5 bar CO pressure (Reaction 6) (32, 33). This was the first example of such a transformation in an aqueous medium. FFA is a precursor for potentially interesting dicarboxylic acid monomers based on a renewable resource. We subsequently applied this method to the carbonylation of benzyl alcohol (Reaction 7) (33) and 1-(4-isobutylphenyl)ethanol (IBPE) (Reaction 5) affording phenylacetic acid and ibuprofen, respectively (34).

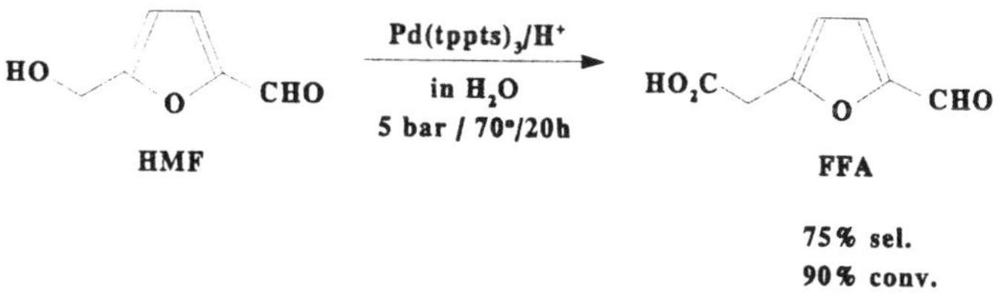

$$\text{HO} \underset{\text{HMF}}{\underset{O}{\diagdown}} \text{CHO} \xrightarrow[\substack{\text{in H}_2\text{O} \\ \text{5 bar / 70°/20h}}]{\text{Pd(tppts)}_3\text{/H}^+} \text{HO}_2\text{C} \underset{\text{FFA}}{\underset{O}{\diagdown}} \text{CHO} \qquad (6)$$

75% sel.
90% conv.

$$\text{PhCH}_2\text{OH} + \text{CO} \xrightarrow[\substack{\text{in H}_2\text{O} \\ \text{60 bar / 100°/10h}}]{\text{Pd(tppts)}_3\text{/H}^+} \text{PhCH}_2\text{CO}_2\text{H} \qquad (7)$$

100% sel.
77% conv.

To account for the observed results in biphasic carbonylations of alcohols catalyzed by Pd(tppts)$_3$ we proposed (32-34) that the Brønsted acid cocatalyst generates a benzylic cation from the alcohol substrate. The cation subsequently reacts with the nucleophile, Pd(tppts)$_3$, to afford the oxidative addition product, a benzylpalladium(II) species. The latter undergoes carbon monoxide insertion to give an acylpalladium(II) intermediate which reacts with water affording the carboxylic acid (Figure 2).

$$\text{ArCH}_2\text{OH} \xrightarrow[\text{H}_2\text{O}]{\text{H}^+} \text{ArCH}_2^+ \xrightarrow{\text{Pd•L}_3} \left[\text{ArCH}_2\text{Pd}^{II}\text{L}_3 \right]^+$$

$$\xrightarrow{\text{CO}} \left[\text{ArCH}_2\text{COPdL}_3 \right]^+ \xrightarrow{\text{H}_2\text{O}} \text{ArCH}_2\text{CO}_2\text{H} + \text{Pd•L}_3 + \text{H}^+$$

Fig. 2. Pd(tppts)$_3$-catalyzed carbonylation of benzylic alcohols

When the benzylic cation contains β-hydrogens this leads to the formation of the corresponding styrene via an El elimination or via β-elimination from the benzylpalladium(II) intermediate. It is difficult to distinguish between these two possible pathways. In either case it leads to the formation of a mixture of the branched and linear acids as observed in the carbonylation of IBPE. The branched/linear ratio was dependent on the temperature and CO pressure (34).

During our investigations of IBPE carbonylation we observed the formation of substantial amounts of p-isobutylstyrene (IBS) at short reaction times and elevated temperatures. This suggested that IBS is a reaction intermediate (probably formed via El elimination) and that the low activity (TOF = 2.3 h^{-1}) observed was due to the low rate of formation of IBS at 90 °C. This led us to study the Pd(tppts)$_3$-catalyzed biphasic carbonylation of olefins (35). A comparison of the carbonylation of IBPE with the hydrocarboxylation of IBS (Reaction 8) under the same conditions - Pd(tppts)$_3$ and p-toluenesulfonic acid at 90 °C with no solvent - showed that the latter reaction proceeds eight times faster than the former. The activity remained low, however, probably due to the low solubility of IBS in the aqueous phase. This led us to study the hydrocarboxylation of styrene (Reaction 8; R = Ph) and propylene (Reaction 8; R = Me), which are more water soluble. Since propylene is less reactive than styrene and less susceptible to polymerization, carbonylation of propylene was performed at higher temperatures (110-120 °C). Under these conditions exceptionally high turnover frequencies (TOF > 2500) were observed. The n/iso ratio of the carboxylic acid products was about 60/40.

$$R\diagdown\diagup + CO/H_2O \xrightarrow{\text{Pd(tppts)}_3/\text{H}^+} R\diagdown\diagup_{CO_2H} + R\diagdown\diagdown\diagup CO_2H \quad (8)$$

R	selectivity (%)	
	iso-	n-
CH$_3$	43	57
C$_6$H$_5$	56	33
4-i-BuC$_6$H$_4$	82	18

More recently, we have shown that analogous palladium complexes of sulfonated chelating diphosphines such as DPPPr-S [1,3-bis-di-m-sulfonatophenylphosphino)propane] in combination with a Brønsted acid catalyze the alternating copolymerization of ethylene and carbon monoxide (Reaction 9) in an aqueous medium (36). Activities were of the same order as those observed with the corresponding organic soluble phosphine ligands in methanol as solvent (37).

$$nC_2H_4 + nCO + H_2O \xrightarrow[\substack{H^+/H_2O \\ 69 \text{ bar}/50°/22h}]{Pd/DPPPr-S}$$

$$CH_3CH_2\!-\!\!\left(\!COCH_2CH_2\!\right)\!\!-\!CO_2H \qquad (9)$$
$$n-1$$

$$n = 100 - 200 \; ; \text{TOF} = 7500 \; h^{-1}$$

The above studies lead us to conclude that a variety of environmentally and economically attractive catalytic methodologies can be developed based on palladium-catalyzed reactions in biphasic media.

Selective oxidations mediated by peroxidases

Another major goal in the development of selective conversions for fine chemicals synthesis is to harness the catalytic potential of enzymes. Biocatalysis has many potential benefits: reactions proceed under mild conditions in water as solvent often with high degrees of chemo-, regio- and enantioselectivity. Our interest in oxidations and, in particular, enantioselective oxidations led us to study biocatalytic oxidations. We were particularly attracted to peroxidases (38), a group of enzymes which catalyzes selective oxidations with H_2O_2 without the stoichiometric cofactor requirement characteristic of oxygenases and dehydrogenases.

Within this group of enzymes we were especially intrigued by one enzyme: chloroperoxidase (CPO; E.C.1.11.1.10) from the marine fungus, *Caldariomyces fumago* (37). CPO is a heme-dependent enzyme which, *in vivo*, catalyzes the oxidation of chloride ion, by H_2O_2, to give hypochlorite. This forms the basis for chlorination of organic compounds *in vivo*. The active oxidant in the CPO-catalyzed oxidation of chloride ion is a high-valent oxoiron porphyrin species.

We reasoned that, in the absence of chloride ion, this oxoiron intermediate should be capable of undergoing (enantio)selective oxygen transfer reactions with organic substrates. Indeed, we (39, 40) and others (41) found that CPO catalyzes the highly enantioselective oxidation of arylmethyl sulfides to the corresponding (R)-sulfoxides (Reaction 10). Furthermore, we showed that CPO performs admirably in tert-butanol/water mixtures which is a better solvent medium for organic substrates. CPO was also shown to catalyze the enantioselective epoxidation of olefins (Reaction 11) (42) and the regioselective oxidation of indoles to the corresponding 2-oxindoles (Reaction 12) (43).

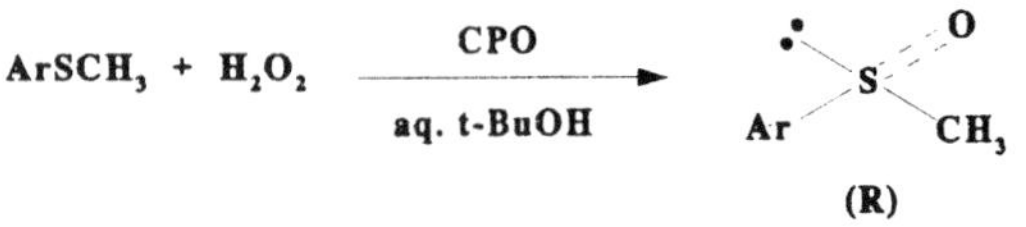

$$ArSCH_3 + H_2O_2 \xrightarrow[\text{aq. t-BuOH}]{\text{CPO}} \quad \underset{(\mathbf{R})}{Ar\overset{\cdot}{\underset{CH_3}{\overset{O}{\,S\,}}}} \qquad (10)$$

>99% yield
>99% ee

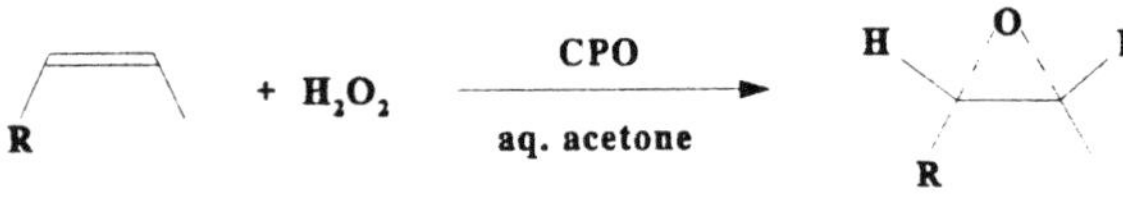

$$\underset{R}{\diagup\!\!=\!\!\diagdown} + H_2O_2 \xrightarrow[\text{aq. acetone}]{\text{CPO}} \quad \underset{R}{H\diagdown\!\!\underset{\triangle}{\overset{O}{}}\!\!\diagup H} \qquad (11)$$

R	Yield (%)	ee (%)
C_5H_{11}	82	96
C_6H_5	67	96

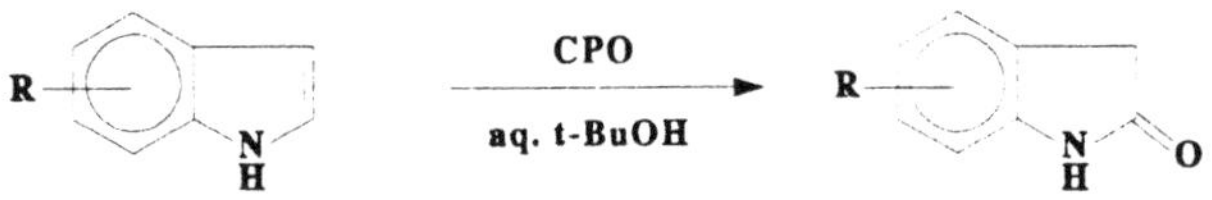

$$R\!-\!\!\!\underset{\overset{|}{H}}{\diagup\!\!\bigcirc\!\!\diagdown N} \xrightarrow[\text{aq. t-BuOH}]{\text{CPO}} \quad R\!-\!\!\!\underset{\overset{|}{H}}{\diagup\!\!\bigcirc\!\!\diagdown N}\!\!\diagdown O \qquad (12)$$

86 - 97% yeild

So far so good, CPO catalyzes a variety of synthetically interesting oxidations with H_2O_2. However, in common with other heme-dependent peroxidases, CPO has a major drawback: limited stability owing to rapid oxidative degradation by H_2O_2. We have been able to increase the stability of CPO by employing aqueous tert-butanol as the reaction medium and slow, feed-on-demand addition of H_2O_2 regulated by means of a H_2O_2 stat (43).

Although this has led to substantial improvements (44) the productivities (g product per g catalyst) of CPO-catalyzed oxidations are still not economically attractive. This is illustrated in Table 4 which compares CPO-catalyzed enantioselective epoxidations with corresponding reactions using the Mn salen/NaOCl system developed by Jacobsen (45). Catalyst costs are directly related to productivities which is illustrated by comparing CPO-catalyzed epoxidation and indole oxidation (see Figure 3) (46).

Chemocatalytic : Mn(Salen*)Cl / NaOCl
Biocatalytic : CPO / H$_2$O$_2$ (NB : product is trans-diol)

	Chemo catalyst	Biocatalyst (CPO)	CPO oxindole synth.
Yield %	90	85	96
ee %	88	97	–
S/C/(mol/mol)	400	2.200	900.000
S/C/ (g/g)[a]	60	6	2.700
Catalyst costs[b] (\$/kg product)	13	ca. 8.000	20

a. M.W. of Mn (salen*)Cl = 635 ; MW of CPO = 42.000
b. Assuming a bulk price of \$ 1000/kg for Mn (salen*)Cl
   and \$ 50.000/kg for CPO

Figure 3. A comparison of biocatalytic and chemocatalytic oxidations

Since the stability problem appears to be an inherent feature of heme-dependent peroxidases it may be more profitable to look for more stable enzymes. Vanadium haloperoxidases (47), for example, are much more stable under oxidizing conditions and tolerate organic cosolvents. Unfortunately, accessibility to the active site is limited to very small substrates such as halide ion. Hence, there is a definite need for the design of peroxidases with increased operational stability and better accessibility to the active site, e.g. by site-directed mutagenesis or by isomorphous substitution of redox metal ions in metallohydrolases (48).

Concluding remarks

The key to waste minimization in chemicals manufacture is clearly the widespread integration of catalytic methods in organic synthesis. The challenge is to develop catalysts with high activities and selectivities, that are stable and easily recycled, and that have a broad scope in organic synthesis. A marriage of convenience after a protracted engagement that lasted more than 150 years.

References

1. P.T. Anastas and C.A. Ferris (eds.), Benign by Design. Alternative Synthetic Design for Pollution Prevention, ACS Symp. Ser. nr. 577, ACS, Washington DC, 1994.
2. J.H. Clark (ed.), The Chemistry of Waste Minimization, Blackie Academic, London, 1995.
3. R.A. Sheldon, *Chem. Ind. (London)*, 12 (1997) and 903 (1992).
4. R.A. Sheldon, *CHEMTECH*, **38** (1994); R.A. Sheldon, *J. Mol. Catal. A: Chemical*, **107**, 75 (1996).
5. R.A. Sheldon, *J. Chem. Tech. Biotechnol.*, **68**, 381 (1997).
6. R.A. Sheldon, in D.T. Sawyer and A.E. Martell (eds.), Industrial Environmental Chemistry, Plenum, New York, 1992, p. 99.
7. B.M. Trost, *Science*, **254**, 1471 (1991); B.M. Trost, *Angew. Chem. Int. Ed. Engl.*, **34**, 259 (1995).
8. P.N. Rylander, Catalytic Hydrogenation over Platinum Metals, Academic Press, New York, 1967.
9. R.A. Sheldon and J.K. Kochi, Metal Catalyzed Oxidations of Organic Compounds, Academic Press, New York, 1981.
10. R.A. Sheldon, Chemicals from Synthesis Gas, Reidel, Dordrecht, 1983.
11. For a recent review see: C.F. de Graauw, J.A. Peters, H. van Bekkum and J. Huskens, *Synthesis*, **10**, 1007 (1994).
12. E.J. Creyghton, S.D. Ganeshie, R.S. Downing and H. van Bekkum, *J. Mol. Catal. A: Chemicals*, **115**, 457 (1997).
13. R. Landau, *Hydrocarbon Process.*, **46**, 141 (1967); M.N. Sheng and J.G. Zajacek, *Advan. Chem. Ser.*, **76**, 418 (1968).
14. R.A. Sheldon and J.A. van Doorn, *J. Catal.*, **31**, 427 (1973); R.A. Sheldon, *J. Mol. Catal.*, **7**, 107 (1980).
15. R.A. Sheldon, *Recl. Trav. Chim. Pays-Bas*, **92**, 253 and 367 (1973).
16. B. Notari, *Stud. Surf. Sci. Catal.*, **37**, 413 (1988); U. Romano, A. Esposito, F. Maspero, C. Neri and M.G. Clerici, *Chim. Ind. (Milan)*, **72**, 610 (1990).
17. M.G. Clerici and P. Ingallina, *J. Catal.*, **140**, 71 (1993).
18. For a recent review see: I.W.C.E. Arends, R.A. Sheldon, M. Wallau and U. Schuchardt, *Angew. Chem. Ind. Ed. Engl.*, **36**, 1144 (1997).
19. A. Corma, M.A. Camblor, P. Esteve, A. Martinez and J.J. Perez-Pariente, *J. Catal.*, **145**, 151 (1994).
20. T. Sato, J. Dakka and R.A. Sheldon, *Stud. Surf. Sci. Catal.*, **84**, 1853 (1994).
21. T. Sato, J. Dakka and R.A. Sheldon, *J. Chem. Soc., Chem. Commun.*,

1887 (1994).
22. J.C. van der Waal and H. van Bekkum, *J. Mol. Catal. A: Chemical*, **124**, 137 (1997).
23. A. Corma, M. Iglesias and F. Sanchez, *J. Chem. Soc., Chem. Commun.*, 1635 (1995).
24. M. van Klaveren and R.A. Sheldon, *Stud. Surf. Sci. Catal.*, **110**, 567 (1997).
25. J.M. Thomas, *J. Mol. Catal. A: Chemical*, **115**, 371 (1997).
26. R. Hutter, T. Mallat and A. Baiker, *J. Catal.*, **135**, 177 (1995).
27. W.F. Maier, J.A. Martens, S. Klein, J. Heilman, R. Parton, K. Vercruyse and P.A. Jacobs, *Angew. Chem. Int. Ed. Engl.*, **35**, 180 (1996).
28. See in J.N. Armor, *Appl. Catal.*, **78**, 141 (1991).
29. R.A. Sheldon, *Curr. Opin. Solid State Mat. Sci.*, **1**, 101 (1996).
30. G. Papadogianakis and R.A. Sheldon, *New J. Chem.*, **20**, 175 (1996).
31. C. Moreau, R. Durand, C. Pourcheron and S. Razigade, *Ind. Crop. Prod.*, **3**, 85 (1994).
32. G. Papadogianakis, L. Maat and R.A. Sheldon, *J. Chem. Soc., Chem. Commun.*, 2659 (1994); see also G. Papadogianakis, J.A. Peters, L. Maat and R.A. Sheldon, *J. Chem. Soc., Chem. Commun.*, 1105 (1995).
33. G. Papadogianakis, L. Maat and R.A. Sheldon, *J. Mol. Catal. A: Chemical*, **116**, 179 (1997).
34. G. Papadogianakis, L. Maat and R.A. Sheldon, *J. Chem. Tech. Biotechnol.*, **70**, 83 (1997).
35. G. Papadogianakis, G. Verspui, L. Maat and R.A. Sheldon, *Catal. Lett.*, **47**, 43 (1997); this reaction was independently discovered by another group, see S. Tilloy, E. Monflier, F. Bertoux, Y. Castagnet and A. Mortreux, *New J. Chem.*, **21**, 529 (1997).
36. G. Verspui, G. Papadogianakis and R.A. Sheldon, *J. Chem. Soc., Chem. Commun.*, accepted for publication.
37. E. Drent, J.A.M. van Broekhoven and M.J. Doyle, *J. Organometal. Chem.*, **417**, 235 (1991).
38. For a recent review see: M.P.J. van Deurzen, F. van Rantwijk and R.A. Sheldon, *Tetrahedron Report Nr. 427*, **53**, 13183 (1997).
39. M.P.J. van Deurzen, B.W. Groen, F. van Rantwijk and R.A. Sheldon, *Biocatalysis*, **10**, 247 (1994).
40. M.P.J. van Deurzen, I.J. Remkes, F. van Rantwijk and R.A. Sheldon, *J. Mol. Catal. A: Chemical*, **117**, 329 (1997).
41. S. Colonna, N. Gaggero, L. Casella, G. Carrea and P. Pasta, *Tet. Asymm.*, **3**, 95 (1992).

42.   E.J. Allain, L.P. Hager, L. Deng and E.N. Jacobsen, *J. Am. Chem. Soc.*, **115**, 4415 (1993).
43.   M.P.J. van Deurzen, K. Seelbach, F. van Rantwijk, U. Kragl and R.A. Sheldon, *Biocat. Biotrans.*, **15**, 1 (1997).
44.   K. Seelbach, M.P.J. van Deurzen, F. van Rantwijk, R.A. Sheldon and U. Kragl, *Biotechnol. Bioeng.*, **55**, 283 (1997).
45.   E.N. Jacobsen, in I. Ojima (ed.), Catalytic Asymmetric Synthesis, VCH, Berlin, 1993, p. 159.
46.   Data for Mn(Salen*) in Figure 3 were taken from C.H. Senanayake, G.B. Smith, K.M. Ryan, L.E. Fredenburgh, J. Liu, F.E. Roberts, D.L. Hughes, R.D. Larsen, T.R. Verhoeven and P.J. Reider, *Tetr. Lett.*, **37**, 3271 (1996).
47.   A. Butler, in J. Reedijk (ed.), Bioinorganic Catalysis, Marcel Dekker, New York, 1993, p. 425.
48.   M. Bakker, F. van Rantwijk and R.A. Sheldon, to be published.

Novel Recyclable Catalysts for Atom Economic Aromatic Electrophilic Nitration

F. J. Waller[a], D. Ramprasad[a], A. G. M. Barrett[b] and D. C. Braddock[b]

a: Air Products and Chemicals, Inc., Allentown, Pa. 18195,
in Joint Strategic Alliance with
b: Imperial College of Science, Technology and Medicine, London, UK.

Abstract

Catalytic quantities of lanthanide(III) triflates (1-10 mol%) were found to mediate the nitration of arenes to excellent conversions using a single equivalent of 69% nitric acid where the only side product is water. Additionally, the catalysts could be recycled and re-used repeatedly with little or no detriment to rate, yield or isomer ratios. A mechanistic understanding of the process was obtained by structural analysis of the putative mononitrate lanthanide intermediates $[Ln(NO_3)(H_2O)_x](OTf)_2$. This enabled design of more efficient catalysts based on tetrapositive metal centres and/or the use of triflamide and triflide counterions in lieu of triflate and their successful application to the nitration of strongly electron deficient substrates such as o-nitrotoluene.

Introduction

Nitration of aromatic compounds is an immensely important industrial process (1). The nitroaromatic compounds so produced are themselves widely utilized and act as chemical feedstocks for a great range of useful materials such as dyes, pharmaceuticals, perfumes and plastics. Unfortunately nitrations typically require the use of potent mixtures of concentrated or fuming nitric acid with sulfuric acid leading to excessive acid waste streams and added expense. Alternatively, nitric acid may be used in conjunction with strong Lewis acids such as boron trifluoride (2). The Lewis acid is used at or above stoichiometric quantities and is destroyed in the aqueous quench liberating large amounts of strongly acidic by-products. With chemists under increasing pressure to perform atom economic processes (3), creating minimal or no environmentally hazardous by-products, development of novel catalyst systems that facilitate aromatic nitrations in this manner should be of great importance (4).

Lanthanides have found increasing use as mild and selective reagents in organic synthesis (5). In particular, lanthanide(III) triflates (6) have been used to good effect as Lewis acids in Diels-Alder (7), Friedel-Crafts (8), Mukaiyama aldol (9) and other (10) reactions. For the Mukaiyama reaction the optimum solvent system was found to be aqueous THF; the catalyst was recycled *via* aqueous work up and used repeatedly with little detriment to rate or yield. The compatibility of lanthanide(III) triflate salts with water and other protic solvents and yet their apparent ability to function as strong Lewis acids is somewhat paradoxical. We sought to harness this water tolerant Lewis acidity and have instigated a program in the area of clean technology using lanthanide(III) triflates for atom economic transformations (11). Herein we report on the use of catalytic quantities of lanthanide(III) (12) and group IV metal (13) triflates for the nitration of a range of simple aromatic compounds in good to excellent yield using a stoichiometric amount of 69% nitric acid wherein the only by-product is water. Furthermore the catalysts are readily recycled by a simple evaporative process.

Results and Discussion

Ytterbium(III) triflate as a recyclable catalyst for the nitration of arenes with nitric acid. Our investigations began with the commercially available hydrated ytterbium(III) triflate (14). A range of simple aromatic compounds, both electron rich (quantified by a negative Hammett coefficient, σ_p^+) and electron poor (positive coefficient), were treated with 1 equivalent of 69% nitric acid in refluxing 1,2-dichloroethane in the presence of 10 mol%

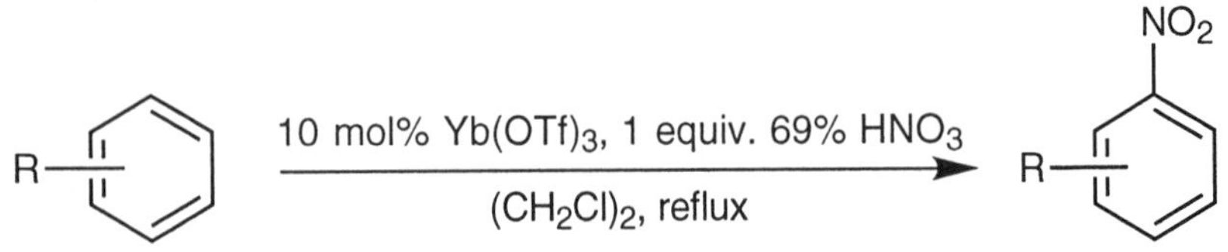

$$R-\text{C}_6\text{H}_5 \xrightarrow[\text{(CH}_2\text{Cl)}_2,\ \text{reflux}]{\text{10 mol% Yb(OTf)}_3,\ 1\ \text{equiv. 69% HNO}_3} R-\text{C}_6\text{H}_4-\text{NO}_2$$

Scheme 1. Nitration of simple arenes with Yb(OTf)₃

Table 1. Nitration of aromatics with catalytic quantities of Yb(OTf)$_3$[a]

Entry	Arene	% Conversion[b,c]	% product distribution[c]		
			ortho	*meta*	*para*
1	Benzene	>75 (75)		n/a	
2	Toluene	>95 (95)	52	7	41
3	Biphenyl	89	38	trace	62
4	Bromobenzene	92	44	trace	56
5	Nitrobenzene	0	-	-	-
6	*p*-Xylene	>95		n/a	
7	*p*-Dibromobenzene	8		n/a	
8	*m*-Xylene	>95	4-NO$_2$: 85 2-NO$_2$: 15		
9	Naphthalene	>95	1-NO$_2$: 91 2-NO$_2$: 9[d]		

a All reactions carried out on a 3 mmol scale with 10 mol% ytterbium(III) triflate and 1.0 equivalent of 69% nitric acid in refluxing 1,2-dichloroethane (5 ml) for 12 h; *b* Isolated yields in parenthesis; *c* Determined by GC and/or ^{1}H NMR analysis; *d* **Care**: nitronaphthalenes are potent human carcinogens.

ytterbium(III) triflate (Scheme 1) (15). Initial work utilized anisole (Hammett coefficient σ_p^+ = -0.78), however this led to extensive polymerisation with the formation of intractable organic tars even at room temperature. Alkyl bearing aromatic compounds (Table 1, entries 2, 6, 8) and naphthalene (Table 1, entry 9) were nitrated smoothly and efficiently in refluxing 1,2-dichloroethane and this represents the effective upper limit in reactivity for the arenes (toluene σ_p^+ = -0.31). The lower reactivity limit was next explored. Biphenyl (entry 3, σ_p^+ = -0.18), and bromobenzene (entry 4, σ_p^+ = +0.15) were nitrated successfully (Table 1) but little success was achieved with benzoic acid (σ_p^+ = +0.42), acetophenone (σ_p^+ = +0.47), ethyl benzoate (σ_p^+ = +0.48) or benzonitrile (σ_p^+ = +0.70). No dinitrated products were observed in any case and in accord with this the system failed to nitrate nitrobenzene (Table 1, entry 5, σ_p^+ = +0.79) and the lower reactivity limit is set at approximate Hammet values of $\sigma_p^+ \approx$ +0.3. In the control experiments with no catalyst, only slow reaction occurred and no more than 10% of nitrated products were observed under these conditions.

It is important to note that the only side product from these nitrations is water. With the additional benefit that the catalyst can be recycled (*vide infra*) this simple methodology represents an efficient atom economic process.

Kobayashi has demonstrated the feasibility of recycling lanthanide(III) triflates for a range of reactions (6b,16). Consequently, ytterbium(III) triflate could be recovered from the reaction mixture at the completion of any particular

Table 2. Recycled Ytterbium(III) triflate for the nitration of *m*-xylene[a]

Run	Conversion (%)[b]	Mass of catalyst (mg)[c]
1	89	190 (>100)
2	81	152 (82)
3	90	127 (68)
4	88	115 (62)

a All runs performed with 3 mmol *m*-xylene, 10 mol% ytterbium triflate (run 1) and 1 equivalent of 69% nitric acid in refluxing 1,2-dichloroethane (5 ml) for 5 h; *b* Determined by GC analysis. The isomeric ratio of 4- and 2-nitroxylene was unchanged throughout (85:15 respectively); *c* Mass of ytterbium(III) triflate recovered from each run. The figures in parenthesis indicate the percentage recovery which were not optimised.

nitration run *via* simple partition work-up and isolated by evaporation of the aqueous phase (15). The resulting free-flowing white solid was found to have an identical IR spectrum to that of the commercially available material and could be re-used as the catalyst for futher nitration runs with no loss in rate or yield or change in isomer distributions. The results for four successive nitrations of *m*-xylene with recycled ytterbium(III) triflate are shown above (Table 2).

Screening the lanthanide(III) triflates. With a view towards rate optimisation a range of lanthanide(III) triflates were examined as potential catalysts. All were found to exhibit catalytic competence but marked differences were apparent (Table 3).

Inspection of the data reveals a clear inverse correlation (with the exception of a few scattered data points) between the ionic radii of the various tripositive lanthanide ions and the extent of nitration whereby the the heavier congeners are the most effective. Thus lanthanum(III) (Z = 57) triflate gave a 64% conversion of naphthalene to mononitronaphthalenes over 1h, whereas the ytterbium(III) (Z =70) triflate catalysed reaction gave a >95% conversion over the same time period.

Postulated mechanism and structural analysis. The isomer distributions from the nitrations of various arenes (Table 1) are consistent with direct electrophilic attack by nitronium ion or, more probably, a nitronium "carrier" of some description (1). The inverse correlation of ionic radii (which should more properly be expressed as charge-to-size ratio, Z/r) and catalytic competence (Table 3) is indicative of interplay between the lanthanide ion and nitric acid where evidently an increasing electrostatic interaction leads to greater reactivity. On this basis a working mechanism can tentatively be proposed. Firstly, nitric acid binds to the lanthanide metal *via* displacement of water from its inner co-ordination sphere (the reactions are performed with 69% nitric acid and the catalyst resides predominately in the aqueous phase as judged by solubility studies). The triflate counterions are outer sphere and effectively

Table 3. Effect of various Ln(OTf)$_3$ for the nitration of naphthalene

Ln(OTf)$_3$[a,b]	Ionic radii/Å[c]	Charge-size (Z/r) ratio	%Conversion[d]
Control	n/a	n/a	38
La (57)	1.172	2.56	64
Ce (58)	1.15	2.61	-
Pr (59)	1.13	2.65	73
Nd (60)	1.123	2.67	67
Pm (61)	(1.11)	(2.70)	radioactive
Sm (62)	1.098	2.73	-
Eu (63)	1.087	2.76	>95
Gd (64)	1.078	2.78	82
Tb (65)	1.063	2.82	93
Dy (66)	1.052	2.85	64
Ho (67)	1.041	2.88	93
Er (68)	1.033	2.90	95
Tm (69)	1.020	2.94	-
Yb (70)	1.008	2.98	>95
Lu (71)	1.001	3.00	-

a All reactions performed on a 3 mmol scale in 5ml refluxing 1,2-dichloroethane with 1.05 equivalents of 69% nitric acid and 10 mol% catalyst; *b* the atomic number is shown in parentheses; *c* taken from *lanthanides in organic synthesis*, T. Imamoto, Academic press, 1994, p. 4; *d* Determined by GCMS: 1-nitro:2-nitronaphthalene were obtained in a 9:1 ratio.

spectator ions (Scheme 2, eqn 1). The resulting strong polarisation due to the metal results in proton liberation affording a lanthanide bound nitrate species **1**

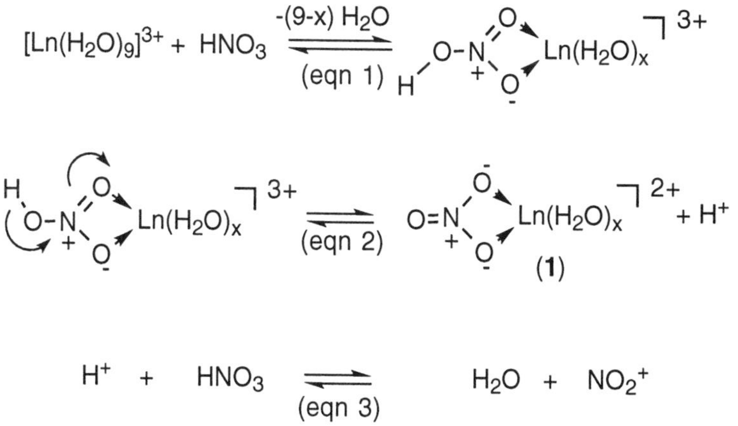

$$[Ln(H_2O)_9]^{3+} + HNO_3 \xrightleftharpoons[\text{(eqn 1)}]{-(9-x) H_2O} \left[\begin{array}{c} H' \overset{O-N}{\underset{+}{\overset{\nearrow O}{\diagdown}}} \overset{O}{\underset{O^-}{\diagup}} Ln(H_2O)_x \end{array} \right]^{3+}$$

$$\left[\begin{array}{c} H \overset{O}{\underset{+}{\overset{O-N}{\diagdown}}} \overset{\nearrow O}{\underset{O^-}{\diagup}} Ln(H_2O)_x \end{array} \right]^{3+} \xrightleftharpoons[\text{(eqn 2)}]{} \left[\begin{array}{c} O=N \overset{\overset{O^-}{\diagup}}{\underset{+}{\underset{O^-}{\diagdown}}} Ln(H_2O)_x \end{array} \right]^{2+} + H^+ \quad \textbf{(1)}$$

$$H^+ + HNO_3 \xrightleftharpoons[\text{(eqn 3)}]{} H_2O + NO_2^+$$

Scheme 2. Proposed mode of action

$$2HNO_3 \rightleftharpoons NO_3^- + NO_2^+ + H_2O$$

The Lanthanide salt 'captures' nitrate ions increasing the equilibrium

concentration of the de facto nitrating agent: NO_2^+

Scheme 3. Effect of lanthanide on classical nitronium ion equilibrium

(Scheme 2, eqn 2) and the proton goes on to liberate NO_2^+ in the classical manner (Scheme 2, eqn 3). Thus, the experimentally observed correlation between increasing charge-to-size ratios (*i.e.* decreasing ionic radii) and extent of conversion is rationalised by noting that the release of the "catalytic proton" is more facile as the metal becomes more polarising.

It becomes clear that the lanthanide salt is acting as a "sink" for nitrate ions, displacing the classical equilibrium process (Scheme 3) invoked for nitrations to the right hand side; the stronger the binding, the greater the equilibrium concentration of nitronium ion and hence the increased rate of nitration.

Inspection of the proposed nitration mechanism (Scheme 2) reveals that the mononitrate dipositive lanthanide species $[Ln(H_2O)_x(NO_3)](OTf)_2$ **1** is the key intermediate. An independent preparation and characterisation of such a species enables possible indentification of **1** directly *in situ* in the reaction mixture. Additionally, spectroscopic examination of these salts may provide some evidence for our working model. We have developed a novel preparation of these mixed salts by simple metathesis of lanthanide chlorides with the requisite quantities of silver nitrate and silver triflate in water (Scheme 4) (17). The resulting hydrated salts were white or lightly coloured (pink, green or yellow) solids which were found to be stable indefinitely at room temperature in the solid state.

Characterisation was accomplished by IR spectroscopy. These materials exhibited the six stretches in the IR spectrum indicative of a symmetrically bound (*i.e.* inner sphere) bidentate nitrate species (18) as well as the requisite (outer sphere) triflate stretches. For example, ytterbium(III) based analogue displayed characteristic nitrate absoptions at 1492, 1384, 1326, 814, 769 and 751 cm^{-1}. The stretch at 1492 cm^{-1}, assigned as the A_1 symmetrical stretch, is known to be critically dependent on the polarising power of the metal

$$LnCl_3 + 2AgOTf + AgNO_3 \xrightarrow{H_2O} [Ln(H_2O)_x(NO_3)] (OTf)_2 + 3AgCl\downarrow$$

(**1**) Ln = La to Lu

Scheme 4. Preparation of putative intermediate **1**

Table 4. Characterisation of $[Ln(H_2O)_x(NO_3)](OTf)_2$ by IR spectroscopy

Ln	Ionic radius (+3) / Å	charge-to-size ratio (Z/r)	IR stretch[a] / cm⁻¹
La	1.172	2.56	1450
Ce	1.15	2.61	1461
Pr	1.13	2.65	1469
Nd	1.123	2.67	1477
Pm	(1.11)	(2.70)	(Radioactive)
Sm	1.098	2.73	1478
Eu	1.087	2.76	1482
Gd	1.078	2.78	1484
Tb	1.063	2.82	1480
Dy	1.052	2.85	1490
Ho	1.041	2.88	1489
Er	1.033	2.90	1492
Tm	1.020	2.94	1496
Yb	1.008	2.98	1492
Lu	1.001	3.00	1497

a The bands for these stretches were fairly broad and at times split into two badly resolved signals - the corresponding error due to peak picking is estimated to be *ca.* 5 cm⁻¹.

centre where increased polarisation of the nitrate leads to the observation of stretches at increased wavenumbers (18). The A_1 stretching frequencies were found to increase steadily in magnitude across the lanthanide period for the $[Ln(H_2O)_x(NO_3)](OTf)_2$ salts starting at 1450 cm⁻¹ for lanthanum (r^{3+} = 1.172 Å, Z/r = 2.56) and ending at a value of 1497 cm⁻¹ for lutetium (r^{3+} = 1.00, Z/r = 3.00) (Table 4). This structural data correlates extremely well with the observed reactivity of the respective lanthanide(III) triflates for nitration and can be taken as strong evidence for the proposed mode of action.

A plot of the A_1 IR stretching frequencies versus the charge-to-size ratio of tripositive lanthanide ions reveals two straight lines with an intersection point located around atomic number Z = 60 (Figure 1). This can be interpreted as a change in co-ordination number where the hydration sphere is somewhat more compact for the heavier (and thus smaller) lanthanide ions in line with literature precedent (19).

Group IV metal triflates as superior nitration catalysts. From the above structural analysis it becomes clear that metal triflates with charge-to-size ratios greater than "3" (*i.e.* greater than that of the smallest lanthanide: lutetium) should be more effective nitration catalysts. We considered that the group IV metals hafnium (r^{4+} = 0.78 Å, Z/r = 5.13) and zirconium (r^{4+} = 0.79 Å, Z/r =

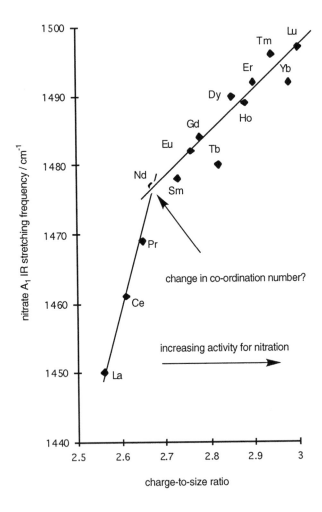

Figure 1. Plot of A₁ IR nitrate stretches for the $Ln(H_2O)_x(NO_3)(OTf)_2$
complexes

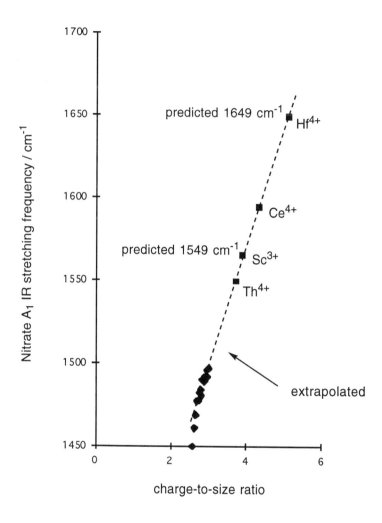

Figure 2. Extrapolation of A_1 stretching frequencies for metals with increased charge-to-size ratios

5.06) might be suitable for such a purpose. In line with this reasoning we noted that hafnium(IV) triflate has been shown to be an effective catalyst for Friedel-Crafts acylations and alkylations where the corresponding lanthanide salts were less active (20).

Extrapolation of the plot shown in Figure 1 indicates that a metal cation with a charge-to-size ratio of approximately "5.1" might be expected to show an A_1 nitrate stretching frequency in its IR spectrum for salts of the type $[M(H_2O)_x(NO_3)](OTf)_3$ in the vicinity of 1650 cm^{-1} (Figure 2) suggestive of a very tightly bound nitrate and hence a very active nitration catalyst (for $M(OTf)_4$).

Hafnium and zirconium mononitrate tris(triflate), $[M(H_2O)_x(NO_3)](OTf)_3$, were prepared from their tetrachlorides in analogous fashion to the lanthanide salts. Much to our delight these (deliquescent) salts displayed A_1 nitrate stretching frequencies in their IR spectra at 1651 and 1650 cm^{-1} respectively. Armed with this pleasing information and with a specific programme aim of nitrating *o*-nitrotoluene (ONT) to dinitrotoluenes (DNTs) with these catalyst types, (catalytic quantities of ytterbium(III) triflate were essentially ineffective for this transformation) hafnium(IV) and zirconium(IV) triflate were prepared as their hydrated salts *via* metathesis of their tetrachlorides with silver triflate in water.

The freshly prepared hafnium salt was employed at a 10 mol% loading for the nitration of ONT with nitric acid in refluxing 1,2-dichloroethane (Scheme 5). After 24h no ONT remained and 2,4- and 2,6-DNT were isolated in 92% yield as a 65:35 mixture after aqueous work-up. Similarly, zirconium(IV) triflate (10 mol%) was found to have comparable catalytic activity for the nitration of ONT; essentially complete conversion was obtained after 24h, and 2,4- and 2,6-DNT were isolated in 87% yield in a 66:34 ratio. The catalyst could be recovered by the usual aqueous work-up regime. This recovered material had an identical IR spectrum to that of the freshly prepared catalyst with additional minor signals for (presumably) nitro containing compounds and was utilized for a futher two nitration runs (Table 5). These results demonstrate that our model based on charge-to-size ratios successfully predicted the activities of various metal triflates for nitrations.

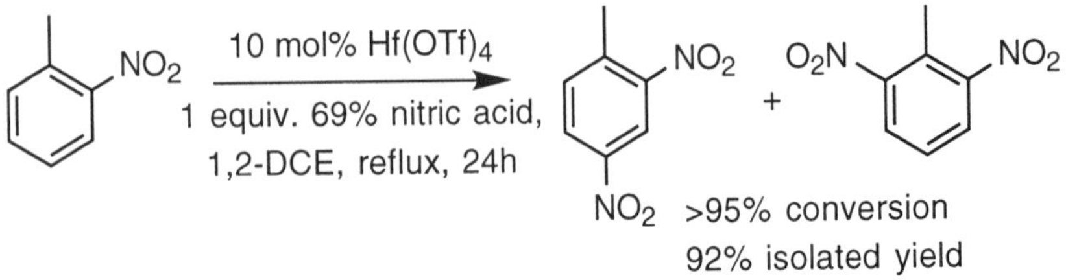

Scheme 5. Nitration of *o*-nitrotoluene with Hf(OTf)$_4$

Table 5. Recycled Zirconium(IV) triflate for the nitration of *o*-nitrotoluene

Run[a]	time / h	% conversion[b]	mass recovery / mg[c]
1	24	> 95 (87)	173 (86)
2	27	> 95 (80)	134 (67)
3	23	88 (76)	121 (61)

a All runs performed with 3 mmol *o*-nitrotoulene, 10 mol% zirconium(IV) triflate (200 mg) (run 1) and 1 equivalent of 69% nitric acid in refluxing 1,2-dichloroethane (2 ml); *b* Determined by GC analysis. The isomeric ratio of 2,4- and 2,6-DNT was unchanged throughout (65:35 respectively). The isolated yields are shown in parenthesis; *c* Mass of zirconium(IV) triflate recovered from each run. The figures in parenthesis indicate the percentage recovery which was not optimised.

Effect of added water. The compatibility of lanthanide(III) triflates with water and yet their ability to function as strong Lewis acids has been noted previously. The effect of water on the rate of mononitration of toluene was studied by varying the composition of nitric acid in the presence of scandium(III) triflate. The nitric acid compositions were 69, 50 and 30%. However, in these experiments toluene was used both as the arene and solvent. Nitration runs (Table 6) demonstrate that the initial rate of formation of nitrotoluenes is rapid in the first hour for both the 69 and 50% nitric acid, but slowed considerably after 5 h. In contrast the 30% nitric acid exhibits an induction time until 2.5 h after which there is a rate increase. Between 5-6 h of reaction time the nitric acid composition of each run levels out to approximately a constant value in the low 20% bracket. No major change in the isomer distribution of the nitrotoluenes was seen as the water content of the nitric acid was varied.

Solvent and counterion effects. The two phase nature of the reaction mixture (aqueous nitric acid/lanthanide salt and solvent/substrate) poses a number of questions. Foremost amongst these is the following: in which phase does the actual nitration occur? Comparison of the nitration rates using 1,2-dichloroethane (b.p. 83 °C) versus cyclohexane (b.p. 80 °C) as the solvents (both reactions performed at reflux) allows speculation on this matter. For the nitration of naphthalene with 10 mol% ytterbium(III) triflate a 78% conversion of naphthalene to mononitronaphthalenes occurred over 0.5h in 1,2-dichloroethane whereas for cyclohexane only a 24% conversion was observed. Based on this result it seems reasonable to conclude that the electrophilic substitution process transpires in the organic phase.

We have previously surmised (*vide supra*) that the *de facto* nitrating agent in this system is the nitronium ion. Evidently the nitronium ion diffuses into the organic phase from the aqueous layer where it has been generated. Moreover, it becomes apparent that in order to maintain overall charge

Table 6. Effect of nitric acid composition on the nitration of toluene

Run	rate[d]	Final HNO$_3$ composition[e]
69%[a]	5	21
50%[b]	1.7	27
30%[c]	0.7	23

a Sc(OTf)$_3$ (0.8 g) was dissolved in 69% nitric acid (0.75 g). The reaction was heated to 60°C and toluene (15g) was added. Aliquots were withdrawn periodically over the reaction time of 6 h and analysed by GC; *b* The run was repeated with 50% nitric acid (1.05g); *c* repeated as in a but with 30% nitric acid; *d* mmol of nitrotoluenes per hour; *e* % by weight.

neutrality, the nitronium ion must be accompanied by an anionic counterion (*i.e.* triflate). Thus, the triflate conterion fulfills two roles; it is the conjugate base of a sufficiently strong acid such that nitric acid is preferentially protonated, producing the nitronium ion *and* acts as a "shuttle" for the transportation of the nitronium ion into the organic phase where nitration occurs (Scheme 6) and where it is thought that the solubility of the nitronium salt (NO$_2^+$ OTf$^-$) is a key factor (21).

  Evidence to support this "counterion modified" model comes from the investigation of hydrated Ln(NO$_3$)Cl$_2$ prepared in the usual way in water (LnCl$_3$, AgNO$_3$). The lanthanide(III) chlorides themselves show little or no catalytic activity for nitrations. A possible rationale for this is obtained by noting that HCl is a poor activator of nitric acid in nitration chemistry. Alternatively, any NO$_2^+$ produced may be trapped by chloride ions to generate NO$_2$Cl which is known to be relatively inactive for nitrations (compared to nitronium salts). However, the IR spectra of the Ln(NO$_3$)Cl$_2$ are essentially

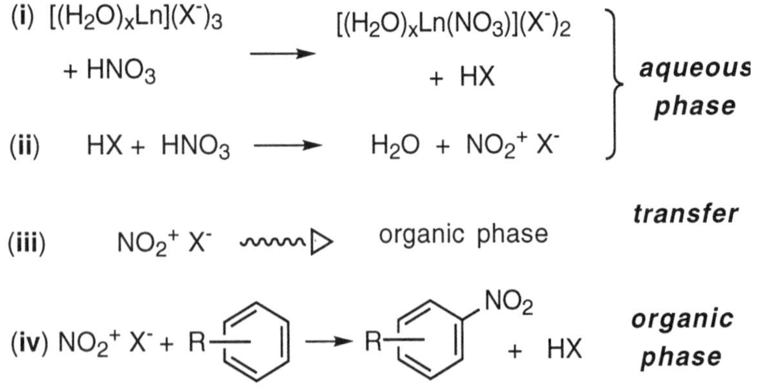

(i) $[(H_2O)_x Ln](X^-)_3$ $\longrightarrow$ $[(H_2O)_x Ln(NO_3)](X^-)_2$

   $+ \; HNO_3$ $+ \; HX$ *aqueous phase*

(ii) $HX + HNO_3 \longrightarrow$ $H_2O \; + \; NO_2^+ \, X^-$

(iii) $NO_2^+ \, X^-$ $\longrightarrow$ organic phase   *transfer*

(iv) $NO_2^+ \, X^- + R\!-\!\!\langle \;\rangle \longrightarrow R\!-\!\!\langle \;\rangle\!-\!NO_2 \; + \; HX$ *organic phase*

Scheme 6. The dual rôle of the counterion

identical to those of $Ln(NO_3)(OTf)_2$ indicating that the chloride ions are outer sphere in these complexes. Additionally the nitrate bands show the same trend as demonstrated for the triflate series (*e.g.*, for $La(NO_3)Cl_2$ the characteristic nitrate stretch is observed at 1459 cm^{-1} and for $Yb(NO_3)Cl_2$ the band appears at 1497 cm^{-1}). This indicates that the lanthanide chlorides are capable of activating nitric acid (*via* metal-nitrate interactions) but critically, the counterion (*i.e.* chloride) is incapable of fulfilling its role (in whatever capacity that may be) and hence no nitration occurs.

We considered various other counterions for the lanthanide salts with a view to producing more active nitration catalysts. The use of trifluoroacetate and pentafluorobenzene sulfonate were ineffective presumably on acidity grounds. The use of the classical conjugate bases of strong acids (BF_4^-, PF_6^-, SbF_6^-) was precluded by their known tendency to undergo fluoride abstraction by strong electrophiles and their penchant for hydrolysis (22).

Yamamoto *et al.* have demonstrated the use of scandium triflamide, $Sc(NTf_2)_3$, for acetylation and acetalisation chemistry where this material is more active than scandium triflate (23,24). We recognised that Tf_2NH is itself a very strong acid (gas phase acidity measurements have shown it to be a stronger acid than triflic acid (25)) and as such the triflamide anion may well be successful as a counterion for nitrations. Additionally it was hoped that the proposed *in situ* nitrating agent $Tf_2N^- NO_2^+$ would have enhanced solubility over nitronium triflate thus leading to an additional rate acceleration. Interestingly, $Tf_2N^- NO_2^+$ is known (26), as is $TfO^- NO_2^+$ (27), but does not appear to have been utilized for nitrations. Indeed, preliminary experiments have shown that catalytic quantities of $Yb(NTf_2)_3$ (28) are 2-3 times more effective than $Yb(OTf)_3$ for the nitration of toluene. In addition, similar experiments have shown that $Sc(NTf_2)_3.8H_2O$ (29) is 2-3 times more effective for toluene nitration than anhydrous $Sc(OTf)_3$ to which an equivalent amount of water has been added. However at this stage it is not clear whether this rate enhancement is due to the triflamides increased inherent acidity or due to a solubility effect or both.

A logical progression along the series TfO^-, Tf_2N^- leads us to Tf_3C^- ("triflide"). Tf_3CH is a known compound (30). Remarkably, this material has been shown to be an even stronger acid than either $HOTf$ or $HNTf_2$ (in the gas phase) (25) and has been employed as an electrolyte in non-aqueous high voltage batteries (31), displaying high anodic stability (32). The Tf_3C^- counterion would seem to be an attractive target on two counts for lanthanide catalysed nitrations: an inherent acidity increase would lead to faster rates and it seems probable that $Tf_3C^- NO_2^+$ would have increased solubility in organic phases compared to the triflate and triflamide salts.

Accordingly, we have recently developed a novel and convenient synthesis of the triflide anion, isolated as its cesium salt (33), and its use as a counterion in conjunction with lanthanides is presently being investigated. These findings will be reported in due course.

Conclusion

We have demonstrated the use of lanthanide(III) salts for the nitration of arenes where the only side product is water and the catalysts can be readily recycled. A mechanism has been presented consistent with the observed product ratios which delineates both the role of the metal centre and the counterion.

Acknowledgements

The Imperial college group thank Air Products and Chemicals Inc. for the support of our research under the auspices of the joint strategic Alliance, the EPSRC, Glaxo Group Research Ltd. for the most generous endowment (to A. G. M. B.) and the Wolfson Foundation for establishing the Wolfson Centre for Organic Chemistry in Medical Science at Imperial College.

References and notes:

1. (a) Olah, G. A.; Malhotra, R.; Narang, S. C. *Nitration: Methods and Mechanisms*, VCH, New York, 1989. (b) Ingold, C. K. *Structure and Mechanism in Organic Chemistry*, 2nd ed., Cornell University Press, Ithaca, New York, 1969. (c) Olah, G. A.; Kuhn, S. J. in *Friedel-Crafts and Related Reactions*, Olah, G. A. ed., Wiley-Interscience, New York, Vol. 2, 1964. (d) Schofield, K. *Aromatic Nitrations*, Cambridge University Press, London, 1980.
2. (a) Thomas, R. J.; Anzilotti, W. F.; Hennion, G. F. *Ind. Eng. Chem.* **1940**, *32*, 408. (b) Hennion, G. F. *U. S. Patent* **1943**, 2,314,212.
3. Trost, B. M. *Angew. Chem., Int. Ed. Engl.* **1995**, *34*, 259.
4. For a recent nitration using clean methodology see: Smith, K.; Musson, A.; DeBoos, G. A. *J. Chem. Soc., Chem. Commun.* **1996**, 469. However this method suffers from the disadvantage of stoichiometric quantities of acidic by-products. for the use of clays and other such solid supported acid sources see reference 1a.
5. Imamaoto, T. in *Lanthanides in Organic Synthesis*, Academic Press, London, 1994.
6. Reviews: (a) Marshmann, R. W. *Aldrichimica Acta* **1995**, *28*, 77. (b) Kobayashi, S. *Synlett* **1994**, 689. (c) Engberts, J. B. N. F.; Feringa, B. L.; Keller, E.; Otto, S. *Recl. Trav. Chim. Pays Bas* **1996**, *115*, 457.
7. (a) Ishitani, H.; Kobayashi, S. *Tetrahedron Lett.* **1996**, *37*, 7357. (b) Kobayshi, S.; Ishitani, H.; Hachiya, I.; Araki, M. *Tetrahedron* **1994**, *50*, 11623. (c) Marko, I. E.; Evans, G. R. *Tetrahedron Lett.* **1994**, *35*, 2771. (d) Saito, T.; Kawamura, M.; Nishimura, J.-I. *Tetrahedron Lett.* **1997**, *38*, 3231.

8. (a) Kawada, A.; Mitamura, S.; Kobayashi, S. *J. Chem. Soc., Chem. Commun.* **1996**, 183. (b) Kawada, A.; Mitamura, S.; Kobayashi, S. *J. Chem. Soc., Chem. Commun.* **1993**, 1157.

9. (a) Kobayashi, S.; Nagayama, S. *J. Org. Chem.* **1997**, *62*, 232. (b) Kobayashi, S.; Hachiya, I. *J. Org. Chem.* **1994**, *59*, 3590 and references cited therein.

10. (a) Kobayashi, S.; Ishitani, H.; Ueno, M. *Synlett* **1997**, 115. (b) Kobayshi, S.; Ishitani, H.; Komiyama, S.; Oniciu, D. C.; Katritzky, A. R. *Tetrahedron Lett.* **1996**, *37*, 3731. (c) Forsberg, J. H; Spaziano, V. T.; Balasubramanian, T. M.; Liu, G. K.; Kinsley, S. A.; Duckworth, C. A.; Poteruca, J. J.; Brown, P. S.; Miller, J. L. *J. Org. Chem.* **1987**, *52*, 1017. (d) Meguro, M.; Yamamoto, Y. *Heterocycles* **1996**, *43*, 2473. (e) Diana, S.-C. H.; Sim, K.-Y.; Loh, T.-P. *Synlett* **1996**, 263. (f) Meguro, M.; Asao, N.; Yamamoto, Y. *J. Chem. Soc., Perkin Trans. 1* **1994**, 2579. (g) Annunziata, R.; Cinquini, M.; Cozzi, F.; Molteni, V.; Schupp, O. *J. Org. Chem.* **1996**, *61*, 8293. (h) Harrington, P. E.; Kerr, M. A. *Synlett* **1996**, 1047. (i) Makioka, Y.; Shindo, T.; Taniguchi, Y.; Takaki, K.; Fujiwara, Y. *Synthesis* **1995**, 801. (j) Fukuzawa, S.-I.; Tsuchimoto, T.; Kanai, T. *Bull. Chem. Soc. Jpn.* **1994**, *67*, 2227. (k) Hosono, S.; Kim, W.-S.; Sasai, H.; Shibasaki, M. *J. Org. Chem.* **1995**, *60*, 4. (l) Chini, M.; Crotti, P.; Favero, L.; Macchia, F.; Pineschi, M.; *Tetrahedron Lett.* **1994**, *35*, 433. (m) Aspinall, H. C.; Browning, A. F.; Greeves, N.; Ravenscroft, P. *Tetrahedron Lett.* **1994**, *35*, 4639. (n) Matsubara, S.; Yoshioka, M.; Utimoto, K. *Chem. Lett.* **1994**, 827. (o) Cozzi, P. G.; Di Simone, B.; Umani-Ronchi, A. *Tetrahedron Lett.* **1996**, *37*, 1691. (p) Jenner, G. *Tetrahedron Lett.* **1996**, *37*, 3691. (q) Hanamoto, T.; Sugimoto, Y.; Yokoyama, Y.; Inanaga, J. *J. Org. Chem.* **1996**, *61*, 4491. (r) Keller, E.; Feringa, B. L. *Synlett*, **1997**, 842.

11. For the atom economic acylation of alcohols using acetic acid as the acetyl source where the only side product is water and the catalyst is readily recyclable see: Barrett. A. G. M.; Braddock, D. C. *Chem. Commun.* **1997**, 351.

12. Waller, F. J.; Barrett, A. G. M.; Braddock, D. C.; Ramprasad, D. *Chem. Commun.* **1997**, 613. For an interesting report utilizing La(NO3)3/HCl/NaNO3 for the nitration of phenols (but not apllicable to any less electron rich aromatics) see: Ouertani, M.; Girard, P.; Kagan, H. B. *Tetrahedron Lett.* **1982**, *23*, 4315.

13. Waller, F. J.; Barrett, A. G. M.; Braddock, D. C.; Ramprasad, D. *Tetrahedron Lett.* **1998**, *39*, in press.

14. All the lanthanide(III) triflates are commercially available from the Aldrich Chemical Co. bar Promethium (radioactive) and Cerium (available in its +4 oxidation state).

15. The following procedure is representative: Nitric acid (69%; 192 μl, 3.0 mmol) was added to a stirred suspension of ytterbium(III) triflate (186 mg, 0.30 mmol) in 1,2-dichloroethane (5 ml). The suspension

dissolved to give a two phase system in which the aqueous phase was the more dense. Toluene (240 µl, 3.0 mmol) was added and the stirred mixture was heated at reflux for 12 h. During the reaction a white solid precipitated and the organic phase became yellow, and after 12 h no phase boundary was apparent. The solution was allowed to cool and diluted with water. The yellow organic phase was dried (MgSO4) and evaporated to give nitrotoluene (390 mg, 95%). The colourless aqueous phase was evaporated to give ytterbium(III) triflate as a white free-flowing solid (183 mg, 98%).

16.   (a) Kobayshi, S.; Hachiya, I.; Yamanoi, Y. *Bull. Chem. Soc. Jpn.* **1994**, *67*, 2342. (b) Kobayashi, S.; Hachiya, I.; Takahori, T. *Synthesis*, **1993**, 371. (c) Kobayashi, S.; Hachiya, I.; Takahori, T.; araki, M; Ishitani, H. *Tetrahedron Lett.* **1992**, *33*, 6815.

17.   The following procedure is representative: ytterbium(III) chloride (775 mg, 2 mmol) as a solution in water was added to a solution of silver nitrate (340 mg, 2 mmol) and silver triflate (1.03 g, 4 mmol) in water. A white precipitate (AgCl) was formed immediately which was filtered at the pump giving a colourless solution. The solution was evaporated under reduced pressure to give a white solid (1.36 g, 99%).

18.   Addison, C. C.; Logan, N.; Wallwork, S. C. *Quarterly Rev.* **1971**, *25*, 289.

19.   (a) A. Chatterjee, E. N. Maslen and K. J. Watson, *Acta Cryst.*, **1988**, *B44*, 381. (b) Lu, T.; Ji, L.; Tan, M.; Liu, Y.; Yu, K. *Polyhedron* **1997**, *16*, 1149. (c) Semenova, L. I.; Skelton, B. W.; White, A. H. *Aust. J. Chem.* **1996**, *49*, 997. (d) Faithfull, D. L.; Harrowfield, J. M.; Ogden, M. I.; Skelton, B. W.; Third, K.; White, A. H. *Aust. J. Chem.* **1992**, *45*, 583. (e) Harrowfield, J. M.; Lu, W. M.; Skelton, B. W.; White, A. H. Aust. J. Chem. **1994**, *47*, 321.

20.   Hachiya, I.; Moriwaki, M.; Kobayshi, S.; *Bull. Chem. Soc. Jpn.* **1995**, *68*, 2053.

21.   Nitronium triflate is approximately 150-fold more reactive than the "more-ordered" nitronium tetrafluoroborate and hexafluorophosphate salts in chlorinated solvents and this is attributed to increased solubility conferred to the nitronium ion by the triflate counterion. (ref. 27a)

22.   Strauss, S. H. *Chem. Rev.* **1993**, *93*, 927.

23.   (a) Ishihara, K.; Kubota, M.; Yamamoto, H. *Synlett* **1996**, 265. (b) Ishihara, K.; Karumi, Y.; Kubota, M.; Yamamoto, H. *Synlett* **1996**, 839.

24.   For the use of bis(trifluorosulfonylmethane)imide as a alternative counterion in conjunction with lanthanides, see: (a) Kobayashi, H.; Nie, J.; Sonada, T. *Chem. Lett.* **1995**, 307. (b) Mikami, K.; Kotera, O.; Motoyama, Y.; Sakaguchi, H.; Maruta, M. *Synlett* **1996**, 171. For the use of perfluorooctanesulfonate see: (c) Hanamoto, T.; Sugimoto, Y.; Jin, Y. Z.; Inanaga, J. *Bull. Chem. Soc. Jpn.* **1997**, *70*, 1421. For the use of a sulphonated resin as a 'counterion' see: (d) Yu, L.; Chen, D.; Li, J.; Wang, P. G. *J. Org. Chem.* **1997**, *62*, 3575.

25. Koppel, I. A.; Taft, R. W.; Anvia, F.; Zhu, S-Z.; Hu, L.-Q.; Sung, K.-S.; DesMarteau, D. D.; Yagupolskii, L. M.; Yagupolskii, Y. L.; Ignat'ev, N. V.; Kondratenko, N. V.; Volkonskii, A. Y.; Vlasov, V. M.; Notario, R.; Maria, P.-C. *J. Am. Chem. Soc.* **1994**, *116*, 3047.

26. Foropoulos, J., Jr.; DesMarteau, D. D. *Inorg. Chem.* **1984**, *23*, 3720.

27. (a) Coon, C. L.; Blucher, W. G.; Hill, M. E. *J. Org. Chem.* **1973**, *38*, 4243. (b) Yagulpol'skii, L. M.; Maletina, I. I.; Orda, V. V. *Zh. Org. Khim.* **1974**, *10*, 2226. (c) Effenberger, F; Geke, J. *Synthesis* **1975**, 40.

28. The ytterbium salt was prepared from commercially availlable lithium triflamide (3M chemical company). A representative procedure is given below (ref. 29)

29. Sc(NTf$_2$)$_3$.8H$_2$0 was prepared as follows: A column (2cm diameter and 21 cm in length) was loaded with an aqueous slurry of Amberlyst A-26 in the chloride form. Lithium triflamide (48.8g, 0.166 mol) dissolved in 1100 mL of water was eluted through the column. A halide test was negative in the last collected fractions. The column was washed with 50 0mL water followed by 200 mL methanol. Scandium triflate (0.9g, 1.82 mmol) was dissolved in 100 ml methanol and eluted through the column. The column was washed with 100 mL methanol and all washings were collected and evaporated in air to yield an oil. Under vacuum, the oil became a light tan solid (1.5 g). A ^{13}C NMR in D$_2$O showed a characteristic quartet. A ^{19}F NMR spectrum showed a singlet at 79.29 ppm. The solid had 15% water by Karl-Fischer analysis. The best fit from all the data is Sc[N(SO$_2$CF$_3$)$_2$.8H$_2$O with a theoretical Sc of 4.37% (found 4.97%).

30. Turowsky, L.; Seppelt, K. *Inorg. Chem.* **1988**, *27*, 2135.

31. (a) Dominey, L. A. *U. S. Patent* **1993**, 5,273,840. (b) Dominey, L. A.; Koch, V. R.; Blakley, T. J. *Electrochim. Acta* **1992**, *37*, 1551. Benrabah, D.; Baril, D.; Sanchez, J. Y.; Armand, M.; Gard, G. G. *J. Chem. Soc., Faraday Trans.* **1993**, *89*, 355. (c) Koch, V. R.; Nanjundiah, C.; Appetecchi, G.; Battista, G.; Scrosati, B. *J. Electrochem. Soc.* **1995**, *142*, L116. (d) Aurbach, D.; Chusid, O.; Weissman, I.; Dan, P. *Electrochem. Acta* **1996**, *41*, 747.

32. (a) Koch, V. R.; Dominey, L. A.; Nanjundiah, C.; Ondrechen, M. J. *J. Electrochem. Soc.* **1996**, *143*, 798. (b) Walker, C. W., Jr; Cox, J. D.; Salomon, M. *J. Electrochem. Soc.* **1996**, *143*, L80.

33. Waller, F. J.; Barrett, A. G. M.; Braddock, D. C.; Ramprasad, D. *J. Org. Chem.* **1998**,*63*, submitted.

Environmentally Friendly Catalysts for Acylation Reactions

M. Campanati, F. Fazzini, G. Fornasari, A. Tagliani and A. Vaccari.

Dip. Chimica Industr. e Materiali, Viale Risorgimento 4, 40136 Bologna (Italy)

O. Piccolo.

Chemi Spa, Via dei Lavoratori 55, 20092 Cinisello Balsamo MI (Italy).

Abstract

Solid acid catalysts were investigated in the acylation of activated organic substrates in order to study the possibility of replacing $AlCl_3$, due to its environmental constraints. Unlike with toluene, in the acylation of mesitylene using catalytic amounts of solid acid catalysts yields comparable to that with $AlCl_3$ were observed using benzoyl chloride, while with the other acylating agents (propionyl, 2-chloropropionyl, acetyl chloride or acetic anhydride) lower values were obtained as a function of the strength and stability of the carbocation formed. However, in comparison to $AlCl_3$, in all cases the solid acid catalysts considerably reduced the formation of by-products due to mesitylene or polymerization reactions. For the catalysts investigated, the following scale of reactivity was detected: acid treated clays > pillared clays > clays > zeolites, as a function of the accessibility and acidity of the active sites, surface area and nature of the pillar constituent. The interesting behaviour of commercial acid treated clays was confirmed by the acylation of thiophene with p-fluorobenzoyl chloride, with almost complete formation of the 2-isomer, useful intermediate for a pharmaceutical active ingredient.

1. Introduction

Homogeneous Lewis acid catalysts such as $AlCl_3$ are employed worldwide in many chemical processes of considerable relevance; in particular the Friedel-Crafts acylation of aromatics is a key step in the production of pharmaceuticals, fragrances, dyes, pesticides, etc. (1-3). However these systems have many important drawbacks such as use in higher than stoichiometric amounts, low selectivity, no possible reuse, troublesome work-up, toxicity and corrosion, significant waste and HCl production. Over the past few years, the development of new chemical processes that are less hazardous to human health and the environment have received extensive attention (4). In particular the substitution of liquid catalysts with solid acid catalysts appears to be very attractive (5-13). Even though their acid sites can be very strong, solid acid catalysts (zeolites, clays and pillared clays) hold their acidity internally and, thus, are safe to handle. Furthermore, most of them are derived from aluminum and silicon oxides and therefore can be disposed of harmlessly. On the whole, solid acid catalysts are environmentally friendly catalysts (14). In addition they can be used in catalytic amounts, are simple to recover and/or reuse and are commercially available at relatively low prices.

The objective of the present research was to shed light on acylation reactions of aromatic substrates in the liquid phase, activated towards electrophilic substitutions, using commercially available solid acid catalysts, and more specifically to study the main reaction parameters as a possible basis for further catalyst improvements and/or applications.

2. Experimental

Different solid acid catalysts were investigated [mainly commercial catalysts or reference materials of the Concerted European Action-Pillared Layered Structures (CEA-PLS)]:

i) zeolites [H-ZSM5 with Si/Al ratio= 55 and 90 (prepared in our laboratory) and HY with Si/Al= 1.5 (University of Cosenza, Italy)];

ii) clays [a montmorillonite Cat-C (University of Modena, Italy) and two saponites EMV-351 and EMY (Tolsa, Spain)];

iii) acid treated clays [K-10 (Aldrich, Germany) and F-13, F-20, and F-20X (Engelhard)];

iv) pillared clays:

    a. montmorillonites, having different pillar constituents: Al [AZA (CEA-PLS) or EXM-534 (Süd-Chemie, Germany)]; Al/Fe (1:9 w/w) [FAZA (CEA-PLS) or CAT-9317 (University of Modena, Italy)]; Zr [EXM-551 (Süd-Chemie, Germany)];

    b. saponites, having Al as the pillar constituent [ATOS (CEA-PLS) or PAL-EV and PAL-EY (Tolsa, Spain)];

v) $ZnCl_2$ (clayzic) or $FeCl_3$ supported on the acid treated clay K-10 (15).

For comparison a homogeneous catalyst $AlCl_3$ (Aldrich, Germany) was also investigated. Before the tests, the solid catalysts were calcined overnight at 573K. The catalytic tests were carried out by mixing under constant stirring 10^{-3} moles of the acylating reagent (acetyl, benzoyl, propionyl, and 2-chloro-propionyl or p-fluorobenzoyl chloride or acetic anhydride) in a flask containing a large excess (about 10 times) of toluene, mesitylene or thiophene, the solid catalyst (0.1 g) or

Table 1. Catalytic behavior of some solid acid catalysts in the acylation of mesitylene with benzoyl chloride (T= 473K; reaction time 15 min). For comparison the result for $AlCl_3$ is also reported.

Catalyst	Conversion (%) of benzoyl chloride	Yield (%) in 2,4,6-trimethylbenzophenone
$AlCl_3$	100	100
H-Y	100	98
K-10	100	99
AZA	100	98
PAL-EY	100	88

$AlCl_3$ ($2 \cdot 10^{-3}$ moles) and an internal standard (tridecane or dodecane depending on the acylating agent employed). The tests were carried out at the reflux temperature of the aromatic substrate operating under an argon atmosphere in order to avoid radical oxidation of the substrate. The products, separated by filtration from the solid catalyst, were tentatively identified by GC-MS using a Hewlett-Packard GCD 1800 system. The structure of the main products, preliminarily isolated by preparative layer chromatography on commercial plates of silica gel, were determined on the basis of the ^{1}H-NMR spectra recorded in CDCl$_3$ by a 200MHz spectrometer Varian Gemini 200.

3. Results and Discussion

The nature of the organic substrate plays a key role in acylation reactions. For example, when toluene (i.e. a slightly activated substrate) was reacted with acetyl chloride negligible yields (< 4%), in comparison to that with $AlCl_3$ (53%), were observed with all the solid catalysts investigated, without any appreciable change in selectivity, unlike that previously reported for La-exchanged Y (5) or, more recently H-ZSM5 zeolites (16). This low activity cannot be attributed to catalyst deactivation due to HCl formation, considering that similar low values also were observed using acetic anhydride.

On the other hand, in the acylation of mesitylene (i.e. an activated substrate) with benzoyl chloride HY zeolite, acid treated and pillared clays gave rise to yields in 2,4,6-trimethylbenzophenone comparable to that obtained using $AlCl_3$ (Table 1), whereas the reaction failed using medium-size pore zeolites H-ZSM5, regardless of the Si/Al ratio, probably due to steric hindrance limitations. With all solid catalysts investigated negligible amounts of by-products due to mesitylene disproportionation (17,18), dimerization or, mainly, polymerization were formed, unlike that observed with the homogeneous catalyst. However, for the solid acid catalysts a key point is careful removal of the water during the activation step, in order to reduce significantly   the hydrolysis of the acylating agent to the corresponding acid.

Together with the nature of substrate, the properties of the acylating agent in the feed are also critical. The yield in the acylation product decreases using the following acyl chlorides: benzoyl > propionyl > acetyl > 2-chloropropionyl (Fig.1). This scale may be justified considering both the activity and the stability of the electrophilic species RCO^+ previously formed. In the case of acetyl chloride, the carbocation formed is very active but poorly selective, easily forming other compounds, such as ketene (1,19). On the other hand, the presence of more electron-donating substituents (such as propionyl and benzoyl) increases the stability of the carbocation favoring the acylation reaction. Finally, in the case

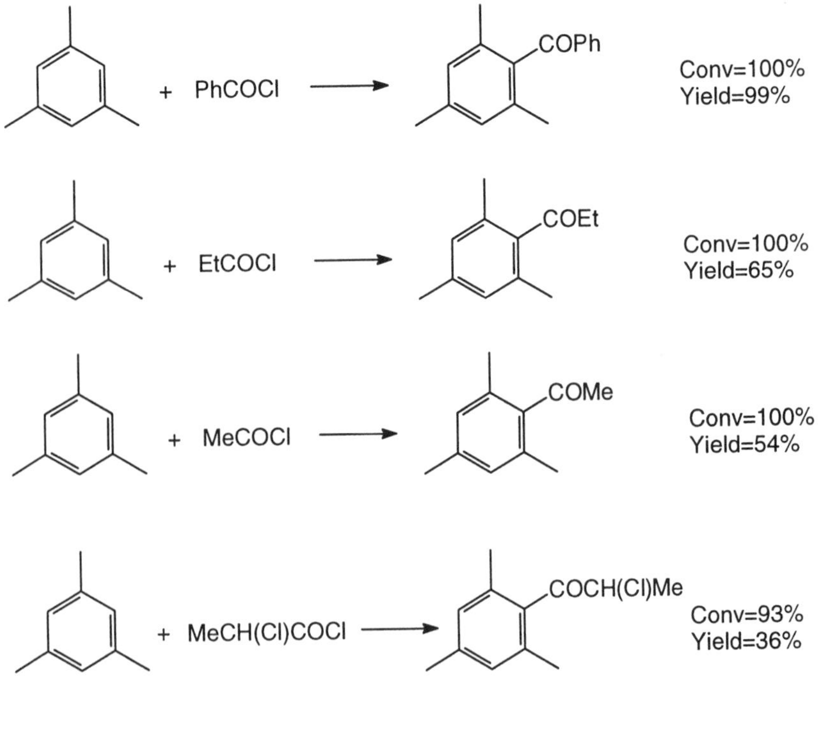

Fig. 1. Role of the nature of the acyl chloride in the reaction with mesitylene (Catalyst = K-10; T= 473K; reaction time = 15 min). Conversion is based on the acyl chloride.

of the 2-chloro-propionyl chloride, consideration must be made of the deactivating effect due to the presence of an electron-withdrawing substituent, evidenced also by the incomplete conversion of the acyl chloride observed only in this case.

On the basis of the previous results, acetyl chloride was chosen for further study of the catalytic performance of the different classes of solid acids investigated. First of all, it must be pointed out that the best yield observed using solid acid catalysts was 30% *ca.* lower than that obtained with AlCl$_3$ (Table 2); however, with the former catalysts mesitylene polymers were not formed and a significant reduction (40% ca.) in the by-products due to its disproportionation or dimerization was observed. Furthermore, an increase in yield (70 instead of 62%) was obtained using acetic anhydride, without increasing the amount of the catalyst F-20 used (unlike that required by AlCl$_3$). For comparison, a yield of 75% has been reported in the reaction of mesitylene with acetyl chloride (T= 423K; reaction time 21h) for the commercial catalyst Envirocat EPZG (Contracts Catalysts, UK) (20). According to Cornelis and Laszlo (6), the preparation of Envirocat catalysts is based on a previous study of the exchange with transition

Table 2. Catalytic behaviour of some solid acid catalysts in the acylation of mesitylene with acetyl chloride (T= 473K; reaction time 15 min). For comparison the result for AlCl$_3$ is also reported.

Catalyst	Conversion (%) of acetyl chloride	Yield (%) in 2,4,6-trimethylacetophenone
AlCl$_3$	100	91
H-ZSM5[a]	100	< 1
H-ZSM5[b]	100	< 1
H-Y	100	1
EMV-351	100	34
EMY	100	30
Cat-Ca	100	20
K-10[c]	100	54
F-13[d]	100	61
F-20[d]	100	62
F-20X[e]	100	34
ZnCl$_2$ on K-10	100	54
FeCl$_3$ on F-13	100	34

a) Si/Al = 55. b) Si/Al = 90. c) Surface area (S.A.) = 220-270m^2/g. d) S.A.= 400m^2/g, acidity = 22mg$_{KOH}$/g; e) S.A.= 275m^2/g, acidity = 13mg$_{KOH}$/g

metal ions of K-10 montmorillonite (21). In our tests, the addition to acid treated clays of $ZnCl_2$ (clayzic) or $FeCl_3$ gave rise, respectively, to no improvement in activity or to a significant decrease, unlike that previously reported by Laszlo and coworkers (15). The negative result observed with $FeCl_3$ can be attributed to the formation of highly crystalline iron oxide particles, showing the key role of the preparation conditions.

The catalytic activity decreases in the order: acid treated clays > clays > zeolites, as a function of the nature and, mainly acidity of the catalysts. The acid treated clays (K-10 and, mainly, F-13 and F-20) show the best catalytic performances, with a direct correlation with acidity and surface area values (compare for example catalysts F-20 and F-20X). Furthermore, it is worth noting that the two saponites (EMV-351 and EMY) show higher activity than the montmorillonite Cat-Ca. These clays all belong to the smectite group of the phyllosilicate family, i.e. are of the 2:1 type, having a sheet of octahedra sandwiched between two sheets of Si-tetrahedra (22-24). On the other hand, it is well known that as a function of their small particle size, clays may exhibit both Brönsted and Lewis acid sites. The former are the external OH-groups, while the Lewis sites are the exposed three-fold coordinated Al^{3+} ions (such as any transition ion), substituting for the Si^{4+} ions in the tetrahedral sheets. Therefore, the different catalytic performances can be interpreted considering that in saponite there is isomorphous substitution of Al^{3+} ions for some of the Si^{4+} ions in the tetrahedral sheets, while in montmorillonite Mg^{2+} ions partially substitute the Al^{3+} ions in the octahedral sheets, emphasizing the important role in the acylation reactions of Lewis acid sites on the catalyst surface (6,7,12,13,15,25).

This last finding is confirmed by comparing the catalytic activity of some clays and the corresponding pillared clays in the acylation of mesitylene with acetyl chloride (Fig. 2). The pillared clays always show (i) higher catalytic activities, related to the presence of additional Lewis acid sites in the pillars (23,25-27), regardless of the nature of the starting clays (saponite or montmorillonite) and (ii) considerable improvement related to the presence of Fe^{3+} ions inside the pillar. This last result stimulated further study on the role of the pillaring species. The following scale of activity: $Fe^{3+} > Al^{3+} > Zr^{4+}$ was found by comparing pillared clays obtained from the same starting clays (AZA and FAZA or EXM-534 and EXM-551, respectively) (Fig. 3). This scale, which differs from the normal sequence of catalytic activity of the metallic chlorides as Lewis acids for the Friedel-Crafts reactions (19), has already been reported in the literature for clays modified through exchange of the interlamellar cations or through impregnation by metallic chlorides, used both in acylation and alkylation reactions (15,21). As previously reported (28), the catalytic activity increases at first as the amount of Fe^{3+} ions inside the pillars increases and then decreases at higher Fe-content due to the segregation of consistent amounts of iron oxide

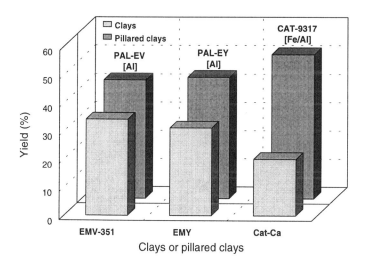

Fig. 2. Comparison of the catalytic activity of some clays and the corresponding pillared clays in the acylation of mesitylene with acetyl chloride (T= 473K, reaction time = 15 min.). The pillaring element is reported in brackets.

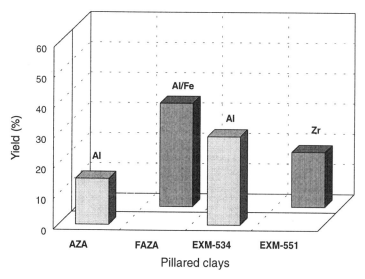

Fig. 3. Catalytic activity as a function of the nature of the pillaring element for some pillared clays in the acylation of mesitylene with acetyl chloride (T= 473K, reaction time = 15 min.)

articles, according to that discussed above for $FeCl_3$ on F-13 acid treated clay. Comparing samples prepared from the same clay-type (montmorillonite) and with the same pillaring element or mixture (Al or Fe/Al), i.e. AZA and EXM-534 or FAZA and CAT-9317, significant differences in activity can be detected, attributable either to possible differences in the preparation method or to the presence of small amounts of transition metals in the clay lattice or as side phases (29). Similar considerations may also be applied to explain the different activities observed for acid treated clays (K-10, F-13 and F-20) or for Al-pillared saponites, for which a yield of 38% was observed using ATOS, lower than the 43% obtained with PAL-EV and PAL-EY.

To confirm the interesting behavior of the commercial acid treated clays, the acylation of thiophene with p-fluorobenzoyl chloride, to produce a useful intermediate for a pharmaceutical active ingredient, was also investigated (Fig. 4). The reaction gives rise to complete conversion of the acylating agent with almost complete formation of the 2-isomer and only traces of the 3-isomer. In this case the main by-product observed is p-fluorobenzoic acid, due to hydrolysis of the acylating agent by residual water present on the catalyst, while other by-products due to side reactions of thiophene dimerization and/or polymerization do not seem to be formed, unlike that observed with $AlCl_3$. On the other hand, acid treated clays have also been very recently studied in the liquid-phase alkylation of naphthalene by isopropanol (30). The clay catalysts showed high stability and activity, with high naphthalene conversion, although on the basis of selectivity strongly dealuminated mordenite was found to be the best catalyst, giving only β- and β,β-isopropylated naphthalenes.

On the basis of the data reported in this paper, clays and pillared clays appear to be very interesting substitutes for $AlCl_3$ with evident advantages for the environment. Considering the very interesting data obtained with commercially available acid treated clays, we also studied the possibility of reducing the amount used. No change in activity was found up to a reduction of 50%. Therefore, the advantages of clay catalysts are evident on the basis of the catalyst/acylating ratio

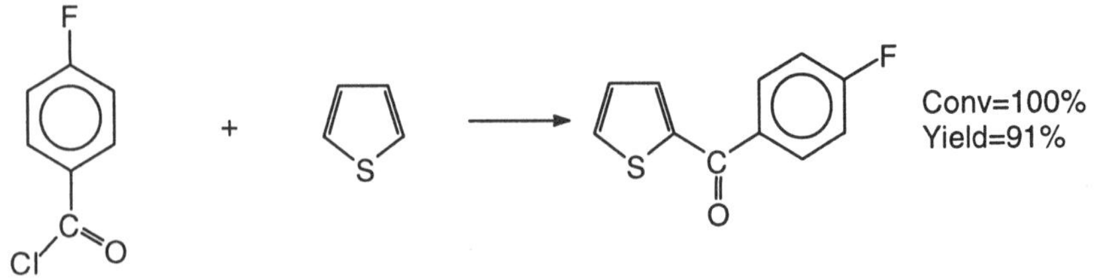

Conv=100%
Yield=91%

Fig. 4. Acylation of tiophene with p-fluorobenzoyl chloride (Catalyst = F-13; T= 357 K; reaction time = 20 min.)

(Fig. 5), calculated for $AlCl_3$ assuming the theoretical amounts necessary for reaction with acetyl chloride or anhydride (10% ca. higher than the stoichiometric amounts), although it is customary to use higher amounts in order to take into account the possible aging of $AlCl_3$, with its partial hydrolysis. Thus on the basis of these considerations, the data reported in Figure 4 can be considered even more favorable than the use of heterogeneous catalysts. Furthermore, using these catalysts it was possible to further add 2-3 consecutive amounts of acetyl chloride without any appreciable deactivation phenomena, in agreement with that previously reported in the literature (15). On the other hand, these catalysts can be easily separated and recycled, after further washing and, mainly, calcination steps, although this may not be economical on the basis of their relatively low prices.

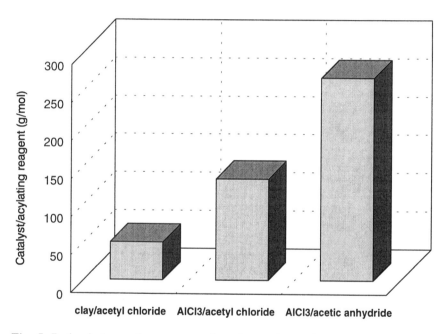

Fig. 5. Ratios between the amounts of catalyst and acetyl chloride or anhydride for homogenous and heterogeneous catalysts (for the latter the same values were obtained with both acylating reagents)

Conclusions

Homogeneous acid catalysts may be suitably replaced with heterogeneous catalysts in the Friedel-Crafts acylation of mesitylene or other activated organic substrates. The yield in the product depends on the nature of the acyl chloride and catalyst employed. The main advantages in the use of solid acid catalysts may be summarized as follow:

i) use in catalytic amounts and relatively low cost;
ii) possible recovery and reuse of the catalyst;
iii) absence of pollution and corrosion effects;
iv) significant reduction in the by-product formed (for example no mesitylene polymer formation unlike that observed with $AlCl_3$).

However, in these catalysts water has to be carefully removed, to avoid hydrolysis of the acylating agent, with a consequent decrease in the reaction yield. Finally, the use of solid acid catalysts may also have very interesting applications in vapor phase reactions, such as, for example, recently reported for the vapor phase synthesis of quinolines or alkylquinolines (31,32). Interesting conversion and selectivity values were achieved using widely available feedstocks and reaction conditions more safe than those required by Skraup's synthesis (high amount of sulfuric acid and significant violence of the initial reaction) (33).

Acknowledgments

The partial financial support from the Ministero Italiano per la Ricerca Scientifica e Tecnologica [MURST, Rome (I)] is gratefully acknowledged. Thanks are due to Dr. M.G. Cramarossa [University of Modena (I)], Prof. G. Giordano [University of Cosenza (I)], CEA-PLS, Engelhard, Süd-Chemie (D) and Tolsa (E) for supplying samples.

References

1. G.A. Olah, *Friedel-Crafts and Related Reactions*, Part I, Interscience, New York, 1964.
2. W. Keim and M. Röper, in *Ullmann's Encyclopedia of Industrial Chemistry* (F.T. Campbell, R. Pfefferkorn and J.F. Rounsaville, Ed.s), Vol. A1, VCH, Weinheim (D), 1993, p. 185.
3. H. -G. Frank and J.W. Stadelhofer, *Industrial Aromatic Chemistry*, Springer and Verlag, Berlin, 1988.
4. P.T. Anastas and T.C. Williamson (Ed.s), *Green Chemistry. Designing Chemistry for the Environment*, ACS, Washington DC, 1996.

5. D.E. Akporiaye, K. Daasvatn, J. Solberg and M. Stöcker, in *Heterogeneous Catalysis and Fine Chemical III* (M. Guisnet, J. Barbier, J. Barrault, C. Bouchoule, D. Duprez, G. Perot and A. Montassier, Ed.s), Elsevier, Amsterdam, 1993, p. 521.

6. A. Cornélis and P. Laszlo, *Synlett.* **3**, 155 (1994).

7. D.R.. Brown, *Geol. Carpathica Clays* **1**, 45 (1994).

8. Y. Izumi, *Aromatikkusu* **46**, 119 (1994); *C.A.* **122**, 239228 (1994).

9. T.J. Pinnavaia, in *Proc. 10th Int. Clay Conference* (G. Churchman, G. Jock, R.W. Fitzpatrick and R.A. Eggleton, Ed.s), CSIRO, Melbourne, 1995, p. 3.

10. M. Onaka, *Yuki Gosei Kagaku Kyokaishi* **53**, 392 (1995); *C.A.* **123**, 82545 (1995).

11. K.G. Ione, *React. Kinet. Catal. Lett.* **57**, 275 (1996).

12. P. Laszlo, *Chem. Ind. (Dekker)* **54**, 429 (1994).

13. J.H. Clark and D.J. Macquarrie, *Org. Proc. Res. & Developm.* **1**, 149 (1997).

14. J.M. Thomas, *Scientific American* **266**, 112 (1992).

15. A. Cornélis, A. Gerstmans, P. Laszlo, A. Mathy and I. Zieba, *Catal. Letters* **6**, 103 (1990).

16. R. Sreekumar and R. Padmakumar, *Synth. Commun.* **27**, 777 (1997).

17. E. Kikuchi, T. Matsuda, H. Fujiki and Y. Morita, *Appl. Catal.* **11**, 331 (1984).

18. E. Kikuchi, T. Matsuda, J. Ueda and Y. Morita, *Appl. Catal.* **16**, 401 (1985).

19. G. A . Olah, *Friedel-Crafts Chemistry*, Wiley, New York, 1973.

20. *Envirocats Supported Reagents*, Technical Bulletin, Contract Catalysts, Prescot (UK), 1995.

21. P. Laszlo and A. Mathy, *Helv. Chim. Acta* **70**, 577 (1987).

22. L. Fodwen, R.A. Barrer and P.B. Tinker (Ed.s), *Clay Minerals: their Structure, Behaviour and Use*, The Royal Society, London, 1984.

23. R. Burch (Ed.), *Catal. Today* **2**, 185 (1987).

24. A.C.D. Newman, *Chemistry of Clays and Clay Minerals*, Monograph 6 of the Mineralogical Society, Longmans, London, 1987.

25. A. Vaccari, in Actas del XV Simp. Iberoamericano de Catalisis (E. Herrero, O. Annunziata and C. Perez, Ed.s), Vol. 1, Univ. Nac. Cordoba, Cordoba (Arg), 1996, p. 37.

26. R.A. Schooheydt, in *Introduction to Zeolite Science and Practice* (H. van Bekkum, E.M. Flanigen and J.C. Jansen, Ed.s), Elsevier, Amsterdam, 1991, p. 201.

27. F. Figueras, *Catal. Rev. -Sci. Eng.* **30**, 457 (1988).

28. F. Fazzini, G. Fornasari, A. Vaccari, L. Valdiserri and P. Carrozza, in *Intern. Symposium on Acid-Base Catalysts III*, Rolduc (NL), April 20-24, 1997.

29. K.R. Sabu, R. Sukumar and M. Lalithambika, *Bull. Chem. Soc. Jpn.* **66**, 3535 (1993).

30. I. Ferrino, E. Rombi, R. Monaci, V. Solinas and A. Vaccari, in *XII Congress of the Division of Industrial Chemistry of the Italian Chemical Society*, Giardini Naxos CT (I), June 22-25, 1997.

31. M. Campanati, P. Savini, A. Tagliani, A. Vaccari and O. Piccolo, *Catal. Letters* **47**, 247 (1997).

32. C.H. MCaater, R.D. Davies and J.R. Calvin, *World Patent* 03,051 (1997); assigned to Reilly Industries Inc. (USA).

33. R.H.F. Manske and M. Kulka, in *Organic Reactions* (R. Adams, Ed.), Vol. 7, Wiley, New York, 1953, p. 59.dqcx

Oxidative anti-Markovnikov-Amination of Aromatic Olefins

Matthias Beller*, Martin Eichberger, Harald Trauthwein
Anorganisch-chemisches Institut, Technische Universität München,
D-85747 Garching

Abstract

Cationic rhodium-complexes catalyze the oxidative *anti*-Markovnikov-amination of aromatic olefins to enamines by simultaneous formation of one equivalent ethylbenzene. Kinetic and mechanistic studies reveal that the yield and the rate of the reaction grow up by increasing styrene : amine-ratio. Furthermore the type of phosphane greatly influences the reaction. The formation of cationic rhodium-olefin-amine-complexes is supposed to be the first step towards the active catalytic species.

Introduction

Amines and their derivatives are of fundamental importance as natural products, pharmacological agents, fine chemicals, and dyes. Hence, there is considerable interest in the development of new more efficient synthetic protocols (1). In this respect, the transition-metal catalyzed hydroamination of olefins, is a particularly convenient method for the synthesis of amines (Scheme 1) (2).

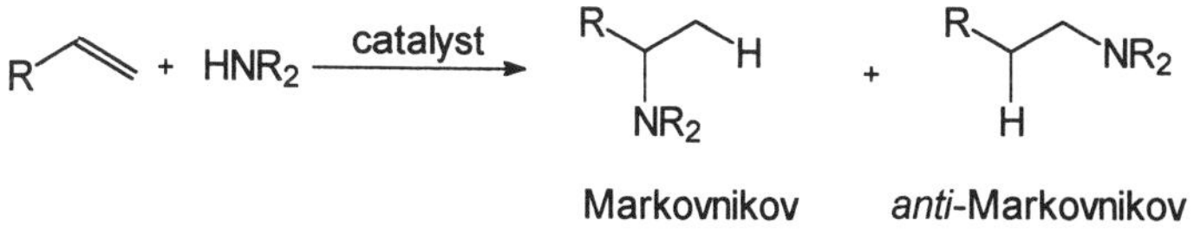

Scheme 1: Hydroamination of olefins.

The enviromentally friendly procedure is straightforward and to 100% atomeconomic, i.e. each atom from the starting material is present in the product molecules and no byproducts are formed. Furthermore, the olefin and amine educts are inexpensive and readily available.

The negative entropy balance - two reagents form one product- does not permit too high temperatures for this reaction, so that a catalyst is required. As catalyst for hydroaminations (3) strong bases, various acids and transition-metal complexes have been used. Due to the possibility of changing catalyst properties by simple ligand exchange reactions, transition-metal complexes offer probably the most promising route to develop for the first time an efficient catalytic hydroamination process.

Aminations of olefins catalyzed by transition-metals can follow two different paths. The first, a), shown in Scheme 2 involves coordinative binding of an olefin to an electron-poor metal. Nucleophilic addition of amine upon the ligated alkene leads to a σ-alkyl complex. Subsequent reductive elimination gives the hydroamination-product. The second reaction path, b) in Scheme 2, involves an oxidative insertion of an electron-rich metal into the N-H bond to form an amidohydrido complex (4). Subsequent insertion of the olefin into the metal-nitrogen bond generates the same intermediates as in a).

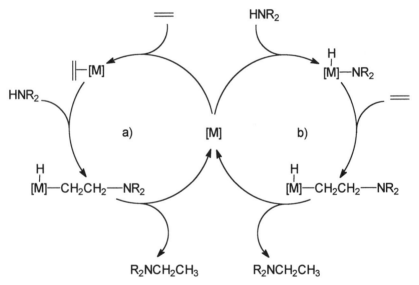

Scheme 2: Possible pathways for the hydroamination of olefins.

Industrially important terminal olefins provide, in principle, two regioisomers, the Markovnikov and the *anti*-Markovnikov product. The Markovnikov-product should be favored based on the stability of the intermediate carbocation. Indeed, in all hydroamination reactions catalyzed by transition-metals to date, the Markovnikov-product is preferentially formed (5). However, the linear *anti*-

Markovnikov-products are important for many practical applications. Hence, the development of a catalytic method to functionalize olefins with amines in *anti*-Markovnikov-regiochemistry has been mentioned as one of the ten most important challenges for new catalytic methods (6).
Whereas work on hydroamination is done since the seventies, publications on the *anti*-Markovnikov-problem of inactivated olefins are rare (7). A summary of important previous achievements in the area of aminations is shown in Table 1.
The first transition-metal catalysts for hydroamination of olefins were rhodium salts introduced by Du Pont (8). Later on Taube et al. could synthesize more active rhodium-catalysts for the hydroamination of ethylene (9).
Hegedus et al. have shown the catalytic activity of palladium complexes for intramolecular oxidative amination (10).

Table 1: Survey of transition-metal catalyzed aminations of olefins.

Reference	Reaction	Yield [%]	TON
Du Pont (8)	C_2H_4 + RNH_2 $\xrightarrow{RhCl_3}$ R_2NEt	70	70
Taube (9)	C_2H_4 + (piperidine) $\xrightarrow{[RhCl(C_2H_4)pip_2]}$ (N-Et piperidine)	70	70
Hegedus (10)	(2-allylaniline NHX) $\xrightarrow[O=\bigcirc=O]{2 \% PdCl_2(MeCN)_2}$ (indole N-X) X = H, Ac, Ts	90	50
Milstein (11)	(norbornene) + $Ph-NH_2$ $\xrightarrow{Ir(PEt)_3C_2H_4Cl}$ (aminonorbornane N-Ph)	-	6
Marks (12, 13)	$\diagup\!\diagdown\!\diagup NH_2$ $\xrightarrow{Cp^*_2LaR}$ (2-methylpyrrolidine)	86	170
Brunet (18)	(styrene) + $PhNH_2$ $\xrightarrow[PhNHLi]{[Rh(PEt_3)_2Cl]_2}$ (α-methylbenzyl NPh) +	65	21
	(phenethyl NHPh)	30	

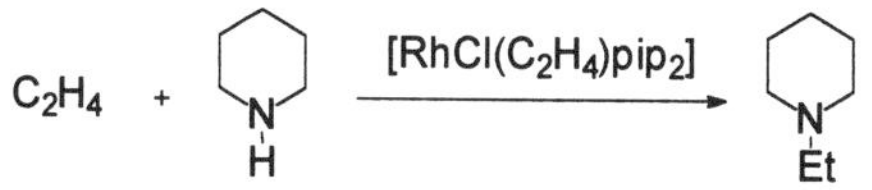

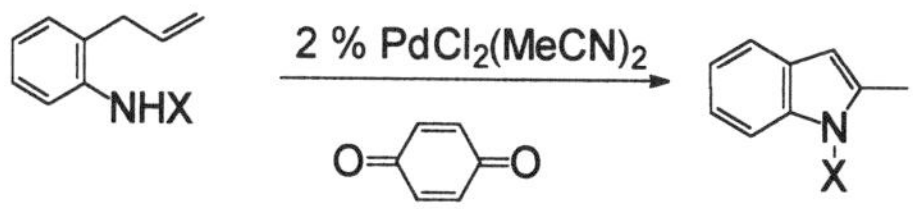

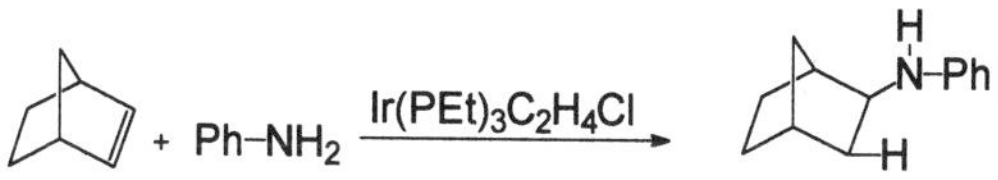

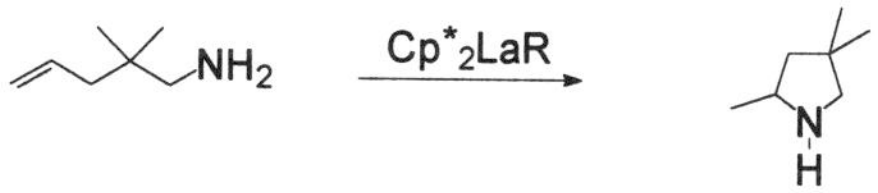

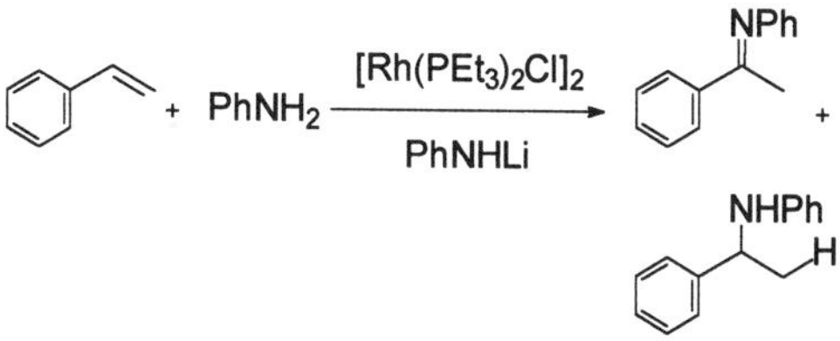

In case of iridium Milstein was able to provide evidence for an oxidative N-H addition pathway for the reaction of aniline and norbornene corresponding to path b) in Scheme 2 (11). However, turnover numbers (TON = 6) were extremely low for this reaction. A new impulse was recently added by Marks using lanthanide-complexes for intramolecular hydroamination (12). Also an intermolecular hydroamination of olefins was reported by Marks. However, the TOF was disappointing low (TOF $= 2 \ h^{-1}$) (13).

In summary, all the previous employed metal complexes for transition-metal catalyzed amination are not very active and some of the aminations only progress intramolecularly. In addition, the Markovnikov products are always favored when asymmetric olefins are used. The only exception is the intramolecular oxidative amination of *o*-aminostyrene in which, for reasons of ring strain in the product, only attack at a terminal position is possible (14).

The catalytic *anti*-Markovnikov oxidative amination reaction

We started our work for *anti*-Markovnikov amination by a catalyst-screening choosing styrene as olefin and piperidine as amine. Styrene is an neutral terminal olefin. Hence, two regioisomers are possible. At the beginning of our investigation we tested several late transition-metal complexes, like Ru, Rh, Ir and Pd in different oxidation states. None of these metal complexes worked as a hydroamination catalyst in the test reaction. However, we discovered that cationic Rh-complexes like $[(cod)_2Rh]^+$ (15) catalyzes in the presence of phosphane the oxidative amination of styrene to the terminal enamine (Scheme 3) (16). This reaction is completely unknown so far.

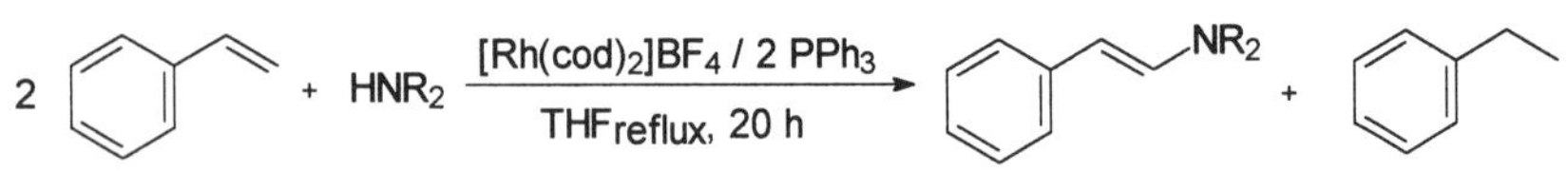

$$2 \quad \text{Ph-CH=CH}_2 \quad + \quad HNR_2 \quad \xrightarrow[\text{THF}_{reflux}, \ 20 \ h]{[Rh(cod)_2]BF_4 \ / \ 2 \ PPh_3} \quad \text{Ph-CH=CH-NR}_2 \quad + \quad \text{Ph-CH}_2CH_3$$

Scheme 3: Catalytic oxidative amination of styrene.

The ^{1}H-NMR spectrum indicated, that the *E*-isomer of the enamine is formed exclusively ($^3J = 14$ Hz). Surprisingly, this reaction is totally regiospecific towards the *anti*-Markovnikov-isomer since the other regioisomer, the Markovnikov-product, can not be detected at all. This is in opposite to other hydroaddition-reactions of styrene, like the hydroboration, where a mixture of both regioisomers is usually generated (17). An equimolar amount of ethylbenzene is concomitantly formed by hydrogenation of styrene.

The above transformation can be seen as a disproportion of the olefin to an enamine and ethylbenzene. The oxidative part of this reaction is related to the Wacker-reaction, as the hydrolysis of the enamines produces the corresponding aldehydes, but in contrary to the Wacker-oxidation the opposite regioisomer is observed.

Interestingly, Brunet detected by the conversion of styrene and aniline, with a mixture of $[(PEt_3)_2RhCl]_2$/lithiumanilide as catalyst, the Markovnikov-isomer (Table 1) (18).

As shown in Figure 1 enamine and ethylbenzene are simultaneously formed in a sigmoid-shaped curve. Furthermore at least one equivalent of the cod-ligand is converted to cyclooctene and cyclooctane, as verified by GC-experiments.

As shown in Figure 2, the yield and reaction rate were highly dependent on the styrene : amine ratio. The higher the amount of styrene compared to amine is, the higher the yield is observed. The rate of the reaction is also proportional to this styrene : amine-ratio. The amount of the derivative in the turning point as a measurement for the highest velocity of the reaction rises also by increasing styrene : amine-ratio. Thus, we infer that the olefin influences greatly the rate determining-step of the reaction.

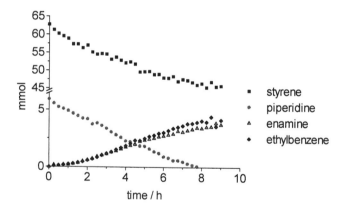

Figure 1: Time-yield-diagram for the oxidative amination of styrene (10 eq) with piperidine (1 eq) and 2.5 mol % $[Rh(cod)_2]BF_4$ / 2 PPh_3.

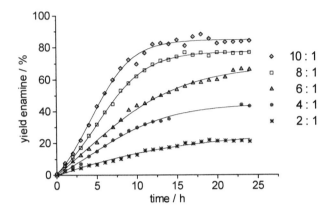

Figure 2: Dependence of the yield in respect to the styrene : piperidine-ratio, 2.5 mol % [Rh(cod)$_2$]BF$_4$ / 2 PPh$_3$.

The slightly higher consumption of styrene than needed to form the products enamine and ethylbenzene is likely due to a small extent of polymerization of styrene. So far, we were not able to detect polystyrene by standard analytical methods. However styrene-dimeres were observable by GC. Furthermore, the rate of oligomerization increases with increasing styrene : amine- ratio.

Scope and limitations

In order to investigate the steric and electronic influences of this new reaction, a variety of styrene derivatives were examined under standard reaction conditions (Table 2).
As shown in Table 2, the best result was achieved with 2-vinylnaphthalene emphasizing the influence of the aromatic system. In general, electron-donating groups attached to the aromatic ring increase the yield. Contrastingly, electron-withdrawing groups lower as expected the yield. Electronic reasons also seem to be responsible for the low reactivity of 4-vinylbiphenyl. Steric factors also play an important role. For instance, the *p*-and *m*-isomer of vinyltoluene are much more reactive than the sterically more hindered *o*-isomer.
Furthermore, the non-aromatic olefin acrylamide, which normally give the Michael-product react in this oxidative amination with 18% yield.

Table 2: Variation of olefins in the reaction with piperidine. [a]

olefin	enamine [%] [b]	ethylbenzene [%] [b]
styrene	55	57
2-vinylnaphthalene	99	99
4-methylstyrene	75	69
3-methylstyrene	75	94
2-methylstyrene	12	18
4-methyoxystyrene	55	49
3,4-dimethoxystyrene	44	64
4-vinylbiphenyl	24	8
4-chlorostyrene	13	8
4-fluorostyrene	18	13
4-trifluoromethylstyrene	14	6
acrylamide	18	-

[a] Ratio of styrene : amine 4 : 1; 2.5 mol % $[Rh(cod)_2]BF_4$ / 2 PPh_3 relative to amine; 20 h reflux in thf.
[b] The yields (GC) are referred to amine.

Table 3: Oxidative aminations of styrene. [a]

amine	enamine [%][b]	ethylbenzene [%][b]
diethylamine	40	54
di-n-butylamine	48	44
piperidine	55	57
hexahydroazepine	45	80
N-methylaniline	9	9
di-iso-propylamine	<1	<1
thiomorpholine	<1	<1
N-methylpiperazine	4	5
n-butylamine	<1	<1

[a] Ratio of styrene : amine 4 : 1; 2.5 mol % $[Rh(cod)_2]BF_4$ / 2 PPh_3 relative to amine; 20 h reflux in thf.
[b] The yields (GC) are referred to amine.

Apart from different styrenes various amines have been tested in the oxidative amination reaction (Table 3). Here, a great number of secondary amines can be converted into styrylamines, especially secondary aliphatic ones. Even deactivated *N*-methylaniline undergoes the reaction, despite its attenuated nucleophilicity due to the phenyl ring. Until now steric demanding amines like di-*i*-propylamine and bidentate amines react sluggishly or not at all.

Ligand-influences and mechanistic approaches

The reaction is very sensitive to the cationic character of the rhodium complex. When halogen anions were added, no reaction occured. Furthermore, the yield of the reaction was dependent on the quantity and type of phosphane. As depicted in Figure 3, the optimum rhodium : phosphane-ratio using PPh$_3$ was 1 : 2. The presence of phosphane has a stabilizing influence on the catalyst, as without phosphane the reaction also takes place but in a much smaller extend. A great surplus of phosphane completely inhibits the reaction. Thus, excess of phosphane most likely blocks the coordination site on the rhodium center rendering the catalyst inactive.

As shown in Table 4, the same result as the 1 : 2-in-situ mixture of [Rh(cod)$_2$]BF$_4$ and PPh$_3$ is obtained by the use of the [Rh(cod)(PPh$_3$)$_2$]BF$_4$ as catalyst. Chelate forming phosphanes like dppe completely inhibit the reaction. This supports the thesis that the phosphane has only a stabilizing function. More specifically, it means that the phosphane occupies vacant coordination-sites to protect the rhodium-complex against decomposition, but has to be displaced by substrate-molecules in order to act as catalyst. The higher homologs of triphenylphosphane AsPh$_3$ and SbPh$_3$, as well as strong nitrogen donors such as pyridine are inactive ligands, which means that no reaction occured.

One step towards the mechanistic explanation of the oxidative amination is the reaction of the rhodium precatalyst with amines. Unfortunately, there are only very few reports known in the literature dealing with cationic rhodium amine complexes (19).

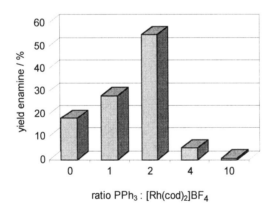

Figure 3: Influence of the phosphane : rhodium-ratio on the oxidative amination

Table 4: Variations on the catalysts of the oxidative amination of styrene with piperidine.

catalyst	enamine [%]	ethylbenzene [%]
[Rh(cod)$_2$]BF$_4$	18	15
[Rh(cod)$_2$]BF$_4$ / 2 PPh$_3$	55	57
[Rh(cod)(PPh$_3$)$_2$]BF$_4$	55	56
[Rh(cod)(dppe)]BF$_4$	<1	<1
[Rh(cod)$_2$]BF$_4$ / 2 AsPh$_3$	<1	<1
[Rh(cod)$_2$]BF$_4$ / 2 SbPh$_3$	<1	<1
[Rh(cod)(piperidine)$_2$]BF$_4$	8	8
[Rh(cod)(piperidine)$_2$]BF$_4$ / 2 PPh$_3$	45	49

An exception is the work of Denise and Pannetier, who treated [Rh(cod)$_2$] $^+$ with piperidine and detected by in-situ-NMR-experiments (20), that the yellow colored [Rh(cod)(piperidine)$_2$]$^+$ is formed.

In view of the limited knowledge of rhodium-olefin-amine-complexes and the lack of general and convenient synthetic routes for the preparation of these complexes, we started to explore the possibility to synthesize these complexes starting with [Rh(cod)Cl]$_2$, silver salt and subsequent addition of the amine (Scheme 4).

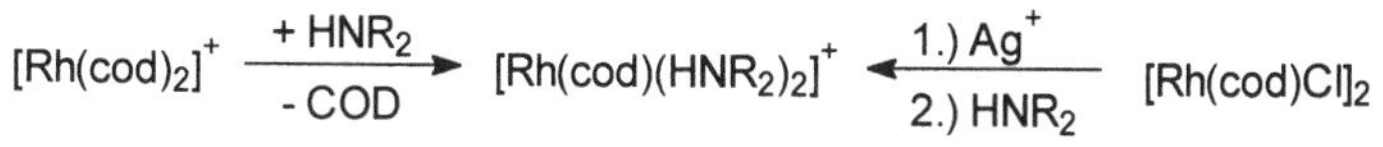

$$[Rh(cod)_2]^+ \xrightarrow[- COD]{+ HNR_2} [Rh(cod)(HNR_2)_2]^+ \xleftarrow[2.) HNR_2]{1.) Ag^+} [Rh(cod)Cl]_2$$

Scheme 4: Synthesis of cationic rhodium-amine-complexes.

Indeed it was possible to synthesize cationic Rh-amine complexes via this route. For example [(1,2,5,6-η)-1,5-cyclooctadien]bis(piperidine)rhodium(I)tetrafluoroborate and [(1,2,5,6-η)-1,5-cyclooctadien]-(N,N')-N-methylpiperazine-rhodium(I)tetrafluoroborate were prepared and completely characterize for the first time (21). In contrary to the monodentate amines the bidentate amines piperazine, N-methylpiperazine and thiomorpholine form chelate complexes (Scheme 5). These complexes precipitate from the catalytic reaction, and deactivate the catalyst (Table 2).

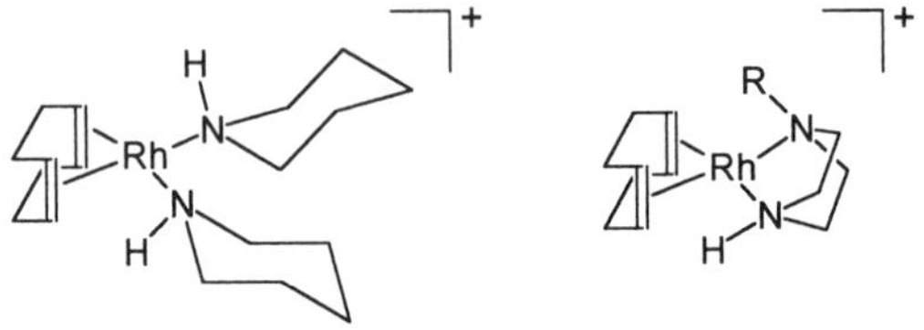

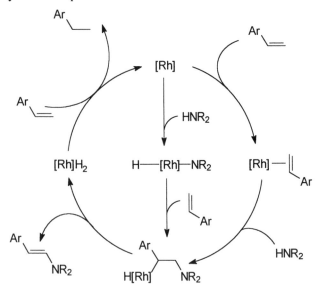

Scheme 5: Monodentate and chelate cationic rhodium-amine-complexes.

This is a further evidence for the necessity of free coordination positions on the catalytic active rhodium-atom. The interesting feature of the cationic rhodium-amine-olefin complexes is their catalytic activity. These complexes show nearly the same, only slightly diminished, catalytic activity as the $[Rh(cod)_2]^+$, both with and without phosphanes, (Table 4). Hence, we conclude that the formation of the $[Rh(cod)(amine)_2]^+$-cation is the first step towards the formation of the catalytic active species.

Ar = Aryl; R'= Alkyl, Aryl

Scheme 6: Possible mechanism for the oxidative regioselective amination of styrene.

Two mechanisms are possible for the oxidative amination of styrene. Both are, in principle, similar to oxidative silylation (22). As shown in Scheme 6, one possibility is the olefin-activation route where the catalytic active rhodium species reacts with the olefin, forming a rhodium-olefin complex. Nucleophilic attack of the amine on the activated olefin affords a Rh-alkylamine compound. β-H-elimination generates the Rh-H species and the enamine. The active catalyst is then reformed by reduction of a further olefin-molecule. The second possibility is the oxidative addition of the amine and subsequent insertion of the olefin in the Rh-N-bond. The following procedure is identical with the olefin-activation-path. So far, our data does not allow us to distinguish which mechanism is operable. In both mechanisms it may occur that the insertion or nucleophilic attack takes place on both sides of the olefin, however only the *anti*-Markovnikov precursor will undergo further reaction.

Outlook

This oxidative reaction provides a new method for the preparation of aromatic enamines. Enamines are reactive intermediates in organic chemistry and precursors to a variety of important functional groups (Scheme 7) (23).

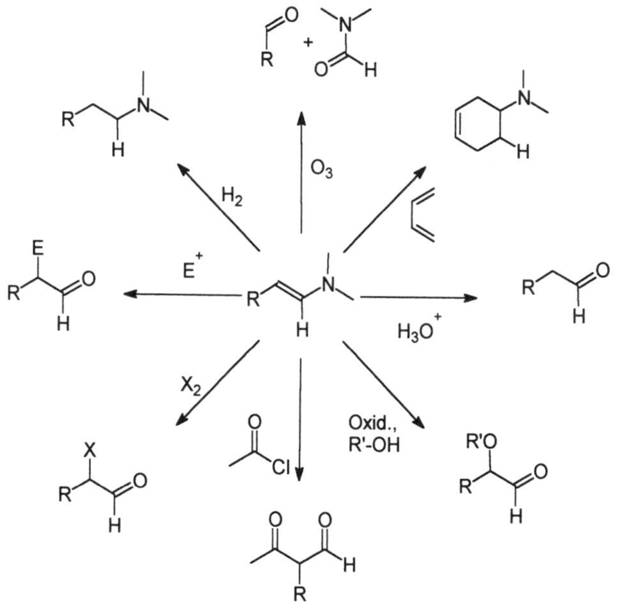

Scheme 7: Enamines as versatile reactive intermediates.

One important transformation among the various reactions of enamines is the hydrogenation resulting in alkylamines of the structural unit $Ph-CH_2-CH_2-NR_2$. This represents the basic structure of several important pharmacological agents (24). Furthermore, the reaction of enamines with electrophiles like H^+ or alkylhalogenides leads to substituted aldehydes. These aldehydes which are normally difficult to synthesize, provide starting materials for the production of non-natural phenylalanines via amidocarbonylation (25).

Conclusion

In conclusion, we have shown for the first time, that cationic rhodium-complexes catalyze the regioselective *anti*-Markovnikov oxidative amination of aromatic olefins, whereby the corresponding arylethane is concurrently formed. This new reaction is of special importance to metal-catalyzed amination reactions due to mainly two reasons. First of all it is the first example of a new synthetic route to enamines from olefins.

For the second and more important reason, it is an intermolecular functionalization of olefins with nucleophiles leading to *anti*-Markovnikov-products, which shows a possibility to solve this problem. The reaction is influenced by electronic and steric factors of both reactants as well as the quantity and the type of the different ligands like phosphanes. Two equivalents of PPh_3 per rhodium lead to the best results. Mechanistic investigations provide evidence that cationic rhodium-amine complexes are involved in formation the active catalyst species. In regard to further applications, this reaction opens a new access to enamines and phenylacetaldehydes, which can be functionalized in pharmacological interesting molecules.

Acknowledgment

We thank the DFG for financial support of this research. Dr. J. Herwig (Hoechst AG), Prof. K. Kühlein (Hoechst AG) and Dr. G. Stark is thanked for valuable discussions. H. T. acknowledges founding form the Fonds der Chemischen Industrie.

Oxidative anti-Markovnikov-Amination

References

1. D. M. Roundhill, *Chem. Rev.*, **92**, 1 (1992).
2. R. Taube in *Applied Homogeneous Catalysis with Organometallic Compounds* (Ed.: B. Cornils, W. A. Herrmann), VCH, Weinheim, p. 507 (1996).
3. T. E. Müller, M. Beller, *Chem. Rev.* (1998) in press.
4. D. Steinborn, R. Taube, *Z. Chemie*, **26**, 349 (1986).
5. V. Markovnikov, *C. R. Acad. .Sci.*, **85**, 668 (1875).
6. Chem. Eng. News, **31 May**, 23 (1993).
7. C. M. Jensen, W. C. Trogler, *Science*, **233**, 1069 (1986).
8. a) D. R. Coulson (Du Pont), US 3.758.586, (1973) [Chem. Abstr., **79**, 125808g, (1973)];
   b) D. R. Coulson, *Tetrahedron Lett.*, 429 (1971).
9. D. Steinborn, B. Thies, I. Wagner, R. Taube, *Z. Chem.*, **29**, 333 (1989).
10. A. L. Casalnuovo, J. L. Calabrese, D. Milstein, *J. Am. Chem. Soc.*, **110**, 6738 (1988).
11. a) L. S. Hegedus, *Angew. Chem.*, **100**, 1147 (1988); *Angew. Chem., Int. Ed. Engl.*, 27, 1113 (1988).
    b) L. S. Hegedus, G. F. Allen, J. J. Bozell, E. L. Waterman, *J. Am. Chem. Soc.*, **100**, 5800 (1978).
    c) B. Åkermark, J. E. Bäckvall, L. S. Hegedus, K. Zetterberg, K. Siirala-Hansén, K. Sjöberg, *J. Organomet. Chem.*, **72**, 127 (1974).
    d) L. S. Hegedus, J. M. McKearin, *J. Am. Chem. Soc.*, **104**, 2444 (1982).
12. a) M. R. Gagné, S. P. Nolan, T. J. Marks, *Organometallics*, **9**, 1716 (1990).
    b) M. R. Gagné, C. L. Stern, T. J. Marks, *J. Am. Chem. Soc.*, **114**, 275 (1992).
    c) M. R. Gagné, L. Brard, V. P. Conticello, M. A. Giardello, C. L. Stern, T. J. Marks, *Organometallics*, **11**, 2003 (1992).
    d) M. A. Giardello, V. P. Conticello, L. Brard, M. R. Gagné, T. J. Marks, *J. Am. Chem. Soc.*, **116**, 10241 (1994).
    e) Y. Li, T. J. Marks, *J. Am. Chem. Soc.*, **118**, 9295 (1996).
13. Y. W. Li, T. J. Marks, *Organometallics*, **15**, 3370 (1996).
14. R. C. Larock, T. R. Hightower, L. A. Hasvold, K. P. Peterson, *J. Org. Chem.*, **61**, 3584 (1996).
15. R. R. Schrock, J. A. Osborn, *J. Am. Chem. Soc.*, **93**, 3089 (1971).
16. M. Beller, M. Eichberger, H. Trauthwein, *Angew. Chem.* (1997), in press.
17. a) S. A. Westcott, H. P. Blom, T. B. Marder, R. T. Baker, *J. Am. Chem. Soc.*, **114**, 8863 (1992).
    b) K. Burgess, W. A. van der Donk, S. A. Westcott, T. B. Marder, R. T. Baker, J. C. Calabrese, *J. Am. Chem. Soc.*, **114**, 9350 (1992).

18. J.-J. Brunet, D. Neibecker, K. Philippot, *Tetrahedron Lett.*, **34**, 3877 (1993).
19. a) M. A. Garralda, R. Hernández, L. Ibarlucea, M. I. Arriotua, M. K. Urtiaga, *Inorg. Chim. Acta*, **232**, 9 (1995).
   b) E. Anzuela, M. A. Garralda, R. Hernández, L. Ibarlucea, E. Pinilla, M. A. Monge, *Inorg. Chim. Acta*, **185**, 211 (1991).
   c) P. Pertici, F. D'Arata, C. Rosini, *J. Organomet. Chem.*, **515**, 163 (1996).
20. B. Denise, G. Pannetier, *J. Organomet. Chem.*, **161**, 171 (1978).
21. M. Beller, H. Trauthwein, M. Eichberger, C. Breindl, T. Müller, A. Zapf, *Organometallics* submitted.
22. R. Takeuchi, H. Yasue, *Organometallics*, **15**, 2098 (1996).
23. P. W. Hickmott, *Tetrahedron*, **38**, 1975 (1982).
24. a) E. Mutschler, *Arzneimittelwirkungen*, 6th ed., Wiss. Verlagsges., Stuttgart (1991).
   b) W. Forth, D. Henschler; W. Rummel, K. Starke, *Pharmakologie und Toxikologie*, 6th ed, Bibliographisches Institut, Mannheim (1992).
25. M. Beller, M. Eckert, F. Vollmüller, S. Bogdanovic, H. Geissler, *Angew. Chem.*, **109**, 534 (1997); *Angew. Chem., Int. Ed. Engl.*, **36**, 1494 (1997).

Catalytic Carbonylations in Supercritical Carbon Dioxide

Iwao Ojima*, Maria Tzamarioudaki, Chih-Yuan Chuang, Donna M. Iula, and Zhaoyang Li

Department of Chemistry; State University of New York at Stony Brook; Stony Brook, New York 11794-3400

Abstract

The Rh-catalyzed hydroformylation of simple alkenes and the cyclohydrocarbonylation of amino- and hydroxydienes in supercritical CO_2 are described.

It has been shown that supercritical fluids are useful as media for extraction and chromatography in academic laboratories as well as industrial processes.[1] Supercritical fluids also provide advantages as unique reaction media since the density, polarity, viscosity, diffusivity, and solvating ability of supercritical fluids can be dramatically changed by small variation of the pressure and/or temperature.[2,3] It has recently been demonstrated that higher reaction rate and selectivity can be achieved by replacing conventional organic solvents by supercritical fluids such as supercritical carbon dioxide ($scCO_2$) and supercritical water (scH_2O).[1,4,5,6] As a medium for organic synthetic reactions and homogeneous catalysis, $scCO_2$ appears to be most suitable because of its mild critical point, i.e., $T_c = 31$ °C, $P_c = 72.9$ atm. In fact, free radical, polymerization, hydrogenation, hydroformylation, and carboxylation reactions have been successfully carried out in $scCO_2$.[7-17] However, in spite of the recent recognition of these very useful features of $scCO_2$, only a very limited number of homogeneous catalytic reactions have been investigated using $scCO_2$ as the medium.[1,10,11,13,15] We report here the results of our preliminary study on the hydrocarbonylation reactions of alkenes and amino- or hydroxyalkadienes in $scCO_2$ using BIPHEPHOS [18,19] or (R,S)-BINAPHOS [20] as the ligand for $Rh(acac)(CO)_2$.

To investigate the characteristics of the hydroformylation reaction in $scCO_2$ medium, the reactions of simple alkenes, i.e., 1-octene, 3-phenyl-1-propene, styrene, cyclohexene, and vinyl acetate were carried out using $Rh(acac)(CO)_2$ (6.67×10^{-5} M) and BIPHEPHOS or (R,S)-BINAPHOS (1.33×10^{-4} M) under 14 atm of CO/H_2 (1/1) in $scCO_2$ (85 atm of CO_2) at 65 °C in a 300 mL autoclave with two sapphire windows. The results are summarized in Table 1. As Table 1 shows, the hydroformylation of 1-octene in the presence of

Table 1. Hydroformylation of alkenes[a]

Entry	Substrate	R	Ligand	Solvent	Time (h)	Conversion (%)	GC Yield (%)	mol ratio 2/3
1	1a	n-C_6H_{13}	BIPHEPHOS	$scCO_2$	24	100	60	6/1[b]
2	1a	n-C_6H_{13}	BIPHEPHOS	hexane	12	100	100	16/1
3	1a	n-C_6H_{13}	BIPHEPHOS	$scCO_2$	12	88	71	175/1[b,c]
4	1a	n-C_6H_{13}	BIPHEPHOS	$scCO_2$	12	69	44	120/1[b,d]
5	1b	$PhCH_2$	BIPHEPHOS	$scCO_2$	12	99	74	150/1[b,e]
6	1c	Ph	(R,S)-BINAPHOS	$scCO_2$	12	100	100	1/8[f,g]
7	1d	AcO	BIPHEPHOS	$scCO_2$	40	85	85	1/10[f]
8	1d	AcO	(R,S)-BINAPHOS	$scCO_2$	43	74	74	1/8.5[f]
9	1d	AcO	BIPHEPHOS	Et_2O	44	>95	95	1/9.2
10	1a	n-C_6H_{13}	none	$scCO_2$	40	87	64	2/1

[a]The reaction was run with $Rh(acac)(CO)_2$ (6.67×10^{-5} M) and ligand (1.33×10^{-4} M) under 14 atm of CO/H_2 (1/1) and 85 atm of CO_2 at 65 °C unless otherwise noted. [b]Products from the isomerization of 1-octene were observed. [c]71 atm of CO_2 was used. [d]3 eq BIPHEPHOS (2×10^{-4} M) was used. [e]1 mmol of $CF_3(CF_2)_6CH_2OH$ was added (6.67×10^{-3} M). [f]78 atm of CO_2 and 21 atm of CO/H_2 (1:2) were used. [g](R)-2-phenylpropanal, was obtained with 92% ee.

Rh-BIPHEPHOS under 71 atm of scCO$_2$ gives a 175:1 mixture of aldehydes in favor of the linear product. Increasing the CO$_2$ pressure results in a significant drop of the linear selectivity which can be improved by increasing the amount of BIPHEPHOS to 2×10^{-4} M. A 17.5:1 mixture of linear and branched aldehydes was formed in the reaction of 3-phenyl-1-propene, while branched aldehydes were predominant for the reactions of styrene and vinyl acetate (eq. 1). (*R*)-2-Phenylpropanal resulting from the reaction of styrene catalyzed by the Rh-(*R,S*)-BINAPHOS was found to possess high enantiomeric purity upto 92% ee. The hydroformylation of cyclohexene gave cyclohexanecarboxaldehyde exclusively although the conversion was low under the given conditions (eq. 2). Accordingly, these results indicate that scCO$_2$ is clearly different from THF and provides a particular advantage for the regioselective syntheses of aldehydes from simple alkenes in addition to the fact that scCO$_2$ is an environmentally benign reaction media. It is interesting to note that when the hydroformylation of 1-octene in THF was run in the absence of bisphosphite ligand, less then 10% conversion was observed whereas in scCO$_2$ medium 87% conversion was obtained.

$$
R \diagdown\!\!\!\equiv \quad \underset{\substack{scCO_2 \\ (60\ °C,\ 80\ atm)}}{\overset{\substack{Rh(acac)(CO)_2 \\ BIPHEPHOS \\ CO/H_2\ (14\ atm)}}{\xrightarrow{\hspace{2cm}}}} \quad R\diagup\!\!\diagdown CHO \;+\; \overset{CHO}{R\diagup\!\!\diagdown} \qquad (1)
$$

$$
\underset{\mathbf{1}}{} \qquad\qquad \underset{\mathbf{2}}{} \qquad\qquad \underset{\mathbf{3}}{}
$$

R = n-C$_6$H$_{13}$, PhCh$_2$, Ph, AcO

$$
\underset{\mathbf{1e}}{\bighexagon\!\!\parallel} \quad \underset{\substack{scCO_2 \\ (60\ °C,\ 80\ atm)}}{\overset{\substack{Rh(acac)(CO)_2 \\ BIPHEPHOS \\ CO/H_2\ (14\ atm)}}{\xrightarrow{\hspace{2cm}}}} \quad \underset{\mathbf{3e}}{\bighexagon\!\!-CHO} \qquad (2)
$$

Next, we looked at the cyclohydrocarbonylation reaction of *N-t*-Boc-allylglycinate **4** [21], 4-amido-1,6-heptadienes, **10a** [22] and **10b**, and 1,6-heptadien-4-ol (**16**) in scCO$_2$ as an extension of our continuous study on the synthesis of functionalized nitrogen- and oxygen-heterocycles that are useful intermediates for the synthesis of natural and synthetic products.[23]

The cyclohydrocarbonylation of **4** proceeded cleanly to yield 5,6-didehydropipecolate **5** [21] as the sole product in quantitative yield (eq. 3). The same result was observed when the reaction was carried out in THF, toluene, ethyl acetate, chloroform or hexane. Thus, it is apparent that scCO$_2$ can replace these traditional organic solvents for this reaction. The reaction in methanol or ethanol has been shown to give the corresponding O-alkylhemiamidal **9**.[21]

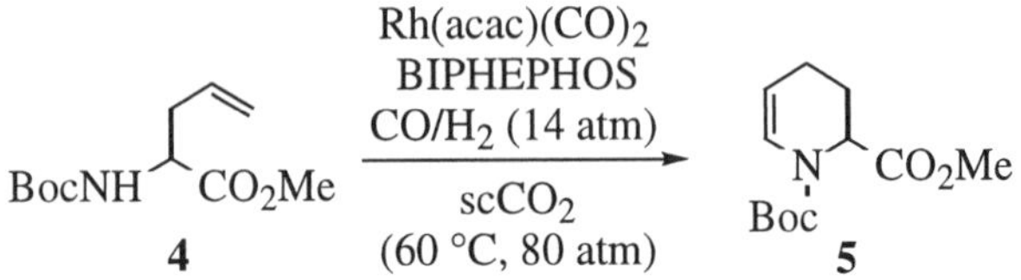

$$\text{(3)}$$

The most likely mechanism for the intramolecular cyclohydrocarbonylation is proposed in Scheme 1. The reaction is believed to proceed via linear aldehyde **6** [21] followed by cyclization to form **7**, which eliminates a hydroxyl group, generating a reactive acyliminium ion intermediate **8**, and **8** gives either 5,6-dehydropiperidine **5** or 6-alkoxylpipecolate **9**. Since no alcohol is present in the current reaction system, only deprotonation and double bond migration takes place to give **5**.

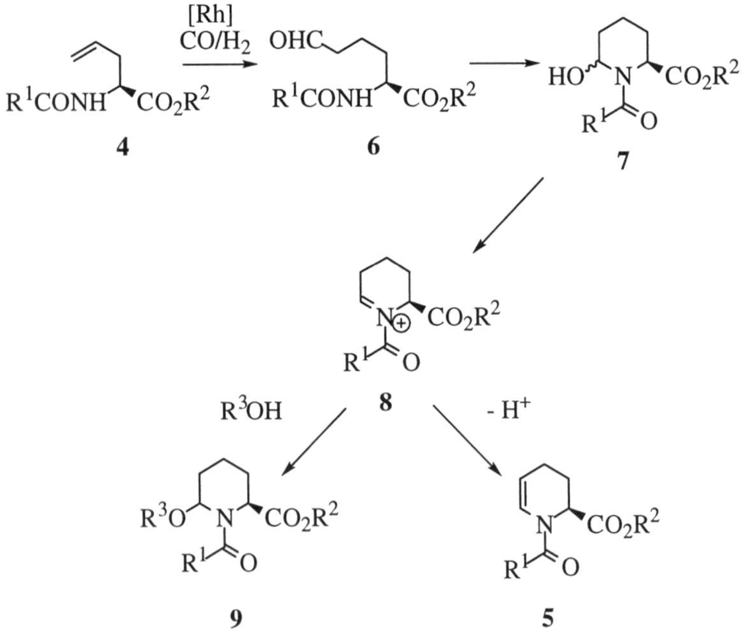

Scheme 1

Reaction of 4-tosylamino-1,6-heptadiene (**10a**) gave didehydropiperidine-aldehyde (**11a**) [22] as the major product (65 %) accompanied by side products (eq. 4).

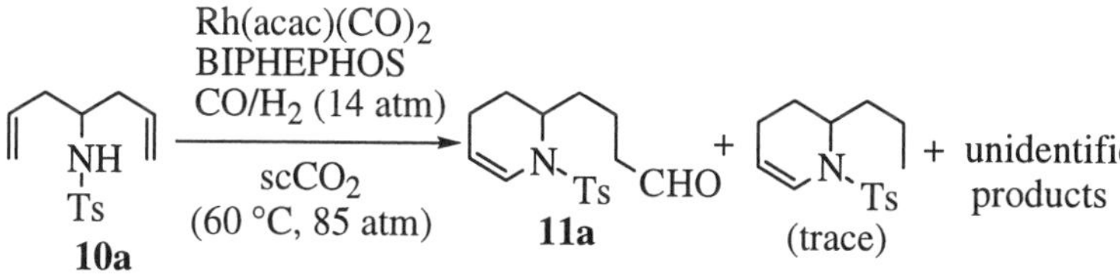
Rh(acac)(CO)2
BIPHEPHOS
CO/H2 (14 atm)
scCO2
(60 °C, 85 atm)
NH
Ts
10a
N
Ts
CHO
11a
+
N
Ts
(trace)
+ unidentified
products

(4)

Extremely high selectivity was accomplished, however, by using 4-(*t*-butoxycarbonylamino)-1,6-heptadiene **10b** as the substrate under the same conditions as those for **10a** , i.e., the reaction gave **11b** exclusively in quantitative yield (eq. 5). Accordingly, $scCO_2$ is an excellent reaction medium for this reaction.

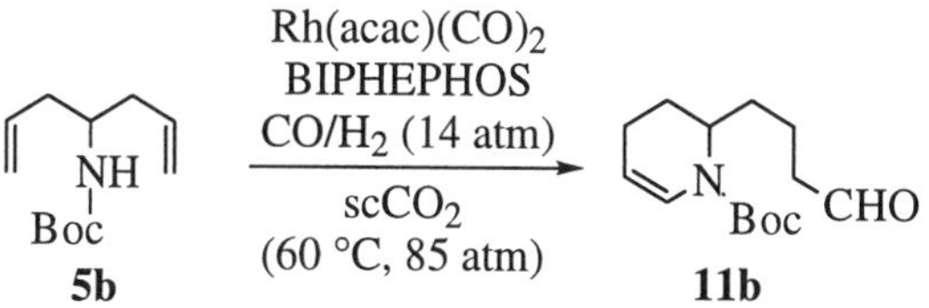
Rh(acac)(CO)2
BIPHEPHOS
CO/H2 (14 atm)
scCO2
(60 °C, 85 atm)
NH
Boc
5b
N
Boc
CHO
11b

(5)

The most likely mechanism for the formation of **11** is shown in Scheme 2. [22] The first step is the linear-selective hydroformylation of one of the two double bonds giving **12** which is followed by cyclization to give alkenyl-hemiamidal **13**. The hydroformylation of the remaining double bond takes place to afford hemiamidal-aldehyde **14**. Then, **14** undergoes dehydration via an iminium ion intermediate **15** to yield **11**.

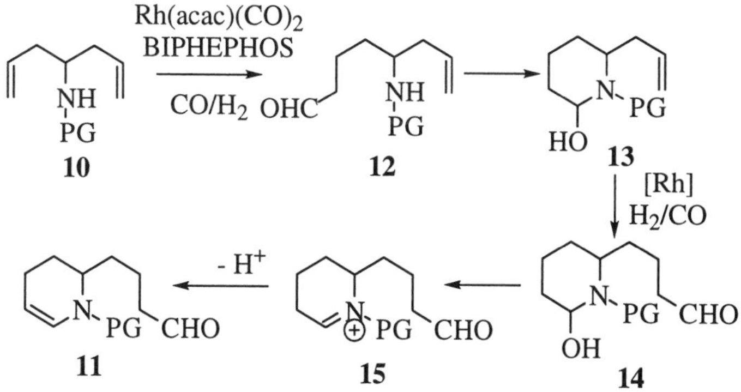
Rh(acac)(CO)2
BIPHEPHOS
CO/H2
NH
PG
10
OHC
NH
PG
12
N
PG
HO
13
[Rh]
H2/CO
N
PG
CHO
OH
14
N
⊕
PG
CHO
15
- H+
N
PG
CHO
11

Scheme 2

Formation of oxygen heterocycles via cyclohydrocarbonylation was also successful. The reaction of 1,6-heptadien-4-ol (**16**) in $scCO_2$ under the same conditions used in the reactions mentioned above afforded lactol-aldehyde **17** in quantitative yield as a mixture of *cis* and *trans* diastereomers (eq. 6).

$$\begin{array}{c} \text{Rh(acac)(CO)}_2 \\ \text{BIPHEPHOS} \\ \text{CO/H}_2 \text{ (14 atm)} \\ \xrightarrow{\hspace{2cm}} \\ scCO_2 \\ \text{(60 °C, 85 atm)} \end{array}$$

16 **17** (6)

Further investigation into a variety of homogeneous catalysis in $scCO_2$ is actively underway in these laboratories.

Experimental

General methods. The ^1H NMR spectra were measured with a Varian Genmi 2300 or a Brüker AC-250 spectrometer with tetramethylsilane as the internal standard. ^{13}C NMR spectra were measured with a Varian Genmi 2300 or a Brüker AC-250 spectrometer. IR spectra were recorded on a Perkin Elmer 1600 FT-IR spectrophotometer with a Hewlett-Packard HP 7470A Plotter, using neat samples. Analytical gas chromatography was performed on a Perkin-Elmer 3920 gas chromatograph equipped with a Hewlett-Packard 3393A integrator, using columns packed with 3% OV-1. Elemental analyses were performed at Quantitative Technologies Inc., Whitehouse, NJ or M-H-W Laboratories, Phoenix, AZ. Mass spectra (GC-MS) were obtained at 70 eV on a Hewlett-Packard HP 5980A mass spectrometer equipped with a HP 5710A gas chromatograph and a HP 5933A data system or a Hewlett-Packard HP 5971A mass spectrometer with a HP 5890 gas chromatograph and a HP Vecta QS/20 workstation. High resolution (exact) mass spectrometric analyses were performed at the Mass Spectrometry Facility of the University of California at Riverside, Riverside, CA.

Materials. Rhodium complex Rh(acac)(CO)$_2$ was a gift from Mitsubishi Kasei Corp. and used as received. BIPHEPHOS and (*R,S*))-BINAPHOS were prepared according to the patented or published procedures. [18, 20] 1-Hexene, styrene, 3-phenylpropene, and vinyl acetate were purchased from Aldrich Chemical Co. and distilled over 4 Å molecular sieves. 1,6-Heptadien-4-ol and cyclohexene were purchased from Aldrich Chemical Co. and used as received.

General Procedure for Hydrocarbonylations in scCO$_2$

    A) Hydroformylation: A mixture of Rh(acac)(CO)$_2$ (5.2 mg, 2 x 10^{-2} mmol) and BIPHEPHOS (31 mg, 4 x 10^{-2} mmol) or (*R,S*)-BINAPHOS (30 mg, 4 x 10^{-2} mmol), alkene (2 mmol), and undecane (2 mmol) was stirred for 30 minutes at room temperature under nitrogen. Then the mixture was introduced into a 300 mL stainless steel autoclave with two sapphire windows. The air was replaced by CO through pressurization (2-3 atm) and release for 3 times, and CO (7 atm) was introduced followed by H$_2$ (7 or 14 atm). The reaction vessel was cooled to -60 °C and CO$_2$ (85 atm) was introduced. The mixture was allowed to stir at 65 °C and 100 atm overnight. The autoclave was cooled to -30 °C and all the gasses were carefully released until the pressure reached 4 atm. The remaining gasses were released by gradually warming the temperature to 0°C. The conversion and yield of the reaction was estimated by GLC analysis using an authentic sample of the aldehyde and undecane as the internal standard. (*R*)-2-Phenylpropanal obtained from the reaction of styrene using the Rh-(*R*,*S*)-BINAPHOS) catalyst was reduced by NaBH$_4$ to (*R*)-2-phenylpropanol, which was then converted to the corresponding Mosher ester for ^{19}F and ^{1}H NMR analyses.

B) Cyclohydrocarbonylation: To a 300 mL stainless steel autoclave with two sapphire windows containing 1.5 mg (0.6 x 10^{-2} mmol) of Rh(acac)(CO)$_2$ and 9.4 mg (1.2 x 10^{-2} mmol) of BIPHEPHOS was added 0.6 mmol of alkene or diene introduced. The air was replaced by CO and CO (7 atm) was introduced followed by H$_2$ (7 atm). The reaction vessel was cooled to -60 °C and CO$_2$ (85 atm) was introduced. The mixture was allowed to stir at 65 °C and 100 atm overnight. The autoclave was cooled to 0 °C and all the gasses were carefully released. The reaction mixture was then concentrated under vacuum and the residue was chromatographed on silica gel using hexane/AcOEt or pentane/ether as the eluant.

Cyclohydrocarbonylation in liquid-CO$_2$ near the critical point.

To a 300 mL stainless steel autoclave with two sapphire windows containing 5.2 mg (2 x 10^{-2} mmol) of Rh(acac)(CO)$_2$ and 31 mg (4 x 10^{-2} mmol) of BIPHEPHOS was added 2 mmol of 16. The air was replaced by CO and CO (7 atm) was introduced followed by H$_2$ (7 atm). The reaction vessel was cooled to -60 °C and 74 atm of CO$_2$ was added. The mixture was allowed to stir at 25 °C for 24 h. The autoclave was cooled to 0 °C and all the gasses carefully released. The reaction mixture was then concentrated under vacuum and the residue was chromatographed on silica gel using pentane/ether as the eluant.

Preparation of 4-tosylamino-1,6-heptadiene (10a)

To a 100 mL round-bottomed flask equipped with a distillation head, tosylamine (40.1g, 0.23 mol) and ethyl orthoformate (60 mL) were introduced. The solution was heated at 80 °C with removal of ethanol by concurrent distillation. When no more ethanol was distilled over, the reaction flask was cooled to room temperature and the excess ethanol and orthoformate were evaporated under vacuum to afford *N*-ethoxymethylene-4-methyl-benzosulfonamide [24] as a pale yellow oil (52.8 g; 99%), which crystallized to form a white solid upon standing: ^{1}H NMR (CDCl$_3$) δ 1.30 (t, J = 7.1 Hz, 3 H), 2.41 (s, 3 H), 4.29 (q, J = 7.1 Hz, 2 H), 7.30 (d, J = 8.1 Hz, 2 H), 7.78 (d, J = 8.1 Hz, 2 H), 8.38 (s, 1 H).

To a solution of *N*-ethoxymethylene-4-methyl-benzosulfonamide (5.0 g, 22 mmol) in dry ether (55.0 mL) was added dropwise a solution of allylmagnesium bromide (1 M in ether; 57 mL, 57 mmol) at 0 °C under nitrogen. The reaction mixture was gradually warmed to room temperature and stirred overnight. The cloudy reaction mixture was then heated at reflux for 2 h and saturated aqueous NH$_4$Cl (pH 4) was added at 0 °C. The resulting two phases were separated and the aqueous phase was extracted twice with ether. The organic extracts were combined, washed with water and brine, dried over MgSO$_4$ and concentrated under vacuum to give a dark orange oily residue (6.0 g). This residue was chromatographed on silica gel using hexane/AcOEt/Et$_2$O (3/1/0.6) as the eluant to afford 4-tosylamino-1,6-heptadiene (**10a**)[24] (3.5 g; 60%) as a yellow oil: ^{1}H NMR (CDCl$_3$) δ 2.07 (t, J = 6.7 Hz, 4 H), 2.34 (s, 3 H), 3.23 (dquintet, J = 6.5, 6.5 Hz, 1 H), 4.66 (br s, 1 H), 4.90 (dd, J = 12.6, 0.8 Hz, 2 H), 4.93-4.96 (m, 2 H), 5.42-5.56 (m, 2 H), 7.21 (d, J = 8.2 Hz, 2 H), 7.67 (d, J = 8.2 Hz, 2 H).

4-(*tert*-Butoxycarbonylamino)-1,6-heptadiene (10b)

To a solution of 1,6-heptadien-4-ol (6.00 g, 53.5 mmol) in CH$_2$Cl$_2$ (60 mL) and pyridine (9.5 mL, 0.12 mol) was added a solution of methanesulfonyl chloride (6.2 mL, 80 mmol) in CH$_2$Cl$_2$ (60 mL) at 0 °C under N$_2$. After addition was complete, the reaction mixture was gradually warmed to room temperature and stirred for 20 h. Water was then added and the resulting two phases were separated. The aqueous phase was extracted with ether and the organic extracts were combined, washed with water, 2 N HCl, water, brine, dried over MgSO$_4$ and concentrated on a rotary evaporator. The residual pale yellow oil was subjected to column chromatography on silica gel using pentane/Et$_2$O (3/2) as the eluant to give the mesylate of alcohol **6** (10.1 g; 100%) as a colorless oil.

To a solution of the mesylate (10.5 g, 55 mmol) in DMF (100 mL) was added sodium azide (17.5 g, 0.27 mol) and the reaction mixture was heated at 65-75 °C for 3 h. The reaction mixture was cooled to room temperature and ether (100 mL) was added. The ether solution was washed with water five times to remove DMF. The organic phase was separated and dried over MgSO$_4$ and the solvent was removed on a rotary evaporator to afford 4-azido-1,6-heptadiene as a yellow oil (7.4 g; 98%): ^{1}H NMR (CDCl$_3$, 300 MHz) δ 2.27-2.34 (m, 4 H), 3.43 (dddd, J = 13.2, 7.05, 7.05, 5.7 Hz, 1 H), 5.11-5.19 (m, 4 H), 5.82 (ddt, J = 17.1, 10.2, 7.2 Hz, 2 H).

To a solution of 4-azido-1,6-heptadiene (3.28 g, 23.9 mmol), thus obtained, in THF (50 mL) was added triphenylphosphine (7.97 g, 30.4 mmol) in portions at room temperature under nitrogen. The reaction mixture was allowed to stir at room temperature for 2 h. Water (4.5 mL) was added and the mixture was heated to reflux for 6 h. The reaction mixture was gradually cooled to room temperature and concentrated. Ethyl acetate (50 mL) was added followed by a saturated aqueous solution of NaHCO₃ (45 mL) and di(*t*-butyl) dicarbonate (7.8 g, 36 mmol). The reaction mixture was allowed to stir at room temperature for 14 h. The two phases were separated and the organic phase was washed with brine, dried over MgSO₄ and concentrated under vacuum to give an oily residue. The residue was subjected to column chromatography on silica gel using hexane/AcOEt (10/1) to afford 4-(*t*-butoxycarbonylamino)-1,6-heptadiene **10b** (3.24 g; 64%) as a pale yellow oil: ^{1}H NMR (CDCl₃, 250 MHz) δ 1.34 (s, 9 H), 2.03-2.22 (m, 4 H), 3.62 (br s, 1 H), 4.43 (br s, 1 H), 4.98 (d, J = 11.6 Hz, 2 H), 4.99 (d, J = 15.4 Hz, 2 H), 5.60-5.77 (m, 2 H); ^{13}C NMR (CDCl₃, 60 MHz) δ 27.30, 27.97, 38.65, 49.49, 117.55, 134.37; IR (neat, cm^{-1}) 3300, 2979, 2933.56, 1811.05, 1701.12, 1516.92, 1369.38, 1213.15, 1173.62, 1072.36. HRMS (CI) calcd for C₁₂H₂₁NO₂ ([MH]⁺): 212.1650. Found: 212.1648.

Identification Data for Cyclohydrocarbonylation Products.

Methyl 1-(*tert*-butoxycarbonyl)-5,6-dehydropipecolate (5): Colorless oil; ^{1}H NMR (CDCl₃) (two rotamers) δ [1.43 (s),1.48 (s)] (9 H), 1.85-1.97 (m, 3 H), 2.28-2.34 (m, 1 H), [3.70 (s), 3.71 (s)] (3 H), [4.72 (br s), 4.76-4.80 (m), 4.89 (br s)] (2 H), [6.77 (d, J = 8.6 Hz), 6.89 (d, J = 8.3 Hz)] (1 H); ^{13}C NMR (CDCl₃) δ [18.26, 18.51], [23.44, 23.60], [28.12, 28.20], [52.19, 52.28], [53.07, 54.31], [81.11, 81.30], [104.14, 104.57], [124.37, 124.85], [152.21, 152.37], [171.52, 171.95]; IR (neat, cm^{-1}) 2977, 2846, 1753, 1712, 1655, 1371, 1313, 1168. HRMS Calcd. for C₁₂H₁₉NO₄: 241.1314. Found: 241.1311.

1-Tosyl-6-(3'-formyl-propyl)-2,3-dehydropiperidine (11a): Colorless oil; ^{1}H NMR (CDCl₃) δ 0.78-0.85 (m, 1 H), 1.25-1.81 (m, 7 H), 2.35 (s, 3 H), 2.43 (t, J = 7.0 Hz, 2 H), 3.84-3.87 (br s, 1 H), 4.98 (br t, 1 H), 6.51 (d, J = 8.2 Hz, 1 H), 7.24 (d, J = 8.2 Hz, 2 H), 7.59 (d, J = 8.2 Hz, 2 H), 9.70 (s, 1 H); ^{13}C NMR (CDCl₃) δ 17.23, 18.56, 21.40, 23.15, 30.96, 43.41, 52.56, 109.38, 123.57, 126.99, 129.60, 136.29, 143.34, 201.92; IR (neat, cm^{-1}) 2924, 2868, 2722, 1723, 1645, 1597, 1450, 1401, 1305, 1288, 1258, 1165, 1092, 684. HRMS (EI) calcd for C₁₆H₂₁NSO₃ [M⁺]: 307.1242. Found: 307.1240.

1-(*tert*-Butoxycarbonylamino)-6-(3'-formylpropyl)-2,3-didehydropiperidine (11b): Colorless oil; ^{1}H NMR (CDCl₃, 300 MHz) (two rotamers) δ 1.39-1.77 (m, 17 H), 1.89-2.06 (m, 2 H), 2.47 (br s, 2 H), [4.15 (br s), 4.29 (br s)] (1 H), [4.77 (br s), 4.86 (br s)] (1 H), [6.64 (br d, J = 7.8 Hz), 6.78 (br s)], (1 H), 9.76 (s, 1 H); ^{13}C NMR (CDCl₃, 60 MHz) δ

17.51, 29.97-30.52, 43.67, [48.91, 50.29], 80.49, [104.73, 105.20], [124.06, 124.08], 202.51; IR (neat, cm^{-1}) 2931, 1723, 1697, 1651, 1409, 1361, 1171, 1116. HRMS (CI) calcd for $C_{14}H_{23}NO_4$ ([MH]$^+$): 254.1756. Found: 254.1760.

6-(3'-Formylpropyl)-2-hydroxy-tetrahydropyran (17): ^1H NMR (CDCl$_3$, 250 MHz) (two diastereomers) δ 1.14-2.02 (m, 10 H), 2.43 (t, J = 6.8 Hz, 2 H), 3.16 (br s, 1 H), [3.36-3.44 (m), 3.88-3.96 (m)] (1 H), [4.66 (d, J = 9.1 Hz), 5.26 (br s)] (1 H), 9.74 (s, 1 H); ^{13}C NMR (CDCl$_3$, 100 MHz) δ [17.50, 18.04], 21.95, [30.26, 30.99], [32.55, 32.03], [35.01, 35.18], [43.13, 43.52], [72.96, 75.64], [91.36, 96.21], [201.90, 202.74]; IR (neat, cm^{-1}) 3420, 2942, 2867, 2723, 1724, 1458, 1440, 1412, 1390, 1352, 1184, 1116, 1066, 1035, 977.

Acknowledgement

This research was supported by grants from the National Institutes of Health (NIGMS) and the National Science Foundation. Generous support from Mitsubishi Chemical Corp. is also gratefully acknowledged.

References

1. P. G. Jessop, T. Ikariya, and R. Noyori, *Science*, **269**, 1065-1069 (1995).
2. G. Kaupp, *Angew. Chem. Int. Ed. Engl.*, **33**, 1452 - 1455 (1994).
3. L. Boock, B. Wu, C. LaMarca, M. Klein, and S. Paspek, *CHEMTECH* , 719-723 (1992).
4. J. A. Banister, P. D. Lee, and M. Poliakoff, *Organometallics*, **14**, 3876-3885 (1995).
5. E. Kiran and J. M. H. Levelt Sengers, *Supercritical Fluids Fundamentals for Application*; Kluwer Acad.: Dordrecht, Vol. 273, (1994).
6. J. W. Rathke, R. J. Klingler, T. R. Krause, *Organometallics* , **11**, 585-588 (1992).
7. M. J. Burk, S. Feng, M. F. Gross, and W. Tumas *J. Am. Chem. Soc.*, **117**, 8277-8278 (1995).
8. J. M. DeSimone, Z. Guan, Z., and C. S. Elsbernd, *Science* , **257**, 945-947 (1992).
9. J. M. DeSimone, E. E. Maury, Y. Z. Menceloglu, J. B. McClain, T. J. Romack, and J. R. Combes, *Science* , **265**, 356-359 (1994).
10. P. G. Jessop, Y. Hsiao, T. Ikariya, and R. Noyori, *J. Am. Chem. Soc.* , **118**, 344-355 (1996).

11. P. G. Jessop, T. Ikariya, and R. Noyori, *Organometallics*, **14**, 1510-1513 (1995).
12. P. G. Jessop, Y. Hsiao, T. Ikariya, and R. Noyori, *J. Chem. Soc., Chem. Commun.* , 707-708 (1995).
13. P. G. Jessop, Y. Hsiao, T. Ikariya, and R. Noyori, *J. Am. Chem. Soc.*, **116**, 8851-8852 (1994).
14. P. G. Jessop, T. Ikariya, and R. Noyori, *Nature*, **368**, 231-233 (1994).
15. J. W. Rathke, R. J. Klingler, and T. R. Krause, *Organometallics* , **10**, 1350-1355 (1991).
16. M. T. Reetz, W. Könen, and T. Strack, *Chimia* , **47**, 493 (1993).
17. R. Tanke, and R. H. Crabtree, *J. Am. Chem. Soc.* , **112**, 7984-7989 (1990).
18. E. Billig, A. G. Abatjoglou, and D. R. Bryant, In *U. S. Patents*; (Union Carbide Corp.): pp 4, 668, 651, (1987).
19. G. D. Cuny and S. L. Buchwald, *J. Am. Chem. Soc.*, **115**, 2066-2068 (1993).
20   K. Nozaki, N. Sakai, T. Nanno, T. Higashijima, S. Mano, T. Horiuchi, and H. Takaya, *J. Am. Chem. Soc.*, **119**, 4413-4423, (1997).
21. I. Ojima, M. Tzamarioudaki, and M. Eguchi, *J. Org. Chem.* , **60**, 7078-7079 (1995).
22. I. Ojima, D. M. Iula, and M. Tzamarioudaki, *Tetrahedron Lett.*, 1998; in press.
23. I. Ojima and C.-Y. Tsai, In *Organic Reactions*; John Wiley & Sons, Inc: New York, 1998; pp in press.
24. F. Barbot, *Tetrahedron Lett.*, **30**, 185-186 (1989).

Continuous Hydrogenation in Supercritical Fluids

T. Tacke, C. Rehren, S. Wieland and P. Panster
Degussa AG, Research and Applied Technology Chemical Catalysts and Zeolites, P. O. Box 1345, 63403 Hanau, Germany

S. K. Ross and J. Toler
Thomas Swan & Co, Ltd., Crookhall, Consett, Country Durham, DH8 7ND, England

M. G. Hitzler, F. Smail and M. Poliakoff
Department of Chemistry, University of Nottingham, Nottingham, NG7 2RD, England

Abstract

The low viscosity, good thermal and mass transport properties of supercritical fluids make them ideal for continuous catalytic reactions. We have investigated the continuous hydrogenation of organic compounds (fats and oils, fine chemicals, and intermediates) in flow reactors (up to 400°C and 40.0 MPa) with supported precious metal fixed-bed catalysts. As solvents we used liquid, near-critical or supercritical CO_2 and/or propane. The hydrogenation of cyclohexene and the selective hydrogenation of fats and oils, fatty acid esters and free fatty acids were investigated as model reactions. In all reactions we determined extremely high space-time yields, when compared to classical batch, trickle bed or gas phase reactions. In addition we observed significantly increased selectivities and catalyst productivities.

Introduction

For two decades, carbon dioxide as a supercritical fluid has established a role in 'Clean Production Technology' on an industrial scale. Well known commercial applications are the decaffeination of coffee and tea and the extraction of hops and spices. In order to explain the supercritical state, a phase diagram for carbon dioxide is shown in Figure 1.

345

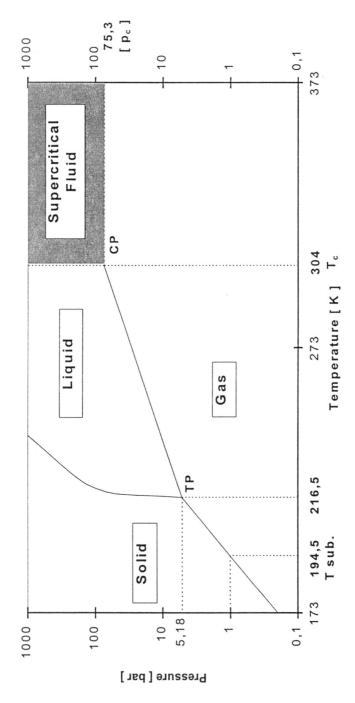

Figure 1   Phase Diagram for Carbon Dioxide (CP= critical point, TP= triple point, T$_C$= critical temperature, p$_c$= critical pressure)

Supercritical fluids are characterized by their unique physical properties which make them suitable for a variety of other applications. Further potential applications are extraction of plant materials (food and pharmaceutical industries), fuel processing, biocatalysis, homogeneous and heterogeneous catalysis, environmental control (Supercritical water oxidation, SCWO), polymerization, and material and chemical synthesis.

In general supercritical fluids are have liquid-like densities and gas-like viscosities. The diffusivities are in between those for liquids and gases. Permanent gases like hydrogen are completely miscible with supercritical fluids. In comparison, the solubility of hydrogen in conventional organic solvents is relatively low. Therefore, supercritical fluids are potentially attractive solvents for hydrogenation reactions. The low viscosities, and good thermal and mass transport properties of supercritical fluids and the ease of separation of product and solvent make them ideal for continuous catalytic reactions in flow reactors (1-5).

In the literature, various catalytic reactions are described, e. g. Fischer-Tropsch synthesis, isomerization, hydroformylation, CO_2 hydrogenation, synthesis of fine chemicals, edible oil hydrogenation, biocatalysis and polymerization. We have investigated the continuous hydrogenation of organic compounds (fats and oils, fine chemicals, and intermediates) in flow reactors (up to 400°C and 40.0 MPa).

The selective and complete hydrogenation of fats and oils, fatty acid esters and free fatty acids was investigated at Degussa whereas the selective hydrogenation of fine chemicals and intermediates was studied at the University of Nottingham in collaboration with Thomas Swan Co. Ltd. and Degussa AG.

The hydrogenation of fats and oils is a very old process, invented in 1901 by Normann in order to increase the melting point and the oxidation stability of fats and oils. For applications in the food industry edible oils are hydrogenated selectively, whereas free fatty acids are completely hydrogenated for oleochemicals (e. g. detergents). This reaction was selected as a model reaction having an important commercial impact. The reaction mechanism in the selective hydrogenation of edible oils is very complex (Figure 2). There are several parallel, consecutive and side reactions. In a cis/trans isomerization reaction elaidic acid (trans C18:1) is formed. From a food diet point of view elaidic acid is an undesirable product.

The current commercial batch process with nickel on kieselguhr catalyst has some disadvantages: discontinuous operation, low space-time-yields, undesirable by-products as a result of strong hydrogen mass-transfer control, and high variable costs (e. g. man-power, energy, filtration). The nickel on kieselguhr catalyst also causes some problems in that undesirable by-products (trans-fatty acids) are formed and these have an impact on health (high cholesterol and high lipid levels in blood), the catalyst is deactivated through formation of nickel soaps in free fatty acid hydrogenation, the costs of product purification are high (e. g. refining of edible oils, distillation of free fatty acids) and disposal of nickel residues is also a problem.

The existing processes in the hydrogenation of intermediates and fine chemicals are also characterized by similiar technical problems relating to the diffusivity, heat transfer and catalyst utilization.

To overcome the existing problems with the state-of-the-art technology in hydrogenation processes, we decided to investigate hydrogenation reactions in liquid, near-critical and supercritical CO_2 and/or propane mixtures with precious metal fixed-bed catalysts on chemical resistant supports.

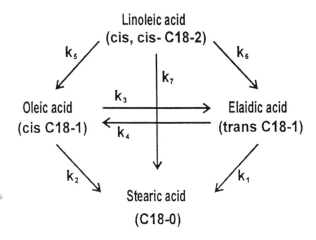

Figure 2   Reaction mechanism for the selective hydrogenation of
edible oils (here: linoleic acid)

Experimental

At Degussa AG catalytic tests in liquid, near-critical or supercritical CO_2 ($T_c = 31.1°C$, $P_c = 7.38$ Mpa) and/or propane ($T_c = 96.8°C$, $P_c = 4.26$ MPa) mixtures were run continuously in oil-heated (200°C, 20.0 MPa) or electrically-heated flow reactors (400°C, 40.0 MPa). The laboratory-scale apparatus for catalytic reactions in supercritical fluids is shown in Figure 3. In hydrogenation experiments we used between 2 and 30 ml of a fixed-bed catalyst. The test unit which is used at the University of Nottingham is described elsewhere (6). Here two different reactors with effective catalyst volumes of 4 and 9 ml were used.

Primarily polysiloxane (DELOXAN®) supported precious metal fixed-bed catalysts were applied as heterogeneous catalysts. We also investigated immobilized, metal-complex fixed-bed catalysts supported on DELOXAN®. DELOXAN® was used because of its unique chemical and physical properties (e. g. high pore volume and specific surface area in combination with a meso- and macro-pore-size distribution, which is especially attractive for catalytic reactions) (7). Reaction products were analyzed by GC and/or NMR.

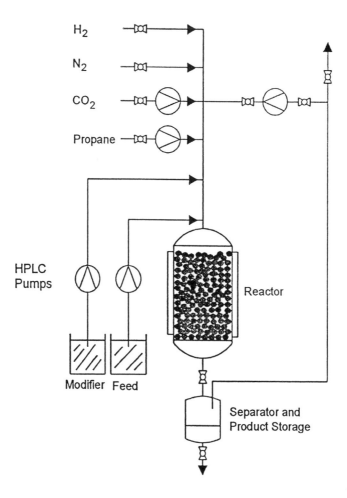

Figure 3   Laboratory scale apparatus for catalytic reactions in supercritical fluids

Results and discussion

A. Edible oil hydrogenation in supercritical fluids

Edible oils and fatty acid esters were selectively hardened in liquid, near critical or supercritical CO_2 and/or propane mixtures at temperatures between 60°C and 120°C and at a total pressure up to 20.0 MPa. In Table 1, results for the selective hydrogenation of edible oils in supercritical CO_2 are shown in comparison to liquid phase hydrogenation reactions in a discontinuous stirred-tank reactor and in a continuous trickle-bed reactor. Space-time-yields in discontinuous hydrogenation reactions with nickel on kieselguhr and activated carbon supported precious metal catalysts are below 1 m^3 oil/(h x m^3 reactor volume). A large amount of undesirable trans fatty acid chains are formed in the triglyceride molecule as a consequence of strong hydrogen mass-transfer control. The continuous hydrogenation with a DELOXAN® AP II-supported 1 wt. % palladium fixed-bed catalyst in the trickle-bed phase results in a higher space-time-yield. However, the linoleate selectivity, a measure of the non-formation of saturated fatty acids, is very low. Therefore large amounts of saturated fatty acids are formed and as a consequence the overall trans fatty acid content remains at a lower level.

Increasing the hydrogen partial pressure results in a further increase in space-time-yield. This result indicates that the reaction is strongly hydrogen mass-transfer controlled. In supercritical CO_2, the hydrogenation activity is even higher. Supercritical CO_2 lowers the viscosity of the reaction medium and increases mass-transfer rate and diffusivity. As a consequence, a higher hydrogenation activity is observed. In addition the linoleate selectivity is increased. With a DELOXAN®-supported palladium complex catalyst, DELOXAN® HK I, the linoleate selectivity is further increased. In comparison to the commercial batch hydrogenation with a nickel on kieselguhr catalyst, the DELOXAN®-supported palladium complex catalyst in combination with supercritical CO_2 as a solvent gives higher space-time-yields, a higher linoleate selectivity and a significantly decreased cis/trans isomerization rate. DELOXAN® AP II-supported platinum catalysts in supercritical CO_2 are less active than DELOXAN® AP II-supported palladium catalysts, but they show an improved linoleate selectivity and a significantly lower cis-trans isomerization rate. The overall yield of undesirable trans fatty acids is 7.5 GC area-% in the edible oil hardening with a DELOXAN® AP II-supported 2 wt. % platinum catalyst.

In a batch hydrogenation using a commercial nickel on kieselguhr catalyst, the undesirable trans fatty acid content was determined to be 40 area-%. Here it should be mentioned that the degree of hydrogenation is very important when linoleate selectivities and trans fatty acid contents are compared. At the

Table 1   Selective hardening of edible oils in liquid phase (slurry, trickle bed) and in supercritical CO2

Catalyst	Process	T [°C]	H_2-pres-sure [bar]	Space-time-yield [m^3 oil/h x m^3 reactor vol.]	Hydro-genation-activity [mol H_2/h x g active metal]	Linoleate-selectivity [-]	max. trans-fatty acid content [GC area-%]
25 % Ni-kieselguhr (powder)	batch/slurry/mass transfer controlled	120	3	< 1	4,1	10,8	40
5 % Pd/C (powder)	"	120	3	< 1	65,4	72,5	60
5 % Pt/C (powder)	"	120	3	< 1	43,1	6,2	37
1 % Pd/Deloxan® AP II	continuous trickle bed	60	5	5	6,3	1,9	20,9
1 % Pd/Deloxan® AP II	continuous trickle bed	60	20	30	28,5	2,1	20,6
1 % Pd/Deloxan® AP II	new continuous fixed bed process	60	100 (H_2 + CO_2)	60	52,3	3	19
1 % Pd/Deloxan® HK1	"	60	100 (H_2 + CO_2)	30	13,3	13,1	15,5
2 % Pt/Deloxan® AP II	"	60	100 (H_2 + CO_2)	30	3,1	7,8	7,5

moment we only achieve good linoleate selectivies and low trans fatty acid contents at a low conversion level (iodine value > 90 g $I_2/100$ g substance). The iodine value of the applied linoleic acid ester was determined to be 138.5 g $I_2/100$ g. The iodine value of the selectively hydrogenated product should be between 60 and 70 g $I_2/100$ g. Compared to CO_2, propane or propane/CO_2 mixtures as liquid, near-critical or supercritical fluids have higher solubilities for fats and oils. The viscosity of the reaction mixture is further decreased, whereas the diffusivity is further increased. In consequence a higher hydrogenation rate is observed. We determined a linear relationship between hydrogenation activity and space velocitiy (LHSV) up to space velocities of 240 h^{-1}. This experiment indicates that there is no limitation with respect to mass-transfer and diffusivity up to measured space velocities. Especially suitable reaction conditions were determined for a homogenous phase formed by a near critical/ supercritical fluid as solvent, the feedstock and hydrogen.

B. Complete hydrogenation of free fatty acids

In the hardening of free tallow fatty acids in supercritical CO_2, we measured iodine values (IV) below 1.0 g $I_2/100$ g product at a space velocity of 15 h^{-1}. In comparison to trickle-bed hardening reactions using activated carbon- and titania-supported 2 wt. % palladium fixed-bed catalysts, between 6 to 15 times higher space-time-yields were obtained in supercritical CO_2. The hydrogen partial pressures in both processes were comparable (2.5 MPa H_2). In supercritical propane the space-time-yield increased even more.

Since the experiments were carried out at a lower temperature compared to the conventional process, the acid value of the fatty acids - a measure for the selectivity of the reaction - remains at a very high level. In catalyst life-time tests, we observed a catalyst productivity that was 3 times higher when using a DELOXAN®-supported 1 wt. % palladium fixed-bed catalyst in supercritical CO_2 compared to the same catalyst in a trickle-bed hardening process. Compared to the commercial applied nickel on kieselguhr catalyst in a batch hydrogenation, we showed that catalyst productivity was 18 times higher (Table 2). All experiments were carried out with the same free fatty acid feedstock.

C. Selective hydrogenation of fine chemicals and intermediates

The continuous hydrogenation in supercritical fluids (CO_2 or propane) using heterogeneous fixed-bed catalysts was investigated at the University of Nottingham in collaboration with Thomas Swan Co. Ltd. and Degussa AG.

Table 2   Hardening of free fatty acids in liquid phase (slurry, trickle bed) and in supercritical CO_2 - Catalyst productivity

Catalyst [-]	Process [-]	Temperature [°C]	H_2-pressure [bar]	Catalyst productivity [kg fatty acid / kg cat.]
25 % Ni-kieselguhr (powder)	discontinuous / slurry / mass transfer controlled	150 - 220	25	333,3
1 % Pd/DELOXAN ® (fixed bed)	continuous / trickle bed / mass transfer controlled	150 - 200	25	2009,2
1 % Pd/DELOXAN ® (fixed bed)	new continuous fixed bed process in supercritical CO_2	140 - 200	140 (H_2 + CO_2)	6086

This reaction can be applied to a wide range of organic compounds including alkenes, alkynes, aliphatic and aromatic ketones and aldehydes, epoxides, phenols, oximes, nitrobenzenes, Schiff Bases and nitriles. The hydrogenolysis of aliphatic alcohols and ethers has also been investigated. The conversion of different organic substrates and the product yield and distribution was investigated as a function of total pressure, H_2 partial pressure, substrate to H_2 ratio, space velocity and the metal content and nature of the applied precious metal fixed-bed catalyst. Under supercritical conditions, it was found that temperature and H_2 partial pressure were the most important reaction parameters. Selected results are summarized in Table 3. In many of the investigated reactions, product yields of higher than 90% are observed at high space velocities. With higher temperatures and a higher excess of hydrogen larger amounts of completely hydrogenated products are formed.

As already reported, it is possible to maximize the yield of any of the hydrogenation products of acetophenone by varying the temperature between 90 and 300°C, and increasing the H_2 : acetophenone ratio (from 2 : 1 to 6 : 1) at a constant pressure of 120 bar (6). The supercritical fluid hydrogenation of isophorone, a functionalized cyclohexene derivative of commercial interest, gives 3,3,5-trimethylcyclohexanone (dihydroisophorone) with high conversion and high selectivity. This is very desirable because the boiling points of substrate, product and by-products are all very close to each other and the purification of dihydroisophorone by distillation is difficult and costly. 7.5 kg of isophorone were hydrogenated with 2 g of DELOXAN®-supported 5 wt. % palladium fixed-bed catalyst (0.3-0.8 mm). The flow reactor system is primarily designed to operate with reactants and products which are liquids at ambient temperature but it can also be used to hydrogenate solids, dissolved in an inert organic solvent.

Conclusion

The combination of DELOXAN®-supported precious metal fixed-bed catalysts together with liquid, near-critical or supercritical CO_2 and/or propane mixtures creates new possibilities for continuous fixed-bed hydrogenations with significantly improved space-time-yields and catalyst life-times. Short residence times and a well-balanced diffusion and desorption of products and reactants results in a decrease in undesirable by-products and thus higher selectivities.

The characteristics of high pressure hydrogenations in near-critical or super-critical fluids can be summarized as follows:

Table 3   Liquid, near-critical and supercritical fluid (scf) hydrogenation of fine chemicals and intermediates

Substrate [% conv.]	Main product [% yield]	Catalyst	Reactor wall temerature [°C]	Total pressure [MPa]	LHSV [1/h]	Solvent	H_2/subst.
m-cresol (100)	3-methylcyclo-hexanone (61.5) 3-methylcyclo-hexanol (26.0)	5% Pd/Deloxan AP II (0.3-0.8 mm)	250	12.0	3.3	scf CO_2	2.5 : 1
benzaldehyde (100)	benzylalcohol (92)	"	95	12.0	7.5	scf CO_2	2 : 1
acetophenone (97.5)	1-phenylethanol (90)	"	90	12.0	7.5	scf CO_2	2 : 1
furan (98)	tetrahydrofuran THF (96)	"	300 – 350	12.0	7.5	scf CO_2	3 : 1
nitrobenzene (100)	aniline (100)	1% Pd/Deloxan AP II (0.2-0.5 mm)	150 – 200	8.0	7.5	scf propane	6 : 1
N-benzylidene-methylamine (99)	benzylmethylamine (99)	5% Pd/Deloxan AP II (0.3-0.8 mm)	40 – 50	12.0	4.5	liquid propane	2 : 1
2-butanone oxime (99)	2-butylamine (80)	"	180 – 200	8.0	15	scf propane	2 : 1
1-octene (100)	octane (100)	"	> 40	12.0 – 14.0	7.5 – 150	scf CO_2	2 : 1
cyclohexene (99.8)	cyclohexane (99.8)	"	> 40	12.0 – 14.0	7.5 – 300	scf CO_2	2-4 : 1
isophorone (100)	dihydroisophorone (100)	"	140 – 200	12.0	7.5 – 30	scf CO_2	2 : 1

- No hydrogen mass-transfer control
  - high space-time-yields
  - avoidance of undesirable by-products
- Continuous processing using stable-fixed bed catalysts
  - adjustable residence time of reactants
  - higher selectivities
  - higher product quality
- Supercritical fluids support the hydrogenation process
  - increase of heat transfer
  - decrease of viscosity
  - high solubility of hydrogen
- In-situ catalyst reactivation gives longer catalyst life-times

References

1. E. Kiran, J. M. H. Levelt Sengers (eds.), Supercritical Fluids, Kluwer Academic Publishers, Dordrecht, The Netherlands, 1994
2. M. Perrut, G. Brunner (Eds.), Proceedings of the Third International Symposium on Supercritical Fluids, 1994
3. P. E. Savage, S. Gopalan, T. I. Mizan, C. J. Martino, E. E. Brock, AIChE Journal, 41, 1723 (1995)
4. K. W. Hutchenson, N. R. Foster (Eds.), Innovations in Supercritical Fluids: Science and Technology, American Chemical Society, Washington, DC, 1995
5. Ph. R. von Rohr, Ch. Trepp (eds.), High Pressure Chemical Engineering, Elsevier, Amsterdam, 1996
6. M. G. Hitzler, M. Poliakoff, Chem. Commun.1997, 1667
7. S. Wieland and P. Panster, in Catalysis of Organic Reactions (M. G. Scaros, M. L. Prunier, eds.), Marcel Dekker, New York, 1995

Novel Catalytic Characteristics of the Co/Mn/Cl/Br Liquid Phase Oxidation Catalyst (1)

Walt Partenheimer
Central Research and Development, DuPont Experimental Station, PO Box 80262, Wilmington DE 19880-0262.

Abstract

The activity and carbon-carbon bond cleavage in metal/halide homogeneous oxidation catalysts (halide = F, Cl, Br, I) is described and rationalized. The catalytic characteristics of the Co/Mn/Cl/Br combination are different than would be anticipated from the Co/Mn/Cl and Co/Mn/Br catalysts. A kinetic model is developed which explains the higher activity of the Co/Mn/Cl/Br catalyst than the Co/Mn/Br catalyst. Benzylic bromides, which are an inactive form of bromide, rapidly form in metal/bromide autoxidation. The addition of chloride causes a displacement reaction: $PhCH_2Br + NaCl ==> PhCH_2Cl + HBr$, which releases the catalytically active bromide. The lack of carbon-carbon bond cleavage in the Co/Mn/Cl/Br catalyst, in contrast to the Co/Mn/Cl catalyst, is because the bromide reacts much faster with Mn(III) than does chloride.

Introduction

Chemical research is the pursuit of knowledge and understanding. It has its joys when unusual phenomena are 'understood' or when the data falls into a rational sequence so it 'makes sense.' In this case, I was working on two apparently unrelated phenomena -- 1) why a Co/Mn/Cl/Br catalyst was more active than a Co/Mn/Br catalyst and 2) that benzylic bromides were inactive forms of bromine during autoxidations with metal/bromide catalysts.   It was Rob Gipe, at a monthly discussion meeting, who made an insightful suggestion that these two phenomena were directly linked. The chloride in the Co/Mn/Cl/Br catalyst reacts with the inactive benzylic bromide to form benzylic chlorides and restore the active form of bromine.   This paper details the subsequent research that substantiated Rob's suggestion.

One of the major industrial processes for producing oxygenates is:

$$\text{hydrocarbon} + O_2 \xrightarrow[\text{acetic acid solvent}]{\text{metal/bromide}} \text{oxygenate} + H_2O \qquad (1)$$

The results of the oxygenation of 251 hydrocarbons using the above method has been summarized (2). The most prominent industrial chemical manufactured with this method is terephthalic acid from p-xylene which is used in polyethylene terephthalate (PET , 'polyester') manufacture.

The different characteristics of autoxidation catalysts of the type $M_1M_2M_3.../X_1X_2...$ where M is a transition metal and X is a halogen is given in table 1. Carbon-carbon bond cleavage is defined as the molar yield of products produced by C-C cleavage per mole of reagent.

Table 1. Some Characteristics of $M_1M_2M_3.../X_1X_2$ Catalysts

	activity	C-C bond cleavage
$M_1M_2M_3.../ F$	weak	little
$M_1M_2M_3.../Cl$	weak	significant amount
$M_1M_2M_3.../Br$	strong	little
$M_1M_2M_3.../ I$	none	---
$M_1M_2M_3.../Cl/Br$	strongest	little

The characteristics of the $M_1M_2M_3.../ Cl/Br$ catalyst is strikingly different from $M_1M_2M_3.../Cl$ and $M_1M_2M_3.../Br$ catalysts. It has higher activity than the $M_1M_2M_3.../Br$ catalyst but does not produce as much C-C bond cleavage as the $M_1M_2M_3.../Cl$ catalyst.

A few illustrative examples will now be given (3). Metal/halogen catalysts were first reported in the patent literature by way of emphasizing the unique high activity of metal/bromide catalysts. The yields of p-xylene to give terephthalic acid:

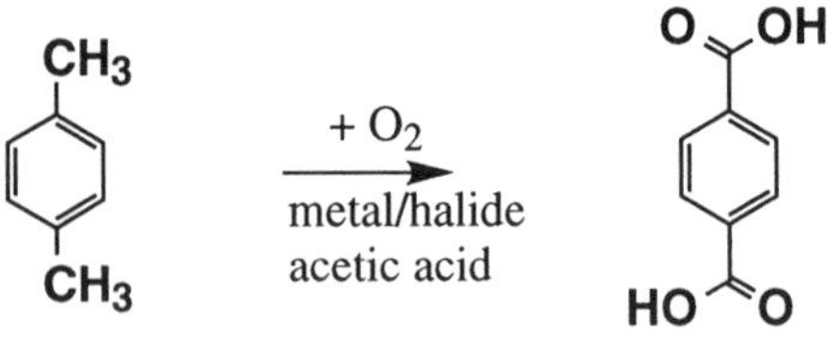

$$\text{(2)}$$

for Co/X catalysts are <19, 19, 75, and <19 for X = F, Cl,Br,I respectively (4). Similarly the autoxidation activity of ethylbenzene to acetophenone for Co/X catalysts:

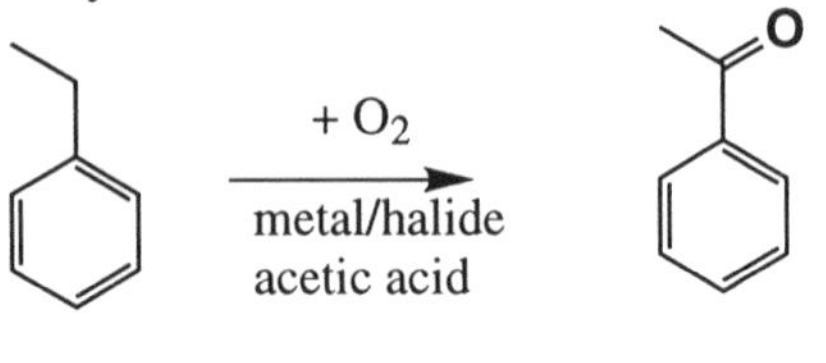

$$\text{(3)}$$

is 1.3, 273, 0.0 for X = Cl, Br, I (5). The lower activity and higher carbon-carbon bond cleavage of Co/Cl vs Co/Br catalysts have been discussed (6). For example the yield to 2,2'-di(p-carboxyphenyl)propane is 49% with Co/Cl and 86% with Co/Br under comparable conditions due to the extensive carbon-carbon cleavage with Co/Cl:

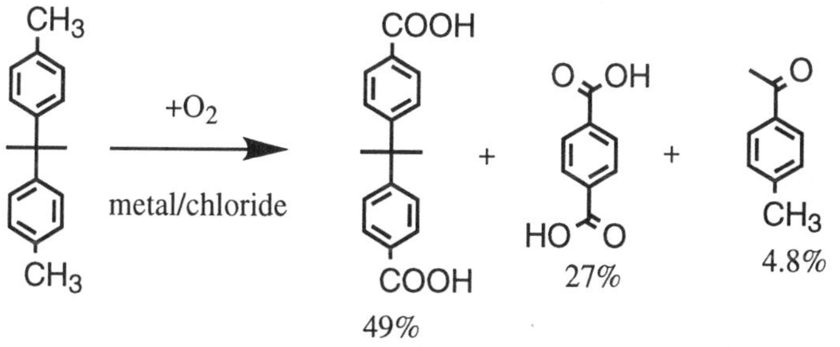

(4)

49%   27%   4.8%

The carbon-carbon cleavage ability of the Co/Mn/Cl catalyst has been used to obtain high yields of 4,4'-dicarboxybiphenyl (7) :

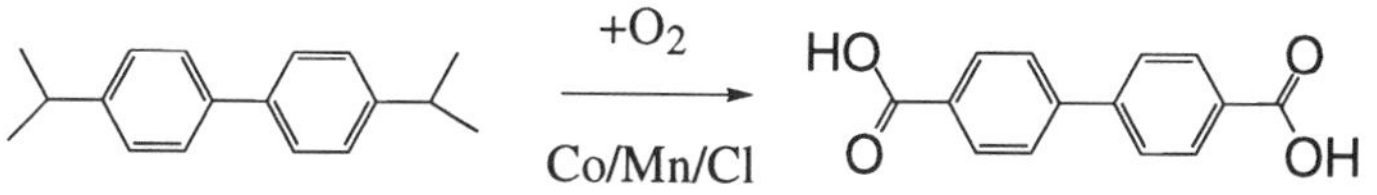

(5)

There are scattered reports on the use of the Co/Mn/Cl/Br catalyst. Holtz reported the reduced carbon-carbon cleavage in the oxidation of p-t-butyltoluene and t-butylbenzene (6). Enhanced activity has been claimed during the oxidation of 1,2,4,5-tetramethylbenzene (durene) to 1,2,4,5-tetracarboxybenzene (pyromellitic acid) (8), with toluene to benzoic acid (9) and bis(p-tolyl)dimethylsilane to bis(p-toluic acid)dimethysilane (10).

Experimental

The catalytic oxygenations were performed in a glass cylindrical reactor with continuous monitoring of the reaction products, oxygen uptake and carbon dioxide and carbon monoxide liberation as previously described (11). The various non-oxidative modeling reactions described herein were measured in this same apparatus.

Results and Discussion

Activation of Reactions due to Chloride Addition to Metal/Bromide Catalysts.- The results of the reaction (at Cl/Br =5.2 mol/mol and Co/Mn = 11.9 mol/mol):

(6)

toluene   benzoic acid

is summarized in table 2. Metals are typically added as the acetates and the halogens as an ionic salt. The rate of toluene disappearance is higher and the rate of benzoic acid formation is faster with chloride present (figure 1). Similar effects are seen with the reactions of p-xylene to p-toluic acid and p-toluic acid to terephthalic acid, see table 2. We do not see activation using p-nitrotoluene. The Co/Mn/Zr catalyst is also activated with chloride, see examples 3 and 4 in table 2. We anticipate this same result with any of the 30 metal/bromide catalysts previously reported (2).

Table 2. Selected Data on the Effect of Chloride Addition to Metal/Bromide Catalysts

	reactant	catalyst	time,hr	rate O_2uptake	Yield, %[b]	CO_x Selec
1	toluene	Co/Mn/Br	1.3	4.2	5	4.6
2	toluene	Co/Mn/Cl/Br	1.3	7.3	8	2.4
3	toluene	Co/Mn/Zr/Br	1.3	6.6	---	1.8
4	toluene	Co/Mn/Zr/Cl/Br	1.3	11.1	---	1.7
5	toluene	Co/Mn/Br/F	1.3	5.7	---	2.5
6	toluene	Co/Mn/Br/I	1.3	0.48	---	---
7	p-xylene	Co/Mn/Br	3.0	---	24	---
8	p-xylene	Co/Mn/Cl/Br	3.0	---	28	---
9	p-toluic acid	Co/Mn/Br	1.3	1.2	12	1.1
10	p-toluic acid	Co/Mn/Cl/Br	1.3	4.8	26	1.6
11	p-nitrotoluene	Co/Mn/Br	1.3	2.1	---	4.4
12	p-nitrotoluene	Co/Mn/Br/Cl	1.3	1.5	---	0.9

a. CO_x Selectivity is defined as the rate of carbon monoxide and carbon dioxide formation divided by the rate of oxygen uptake.
b. To benzoic acid for toluene, to p-toluic acid for p-xylene, and to terephthalic acid for p-toluic acid.

The Cl/Br combination is unique - It cannot be Duplicated by F/Br and I/Br Catalysts. See examples 1,2,5,6 on table 2.
The Addition of Chloride does Not Affect the Metal-Modified Free Radical Chain Reaction (15) - Figure 2 indicates that the formation of intermediates with and without chloride present, are unaffected. This is also suggested by the fact that the selectivity to carbon monoxide and carbon dioxide are, within experimental error, the same.

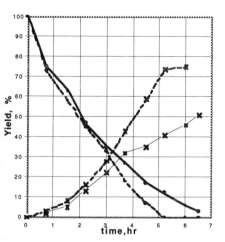

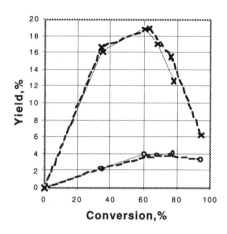

Figure 1. Rate of Oxidation of Toluene (0) to give Benzoic acid (x) with with Co/Mn/Br (solid line) and Co/Mn/Br/Cl (--- line) Catalysts

Figure 2. Oxidation of Toluene to Benzaldehyde (x) and Benzyl acetate (0) with Co/Mn/Br (solid line) and Co/Mn/Br/Cl (--- line) catalysts

A Substantial Amount of the Bromide in Metal/Bromide Catalysts Exist as Benzylic romides- A gc trace of the samples removed from the reactor during the reaction gives minor component which has been identified via gc/ms as α-bromotoluene. The ormation of benzylic bromides has been previously reported for toluene, p-xylene, 2-hlorotoluene, 2,4-dichlorotoluene, m-phenoxytoluene, 4-nitrotoluene and iphenylethane (2). Table 3 extends their formation to a number of other hydrocarbons nd other metal/bromide autoxidation systems. Benzylic bromide formation is not urprising since peroxides in the presence of bromide form benzylic bromides (12). Vhat is important here, is that the yields of benzylic bromides when expressed in terms f toluene is only about 1%, but when the yield is expressed in terms of the bromide resent, is very high. *A majority of the initial added bromide exists as benzylic romide during the reactions. Their formation occurs immediately after initiation of the eaction (table 3 and figure 3).*

Benzylic Bromides Solvolyze During the Reaction- In 10% water/acetic acid, an uthentic sample of a-bromotoluene solvolyzes first to the alcohol which subsequently sterifies to benzyl acetate:

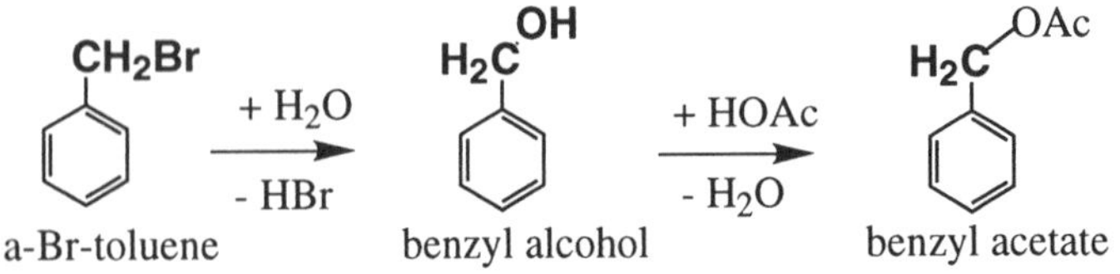

CH₂Br → H₂C-OH → H₂C-OAc

a-Br-toluene   benzyl alcohol   benzyl acetate

$$\text{a-Br-toluene} \xrightarrow[-\ HBr]{+\ H_2O} \text{benzyl alcohol} \xrightarrow[-\ H_2O]{+\ HOAc} \text{benzyl acetate} \qquad (7)$$

It can be seen from the kinetic studies reported in table 4, that during the time frame of the reactions under discussion (typically require 12-24 hrs for complete conversion to the aromatic acid), that substantial amounts of the bromide will solvolyze. The activation energy for solvolysis calculated from table 4 is 22.8 kcal/mol.

Table 3. Benzylic Bromides Yields for Selected Reagents and Catalysts

Reagent, initial conc.	Catalyst,mmol added	time, hr	rate[a]	Conv %	Yield benzylic Br
toluene,1.08M	Mn/Br,2/2	2.0	1.12	13.9	35
toluene,1.08M	Ce/Br,2/2	2.0	0.66	9.6	64
toluene,0.294M	Co/Br,4/1	2.2	6.5	94	100
toluene, 0.29M	Co/Ce/Br,4/0.4/1	2.2	5.4	99	90
toluene, 1.08M	Co/Mn/Br,1/1/2	2.75	6.6	60	96
toluene,2.18M	Co/Mn/Br,2/2/4	2.17	5.3	21	99
toluene,2.18M,5% H₂O	Co/Mn/Br,2/2/4	2.17	4.4	19	79
p-xylene,0.94 M	Co/Mn/Br,2/2/4	1.75	10.6	57	76
p-toluic acid,1.08M	Co/Mn/Br,1/1/2	0.88	7.09	---	54
p-toluic acid,0.74 M	Co/Mn/Br,2/2/4	2.33	4.21	57	76
4-nitrotoluene,1.08M	Co/Mn/Br,1/1/2	1.17	1.58	9.8	11
toluene,1.08M	Co/Mn/Ce/Br,1/1/0.14/2	1.22	7.09	26	88
toluene,1.08M	Co/Mn/Zr/Br,1/1/0.4/2	1.22	6.31	34	98

a. in ml O_2/min. Benzylic bromide yield based on NaBr added.

Table 4. Rates of Solvolysis of a-bromotoluene

[water], wt %	temp.°, C	k,s^{-1}, x 10^6	half-life,hrs	R^2
5	95	13.6	14	0.995
10	95	51.9	3.7	0.998
20	95	169	1.1	0.996
10	85	18.3	11	0.999
10	75	7.96	24	0.999

Benzylic Bromides are An Inactive Form of Bromide- This is suggested by the the rate of initiation given in figure 4. Addition of a ring-brominated compound, in this case 3-bromoanisole, to a mixture of cobalt(II) acetate and toluene, does not initiate the reaction. Cobalt catalyzed oxidations normally take 1-2 hrs to initiate so the Co/toluene/3-Br-anisole is behaving like a Co catalyzed oxidation of toluene. However, if 3-bromoanisole is replaced by an ionic source of bromide, sodium bromide, the reaction immediately initiates and reaches a maximum rate in about 20 minutes. Starting with a-bromotoluene, the rate of initiation is significantly slower than with sodium bromide. We interpret this observation to mean that the benzylic bromide itself is not active, but the slow release of hydrobromic acid via solvolysis results in increased activity.

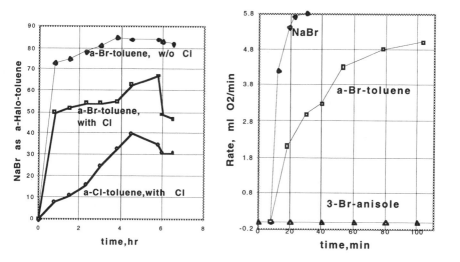

Figure 3. Steady State Conc. of a-Br and a-Cl-toluene during oxidation of toluene with Co/Mn/Br and Co/Mn/Br/Cl catalysts

Figure 4. Rate of Initiation of toluene with different sources of bromide

Chloride Addition to Metal/bromide Catalyzed Autoxidations reduces the steady state concentration of a-bromo-toluene , [PhCH₂Br]ₛₛ, see figure 3 - When chloride is added to a Co/Mn/Br catalyzed reaction, a new gc peak is observable which gc/ms analysis indicates is a-chlorotoluene. The reduction in the $[PhCH_2Br]_{ss}$ is almost matched by the amount of benzyl chloride formed. The reduction in $[PhCH_2Br]_{ss}$ presumably results in an increased concentration of ionic bromide hence a more active reaction.

An Explanation Why Chloride Activates - If we assume the displacement reaction:

$$PhCH_2Br \quad + \quad NaCl \quad \longrightarrow \quad PhCH_2Cl \quad + \quad NaBr \quad (8)$$

a-Br-toluene   a-Cl-toluene

inactive Br   *active Br*

then inactive bromide would be converted to an active form. Since this reaction is independent of the free radical chain mechanism, we would expect the rate of formation of the intermediates to be the same, which is observed. We do need to show that this displacement reaction occurs within the time frame of the oxidation. The observed data in figure 5, performed under identical conditions as the oxidation reactions, is consistent with the following reactions:

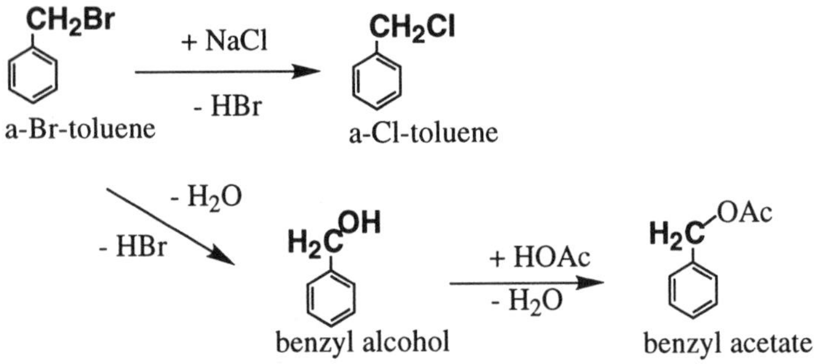

a-Br-toluene   a-Cl-toluene

benzyl alcohol   benzyl acetate

The reaction is quite rapid and consistent with the above network of reactions occuring during autoxidation reactions. A significant amount of solvolysis of the a-bromotoluene also occurs. A high Cl/Br ratio, typically 5, is necessary for the substitution reaction (8), to occur to a significant extent.

Mechanism of Benzylic Bromide Formation - It is not a substitution reaction - Conceivably the formation of a-bromotoluene is via the reactions:

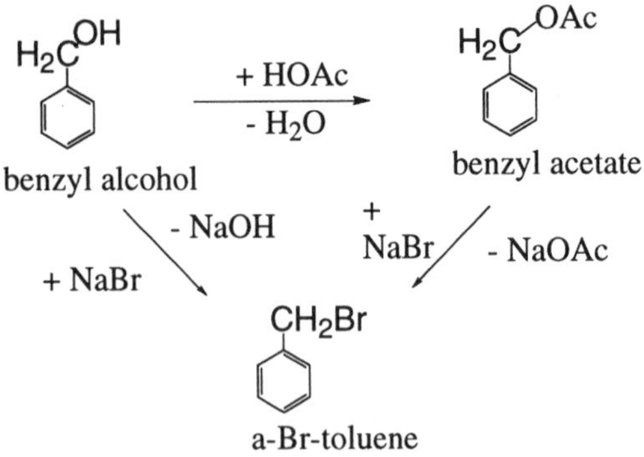

benzyl alcohol   benzyl acetate

a-Br-toluene

This does not occur however upon mixing benzyl alcohol with sodium bromide under identical conditions as the autoxidation reactions were performed. The only observable reaction is the formation of the acetate, figure 6.

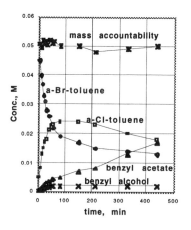

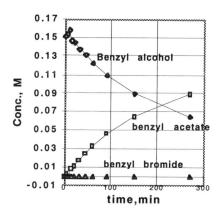

Figure 5. Addition of NaCl to a-Br-Toluene in 10% water/acetic acid at 95 C

Figure 6. Reaction upon mixing benzyl alcohol with NaBr in 10% water/acetic acid at 95 C

The formation of the benzylic bromide is probably free radical in nature. Perhaps via the scheme [13]:

$$\text{Co(III)} + \text{PhCH}_3 \longrightarrow \text{Co(II)} + [\text{PhCH}_3]^{\cdot+} \qquad (9)$$

$$[\text{PhCH}_3]^{\cdot+} \longrightarrow \text{PhCH}_2^{\cdot} + \text{H}^+ \qquad (10)$$

$$\text{PhCH}_2^{\cdot} + \text{Co(III)} \longrightarrow \text{PhCH}_2^+ + \text{Co(II)} \qquad (11)$$

$$\text{PhCH}_2^+ \ \text{Br}^- \longrightarrow \text{ArCH}_2\text{Br} \qquad (12)$$

or perhaps via a ligand transfer reaction (14):

$$\text{Mn(III)(OAc)}_2\text{Br} + \text{PhCH}_2^{\cdot} \longrightarrow \text{Mn(II)(OAc)}_2 + \text{PhCH}_2\text{Br} \qquad (13)$$

Summary of the Activating Effect of Chloride in Metal/Bromide Catalysts.- The primary product of autoxidations are peroxides which form via a chain reaction of benzylic radicals with dioxygen. Cobalt(II) is highly efficient and selective in reacting with these peroxides (15). The cobalt(III) thus produced reacts rapidly with manganese(II) to produce manganese(III), see figure 7 (16). This reaction is important because it reacts with cobalt(III) prior to it forming a lesser reactive form of cobalt and because manganese(III) reacts less rapidly with the acetic acid solvent than cobalt(III) does. The manganese(III) then oxidizes the bromide(-I) to bromine(O) atom which remains coordinated to the metal (15). The latter reacts to further initiate the reaction by forming benzylic radicals. The system is strongly catalytic because of the very rapid reaction of cobalt(II) with the peroxides and because the benzylic radicals are much more rapidly generated with Br(O) then with either Co(III) or Mn(III).

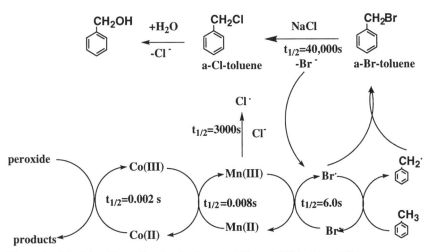

Figure 7. Model to Describe the Activating Effect of Chloride addition to Metal/Bromide Catalysts

Metal/bromide catalysts are quickly deactivated because a substantial amount of the initially reactive ionic bromide (normally added as HBr or NaBr) quickly forms benzylic bromides which are an inactive form of bromide. The presence of sodium chloride regenerates the active form of ionic bromide via a S_N2 substitution reaction to produce a-chloro-toluene. Metal/chloride catalysts promote C-C cleavage while metal/chloride/bromide do not. This is because Co(III) or Mn(III) react 500 times faster with bromide (2) than chloride so that bromine atoms (Br(O)) is preferably generated to Cl(O), see following discussion.

Rationalization of Why Metal/Bromide Catalysts are more Active than Metal/Fluoride, Metal/Chloride, and Metal/Iodide Catalysts (table 1)-
Using a Co/Br combination as an example, the catalysts are efficient if the metal can oxidize the halide and if the resultant halogen(O) can abstract a hydrogen atom from the methylaromatic compound (ligands dropped around cobalt for simplicity):

$$Co(III) + X^- \longrightarrow Co(II) + X^\cdot \qquad (14)$$

$$PhCH_3 + X^\cdot \longrightarrow PhCH_2^\cdot + HX \qquad (15)$$

Using thermodynamic arguments and ignoring the entropy contribution and solvent effects (2), cobalt easily reacts with bromide and iodide but not chloride or fluoride, reaction 14, figure 8. Hence metal/Cl or F catalysts are not very active. Iodide cannot abstract benzylic hydrogen atoms hence metal/iodide catalysts are not active. It is observed that metal/chloride catalysts do promote autoxidations but are much less active than metal/bromide catalysts. This is because Co(III) can react with chloride; however, the reaction is 500 times slower than with bromide (2).

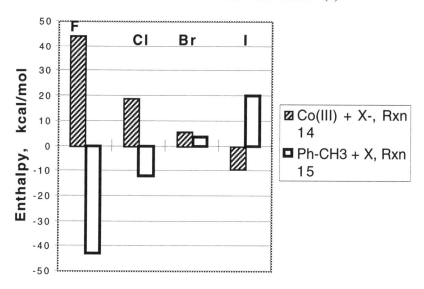

Figure 8. Data from (2).

Rationalization of the Carbon-carbon bond cleavage in Metal/chloride Catalysts-
Metal/chloride catalysts react with t-butyl groups while metal/bromide and metal/chloride/bromide catalysts do not (2),(16). Using bond strength arguments, this is because Br(O) can abstract benzylic C-H bonds but not t-butyl bonds, see rxns 16,17. Cl(O) can react with t-butyl C-H bonds to produce radicals, see rxn 18. This radical initiates a complex sequence of reactions with dioxygen which subsequently yields benzoic acid as the principle product. Little C-C bond cleavage is present in Co/Mn/Cl/Br catalysts because the reaction of chloride with Mn(III) is much slower than with bromide, see figure 7.

$$Br(O) + \underset{\text{CH}_3}{\bigcirc} \longrightarrow HBr + \underset{\text{CH}_2{}^{\cdot}}{\bigcirc} \qquad (16)$$

$$D^{C\text{-}H} - D^{H\text{-}Br} = +335\text{-}339 = -4 \text{ kj/mol}$$

$$Br(O) + \underset{\bigcirc}{H_3C\!-\!\overset{CH_3}{\underset{|}{C}}\!-\!CH_3} \longrightarrow HBr + \underset{\bigcirc}{H_3C\!-\!\overset{CH_2{}^{\cdot}}{\underset{|}{C}}\!-\!CH_3} \qquad (17)$$

$$D^{C\text{-}H} - D^{H\text{-}Br} = +397\text{-}339 = +58 \text{ kj/mol}$$

$$Cl(O) + \underset{\bigcirc}{H_3C\!-\!\overset{CH_3}{\underset{|}{C}}\!-\!CH_3} \longrightarrow HCl + \underset{\bigcirc}{H_3C\!-\!\overset{CH_2{}^{\cdot}}{\underset{|}{C}}\!-\!CH_3} \qquad (18)$$

$$D^{C\text{-}H} - D^{H\text{-}Cl} = +397\text{-}405 = -8 \text{ kj/mol}$$

Acknowledgments- I would like to thank Rob Gipe for his insightful comments and friendship and to Gayle Chany for her careful experimentation.

References

1.   Worked performed at the Amoco R&D Center, Naperville, IL.

2.   W. Partenheimer, Catalysis Today, 23 (2): 69 (1995)

3.   These observations have also been made in the authors laboratory

4.   A. Saffer and R.S. Barker US Patent 2,833,816 (1958)

5.   A.S. Hay, US Patent 3,139,452 (1964)

6.   H.D. Holtz, J. Org. Chem., 37, 2069 (1972)

7.   R.A. Periana, G.F. Schaefer, US Patent 5,068,407 (1991)

8.   H. Shimizu, T. Horei, K. Yoshida, and Y. Katsuyama, Japan Patent
      48/26743(1973)

9.   F.F. Shcherbina and N.P. Belous, Dopov. Aka. Nauk Ukr. RSR,
      Ser. B: Geol., Khim. Biol. Nauki, 3(1984) 54. CA101(1):6404k

10.   H. Mami, Jap Patent LOP Publication No. 310,846 (1988).

11.   W. Partenheimer, J. Mol. Catal., 67 35-46 (1991).

12.   A. Chaintreau, G. Adrian, D. Couturier, Syn. Commun., 11, 669
      (1981)

13.   R.A. Sheldon, J.K. Kochi, "Metal-Catalyzed Oxidations of Organic
      Compounds", Academic Press, New York, N.Y. (1981) pg122

14.   R.A. Sheldon, J.K. Kochi, "Metal-Catalyzed Oxidations of Organic
      Compounds", Academic Press, New York, N.Y. (1981) pg35

15.   W. Partenheimer in S.T. Oyama and J.W. Hightower (Editors), Catalytic
      Selective Oxidation, Am. Chem. Soc., (1993), chapter 7.

16.   Quoted half-lives are from references (2) and (15) at 60 ° C in 10% water/acetic
      acid

The Cativa™ Process for the Production of Acetic Acid

Derrick J. Watson
BP Chemicals, Hull, England

Abstract

Methanol carbonylation to acetic acid is catalysed at very high rates using an iridium carbonyl catalyst. This process is enhanced by the addition of a number of other metals, including as osmium and cadmium. The rate of this reaction has been found to be highly dependant on the concentration of water in the reaction matrix, and a rate maximum has been found to exist at commercially attractive low water concentrations.

Introduction

Acetic acid is an expanding commodity chemical with a world usage of just over 12 thousand million pounds per year.

Since its invention in the late 1960s by Monsanto, the dominant manufacturing route for acetic acid has been via the rhodium catalysed carbonylation of methanol. The rights to this process were acquired by BP Chemicals in 1986. Recent work has focused on the replacement of rhodium with a more effective catalyst package based on an iridium catalytic cycle. This has been commercialised at two plants: the Sterling Chemicals Texas City acetic acid plant in November 1996; and the Samsung-BP acetic acid plant in Ulsan, Republic of Korea in August 1997.

The new process is the most significant breakthrough in this industry in 25 years and gives improvements via a much higher reactivity coupled with lower by-products and lower energy requirements for the purification of the product acid.

Methanol carbonylation via rhodium

The rhodium catalysed production of acetic acid from methanol was developed by Monsanto in the late 1960s and rapidly became the process of choice for new plant builds for this commodity chemical. The previously commercialised route for methanol carbonylation utilised a cobalt catalyst. This process demanded very stringent operating conditions (600 atmospheres pressure and 230°C) and hence high capital and operating costs. The new rhodium process operated at both lower temperature and pressure (30 to 60 atmospheres and 150-200°C). It

very stringent operating conditions (600 atmospheres pressure and 230°C) and hence high capital and operating costs. The new rhodium process operated at both lower temperature and pressure (30 to 60 atmospheres and 150-200°C). It also gave substantially improved selectivity on the major feedstock - from a 90% yield on methanol for cobalt catalysis to over 99% for rhodium (1).

The reaction was investigated in great detail by Forster and his co-workers at Monsanto (2) and the accepted mechanism is shown below (Figure 1). This cycle is a classic example of a homogeneous catalytic process and is made up of six discreet but interlinked reactions.

Methyl iodide is generated by the reaction of added methanol with hydrogen iodide. Infrared spectroscopic studies have shown that the major rhodium species present is $[Rh(CO)_2I_2]^-$. Methyl iodide oxidatively adds to this rhodium species to give a rhodium-methyl complex. The key to the process is that this rhodium-methyl complex can rapidly undergo a methyl shift to a neighbouring carbonyl group, with the rhodium complex becoming locked into this acyl form by the subsequent addition of added carbon monoxide. Reductive elimination of acyl iodide can then occur to liberate the original rhodium-dicarbonyl-diiodide complex. Hydration of acyl iodide is very rapid in the presence of excess water and will result in the formation of acetic acid and hydrogen iodide to complete the cycle.

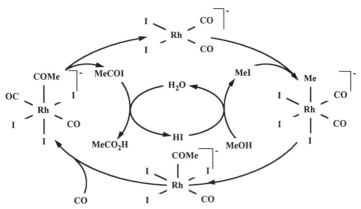

Fig. 1: Cycle for the rhodium catalysed carbonylation of methanol to acetic acid

The rate determining step in this process is the oxidative addition of methyl iodide to the rhodium centre, and so the reaction rate is essentially first order in both catalyst and methyl iodide concentrations and, under normal reaction conditions, largely independent of any other parameters:

$$Rate \propto [catalyst] * [CH_3I]$$

Although a highly selective and efficient process, rhodium catalysed carbonylation does suffer from disadvantageous side reactions. For example, rhodium will also catalyse the water gas shift reaction:

$$CO + H_2O \Rightarrow CO_2 + H_2$$

This not only represents a loss of selectivity with respect to the carbon monoxide raw material, but the gaseous by-products will dilute the carbon monoxide present in the reactor. This lowers the carbon monoxide partial pressure leading to the necessity to vent significant volumes of gas with further loss of yield. Yield on carbon monoxide is good (>85%) (3,4), but does leave room for improvement.

Propionic acid is observed as the major liquid by-product. This may be produced by the carbonylation of ethanol (present as an impurity in the methanol feed), but much more propionic acid is observed than can be accounted for by this route.

The rhodium catalytic system can generate acetaldehyde, and it is proposed that this acetaldehyde, or it's rhodium bound precursor, undergoes reduction by hydrogen present in the system to give ethanol which subsequently yields propionic acid.

One possible precursor for this generation of acetaldehyde is the rhodium-acyl species shown in the preceding mechanism. This species can be observed spectroscopically in the normal reaction mixture. Reaction of this material with hydrogen iodide would yield acetaldehyde and [RhI₄CO]⁻. The latter species is well known in this system and is postulated as the principal cause of catalyst loss by precipitation of inactive rhodium tri-iodide.

$$[RhI_3(CO)(COCH_3)]^- + HI \Rightarrow [RhI_4(CO)]^- + CH_3CHO$$
$$[RhI_4(CO)]^- \Rightarrow RhI_3 + I^- + CO$$

In addition to the propionic acid produced, very small amounts of acetaldehyde condensation products and their derivatives are also observed. Under the normal operating conditions of the original Monsanto process, these trace compounds do not present a problem to either product yield or product purity.

Separation of pure acetic acid product from the reaction medium presents few problems and a schematic of a typical plant configuration is shown below (Figure 2).

Rhodium catalyst is separated from the product acid by conducting a simple flash, the non-volatile catalyst remains in the liquor and can be readily recycled to the reactor. "Lights" separation - principally methyl iodide - is carried out in the first distillation column. That is followed by a drying column. Energy usage in this distillation can be high, depending on the water concentration chosen for the reaction stage. Similarly, high energy usage can be encountered in the third distillation column, particularly if the reactor is operated in a manner that results in the production of high levels of propionic acid.

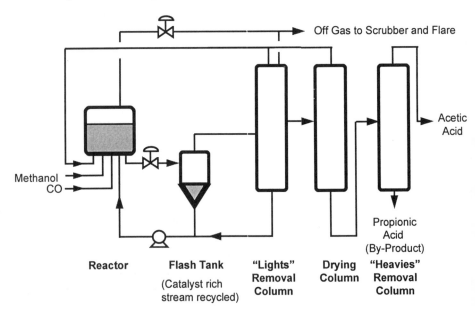

Fig. 2: Schematic of a rhodium catalysed acetic acid plant

Operating experience has shown that high water concentrations and operating regimes that give high propionic acid production, both lead to enhanced catalyst stability. Operation can therefor be a balancing act between energy usage and catalyst loss.

Enhancements to the rhodium process.

Rhodium stability was extensively investigated as part of the original work by workers at Monsanto (5). Among other species, lithium iodide was early identified as a very good agent for imparting an enhanced degree of stability to the rhodium catalyst. Further work in this area has identified that the addition of a substantial quantity of Group I metal iodides also enhances the reactor productivity under some conditions (6,7,8). The potential for use of these metals, especially lithium, as an agent for the chemical sequestration of water within the reaction system has also been identified (9).

At first sight it would appear that this provides a clear path for progressing this technology.

The addition of a significant quantity of a Group I metal iodide speeds up the oxidative insertion of methyl iodide (the rate determining step) and therefor promotes the primary carbonylation reaction. It also has the effect of reducing the rate of the unwanted water-gas shift reaction giving a beneficial effect on yield on carbon monoxide.

However, these two effects in combination significantly change the amount and nature of the liquid by-products observed in the system. The oxidative insertion step is now faster, but the CO insertion into the rhodium-acyl species remains at the same rate, and there is therefor an increase in the standing concentration of the rhodium-acyl species. This moiety will continue to form free acetaldehyde, but at a higher rate than in an unpromoted system.

The primary effect of this change could be to produce more propionic acid, but, because of the lower water-gas shift reaction rate giving less hydrogen present, acetaldehyde condensation products are favoured instead.

Numerous liquid by-products are observed in the promoted system and also, at much lower concentration, in the unpromoted system. The natural fate of acetaldehyde, if not reduced, is to undergo self condensation yielding butenal and higher analogues. These can undergo further reaction as summarised in the network below:

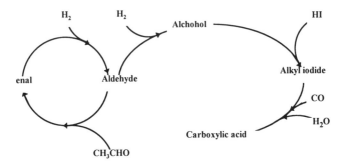

Fig. 3: Generation of higher aldehydes, iodides and carboxylic acids

From this, it would be expected that the homologation observed would be of unsaturates and iodides having an even number of carbon atoms, and long chain carboxylic acids with an odd number of carbon atoms. Some of the intermediates in these chains are readily observable. For example, the alkyl iodides tend to accumulate in the heads material of the first distillation column of the process. Analysis of this material (principally methyl iodide) shows a pattern of C2 additions. Analysis of the base material from the "heavies" removal column shows a similar pattern, but with one more carbon atom in each.

The end products of each chain (carboxylic acids) present no problem apart from a loss of raw materials. They are readily removed in the final "heavies" removal column. However, some of the intermediate compounds have physical properties that make their separation from acetic acid far more difficult.

Particular problems are encountered with the C6 species present. The boiling points of the unsaturated compounds (hexenal and some of it's isomers) are very similar to acetic acid. Further, hexyl iodide is observed to form a constant boiling mixture with acetic acid.

The presence of the unsaturates even in low part per million quantities will cause the product acid to fail a simple permanganate test (a widely respected standard for the quality of the acid).

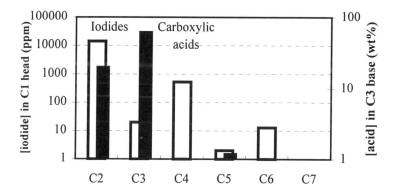

Fig. 4: Concentration of higher molecular weight by-products in a promoted rhodium system.

Iodide compounds are an accumulative poison for processes that use palladium catalysts, so need to be controlled to the low part per billion levels that are normally observed in the unpromoted rhodium system. In the promoted system, iodide levels of several hundred parts per billion are observed in the product acid. In order to remove these problems, extensive further treatment steps may need to be incorporated into the process. These include extraction with water (11), treatment with ozone (13,16), hydrogenation (21), and extra distillation steps (10).

Chemical removal of the long chain alkyl iodides is also required, and proposals for doing this centre around the use of various types of ion-exchange resin (22), often in the presence of added silver salts (14,15,23,24).

Again, additional processing steps need to be included into the operating unit in order to remove this potential source of product contamination.

This approach does not resolve all of the improvement issues and it also gives further complications in terms of extra processing steps being required in order to ensure consistent product quality.

Development of an iridium catalysed methanol carbonylation

The potential use of iridium instead of rhodium was identified as part of the early work done by Monsanto (2), but this was never followed through to commercialisation as, with commonly accepted reactor parameters, the reaction rate with rhodium species is superior to that of the equivalent iridium system.

The case for using iridium was re-opened in late 1991, largely as a response to the problems identified above with the use of a rhodium system - but also because of the then very attractive price difference between rhodium ($3500 per ounce) and iridium ($60 per ounce). Initial experiments showed significant promise, and this development rapidly required the co-ordinated effort of several diverse teams. The resulting new route for the production of acetic acid has been commercialised on two world scale plants to date, and has received wide publicity as the "Cativa™" process.

Methanol carbonylation via iridium

One early result of the investigations conducted at the Hull Research Centre was the extreme robustness of the iridium catalyst species (17). This work has demonstrated that the iridium catalyst remains stable under conditions that would cause the rhodium analogues to decompose completely to inactive and largely irrecoverable rhodium salts. Of particular significance is the robustness of the iridium catalyst at even very low water concentrations, ideal for optimisation of the methanol carbonylation process.

The unique differences between the rhodium catalytic cycle and that of iridium in methanol carbonylation have been investigated in a close partnership between BP Chemicals researchers at Hull and a research group at the University of Sheffield (18). The anionic iridium cycle (Figure 5 below), is similar to that shown before for rhodium, but contains sufficient key differences to yield the major advantages seen with the iridium system.

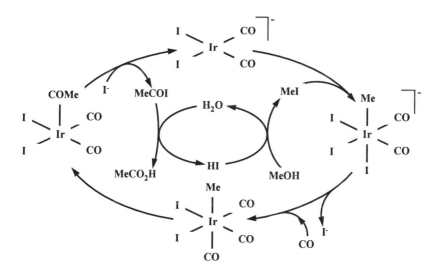

Fig. 5: Catalytic cycle for iridium carbonylation

Model studies (18) have shown that the oxidative addition of methyl iodide to the iridium centre is of the order of 150 times faster than the equivalent reaction with rhodium. This represents a dramatic improvement in the available reaction rates, as this step is now no longer rate determining. The slowest step in this cycle is the subsequent migratory insertion of carbon monoxide to form the iridium acyl species, which involves the elimination of ionic iodide and the co-ordination of an additional carbon monoxide ligand. This would suggest a totally different form of rate law:

$$Rate \propto \frac{[catalyst]*[CO]}{[I^-]}$$

The implied inverse dependence on ionic iodide concentration suggests that very high reaction rates should be achievable by operating at low iodide concentrations. It also suggests that the inclusion of species capable of assisting in the abstraction of iodide should promote this new rate limiting step. Promoters for this system tend to fall within two distinct groups: ruthenium, osmium and rhenium (19,20) and cadmium, mercury, zinc, gallium, indium and tungsten (26). In effect, a proprietary blend of promoters has been found to significantly increase reaction rate and also to virtually eliminate the dependence on the carbon monoxide partial pressure of the system.

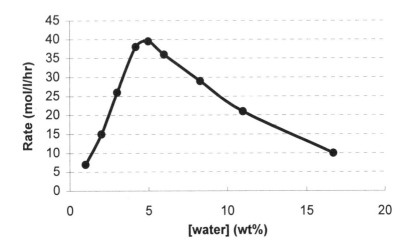

Fig. 6: Rate variation in a promoted iridium system with water.

The above expression does not imply any effect from the water present in the matrix, but water is found to have a very significant effect on rate (25).

Figure 6 shows reaction rates achievable using this catalyst system. Rates of the order of 10 mols/litre/hour have in the past been considered reasonable for commercial operation of such processes. This very high activity at low water concentrations, combined with the inherent stability of the catalyst under such conditions, leads to significant economic improvements in the process.

An equally important feature of the iridium system is that it has been found to be slower at producing acetaldehyde than the rhodium system. Further, what acetaldehyde is produced is rapidly reduced in the iridium system to yield ethanol and hence propionic acid.

The net effect of these two features is that the iridium system makes significantly less propionic acid than the rhodium analogue, and that the production of higher molecular weight derivatives of acetaldehyde (both unsaturates and higher iodides) is vanishingly small. This gives an inherently "clean" product without the need to introduce the additional processing steps required in a promoted rhodium system.

Benefits of the iridium catalysed system

Several targets for an improved methanol carbonylation have been identified above. The Cativa™ process successfully addresses each of the points raised:

Higher reaction rates: 75% increase in throughput has been commercialised on an existing plant. Higher rates have been demonstrated in the laboratory.

Enhanced yield on carbon monoxide: The use of promoters effectively removes the dependence of rate on carbon monoxide partial pressure.

Better catalyst stability: Under almost all envisaged reactor conditions the catalyst remains inherently stable.

Lower selectivity to propionic acid: Iridium has a much lower selectivity to acetaldehyde (the main propionic acid precursor) than does rhodium. Propionic acid production is substantially reduced in the Cativa™ process.

Less acetaldehyde condensation products: Iridium acts as a very effective hydrogen catalyst for unsaturated species present. Unsaturated condensation products and the iodide derivatives of these are virtually eliminated.

Lower standing concentration of water in the reactor: The iridium catalyst species are far more stable than their rhodium counterparts even when exposed to what would normally be perceived as extremely harsh conditions such as very low water concentrations. Substantial reduction in the amount of water in the system is therefor possible.

Acknowledgements.

Much of the above work has been carried out by my co-workers at BP Chemicals at our Hull and Sunbury research facilities. Special acknowledgement is also due to the external parties that have participated in this development. In particular to Professor P. Maitlis and Dr. A.Haynes at the University of Sheffield (for work on catalyst fundamentals), Dr. S.Collard of Johnson Matthey (for catalyst development) and Mr. J.A.Stal of Sterling Chemicals (for process implementation).

Parts of this article were previously published in the Proceedings of the IPMI Precious Metals Catalysis Seminar held in Houston, Texas, USA and are reproduced with their kind permission.

References

1. R. T. Eby and T. C Singleton, *Applied Industrial Catalysts*, 1983, **1**, p. 275
2. T. W. Dekleva and D. Forster, *Adv. Catal.*, <u>34</u>:81 (1986)
3. F. E. Paulik and J. R. Roth, *J. Am. Chem. Soc., Chem. Comm.*, 1578 (1968)
4. R. G. Shultz, *US Pat.* 3,717,670 to Monsanto (1973)
5. T. C. Singleton, W. H. Urry, F. E. Paulik,, *US Pat.* 55,618 (1982)
6. F. E. Paulik, A. Hershman, W. R. Knox, R. G. Shultz, J. F. Roth , *US Pat.* 5,003,104 (1988)
7. B. L. Smith, G. P. Torrence, A. Aguilo, J. S. Adler, *US Pat.* 5,144,068 (1992)
8. H. Koyama, H Kojima, *GB Pat.* 2,146,637 (1987)
9. D. J. Watson, *Eur. Pat. Pub.* 506,240 (1996)
10. G. A. Blay, M. O. Scates, M. Singh, W. D. Picard, *Eur. Pat. Pub.* 497,521 (1992)
11. S. Kimura, N. Araishi, *Eur. Pat. Pub.* 696,565 (1996)
12. M. Shimizu, T. Sato, Y. Morimoto, M Kagotani, *Eur. Pat. Pub.* 687,662 (1995)
13. Y. Harano, Y. Morimoto, *Eur. Pat. Pub.* 645,362 (1994)
14. M. D. Jones, D. J. Watson, B. L. Williams, *Eur. Pat. Pub.* 538,040 (1993)
15. M. O. Scates, R. J. Warner, *Eur. Pat. Pub.* 544,496 (1993)
16. M.O.Scates, R.K.Gibbs, G.P.Torrence, *Eur. Pat. Pub.* 322,215 (1988)
17. K. E. Clode, D. J. Watson, *Eur. Pat. Pub.* 616,997 (1994)
18. P. M. Maitlis, A. Haynes, G. J. Sunley, M. J. Howard, *J. Chem. Soc., Dalton Trans.*, 2187 (1996)
19. C. S. Garland, M. F. Giles, J. G. Sunley, *Eur. Pat. Pub.* 643,034 (1994)
20. C. S. Garland, M. F. Giles, A. D. Poole, J. G. Sunley, *Eur. Pat. Pub.* 728,726 (1994)
21. G.P.Torrence, P.M.Colling, W.D.Picard, *Eur. Pat. Pub.* 372,993 (1988)
22. C.F.Fillers, S.L.Cook, *Eur. Pat. Pub.* 685,445 (1995)
23. C.B.Hilton, *Eur. Pat. Pub.* 196,173 (1988)
24. M.D.Jones, *Eur. Pat. Pub.* 484,020 (1992)
25. M.J.Baker, M.F.Giles, C.S.Garland, M.J.Muskett, *Eur. Pat. Pub.* 752,406 (1995)
26. M.J.Baker, M.F.Giles, C.S.Garland, G.Rafeletos, *Eur. Pat. Pub.* 749,948 (1995)

Synthesis of 2-Coumaranone by Catalytic Dehydrogenation of α-Carboxymethylidene Cyclohexanone

N. Carmona, P. Gallezot*, and A. Perrard

*Institut de Recherches sur la Catalyse-CNRS,
69626 Villeurbanne cédex, France*

L. Carmona[1], G. Mattioda[2] and J.C. Vallejos[2]

Société Française Hoechst, [1] *Laboratoire de Recherches et d'Applications, 60350
Cuise-Lamotte, France.* [2] *Centre de Recherches et d'Applications, Stains, France*

Abstract

The synthesis of 2-coumaranone from a-carboxymethylidene cyclohexanone isomers and enol lactone derivative, was achieved by vapor phase dehydrogenation on alumina-supported palladium catalysts. From a mixture in acetic acid solution, 2-coumaranone was obtained with a 40 % yield. However, the catalyst deactivated rapidly due to the formation of carbonaceous deposits on the catalysts, and up to 35% of the reactants were lost as cracking products. This was attributed to the presence of the thermally unstable *trans*-isomer. In contrast, the *cis*-isomer in acetic acid solution was converted to 2-coumaranone with good initial yields (ca. 75%) with low amounts of cracking products; however, the catalyst still deactivated and would require frequent regenerations. The vapor phase dehydrogenation of pure enol lactone, vaporized without solvent, proceeded continuously with negligible deactivation and yielded 67.3% of 2-coumaranone at a rate of 52 g h^{-1}g$_{Pd}^{-1}$. The final yield in 2-coumaranone could be further increased by recycling the hydrogenation by-products using the same process of catalytic vapor phase dehydrogenation.

Introduction

Dehydrogenation reactions in the presence of oxygen (oxidative dehydrogenation) are conducted on silver catalysts to transform alcohols into the corresponding aldehydes, e.g. ethylene glycol into glyoxal (1). These reaction types can be extended to prepare fine chemicals, thus, dec-9-en-1-ol (rosalva) was dehydrogentaed into dec-9-enal (costenal) on silver catalysts with good yields (2). In contrast, dehydrogenation reactions can be conducted in the absence of oxygen on platinum or palladium catalysts to aromatize substituted cyclohexyl or cyclohexenyl compounds. Thus, in the field of fine chemistry, p-cymene was obtained by dehydrogenation of limonene with a 97% yield on active carbon-supported Pd-catalysts (3). In the same way, substituted phenols were obtained by dehydrogenation of the corresponding cyclic alcohols or cyclic ketones (4). Thus, 2,3,6-trimethyl-2-cyclohexen-1-one was converted with a 85% yield in 2,3,6-trimethyl-phenol on palladium catalysts supported on a Al_2O_3-MgO spinel. A 74 % yield in catechol was obtained from the dehydrogenation of the methylic acetal of hydroxycyclohexanone on active carbon-supported palladium catalysts (5). In these processes, reactants were vaporized and highly diluted in a stream of nitrogen and/or hydrogen and flowed with a very short residence time through a fixed bed of palladium catalysts at 220-280°C.

This investigation was intended to employ vapor phase dehydrogenation on metal catalysts to prepare 2-coumaranone **4**, which is a valuable intermediate in the preparation of fungicides. 2-coumaranone can be prepared by dehydration of *o*-hydroxyphenylacetic acid obtained by different routes, e.g. from *o*-chlorophenylacetic acid (6), or *p*-chlorophenol (7), and by internal acetoxylation of phenylacetic acid (8). In a first stage of this study, an attempt was made to prepare **4** by dehydrogenation on palladium catalysts of a mixture of α-carboxymethylidene cyclohexanone isomers (*cis* **1**, present in the lactonic form **1'**, and *trans* **2**), and of the enol lactone **3** derived by dehydration. This mixture was obtained upon reacting cyclohexanone and glyoxylic acid in acetic acid solution. In a second stage, the dehydrogenation reaction was achieved starting from either the pure *cis*-isomer or enol lactone.

Experimental

Catalysts
Catalysts were in pellet form to minimize the pressure drop resulting from the high flow rate in a continuous flow reactor. A commercial palladium catalyst supported on γ-alumina (100 m^2g^{-1}), 0.5% Pd/Al_2O_3 (Engelhard ESCAT 16), was employed in most reactions. Catalysts 1.4% or 2.5%Pd/α-Al_2O_3 were obtained by impregnating α-Al_2O_3 spheres from Degussa (10 m^2g^{-1}) with the required amounts of H_2PdCl_4. The catalyst was dried at 100°C and reduced under flowing hydrogen at 200°C for 2 h.

Starting materials
α-carboxymethylidene cyclohexanone was obtained by reacting 8 mol of glyoxylic acid in 50% aqueous solution with 10 mol of cyclohexanone and 16.7

mol of acetic acid. The solution was heated at 130°C for 1.5 h. A mixture of **1** (**1'**), **2**, and **3** at 18, 16 and 1 wt%, respectively, in acetic acid solution was obtained. Enol lactone **3** was prepared by heating for 2 h. under reflux 1 mol of glyoxylic acid and 1.5 mol of cyclohexanone in hydrochloric water solutions; the solution was then heated under reduced pressure to eliminate reagents in excess and solvent. The mixture was dissolved in xylene in the presence of *p*-toluenesulfonic acid. After azeotropic distillation, neutralization and distillation under reduced pressure (4 Torr), 0.87 mol of enol lactone **3** were obtained (boiling point 120°C under 4 Torr).

Reaction procedure

The liquid reactant was pumped (12 mLh^{-1}) and injected through an ultrasonic nebulizer (US1 Lechler) that produced an aerosol of 10 mm droplets in a stream of N_2 or N_2+H_2 (200 Lh^{-1}) The aerosol was vaporized and heated at 220°C in a preheater before entering the reactor, which consisted of a 40 cm-long quartz tube with an inner diameter of 15mm. The gas mixture flowed at atmospheric pressure and 250°C through a bed of 5-50 g of palladium catalysts in pellet form. The catalyst bed was supported on quartz wool located at the bottom of the reactor. Reaction products were quenched and condensed in two acetonitrile-filled traps cooled to -10°C. The condensation of the very stable aerosols, formed by droplets of acetic acid and reaction products in nitrogen, was improved by flowing the gas through glass frits immersed in the traps. Reaction products contained in the acetonitrile-filled traps were analyzed by HPLC on a Nucleosil C18 column (7mm) with a $H_2O/CH_3CN/H_3PO_4$ (90:10:1) eluant and a Shimadzu SPD-6AV detector. Products were also analyzed by GC with a DB 5 column mounted on a Fisons GC 8000 chromatograph. The various products detected are presented in figure 1.

Results and Discussion

1. Dehydrogenation of 1, 2, and 3 in acetic acid solution

Dehydrogenation reactions were conducted with the mixture of **1** (**1'**), **2**, and **3** at 18, 16 and 1 wt%, respectively, in acetic acid solution. In a few experiments, product analysis was carried out by HPLC that did not isolate all of the products formed. This resulted in a lower carbon balance and overestimated selectivity to **4**, particularly for reactions conducted in the presence of hydrogen and yielding hydrogenation and hydrogenolysis products. Therefore, selectivities were expressed with respect to transformed reactants rather than to analyzed products. Preliminary experiments conducted in the absence of a catalyst (liquid flow rate: 12 ml/h; nitrogen flow rate: 200 l/h; T(vaporizator): 300°C; T(reactor): 300°C) showed that 90.4% of the reactants were recovered in the two traps (Table 1). The incomplete carbon recovery was due to cracking products deposited in the vaporizer and reactor which were operating at comparatively high temperature (300°C) with respect to the standard conditions, and to products not condensed in the traps. Indeed, reactants, products, and acetic acid formed an aerosol that was difficult to condense completely even with the specially designed traps (vide supra).

Carbon-supported catalysts

The dehydrogenation reaction conducted during 1 h. under standard reaction conditions on 5 g of 1% Pd/C catalyst (Degussa E152 XH/D) gave a 43.5% selectivity to 4 (82% with respect to analyzed products) at 58% conversion (Table 1); the main by- products were phenyl acetic acid 6 and o-cresol 7. However, the carbon balance was very poor because of cracking reactions, and the catalyst deactivated within one hour. Regeneration of the catalyst activity was obtained by mild oxidation treatment at ca. 200°C for 10 h with air diluted with nitrogen. However, this oxidative treatment on active carbon-supported was too risky to be scaled up. Therefore although dehydrogenation reactions were reported on active carbon-supported catalysts [3], the present data suggest that carbon supports are not well suited for industrial applications: cracking reactions are favored by their microporosity, which increases the contact time between molecules and the catalyst, and oxidative regeneration is too dangerous.

Alumina-supported catalysts

Palladium supported on γ-alumina, (0.5% Pd/Al$_2$O$_3$, ESCAT 16) deactivated within a few hours when the dehydrogenation reaction was conducted in the presence of pure nitrogen. The catalyst performance can be recovered by heating under a flow of oxygen diluted in nitrogen (1/200) at 400°C, then re-reducing the catalyst under hydrogen diluted with nitrogen (1/200) at 250°C. The catalyst was less active (assay 3, Table 1) than that supported on active carbon (assay 2),

Table 1 Conversion of 1, 2, and 3 in acetic acid solution

Assay	catalyst	catalyst mass (g)	gas and flow rate (l/h)	conversion (%)	analyzed products[a] (%)	4 (%)	selectivity[b] (%)
1	none		N$_2$ (200)		90.4[1]		
2	1% PdC	5	N$_2$ (200)	58.0	53.0[1]	25.2	43.5
3	0.5%Pd/Al$_2$O$_3$	14	N$_2$ (200)	27.5	60.0[1]	13.9	50.4
4	0.5%Pd/Al$_2$O$_3$	14	N$_2$(200), H$_2$(4)	70.0	48.0[1]	24.9	35.5
5	0.5%Pd/Al$_2$O$_3$	30	N$_2$ (200)	67.0	63.0[1]	34.6	51.6
6	0.5%Pd/Al$_2$O$_3$	30	H$_2$ (200)	84.0	31.0[1]	7.5	9.0
7	0.5%Pd/Al$_2$O$_3$	30	N$_2$(200)	79.3	74.3[2]	43.7	55.1
8	1.4%Pd a-Al$_2$O$_3$	10	N$_2$(200)	24	67.0[1]	14.8	61.6
9	2.3%Pd a-Al$_2$O$_3$	10	N$_2$(200)	70	65.0[1]	35.0	50.0

a: (1) analyzed by HPLC; (2) analyzed by GC (all products analyzed). b: selectivity calculated with respect to transformed reactant.

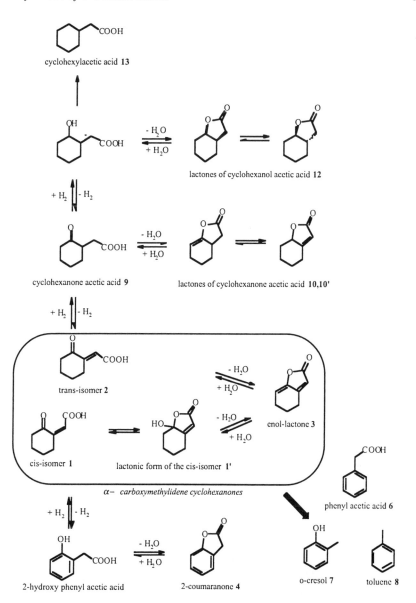

Figure 1 General reaction scheme and detected products

essentially because of lower cracking reactions. In the presence of small amounts of hydrogen in the feed ($H_2:N_2$ = 4:200), the conversion was much higher (70 compared to 27.5%) but the selectivity decreased (assay 4, Table 1).

In order to better evaluate the influence of hydrogen, dehydrogenation experiments were conducted on the same catalyst under pure nitrogen (assay 5) and pure hydrogen (assay 6). The conversion and mass balance were improved under hydrogen, but the selectivity decreased dramatically. The product distribution measured by GC indicated that higher amounts of hydrogenolysis and hydrogenation products were formed at the expense of 2-coumaranone (Table 2). The 0.5% Pd/Al_2O_3 catalyst in the presence of pure hydrogen did not deactivate after 40 h on stream whereas regeneration was needed after a few hours under pure nitrogen. However, the advantage in terms of stability of operating in the presence of hydrogen had to be paid for by a loss of selectivity.

Table 2 Comparison of the product distribution under nitrogen and hydrogen

	product %	
	assay 5	assay 6
toluene **8**	2.4	10.9
o-cresol **7**	11.1	23.1
cyclohexyl acetic acid **13**	0	11.8
phenyl acetic acid **6**	2.5	11.0
2-coumaranone	78.5	9.6
lactones of cyclohexanol acetic acid **12**	1.3	14.7
cyclohexanone acetic acid. and lactones **9, 10, 10'**	4	19.1

The effect of residence time on the performance of 0.5%Pd/Al_2O_3 catalyst was studied by measuring the product distribution with GC, after 1 h on stream, for various nitrogen flow rates and catalyst masses (Table 3). Interestingly, the 2-coumaranone yield increased with conversion; thus, at the highest conversion (79.3%), a 43.7% yield was achieved corresponding to a 55.1% selectivity with respect to the transformed reactant. The amount of products not accounted for in this assay was 25.7%; the carbon balance deficit corresponds essentially to cracking products.

Palladium catalysts supported on α-alumina spheres were also tested under standard conditions. Table 1 show that the activity was lower and the selectivity slightly higher than that of Pd/γ-Al_2O_3. Since, this catalyst was not available commercially and exhibited catalytic performance not markedly higher than that of ESCAT 16, the study was continued with the latter.

Palladium catalysts were also prepared by impregnation of other supports such as Al-Ni or Al-Mg spinels, magnesium oxide, silica, and silico-alumina of various specific surface area and porosity, but none of these catalysts performed as well as Pd/Al_2O_3, both in terms of conversion and selectivity. Alumina-supported Pd-Sn catalysts were tested with the hope of decreasing the cracking reactions, as observed in the reactions of hydrocarbon conversion (9). However, the selectivity was hardly improved and the activity dropped dramatically.

Table 3 Conversion of α-carboxymethylidene cyclohexanone (**1**+**2**+**3**) and product distribution as a function of gas and liquid flow rate and catalyst mass

N_2 (L/h)	liquid (mL/h)	mass (g)	conversion (%)	**4** (%)	**6** (%)	**7** (%)	9+10+10 (%)	D^a (%)
100	6	14	77,6	33,4	4,2	4,1	2,9	33
100	6	30	75,4	30,7	0,2	13,8	0,7	30
100	6	50	65,1	21,6	1,9	25,6	0,7	15,3
200	12	14	56,4	15,3	0,6	0,9	1,5	38,1
200	12	30	79,3	43,7	1,2	6,4	2,3	25,7
200	12	50	70,5	35,3	2,6	2,9	4,6	25,1
600	36	14	25	4,3	0,2	0,6	0,5	19,4
600	36	30	57,9	9,9	0,6	1,6	1	45,7
600	36	50	50,8	17,8	0,7	4	2,1	25,4

a : mol% loss. Standard reaction conditions on 0.5% Pd/Al_2O_3

In conclusion, the studies conducted on acetic acid solutions of **1**, **2**, and **3**. indicated that 2-coumaranone can at best be obtained with a selectivity of ca. 50%, at 75% conversion. The main limitation was the high loss of reactants by conversion into cracking products regardless of the nature and texture of supports or the composition of the active phase. It was inferred that one of the isomers of the mixture might be responsible for the cracking reactions: most likely the *trans*-isomer **2** which may be difficult to convert into 2-coumaranone given its structure. Therefore, subsequent attempts to prepare 2-coumaranone by catalytic dehydrogenation were conducted with the *cis*-isomer and the enol lactone.

2 Dehydrogenation of the *cis*-isomer

The *cis*-isomer was available as a solid that had to be dissolved in acetic acid (18 wt% solution) in order to be vaporized. The dehydrogenation reaction was carried out under standard conditions on the 0.5% Pd/Al_2O_3 catalyst. In contrast with the dehydrogenation reaction conducted with the mixture of isomers (vide supra), the reaction data given in Table 4 show that a carbon balance of almost 100% was obtained. The selectivity to 2-coumaranone attained 90% after 2 h on stream with the lactones of cyclohexanone acetic acid as the main by-product. Still, the catalyst deactivated with time so that frequent oxidative regenerations to eliminate carbonaceous deposits would be needed for practical operation.

Table 4 Conversion and selectivities in the dehydrogenation of the *cis*-isomer.

time (h)	conversion (%)	4	6	7	9, 10, 10'
0.2	90	86	3.8	9	1.6
1	70	82	6	8	4.5
2	· 90	90	1.6	-	8

Reaction conditions: 12 mLh^{-1} of **1'** in 18 wt% acetic acid; T_{vapo} = 220°C, $T_{react.}$= 250°C; N_2 flow rate: 200 Lh^{-1}; 30 g of 0.5% Pd/Al$_2$O$_3$

3. Dehydrogenation of the enol lactone

Pure enol lactone was a liquid that could be vaporized readily in the reactor. The dehydrogenation reaction was carried out under standard conditions over the 0.5% Pd/Al$_2$O$_3$ catalyst; all the products were analyzed and a good carbon balance was attained indicating little loss of reactants via cracking reactions. Table 5 gives the selectivity data after different periods of time on stream. In comparison with the experiments reported in the previous two sections, the most striking feature is the high and almost constant conversion of the reactant. This indicates that the catalyst was not poisoned significantlly by carbonaceous deposits and, therefore, would require infrequent regeneration if used over a long period of time.

Table 5 Conversion and selectivities in the dehydrogenation of enol lactone .

time (h)	conversion (%)	4	6	7	9, 10, 10'
1	98	83	2.7	5	10
2.2	97	81	4.2	0.3	15
3.5	96	78	4.9	-	17
4.3	94	76	4.1	-	19
5.6	93	76	4.1	-	19.5
6.6	93	79	2.9	-	18.5
7.7	91	76	2.6	-	22

Reaction conditions: 12 mLh^{-1} of **3** ; T_{vapo} = 220°C, $T_{react.}$= 250°C; N_2 flow rate: 200 Lh^{-1}; 30 g of 0.5% Pd/Al$_2$O$_3$

During 8 hours on stream under standard conditions, 95% of **3** were converted, and the cumulative yields of reaction products were: 67.3% of 2-coumaranone (productivity: 52 g h^{-1} g$_{Pd}^{-1}$), 14.8 % of **10'**, 2.9% of **6**, 2.1% of **10**, 1.2% of **8**, and less than 1% of **7, 12**, and **12'**. Reactions conducted at lower flow rates of nitrogen gave lower selectivities to 2-coumaranone. In contrast, higher selectivities to **4** were obtained by decreasing the contact time with higher flow rates of N_2 (e.g., 90% selectivity at 800 Lh^{-1}), because the concentration of hydrogen and thus the yield of hydrogenated products decreased. However, this improvement had to be paid for by a decrease of conversion (from 95 to 27%) and impractical nitrogen flow rates.

The main by-products of the dehydrogenation of **3** were the lactones of cyclohexanone acetic acid, particularly **10'**. These hydrogenated derivatives can possibly be recycled by catalytic dehydrogenation to further increase the final 2-coumaranone yield. To verify this possibility, the liquid issued from enol lactone dehydrogenation was submitted to another dehydrogenation reaction. It was found that the amount of coumaranone increased at the expense of **10'**. The dehydrogenation of the lactones of cyclohexanone acetic acid was studied in more details by starting from pure products. A mixture of **10** and **10'** was synthesized from cyclohexanone acetic acid by dehydration in the presence of acetic anhydride and sodium acetate as described by Newman and van der Werf (10). The dehydrogenation conducted under standard conditions on 30 g of 0.5% Pd/Al_2O_3 catalyst yielded 48% of 2-coumaranone (69% selectivity at 70% conversion) after 1 hour on stream. This experiment showed that the lactones of cyclohexanone acetic acid cannot be considered as waste products because they can be dehydrogenated further into 2-coumaranone

Conclusion

This study shows that vapor phase dehydrogenation on alumina-supported palladium catalysts is potentially an efficient one-step process for the synthesis of 2-coumaranone from a-carboxymethylidene cyclohexanone isomers. Starting from a mixture of **1(1')+2+3** in acetic acid solution, 2-coumaranone was obtained with a 40% yield. However, the catalyst deactivated rapidly due to the formation of carbonaceous deposits on the catalysts, and up to 35% of the reactants were lost as cracking products. This was attributed to the presence of the *trans*-isomer that cannot be transformed into useful products and, because it is thermally unstable, accounts for the high amounts of cracking products.

Isomer **1 (1')** in acetic acid solution was converted to 2-coumaranone with good initial yields (ca. 75%) and yielded low amounts of cracking products. However, the catalyst deactivated, possibly because of the presence of acetic acid, thus requiring frequent regenerations which would decrease the economy of the process. In contrast, the enol lactone **3** in pure form can be vaporized without requiring a solvent. Whereas its dehydrogenation in liquid phase batch process yielded a maximum of 50% of **4**, the vapor phase dehydrogenation proceeded continuously exibiting negligible deactivation and yielding 67.3% of 2-coumaranone at a much higher activity of 52 g $h^{-1}g_{Pd}^{-1}$. The final yield in 2-coumaranone can be further increased by recycling the hydrogenation by-products using the same process of catalytic vapor phase dehydrogenation.

References

1.   P. Gallezot, S. Tretjak, Y. Christidis, G. Mattioda, and A. Schouteeten, *J. Catal.*, **142**, 729 (1993).
2.   L. Ceroni, A. Perrard, and P. Gallezot, *J. Mol. Cat.* (in press).
3.   R. Martin, and W. Gramlich (BASF) DE 3,607,448 (1987).
4.   H. Laas, P. Tays, and L. Hupfer (BASF) DE 3,314,372 (1984).
5.   M. Karcher (BASF) EP 0,499,055 A2 (1991).
6   Ihara Chemical Industry, Fr 2,489,312 (1981)
7   J.C. Vallejos, and Y. Christidis (S.F. Hoechst), Fr 2,686,877 and Fr 2,686,880 (1992)
8   T. Fukagawa, Y. Fujiwava, and H. Taniguchi *J. Org. Chem.*, **47**, 2491 (1982)
9.   F.M. Dautzenberg, J.N. Helle, P. Biloen, and W.M.H. Sachtler, *J. Catal.*, **63**, 119 (1980).
10. M.S. Newman, and van der Werf, *J. Am. Chem. Soc.*, **67**, 233 (1945).

Cu/Zn/Zr Catalysts for the Production of Environmentally Friendly Solvents

G.L. Castiglioni[a], C. Fumagalli[a], E. Armbruster[b], M. Messori[c] and A. Vaccari[c]

[a]Lonza Spa, Via Fermi 51, 24020 Scanzorosciate BG (Italy).
[b]Lonza Spa, Furkastrasse 64, 3904 Naters (Suisse).
[c]Dip. Chimica Industr. e Materiali, Viale Risorgimento 4, 40136 Bologna (Italy)

Abstract

New catalytic systems without chromium were investigated in the γ-butyrolactone (GBL) synthesis by vapour phase hydrogenation of maleic anhydride (MA); in particular the activity and selectivity were studied by using Cu/Zn/Zr catalysts, thermally treated at different times and temperatures. It was found that for the dried samples the activity is directly connected to the catalytic composition, in particular it has been noted an increased activity with the samples with higher copper content due to the presence of higher amounts of Cu not strongly interacting with other phases. Moreover the thermal treatment may tune the activity, changing the surface area and the amount of extractable copper and, mainly, according to the formation of a brass-like phase (not detectable by XRD analysis and characterized by a greater reducibility) stabilized by the amorphous ZrO_2. Using the proper catalyst composition and thermal treatment it is possible to achieve GBL yields close to 99% with optimum mechanical properties and carbon balance. Finally a technical-economical analysis of different processes to produce GBL (or N-methyl pyrrolidone) led to the conclusion that the direct vapor phase hydrogenation of MA is the most advantageous technology, also providing a good example of an environmentally friendly industrial process.

Introduction

Because of the increasing interest in the last years in the substitution of the chlorine-based solvents with alternative ones at lower impact on humans and on the environment many companies have been studying the direct hydrogenation of maleic anhydride (MA) to γ-butyrolactone (GBL), a key intermediate for the N-methyl pyrrolidone (NMP) synthesis, an extremely versatile and environment friendly solvent. However the development of heterogeneous catalysts has been led mainly from Cr-containing mixed oxides, mostly with chemical or physical promoters, to Cr-free samples (1-5); the driving force for this evolution must be found in the toxicity and the impact on the environment of the Cr^{6+} (6) that is the reason why of more and more restrictive regulations.

In our previous works we have reported interesting data concerning to the possibility of substituting chromium with gallium and better (especially for economic reasons) with aluminium (7-9); unfortunately the problems connected to the organic adsorption on the catalyst and/or its unsatisfactory mechanical properties which can be only partially avoided with pelletting the catalyst (10). The aim of this paper is to extend the study on more friendly systems using zirconium instead of a trivalent ion other than chromium, focusing the attention on the relationships between the chemical composition, the thermal treatment and the catalytic properties.

Experimental

All the samples were prepared by coprecipitation, adding dropwise a solution of the correspondent chloride salts of copper, zinc and zirconium into an aqueous solution of NaOH under vigorous stirring. The NaOH amount in the last solution was calculated as equimolar to the chloride amount of the former one plus a slight excess. The obtained slurry was filtered, washed with deionized water, dried overnight at 373-393K and eventually calcined at different temperatures and/or for different times. In table 1 are reported the chemical compositions of the various catalysts used in this work and their calcination conditions.

Before the catalytic tests the samples were activated in-situ at atmospheric pressure flowing a mixture of H_2/N_2 (5/95 v/v), while the temperature was progressively increased from room temperature to 603K.

Table 1. Catalytic composition and thermal treatment of the various samples studied.

Sample		30Zr	40Zr	50Zr	60Zr
Composition	(atom. %)	Cu:Zn:Zr=35:35:30	Cu:Zn:Zr=30:30:40	Cu:Zn:Zr=25:25:50	Cu:Zn:Zr=20:20:60
Calc. T	(K)	373-393	373-393	373-393	373-393
Calc. time	(h)	o.n.	o.n.	o.n.	o.n.

Sample		30Zr-653	40Zr-653	50Zr-653	60Zr-653
Composition	(atom. %)	Cu:Zn:Zr=35:35:30	Cu:Zn:Zr=30:30:40	Cu:Zn:Zr=25:25:50	Cu:Zn:Zr=20:20:60
Calc. T	(K)	653	653	653	653
Calc. time	(h)	24	24	24	24

Sample		60Zr-773/3	60Zr-773/8	60Zr-773/17	60Zr-773/24
Composition	(atom. %)	Cu:Zn:Zr=20:20:60	Cu:Zn:Zr=20:20:60	Cu:Zn:Zr=20:20:60	Cu:Zn:Zr=20:20:60
Calc. T	(K)	773	773	773	773
Calc. time	(h)	3	8	17	24

Sample		50Zr-773	50Zr-973		
Composition	(atom. %)	Cu:Zn:Zr=25:25:50	Cu:Zn:Zr=25:25:50	(o.n. = over night)	
Calc. T	(K)	773	973		
Calc. time	(h)	3	3		

X-ray powder diffraction patterns were recorded using a diffractometer (Philips PW1050/81) controlled by a PW1710 unit and Ni-filtered CuK$_\alpha$ radiation (λ=0.15418nm) (40kV, 40mA). The data were processed using an Olivetti M240 computer. A Carlo Erba Sorpty 1750 apparatus with N$_2$ adsorption was used to measure the surface area of the samples. Thermogravimetric analysis (TG) was carried out using a Perkin-Elmer TGS-2 thermobalance, with an N$_2$ flow of 50mL/min and a constant heating rate of 10K/min. The temperature programmed reduction (TPR) tests of the samples were made in the above thermobalance using an H$_2$/N$_2$ flow (6:94 v/v) of 3l/h, with a heating rate of 1K/min, starting from room temperature. The free copper was determined by extracting the samples by a NH$_4$OH/NH$_4$NO$_3$ solution and analysing, after filtration, the obtained solution by using a Perkin-Elmer model 124 spectrophotometer (11). The surface area of copper in the catalysts was determined in-situ, after reduction at room temperature, by the N$_2$O pulse method (12).

The catalytic tests were carried out using about 2g of catalyst in a tubular fixed-bed microreactor (i.d. 2mm, length 520mm), operating at atmospheric pressure in the 485-548K range. Before the catalytic tests the samples were activated in-situ at atmospheric pressure in a flowing H$_2$/N$_2$ (5:95 v/v) stream, while the temperature was increased from room temperature to 603K. The reactor was fed with a stream of either GBL or MA/GBL solution (60:40 w/w) in hydrogen (total flow 5.0l/h, H$_2$/C$_4$ molar ratio equal to 170 or 70); the organic feedstock being introduced by an Infors Precidor 5003 infusion pump. GBL was used as the solvent, taking into account the difficulties to feed a very small amount of liquid MA and considering that most of the usual solvent reacts preferentially with MA or with the hydrogenation products. It must be noted that in this case GBL represents both a reactant and a product, so a molar yield of about 43% must be considered as the theoretically absence of conversion and production. The reaction products were analysed on-line without condensation using a Carlo Erba gas chromatograph model 4300, equipped with FID and two columns (3.2mm x 2.0m) filled with Poropack QS.

Influence of chemical composition

The XRD analysis of the dried catalysts (samples 1-4) showed great differences for what concerns the crystallinity (Table 2): infact, while for systems 50Zr and 60Zr only two very broad bands at 2Θ values of about 30° and 55° have been evidenced, corresponding to amorphous zirconia (13,14). An increase of the copper and zinc contents led to the appearance of highly crystallineCuO, ZnO or Cu(OH)$_2$.

The thermogravimetric analysis evidenced two main weight losses up to 753K, due to the elimination of the physically adsorbed water and to the chemically co-ordinated water respectively. Moreover, increasing the temperature it was possible to note a further small weight loss corresponding to the transition of the amorphous zirconia into a crystalline phase (14). The corresponding temperature of this transition is strictly connected to the Zr amount in the samples, going from 858K for sample 30Zr to 905K for 60Zr.

The TPR analysis showed for each system a peak due to the reduction of CuO, whose intensity and temperature were correlated to the Cu amount. Nevertheless, while the temperature increased proportionally with the increase of the Zr amount, the weight loss of sample 30Zr is more than two times of that of 50Zr and 60Zr, indicating a greater amount of reducible copper. Taking into account the data in Table 2 we may note that in the two samples with lower Zr amount the copper is completely extracted, while this value is only about 50% for 50Zr and 60Zr. This results suggest that, increasing the Zr amount, the copper interacts with others species, mainly the amorphous zirconia previously detected, making disposable to the extraction a lower fraction of copper, with a corresponding increase of the reduction temperature. On the other hand it is not possible to correlate the chemical composition of the systems with their values of surface area or metallic copper area (Table 2).

The catalytic tests were performed on samples 30Zr, 50Zr and 60Zr; the results obtained by using a solution feed of 0.127g/h (Fig. 1) show well the differences between 30Zr and the other two catalysts: while the former system is characterized by a good activity at every temperature (total MA conversion), 50Zr and 60Zr achieve a complete conversion only at 548K, and with lower yields in GBL.

Preliminary tests performed feeding 0.280g/h of organic solution showed the impossibility to obtain total conversion for the samples with Zr≥50% also at 548K confirming the unsatisfactory properties of these systems. Moreover, it must be noted that also the yield in SA increases with the increase of the zirconium amount, strengthening the following scale of activity (15):

$$30Zr \gg 50Zr > 60Zr.$$

Table 2. Physical-chemical characterization of the dried and unloaded samples investigated.

Sample	Identified Phases (XRD)		Surface Area	Cu° Area	Extr. CuO % (g_{CuO}/g_{cat})	
	dried	unloaded	(m^2/g_{cat})	(m^2/g_{Cu})	exp.	calc.
30Zr	ZnO, CuO, Cu(OH)₂	ZnO, Cu°	139	3,2	28,5	29,9
40Zr	CuO	CuO	229	n.d.	22,6	24,5
50Zr	amorphous	amorphous	195	2,5	10,2	19,9
60Zr	amorphous	amorphous	194	4,3	8,0	15,0

ZnO (ICDD 5-664); CuO (ICDD 5-661); Cu(OH)₂ (ICDD 13-420); Cu° (ICDD 4-836).

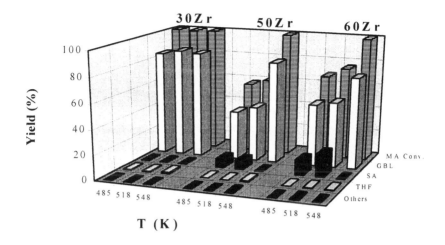

Fig. 1. Catalytic activity of the dried Cu/Zn/Zr samples as a function of the Zr content, in the hydrogenation of the MA/GBL solution (60:40 w/w), (Q=0.127g/h, P=1.0bar, H_2/C_4=170 mol/mol)

As a direct consequence of the highest activity, connected to the highest amount of bivalent ions, the maximum yield in GBL for 30Zr is achieved at 518K; a further increase in the temperature leads to a low decrease of the yield, due to the formation of unknown light products, while only negligible lack to the carbon balance may be justified by organic adsorption on the catalyst.

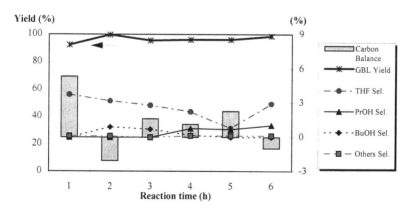

Fig. 2. Catalytic activity and carbon balance for the sample 30Zr, in the hydrogenation of the MA/GBL solution (60:40 w/w), (Q=0.280g/h, P=1.0bar, H_2/C_4=70 mol/mol); total MA conversion.

The sample 30Zr was also studied using a feedstock of 0.280g/h, obtaining at 548K a GBL yield close to the maximum value, with optimum carbon balance. In order to confirm the reliability of this result the test was prolonged for 6h, obtaining good reproducibility for both GBL yields and carbon balance (Fig. 2).

In conclusion it may be said that the activity of the various Cu/Zn/Zr systems is strictly connected to the chemical composition, in particular an increase of the Cu and Zn amount leads to an increase of the activity due to the presence of a greater amount of copper not strongly interacting with other phases present in the system.

Influence of thermal treatment

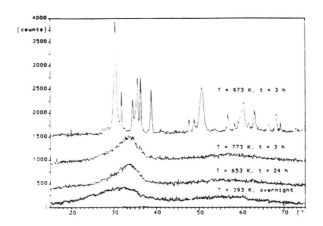

Fig. 3. XRD patterns of the sample 50Zr calcined at different temperatures.

Considering that the samples seen before had been previously only dried and that in the industrial plants it may be advantageous to regenerate the catalyst with a thermal treatment when its activity has become too low, also the influence of the thermal treatment was studied. The XRD patterns reported in figure 3 show well how the system containing 50% of Zr is not affected by the increasing of the temperature until 773K: all the samples are characterized by the presence of an amorphous ZrO_2, whose band at $2\Theta \approx 30°$ became slightly narrower with the increasing of the calcination temperature. Only a further increase of the temperature (973K) it may be detected a crystallization of the system, with the appearance of crystalline CuO, ZnO and tetragonal ZrO_2, in agreement with the results already discussed for the TG analysis.

The crystallization temperature of the ZrO_2 phase, higher than that previously reported by Koeppel and Baiker (773K) (14), may be due to the further presence in the studied samples of the zinc inside the structure that delays the formation of the tetragonal ZrO_2. Notwithstanding that the calcination treatment does not modify the crystallinity up to 773K, it leads to strong decrease of the catalytic surface, going from 195 to 90 and $74m^2/g_{cat.}$ for the 50Zr, 50Zr-653 and 50Zr-773 samples respectively. For what concerns the calcination time neither modifications on the crystallinity degree nor on the appearance of different phases have been detected, with only a small decrease in the surface area.

On the dried 60Zr and the corresponding calcined samples at 773K for 3 (60Zr-773/3) and 24h (60Zr-773/24) a TPR analysis was done: assuming a catalyst composition as CuO, ZnO and ZrO_2, a weight loss of 6.0% may be calculated due to the reduction of both copper and zinc. The TPR profiles show for the calcined systems lower reduction temperature (40-50K) and higher weight losses than the dried one (5.6, 5.1 and 2.8 for 60Zr-773/3, 60Zr-773/24 and 60Zr respectively), suggesting that only CuO may be reduced in the dried sample, while both CuO and ZnO after the calcination treatment. The greater reducibility of these phases may be connected to the formation of amorphous brass-like phases (16), in agreement with the results obtained in determining of the extractable copper. In fact it was found that increasing the calcination temperature and time, the amount of soluble copper in the NH_4OH/NH_4NO_3 solution decreased, in accordance with the formation of a new brass-like phase with lower solubility.

Thus it can be concluded that a mixed Cu-Zn phase may be obtained by thermally treating the samples: this brass-like phase is characterized by a greater reducibility [similar behaviour was already observed with Cu/Zn/Cr systems (17)] and is not detectable by RXD analysis. Moreover, the calcination treatment strongly influences the surface area and the amount of extractable copper.

Catalytic activity as a function of the calcination temperature

In figure 4 are reported the results obtained in the catalytic tests by using the samples calcined at 653K: it is possible to note an equalization of the catalytic activities, with a total MA conversion only at $T \geq 518K$ and similar GBL yields.

Table 3. Physical-chemical characterization of the Cu/Zn/Zr samples calcined at 653K, with atomic ratio Cu/Zn=1.0 and different Zr content.

Sample	Identified Phases (XRD)		Surface Area	Free CuO % (g_{CuO}/g_{cat})	
	fresco	scaricato	(m^2/g_{cat})	exp.	calc.
30Zr-653	ZnO, CuO	ZnO, Cu°	53	28,4	29,9
40Zr-653	CuO	CuO	63	22,2	24,5
50Zr-653	amorphous	amorphous	90	6,1	19,9
60Zr-653	amorphous	amorphous	57	n.d.	15,0

ZnO (ICDD 5-664); CuO (ICDD 5-661); Cu(OH)₂ (ICDD 13-420); Cu° (ICDD 4-836); n.d.= not detected.

Taking into account that the appearent low conversion of sample 50Zr-653 at 485K is due to a lack to the carbon balance (only for some samples and at low temperature adsorption phenomena has been detected), a more accurate analysis of the graph leads to the following activity order:

50Zr-653 ≥ 40Zr-653 > 60Zr-653 > 30Zr-653.

It must be noted that this order corresponds to the surface area values for the calcined samples reported in table 3, although the only surface area does not justify the low yields in by-products of the different systems; on the contrary, if it is considered the synergetic effect of the modified surface area and extractable copper it may justified the best catalytic behaviour for 50Zr-653 and the intermediate hydrogenating power of the catalysts containing 40 and 60% of Zr. A confirmation of these data has been achieved by repeating the tests with a higher organic feedstock (0.280g/h): also under this condition the catalytic behaviours were similar for all the catalysts with, best results using 50Zr-653.

The catalytic study was completed comparing the activity of the sample contained 50% of Zr calcined at 653 and 773K, mainly considering the improvements evidenced passing from the dried system to that calcined at 653K. However, it must be noted that while Cu/Zn/Al systems (10) may be effected by low mechanical strength (that may increase the pressure drop inside the reactor), with both the dried and the various calcined Cu/Zn/Zr samples this problem was never detected.

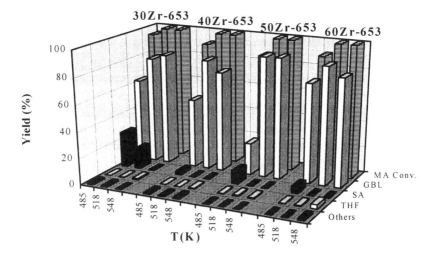

Fig. 4. Catalytic activity of the calcined Cu/Zn/Zr samples as a function of the Zr amount, in the hydrogenation of the MA/GBL solution (60:40 w/w), (Q=0.127g/h, P=1.0bar, H_2/C_4=170 mol/mol)

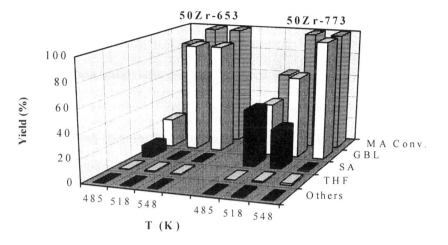

Fig. 5. Catalytic activity of the sample 50Zr calcined at 653 and 773K in the hydrogenation of the MA/GBL solution (60:40 w/w), (Q=0.127g/h, P=1.0bar, H_2/C_4=170 mol/mol)

Fig. 5 shows that a part from the results regarding the system 50Zr-653 at 485K, for which the organic adsorption on the catalyst strongly influenced the activity behaviour, an increase in the calcination temperature leads to a detectable decrease of the catalytic activity (analogous results were also obtained from preliminary tests feeding GBL). In fact, while for the sample calcined at 653K a low amount of SA was detected only at the lowest temperature, for 50Zr-773 SA is still present at 518, with a corresponding lower yield in GBL. However, this decrease in activity leads to an evident improvement of the catalytic behaviour at 548K, at which the sample calcined at 773K reaches a GBL yield next to the maximum theoretical, due to the almost total absence of other by-products (only 1.5% of THF). The decrease in activity increasing the calcination temperature was confirmed by the data obtained feeding 0.280g/h of the MA/GBL solution, showing the presence of both MA and SA for 50Zr-773 at any temperature.

Starting from the good results obtained with 560Zr-773 the test at 548K was repeated for a longer time (8h), obtaining constant GBL yield (about 99%), 1.0% of THF with good carbon balance, suggesting that this system under these reaction conditions is characterized by an optimum activity/selectivity ratio associated with good stability and excellent reproducibility, without adsorption phenomena. Also the catalytic activity of a sample calcined at 973K was investigated, showing that the recrystallization of the system already discussed leads to a dramatic deactivation of the catalyst.

Thus it can be concluded that in order to optimize the activity of the system it is important to tune the calcination treatment with the purpose to obtain the right balance between the amount of copper and zinc free or stabilized by the ZrO_2

phase; furthermore it must be noted that the ZrO_2 does not act as a simple support, but it can lead to a positive synergetic effects on the catalytic activity (18).

Economic evaluation of the GBL production

Taking into consideration the great interest that the direct hydrogenation of MA holds, we conclude this work with a GBL production cost comparison for the main processes, actually on run or of next generation: 1) REPPE process; 2) MA liquid phase hydrogenation via 1,4-butanediol (BDO); 3) MA liquid phase hydrogenation (19); 4) MA vapour phase hydrogenation; 5) hydroformilation of propylene oxide (ARCO process). The Davy McKee process (20) has not been considered because from a preliminary analysis it had appeared competitive only when exploiting both GBL and THF. For what concerns the REPPE and ARCO processes it is reported the BDO cost, as it was the feedstock, without splitting the single costs. The total cost of the REPPE process was set to 100 and all the others were refered to this one.

From the data reported in table 4 it can be clearly seen that the most economical GBL process is the vapour phase hydrogenation of MA: in fact, even if the REPPE process is no more affected by the depreciation cost (it has been running for many years), it is disadvantageous because of the further BDO dehydrogenation stage to GBL. Moreover the economics of the REPPE process (just like all the processes using BDO) is strictly connected to the BDO price that otherwise may be sold by its own.Considering that the main cost in the NMP production is given by the GBL cost, the conclusions above reported may justify also the most economical of the NMP production technology starting from the gas phase hydrogenation of the MA.

Table 4: GBL production costs comparison for the various analyzed processes (it was set the REPPE cost as 100).

	Specific Consumption Kg/Kg GBL	REPPE	Liquid* via BDO	Liquid* WO 92/02298	MA Hydrog. gas phase	Hydrofor***
BDO (Integral Cost)	1,14	77 (1)	=	=	=	72 (1)
n- BUTANE		=	13	12	15	=
HYDROGEN		=	15	8	8	=
BuOH (Credit)		=	-11	-3	=	=
RAW MATERIAL COST		77	17	17	23	72
UTILITIES		4	2	2	5	5
VARIABLE COST		81	19	19	28	77
OTHER COSTS		19	100	70	47	23
TOTAL		100	119	89	75	100

(1) Incl. ROI, without depreciation (old process); (2) Incl. Depreciation 10%, ROI 20%, SAR, Direct cost and Overhead; (*) Liquid phase hydrogenation of MA via BDO; (**) Direct liquid phase hydrogenation of MA; (***) Propylen oxide hydroformilation.

In conclusion it is possible to say that the thecnical-economical analysis identifies that the direct hydrogenation of the MA in gas phase is the most advantageous technology for both GBL and NMP production, because of its optimization in the synthesis steps and better evaluation of the various involved substances, starting from n-butane till the final product NMP (21).

Conclusions

In order to obtain more environmentally friendly catalysts for the GBL synthesis process the Cu/Zn/Zr systems can represent a good substitute of the chromitic systems, due to their main characteristics: i) high GBL yield as a function of the thermal treatment used; ii) almost total absence of organic retention/adsorption (good carbon balance); iii) good mechanical properties; iv) optimum reproducibility of the data obtained in the tests performed in microreactor.

The use of these catalysts joint to the vapour phase hydrogenation of MA technology consists of the most economic way to produce GBL (and NMP) according to an integrated cycle, able to optimize the various synthesis steps and to evaluate all the products, starting from the n-butane to the final NMP.

References

1. J. R. Budge, US Patent 5,055,599 (1991) to The Standard Oil Co..
2. P.D. Taylor, W. De Thomas and D.W. Buchanan, U.S. Patent 5,122,495 (1992) to ISP Investments Inc.
3. W. De Thomas, P.D. Taylor and H.F. Tomfohrde III, U.S. Patent 5,149,836 (1992) to ISP Investments Inc.
4. N.E. Johnson, R.T. Miskinis and R.A. Shafer, U.S. Patent 5,030,773 (1991) to General Electric Co.
5. M. Bergfeld and G. Wiesgickl, German Patent 4,203,527 (1993) to Akzo Nobel N.V.
6. In "Repertorio dati chimico-fisici e tossicologici" 5th ed., **A7**, 69 (1986).
7. G. L. Castiglioni, C. Fumagalli, A. Guercio, M. Messori and A. Vaccari, *Erdol & Kohle, Erdgas Petrochemie*, **48**, 174 (1995).
8.. G. L. Castiglioni, M. Ferrari, C. Fumagalli, A. Guercio, R. Lancia and A. Vaccari, *Catal. Today*, **27**, 181 (1996).
9. G. L. Castiglioni, C. Fumagalli, R. Lancia and A. Vaccari, in "Catalysis of Organic Reactions", Russell E. Malz Jr. Ed., Dekker, New York, 65 (1996).
10. G. L. Castiglioni, A. Guercio, R. Lancia and A. Vaccari, *J. Porous Materials*, **2**, 79 (1995).
11. J. Escard, I. Mantin and R. Sibut-Pinote, *Bull. Soc. Chim. Fr.*, **10**, 3403 (1970).
12. B. Dvorak and J. Pasek, *J. Catal.*, **18**, 108 (1970).

13. Ch. Schild, A. Wokaun, R. A. Koeppel and A. Baiker, *J. Catal.*, **130**, 657 (1991).

14. R. A. Koppel, A. Baiker, Ch. Schild and A. Wokaun, in "Preparation of Catalysts V", G. Poncelet, P.A. Jacobs, P. Grange and B. Delmon, Ed.s, Elsevier, Amsterdam, 59 (1991).

15. E. Armbruster, C. Fumagalli, R. Lancia and A. Vaccari, WO Patent 95/22539; to LONZA S.p.A..

16. T. Van Herwijnen and W. A. De Jong, *J. Catal.*, **34**, 209 (1974).

17. G. L. Castiglioni, A. Vaccari, G. Fierro, M. Inversi, M. Lo Jacono, G. Minelli, I. Pettiti, P. Porta and M. Gazzano, *Appl. Catal.*, **A123**, 123 (1995).

18. G. L. Castiglioni, A. Vaccari, G. Busca, M. Trombetta, to be pubblished.

19. R. E. Ernst and M. J. Byrne, WO Patent 92/0.002.298; to Du Pont.

20 K. Turner, M. Sherif, C. Rathmell, A.B. Carter and J. Scarlett, WO Patent 88/0.000.937; to Davy McKee L.td

21. G. L. Castiglioni, C. Fumagalli and A. Vaccari, *Chim. Ind. (Milan)*, **78**, 575 (1996).

Reusable, Recoverable, Polymeric Supports: Applications in Homogeneous Catalysis

Brenda L. Case, Justine G. Franchina, Yun-Shan Liu, David E. Bergbreiter
Texas A&M University, Department of Chemistry, College Station, TX

Abstract

Copolymers of *N*-isopropylacrylamide (NIPAm) with *N*-acryloxysuccinimide (NASI) exhibit inverse temperature dependent solubility in water. Catalysts bound to these polymers can be recovered by heating the aqueous solution or by adding brine. Catalysts prepared include phosphine-ligated transition metal catalysts that are active in alkene hydrogenation and covalently bound acid catalysts that are active in acetal hydrolysis. NMR studies have shown that substrate activity on such supports can be equivalent to that of low molecular weight analogs if a sufficiently long chain is used to attach the species in question to PNIPAm.

Poly(fluoroacrylates) are exclusively soluble in fluorocarbon solvents. The complete immiscibility of fluorocarbon solvents with water and common organic solvents suggests their utility in bisphasic syntheses and catalyses, where the polymer-bound reagent or catalyst can be recovered simply by separation of the two phases. Here we describe the synthesis of fluorous phase soluble supports that readily bind amine-terminated reagents that are utilized in bisphasic syntheses and catalyses.

Introduction

There is a strong interest in synthesizing highly reactive and highly selective catalysts that can be easily separated from the product and excess reagents after completion of a reaction. Homogeneous catalysts have high activities and selectivities. However, simple and complete separation of the catalyst is an inherent problem in traditional homogeneous catalysis (1). Several strategies have emerged as solutions to the recovery of homogeneous catalysts. Those that have generated the most interest include supported homogeneous catalysts - "heterogeneized homogeneous catalysts," or soluble, polymer-bound catalysts - and two-phase homogeneous catalysts.

The most frequently used organic polymer supports are linear (soluble) and cross-linked (insoluble) polymers. The macroscopic properties of the polymer are what make them so useful in organic chemistry. The relatively simple properties of polymer size and solubility are the basis of most applications of polymers in organic chemistry where the polymer provides a means of separating the desired product or the catalyst from excess reagents (2). There are numerous examples of the use of polymers in the solid-phase synthesis of organic compounds (3). Catalysts bound to insoluble polymers such as DVB cross-linked polystyrene beads (chemistry based on Merrifield's solid phase peptide synthesis) can be easily recovered (4). The use of linear, soluble polymers to effect phase isolation with minimal alteration of a catalyst's activity is also known (5). Catalysts supported on polymers can potentially combine the advantages of homogeneous catalysis (high activity and selectivity, well-defined catalytic sites, and good reproducibility) with the positive aspects of heterogeneous catalysis (long lifetime and ease of separation). The underlying idea is to use the properties of the macromolecules to control the physical properties of the catalyst.

While most workers have used insoluble, cross-linked polymers in phase isolation, there has been significant work with soluble, linear polymers. These linear soluble polymers have either normal or inverse temperature dependent solubility. Polymers that possess inverse temperature dependent solubility have the unusual property of undergoing a temperature dependent phase change wherein they dissolve upon cooling and phase separate on heating above the lower critical solution temperature (LCST). Unlike natural polymers, such as proteins, these synthetic polymers re-dissolve when the temperature is decreased below the LCST (6). The polymer's reversible phase behavior can thus be used to synthesize catalysts whose activity is sensitive to changes in temperature, with reaction proceeding (turned "on") at low temperature and stopping (turned "off") at elevated temperature (7).

Here we describe poly(*N*-isopropylacrylamide) (PNIPAm) bound catalysts whose inverse temperature dependent solubility has a useful effect on the catalyst solubility. First, the PNIPAm derivatives can be isolated and separated from soluble reagents simply by heating and decantation of excess water from the resulting polymer suspension. In addition, the "on/off" behavior (in solution at low temperatures, out of solution at elevated temperatures) of these derivatives allows for self-regulating systems in exothermic reactions.

Homogeneous two-phase catalysis has proven to be another successful method of recovering homogeneous catalysts from solution (8). A classic example is the aqueous-hydrocarbon system used in the Rhone-Poulenc process (9). A second two-phase system, the fluorous biphase system (FBS), has recently been reported (10). This system is based on the limited miscibility of fluorinated compounds with non-fluorinated compounds. Horvath first described fluorous phase chemistry as consisting of three elements: a fluorocarbon solvent, a hydrocarbon solvent, and a fluorous tag that renders the desired reagent fluorous phase soluble. His pioneering work in this area focused on the recovery of the Rh catalyst, $HRh(CO)\{[P(CH_2)_2(CF_2)_5CF_3]_3\}$, from biphasic systems in hydroformylation reacations (10).

In this paper, we describe an alternative approach to the synthesis of fluorous labels. Specifically, we have devised a scheme where reagents or ligands are rendered fluorous phase soluble by attachment to a fluorous acrylate copolymer containing reactive binding sites. These fluorous phase soluble reagents can then be used in synthesis and as catalysts.

Results and Discussion

To achieve chemical linkage between a soluble metal complex catalyst and an organic polymer a suitable functionality must be introduced into the original polymer. A number of methods, most of which are of general applicability, have been used to graft metal complexes onto polymer supports that contain ligands such as phosphines. Common methods include ligand exchange and direct reaction of the functionalized polymer with a metal halide.

Due to their structural similarity with the PNIPAm backbone and their chemical inertness, we chose to use amide bonds to bind the phosphine ligands to the polymer. To introduce new amide bonds we prepared a series of copolymers of NIPAm and *N*-acryloxysuccinimide (NASI). NASI is a very useful monomer in that the activated ester provides a reactive site at which the copolymer can be derivatized. IR spectroscopy (imide peaks at 1813 and 1785 cm^{-1}, carbonyl peak at 1735 cm^{-1}) and ^{1}H NMR (broad singlet at 2.8 ppm in D$_2$O) confirmed the presence of the NASI groups. The M_v of the 13:1 copolymer was determined to be 8.7 x 10^5 Daltons using values of 9.59 x 10^{-3} and 0.65 for K and a, respectively, in THF at 30 °C (11).

Phosphine ligands, such as bis(2-diphenylphosphinoethyl)amine (DPPE), were readily bound to the polymer backbone of p[NIPAm-co-NASI] by a simple substitution reaction with the activated ester (Figure 1). The unreacted active ester units were quenched with a suitable amine. When isopropyl amine was used as the quenching agent, the resulting copolymer was completely insoluble in water in the range of 0 °C to 100 °C (LCST < 0 °C). This was not surprising since the bisphosphine is highly hydrophobic. When ammonia, a more hydrophilic amine, was used as the quenching agent the resulting terpolymer was soluble in water below 19 °C (LCST ~ 18-19 °C). This illustrates how *N*-substitution can be used to tune PNIPAm's LCST. Based on ^{1}H NMR analysis the loading of the bisphosphine was approximately 1%. Reaction of an aqueous solution of the terpolymer with [Rh(COD)]$^+$CF$_3$SO$_3^-$ in a 2:1 phosphorus to rhodium ratio yielded the polymer-bound cationic Rh(I) catalyst as a golden solution. This aqueous catalyst has an LCST of approximately 17 °C.

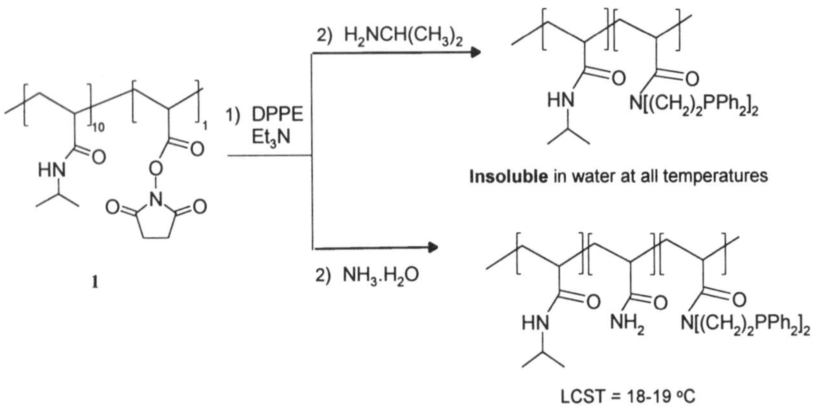

Figure 1. Coupling of bis(diphenylphosphino)ethylamine to p[NIPAm-co-NASI].

The "smart" behavior of the PNIPAm-bound catalyst was observed in the aqueous hydrogenation of allyl alcohol. At 0 °C hydrogenation took place under homogeneous conditions with a rate of approximately 2 mol H_2/mol Rh/h. Heating of the reaction mixture to 40-45 °C resulted in precipitation of the catalyst and almost complete cessation of hydrogenation. According to normal Arrhenius-type kinetics a temperature change from 0 °C to 40 °C should result in an 8-10 fold increase in the reaction rate. Instead, the reaction rate decreased by greater than 50-fold, an approximately 500-fold decrease in activity given the observed temperature change. This rate change was attributed to the solubility changes the polymer-bound catalyst experienced on heating and cooling. When cooled to 0 °C hydrogenation resumed at a rate comparable to that of the original catalyst. This "on/off" behavior was observed through three heating/cooling cycles with only a slight loss in catalytic activity. When isolated from aqueous solutions, the polymer-bound catalyst was recovered in >99% yield.

Although the "on/off" behavior is impressive, the rate of the hydrogenation is relatively low. (For comparison, the rate of hydrogenation of a cyclohexene in benzene with the non-polymer bound Wilkinson's catalyst was found to be 75 mol H_2/mol Rh/h.) (12) This is not true for the activity of all PNIPAm-bound catalysts. We recently described Pd(0) catalysts bound to PNIPAm that are nearly as active as their low molecular weight analogs (13). However, since it is known that binding a reagent to a polymer can affect the bound species' behavior (14), we have further studied this issue. Specifically, NMR spectroscopy studies were undertaken to determine what influence the polymer has on a bound catalyst ligand's behavior. This question is of interest because the slow hydrogenation rates could be explained if it is concluded that the polymer has a profound effect on the bound reagent (i.e. the bound catalyst's behavior resembles that of a polymer rather than a small molecule).

NMR studies are ideal to ascertain whether the polymeric side chains are acting as large or small molecules. The effects of increasing the side chain length can be determined by carbon T_1 experiments and by measuring the linewidths in ^{1}H NMR. To study the effectiveness with which spacers affect side chain bound groups in soluble polymers, we prepared a p[NIPAm-co-NASI] copolymer and used this soluble polymeric reagent to bind two simple amines – hexylamine and octadecylamine. We then measured the T_1 relaxation times of the terminal carbon of the isopropyl groups of the NIPAm and of the C6 and C18 side chains. We also measured the linewidths of the central peak in the triplet of the terminal methyl groups.

Longitudinal relaxation is more efficient at higher molecular weights where the tumbling rate is slower. Thus, polymers typically have shorter T_1 relaxation times than small molecules. Indeed, the T_1 relaxation time of the methyl groups of the homopolymer PNIPAm (0.43 seconds) is much shorter than those of hexyl- or octadecylamine small molecules (6.8 and 4.6 seconds, respectively). The T_1 relaxation time of the terminal $-CH_3$ group of the polymer-bound hexyl chain is measurably different from that of the parent amine (2.7 seconds). However, there is no measurable difference in the T_1 relaxation time values for the $-CH_3$ group of the free and polymer-bound octadecyamine (4.1 seconds). In addition to T_1 relaxation times, we also measured the linewidth of the central peak in the pendant methyl groups in each polymer. Using hexamethyldisilazane as a reference, the PNIPAm methyls were, as expected, very broad. Relatively sharp signals were seen for the octadecyl methyl group.

As a graphic illustration of the effect of spacer length on ^{1}H NMR resolution, allylamine and 11-undecenamine were bound to p[NIPAm-co-NASI]. The region of interest (δ 5-6 ppm) is shown in Figure 2, where it is obvious that extending the side chain by just eight carbon atoms markedly improves the spectral resolution. These T_1 relaxation studies and ^{1}H NMR linewidth measurements confirm the conclusion that soluble polymers require only small spacer groups to significantly change a bound substrate's (or catalyst's) behavior so that it resembles that of a small molecule.

Ongoing studies in our laboratoy involve the design and synthesis of amine and hydroxy-terminated tethers of varying spacer lengths and hydrophobicity. Subsequent studies will examine the utility of such tethers and the effect of the spacer length on the activity of PNIPAm-bound transition metal catalysts and on the utility of these soluble resins in multistep syntheses. Such tethers should also be useful as tools in modifying reactive groups on other vinyl polymers whose potential utility in synthesis and catalysis has recently been described (15, 16).

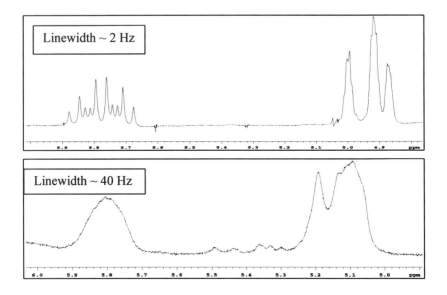

Figure 2. ¹H NMR of pNIPAm-bound 11-undecenamine (top) and allylamine (bottom). (Alkene region only shown.)

As previously mentioned, we have also studied the recovery of catalysts from fluorous biphasic systems. We investigated the reactivity and recoverability of fluoropolymer-bound rhodium catalysts in biphasic hydrogenation reactions. Four fluoropolymer-bound neutral rhodium catalysts were investigated. The fluoropolymers were synthesized with various fluoroacrylates: FX-13 and FX-14 from 3M; Zonyl TAN and TM from DuPont (Figure 3). The fluorinated acrylates were allowed to react with the NASI monomer, giving the copolymers an active site at which the copolymer was derivitized by reaction with primary amines. The copolymerization was run in α,α,α-trifluorotoluene at 100 °C for 48 hours with 2,2'-Azobis(2-methylpropionitrile) (AIBN) as the initiator. The fluorous copolymers were isolated by precipitation in methanol and were characterized by IR spectroscopy and ¹H NMR.

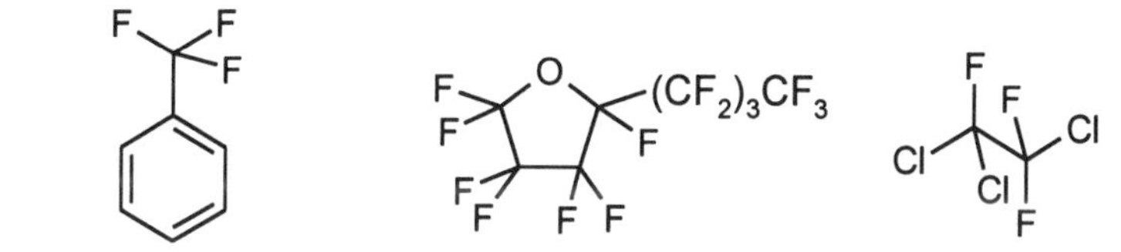

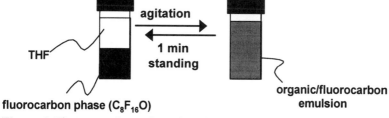

α,α,α-**Trifluorotoluene**   **FC-77**   **1,1,2-Trichlorotrifluoroethane**

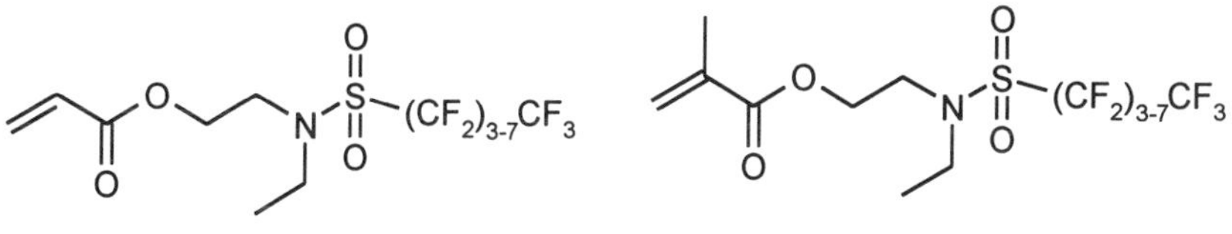

FX-13

FX-14

TAN

TM

Figure 3. Fluorocarbon solvents (top row) and fluoroacrylate monomers.

An amine-terminated dye was attached, in a biphasic reaction, to the fluorous copolymers at the active ester site. The copolymer was dissolved in FC-77 (3M Co., fluorinated cyclic ether, $C_8F_{16}O$), followed by the addition of the amine-terminated *p*-methyl red derivative in THF. The biphasic system was vigorously stirred to form an emulsion. Once the reaction was complete, the product was isolated simply by separation of the two phases and evaporation of the FC-77 solvent. The colored copolymer was placed in a biphasic system and vigorously shaken to determine in which phase(s) the fluorous polymer-bound dye was soluble (Figure 4). The methyl red-containing polymers were not

THF

agitation

1 min standing

organic/fluorocarbon emulsion

fluorocarbon phase ($C_8F_{16}O$)

Figure 4. Fluorous polymer-bound methyl red readily and quantitatively separates from an organic solvent.

soluble in organic solvents such as THF, DMSO, ethanol, methanol, chloroform and toluene. They were soluble in fluorinated solvents such as FC-77; 1,1,2-trichlorotrifluoroethane and α,α,α-trifluorotoluene. Separation of this fluoropolymer-bound dye from organic solvents appeared visually quantitative. A more quantitative estimate of >99.9% of the polymer-bound dye being retained in the fluorous phase was obtained by UV-visible spectroscopy of the THF layer from the biphasic system, which showed no methyl red present in the organic phase.

Once it was determined that 1) the fluoropolymers synthesized were exclusively soluble in the fluorocarbon layer of the biphasic system, and 2) that the emulsion of the two phases resulted in an interfacial reaction, the fluoropolymers were derivatized in such a way as to be used in catalysis.

The monodentate ligand, diphenylphosphinopropylamine (DPPP), was bound to the fluoropolymer in an interfacial reaction in FC-77 and THF. Once the monodentate phosphine was attached, the fluoropolymers were further reacted in a biphasic system (FC-77/THF) with a neutral rhodium catalyst precursor, μ-dichlorotetraethylenedirhodium, $[RhCl(CH_2CH_2)_2]_2$. These fluorous polymer-bound catalysts were subsequently used in the hydrogenation of organic-phase soluble alkenes. The catalysts were recyclable and easily separable from the reactants and products by liquid-liquid extractions. When the hydrogenation was complete (as determined by ^{1}H NMR), the organic layer was removed and the fluorocarbon layer was washed twice with fresh THF. A new organic layer containing an alkene was added to the fluorocarbon phase and the hydrogenation resumed. Full conversion to the alkane was seen in most cases, with reaction rates of 10-40 mmol H_2/mmol Rh/h. The neutral rhodium catalyst $[TM-PPh_2]_3RhCl$ was recycled nine times, with complete conversion of 1-octene each time. The turnover number (moles of alkene reduced per mole of rhodium catalyst) was estimated to be 6000 in this instance. The neutral rhodium catalysts $[TAN-PPh_2]_3RhCl$ and $[FX14-PPh_2]_3RhCl$ were each recycled six times (turnover number~4000), with complete hydrogenation of six different substrates.

Conclusion

We have found that catalysts bound to the PNIPAm backbone are recoverable and also exhibit "smart" behavior in the hydrogenation of simple alkenes in aqueous systems. T_1 relaxation studies and ^{1}H NMR linewidth measurements confirm the conclusion that soluble polymers require only modest spacer groups to significantly change the bound substrate's behavior so that it resembles that of a small molecule.

The methyl red experiments have shown that fluorous polymer-bound substrates and catalysts can be designed such that they are soluble exclusively in the fluorocarbon phase. We have also shown that catalysts bound to fluorous polymers are fully recyclable and have good activity.

Experimental Section

General methods. All reagents and solvents were obtained from commercial sources and used without further purification unless otherwise stated. THF was distilled from sodium benzophenone ketyl. NIPAm was recrystallized from hexanes/benzene (10% benzene). ^{1}H and ^{13}C NMR spectra were recorded on a Varian XL200E spectrometer at 200 MHz (50 MHz for ^{13}C) or a Unity p300 at 300 MHz (75 MHz for ^{13}C). Chemical shifts are reported in ppm using HMDS (0.055 ppm) as an internal reference. ^{31}P NMR spectra were recorded on the Unity p300 at 121MHz using H_3PO_4 (80%) as an external reference. Infrared spectra were recorded as pressed KBr pellets on a Mattson Galaxy 4021 FT-IR spectrometer. The transmission spectra are reported in wavenumbers (cm^{-1}) and were recorded with a resolution of 2 cm^{-1}.

Poly(N-isopropylacrylamide-co-N-acryloxysuccinimide) (1). Preparation of the p(NIPAM-co-NASI) copolymers and derivatization of these copolymers with amines follows an established procedure (11). In a typical example, 11.3 g (0.100 mol) of recrystallized N-isopropylacrylamide and 1.69 g (0.010 mmol) of N-acryloxysuccinimide were dissolved in 80 mL of t-butanol and heated to 70 $^\circ$C under positive nitrogen pressure. AIBN initiator (0.053 g, 0.32 mmol) in 10 mL of t-butanol was added at once and the solution was stirred for 15 hours at 70 $^\circ$C. The solvent was removed under reduced pressure and the polymer **1** was purified by repeatedly dissolving in THF (~250 mL) and precipitating into hexanes (~ 700 mL). The final product was put on a high vacuum system for 24 h to yield 9.5 g (73%) of the copolymer: IR (KBr, cm^{-1}) 3400, 2970, 2920, 1818, 1785, 1735, 1652 (imide peaks at 1813 and 1785 cm^{-1}, carbonyl at 1735 cm^{-1}); ^{1}H NMR (300 MHz, D$_2$O) δ 1.15 (br s, 6H), 1.40-2.09 (m, 3H), 2.93 (br s, 0.4H), 3.96 (br s, 1H). The M$_v$ was determined to be 8.7 x 10^5 Daltons using values of 9.59 x 10^{-3} and 0.65 for K and a, respectively, in THF at 30 $^\circ$C (11). Modification of this copolymer with an amine was carried out by stirring the copolymer with excess amine in THF for 24 h. Subsequent addition of NH$_3$ (as NH$_4$OH or as the gas) and further stirring for 4 h ensured that all of the NASI groups were consumed – a supposition that was verified by ^{1}H NMR spectroscopy and by IR spectroscopy.

Synthesis of Diphenylphosphinopropylamine (DPPP) . The reaction and work-up were done under strict N$_2$ atmosphere. Li wire (3.199 g, 460.9 mmol) was washed in hexanes, dried, cut into 3mm pieces and added to a flame-dried N$_2$-flushed flask. Chlorodiphenylphosphine (35 mL,185 mmol) and 15 mL of dry THF were added via syringe. The reaction was stirred at room temperature overnight to yield a dark red solution. This dark red solution was added by forced cannulation to a flame-dried, N$_2$-flushed flask containing 10.98 (82.7

mmol) 3-chloropropylamine hydrochloride and 50 mL of dry THF cooled to –78 °C. The dark red solution turned yellow within one hour. After stirring for 2 h it was warmed to room temperature stirred for 4 hours and then stirred at 50 °C overnight. The solution was transferred to a separatory funnel and diluted with hexanes. The organic phase was washed with saturated NaCl 2x20 mL. The organic phase was extracted with 1 N H_2SO_4 3x15 mL, the combined aqueous extracts were neutralized with 2 N NaOH and the product extracted with hexanes/ether (70:30) 2x100 mL. The solvent was removed under reduced pressure to give 7.8 g (39% yield) of a colorless oil: ^{1}H NMR: δ 1.30 (br s, 2H), 1.45-1.62 (m, 2H), 2.01-2.09 (m, 2H), 2.76 (m, 2H), 7.3-7.42 (m, 10H); ^{13}C NMR: δ 25.4, 30.05, 43.35, 131, 132.71, 138.67; ^{31}P NMR: δ -21, -20, 34.5 (9% oxidized).

Synthesis of bis(2-Diphenylphosphinoethyl)amine hydrochloride (DPPE). The reaction was done under a strict N_2 atmosphere. Li wire (2.476 g, 356 mmol) was washed in hexanes, dried, cut into 3mm pieces and added to a flame-dried N_2-flushed flask. Dry THF (80 mL) and 22 mL (116 mmol) of freshly distilled chlorodiphenylphosphine were added via syringe. The reaction was stirred at room temperature overnight to yield a dark red solution. This dark red solution was added by forced cannulation to a flame-dried, N_2-flushed flask containing 5.28 g (29 mmol) of oven-dried bis(2-chloroethyl)amine hydrochloride and 150 mL of dry THF. The reaction was stirred under reflux, using a cold water bath to circulate water through the condenser, overnight. The red solution was cooled to room temperature, diluted with 400 mL of hexanes, and washed with 10% NaOH 1x150 mL and saturated NaCl 1x150mL. The hexanes were stirred vigorously with 400 mL of 2 M HCl for 45 minutes, resulting in a white precipitate, which was collected by filtration (11.15 g, 80% crude yield). The white solid was recrystallized from boiling acetonitrile to give 7.7 g (55% pure yield) of a white solid: MP 169.5-170.8 °C; ^{1}H NMR δ: 2.4-2.6 (m, 4H), 3.8-4.1 (m, 4H), 7.2-7.7 (m, 20H), 9.8-10.1 (2 br s, 2H); ^{31}P NMR: δ -21, -20, 34.5 (9% oxidized).

Synthesis of a cationic Rh catalyst bound to p[NIPAM-co-NASI]: Bis-(2-diphenylphosphinoethyl)amine (0.115 g, 0.24 mmol) and p[NIPAM-co-NASI] (1.01 g, 0.78 mmol) were dissolved in a mixed solvent system of dioxane and *t*-butanol (10 mL: 1 mL). Following the addition of 1 mL (7 mmol) of triethylamine, the flask was evacuated and flushed with argon several times. The reaction mixture was stirred for 7 h at 105 °C, then cooled to room temperature and any reacted NASI groups were quenched with NH_4OH. The polymer was isolated by precipitating into hexanes. Reaction of an aqueous solution of the bisphosphine ligated terpolymer with $[Rh(COD)]^+CF_3SO_3^-$ in a 2:1 phosphorus to rhodium ratio afforded a golden solution of the polymer bound cationic Rh(I) catalyst.

Synthesis of poly(TAN-co-NASI). The poly(fluoroacrylates) were polymerized using identical procedures. The monomers from DuPont were used as received.

The monomers from 3M were recrystallized from MeOH before use. In an oven- dried, N_2-flushed flask, 6.64 g (12.8 mmol) of TAN monomer and 0.114 g (0.77 mmol) of NASI were dissolved in 50 mL of α,α,α-trifluorotoluene, with stirring. The flask was evacuated and flushed with nitrogen three times, then fitted with a reflux condenser. After heating to 80 °C, AIBN (0.025 g) in 5 mL of α,α,α -trifluorotoluene was added. The reaction was stirred for 2 days at 100 °C and was then cooled to room temperature. The copolymer product was soluble in α,α,α -trifluorotoluene. The copolymer was precipitated using 75 mL of methanol. The ^1H NMR after the first precipitation showed the presence of monomer so the copolymer was redissolved in 25 mL of α,α,α-trifluorotoluene and reprecipitated in 50 mL of methanol. The copolymer was isolated by filtration and dried under vacuum overnight. ^1H NMR: sample was dissolved in a small amount of 1,1,2-trichlorotrifluoroethane with acetone for locking, all peaks were broad. ^1H NMR (300 MHz): δ 4.5 (CH_2-O), 2.9 (($CH_2)_2$), 2.6 (CH_2-CF_2 and CH of polymer backbone), 1.9- 1.5 (CH_2 of polymer backbone); IR (KBr, cm^{-1}): 3000, 1818-1740 (C=O), 1250-1150 (C-F). Ratio of CH_2-O to $(CH_2)_2$ was 10:1 by ^1H NMR. Yield was 6.29g (96%).

Synthesis of diphenylphosphinopropylamine ligated fluorous copolymers (DPPP copolymers): In an oven-dried, N_2-flushed flask, p[TAN-co-NASI] (0.100 g) was dissolved in 5 mL of FC-77 under nitrogen. After the flask was flushed with nitrogen, 15 mL of THF was transferred via syringe to the flask followed by the addition of 5 drops of diphenylphosphinopropylamine via syringe. The reaction was stirred vigorously overnight at room temperature under N_2. The top layer was removed by syringe. Fresh THF was introduced into the reaction and stirred for ten to fifteen minutes to remove any excess diphenyphosphinopropylamine. This was repeated two times. The lower fluorocarbon layer was then transferred into a N_2-flushed vial and stored under nitrogen until use. ^1H NMR (300 MHz, FC-77 with an acetone-d$_6$ lock): δ 7.3- 7.7 (-P($C_6H_5)_3$), 4.5 (CH_2-O), 2.6 (CH_2-CF_2 and CH of polymer backbone), 1.9- 1.5 (CH_2 of polymer backbone). ^{31}P NMR: δ −16.5, 35 (oxidized).

Synthesis of diphenylphosphinopropylamine (DPPP) neutral rhodium catalysts: The solution of DPPP copolymer in FC-77 was diluted to 10 mL. Five mL (containing 50 mg of copolymer) was transferred via syringe into a N_2-flushed, oven-dried, flask. The neutral rhodium complex, [RhCl($CH_2CH_2)_2$]$_2$, (4 mg) was dissolved in 10 mL of THF and transferred via syringe into the copolymer solution. The reaction was stirred vigorously overnight at room temperature. The following day the stirring was stopped, the lower fluorocarbon layer was removed, transferred to a hydrogenation vessel and stored under nitrogen until use.

Acknowledgments

Financial support from the National Science Foundation (CHE-9707710 and CHE-9222717) and the Robert A. Welch Foundation is greatly appreciated. We are also indebted to 3M for supplying the FX-13 and FX-14 monomers and fluorocarbon solvents and to DuPont for supplying the TM and TAN monomers.

References

1) C. U. Pittman, Jr., in *Comprehensive Organometallic Chemistry*, G. Wilkinson Ed.; Pergamon: Oxford, Vol. 8, p. 553 (1982).

2) E. Bayer and V. Schurig, *Angew. Chem. Int. Ed. Engl.*, **14**, 493 (1975).

3) J. S. Fruchtel and G. Jung, *Angew. Chem. Int. Ed. Engl.*, **35**, 17 (1996). D. J. Gravert and K. D. Janda, *Chem. Rev.*, 489 (1997).

4) N. K. Mathur, C. K. Narang and R. E. Williams, *Polymers as Aids in Organic Chemistry*; Academic: New York, 1980; C.U. Pittman, Jr., in *Polymer-supported Reactions in Organic Synthesis*, P. Hodge and D. C. Sherrington Ed.; Wiley: New York, p. 249 (1980). D. C. Bailey and S. H. Langer, *Chem. Rev.* **81**,109, (1981). F. R. Hartley, *Supported Metal Complexes: A New Generation of Catalysts*, D. Reidel: Amsterdam, 1985. W. T. Ford, CHEMTECH, **17**, 246 (1987), **14**, 436 (1984).

5) D. E. Bergbreiter, *ACS Symp. Ser.*, **308**, 1741 (1986).

6) I. Y. Galaev and B. Mattiasson, *Enzyme Microb. Technol.*, **15**, 354 (1993).

7) I. Y. Galaev, *Russian Chem. Rev.*, **64**, 471 (1995).

8) W. A. Hermann and C. W. Kohlpaintner, *Angew. Chem. Int. Ed. Engl.*, **32**, 1524 (1993). J. J. J. Juliette, I. Y. Horvath and J. A. Gladysz, *Angew. Chem. Int. Ed. Engl.*, **36**, 1610 (1997). M.-A. Guillevic, A. M. Arif, I. T. Horvath and J. A. Gladysz, *Angew. Chem. Int. Ed. Engl.*, **36**, 1612 (1997). S. Kainz, D. Koch, W. Baumann and W. Leitner, *Angew. Chem. Int. Ed. Engl.*, **36**, 1628 (1997).

9) E. G. Kuntz, CHEMTECH, **17**, 570 (1987).

10) I. T. Horvath and J. Rabai, *Science*, **266**, 72 (1994).

11) F. M. Winnik, *Macromol.*, **23**, 233 (1990).

12) J. A. Osborn, F. H. Jardine, J. F. Young and G. Wilkinson, *J. Chem. Soc. (A)*, 1711 (1966).

13) D. E. Bergbreiter, Y.-S. Liu and P. L. Osburn, submitted to *J. Am. Chem. Soc.*

14) D. E. Bergbreiter, T. Kimmel and J. W. Caraway, *Tetrahedron Lett.*, **36**, 4757 (1995).

15) D. E. Bergbreiter and J. G. Franchina, *J. Chem. Soc., Chem. Commun.*, 1531 (1997).

16) D. E. Bergbreiter and Y.-S. Liu, *Tetrahedron Lett.*, **38**, 3703 (1997).

Utilization of the Heterogeneous Palladium-on-Carbon Catalyzed Heck Reaction in Applied Synthesis

A. Eisenstadt
IMI (TAMI) Institute for R&D Ltd.

Israel

Abstract

The last two decades have seen a revival of the facile palladium-catalyzed arylation and alkenylation reactions of alkenes known as the Heck reaction. This only occurred after extensive studies into the reactivity and regioselectivity of the reaction. This paper discusses how palladium-on-carbon has been successfully used to prepare sunscreens, special monomers and pharmaceutical intermediates.

Introduction

Heck discovered in the early 70's several new reactions of considerable potential value to organic chemists. Perhaps the most promising of these reactions as far as synthetic utility is concerned, is the catalytic arylation and vinylation of olefins.

In this reaction, an organic halide first forms, via oxidative addition, an organopalladium halide complex with the catalyst. The complex then adds to an olefin and the adduct decomposes by elimination of a hydrido-palladium halide to form a new olefin in which a vinylic position is substituted by the organic group of the halide

The use of homogeneously-catalyzed Heck arylations in the synthesis of variety of compounds is well documented [1-4]. This includes (among many others) the synthesis of Lilial (fragrance), developed by Givaudan [5], Metroprolol (beta blocker) [6] and Nabumetone (NSAID) [7].

Some of the problems associated with this technology can be obviated by the use of heterogeneous catalysts and there have been reports stating that Pd/C can promote the Heck arylation, but with limited data [8].

It seems that more complete investigation of the usefulness of supported metal as catalysts for this reaction would lead to a better appreciation of the use of heterogeneous catalysts by organic chemists.

Results and Discussion

Sunscreen raw materials [9]

The production of OMC (OctylMethoxyCinnamate), the most common UV-B sunscreen on the market, is an example of a successful industrial technology and demonstrates the novel chemistry and economic attractiveness of the Heck reaction. The main parameters that control the process in the scaled-up industrial plant, will be discussed.

The IMI OMC production process [9] involves two main steps: bromination of commercial anisole at a para-regioselectivity of greater than 97%, and heterogeneously-catalysed Heck coupling between commercially available octyl acrylate (OA), in the presence of a palladium-on-carbon catalyst, using Na_2CO_3 as an HBr sponge, and N-methylpyrrolidone (NMP) as a polar, non-protic solvent. The pathway to OMC is illustrated in Scheme 1.

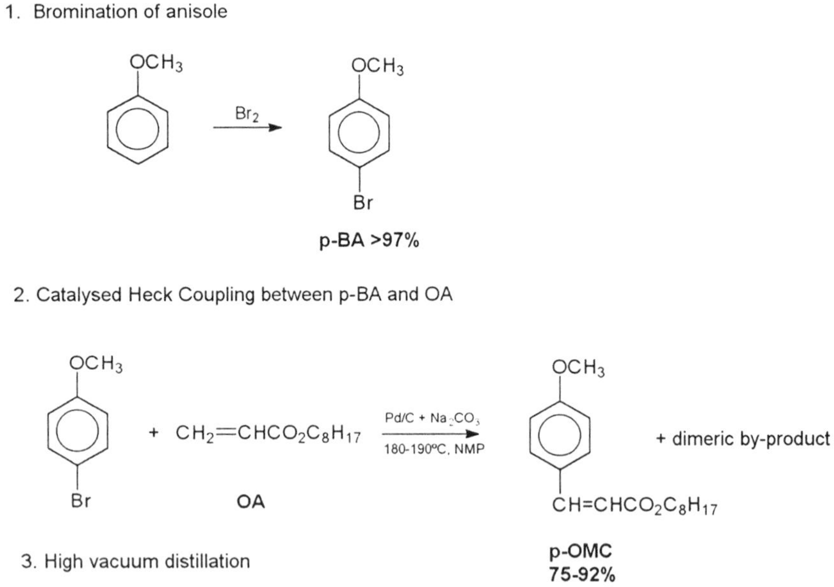

1. Bromination of anisole

p-BA >97%

2. Catalysed Heck Coupling between p-BA and OA

3. High vacuum distillation

p-OMC
75-92%

Scheme 1: Preparation of octylmethoxy cinnamate (OMC)

The identifiable by-products of the OMC preparation are shown in Scheme 2.

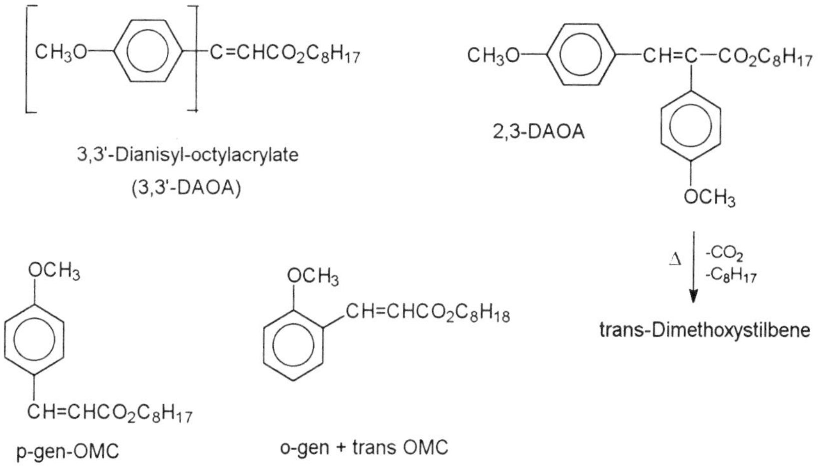

3,3'-Dianisyl-octylacrylate
(3,3'-DAOA)

2,3-DAOA

Δ | -CO$_2$ / -C$_8$H$_{17}$

trans-Dimethoxystilbene

p-gen-OMC

o-gen + trans OMC

Scheme 2: By-products of OMC production

OMC isomers such as ortho-*cis* and ortho-*trans*-OMC are the result of some ortho bromination. The formation of methyl-OMC by-product is due to the presence of methyl anisole (0.11%) in the commercial anisole starting material. *Cis*-para-OMC is a minor side-product which is formed together with the major *trans* para-OMC.

A major by-product, 3,3`-dianisyl-octyl acrylate (DAOA), results from a double Heck coupling between p-BA and OA. The dimeric 2,3-DAOA, present in small amounts in the final reaction mixture, undergoes thermal decomposition to give *trans*-dimethylstilbene (DAMS).

Due to the fact that our OMC process is different from the known one, it was important, from a safety aspect, to prove the complete non-mutagenic nature of our OMC process; a quantitative separation and Ames testing of each of the four isomers and the two by-products was carried out.

The pilot plant runs of the coupling stage were carried out in a 250-L reactor at 70% loading of the reaction mixture, and resulted in excessive OA polymerization and oligomerization i.e., fast consumption of OA relative to the p-BA. This caused a significant decrease (15-20%) in the yield of OMC and increased formation of the double adduct (DAOA) prior to the flash distillation, and to difficulties in the salt-cake filtration. This major problem was overcome by dilution of the reaction medium. Increasing the NMP volume dramatically reduced the reaction time from 5-6 to ca. 2 hrs. and improved the yield to a level of 80-90% when the loading was changed from 70% to 47-58%. The OMC production process can tolerate a content of up to 15% water with some acceleration of the process (to less than 2 hrs) without any foaming or overflooding phenomena. A possible explanation for the "wet" production behaviour is that the presence of water in the heterogeneous, polar, aprotic medium (NMP mixed with insoluble catalyst and sodium carbonate) promotes the solubility of the base in this medium and pushes the proton abstraction from the palladium-intermediate complex.

We also utilized the same technology in making another sunscreen, ODMAC (OctylDimethylAminoCinnamate) which has the potential of being a UVA+B filter reagent (Scheme 3);

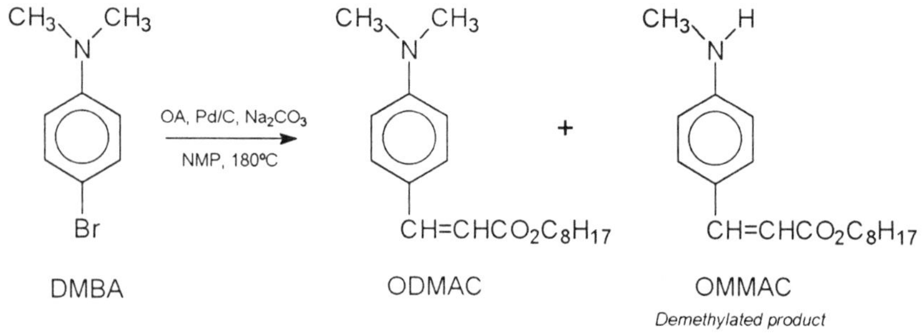

Scheme 3: Octyl Dimethyl Amino Cinnamate (ODMAC) preparation

The γ max of ODMAC is 30,600 at 363 nm. At a ratio of 1:1 with another UV-B component (PABA or OMC), ODMAC can block 90%, and more, of all irradiation at 290-400 nm

Substituted benzophenones

IMI has also developed a general route for synthesizing homo- and hetero-disubstituted benzophenones such as 4,4'-difluorobenzophenone (DFBP), 4,3'-difluorobenzophenone and 4,4'-diphenoxybenzophenone (DPOBP). These substituted benzophenones are important materials having a variety of uses as special monomers (for polyetherketones and polyarylene etherketones) and pharmaceutical intermediates. These benzophenones are prepared by utilizing the Heck technology in conjunction with an oxidative cleavage reaction.

Synthesis of 4,4'-Difluorobenzophenone (DFBP) [10]

A representative example is the synthesis of DFBP starting from p-bromofluorobenzene (BFB). This reaction involves the formation of a double-Heck adduct (Scheme 4).

DFBP is used as a monomer for HP & LC polymers (PEEK & PEK) and as a pharmaceutical intermediate, and is prepared via two steps as follows:
Step 1) Double Heck coupling between p-BFB and an olefinic substrate,
Step 2) Catalyzed oxidative-cleavage of the double adduct.

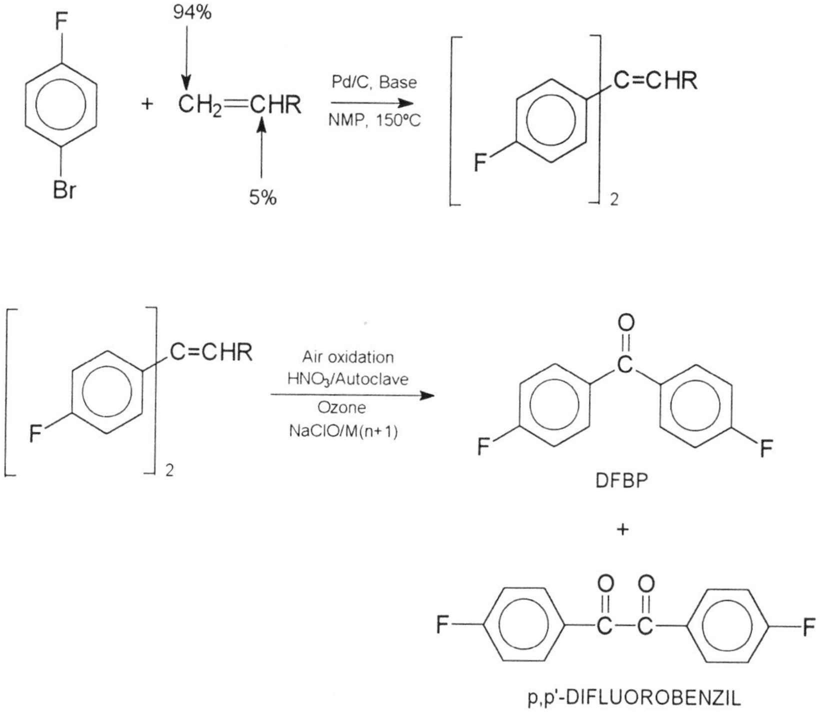

Scheme 4: Synthesis of 4,4'-Difluorobenzophenone (DFBP)

A kinetic follow-up of the double Heck coupling process is shown in Fig. 1.

Heck coupling between p-FBB & OA

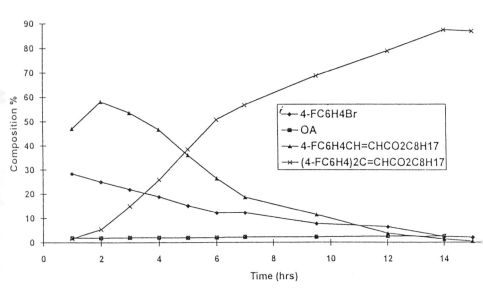

Fig. 1: Kinetic follow-up of the Double Heck Coupling Process

Substituted Benzophenones [3]

The same strategy was applied to the preparation of the following di-heterosubstituted benzophenones (Scheme 5) [11] and the pharma intermediate, dibenzosuberone (Scheme 6):

1. Symmetric disubstituted BP's: 4,4'-Difluoro- and diphenoxy BP's

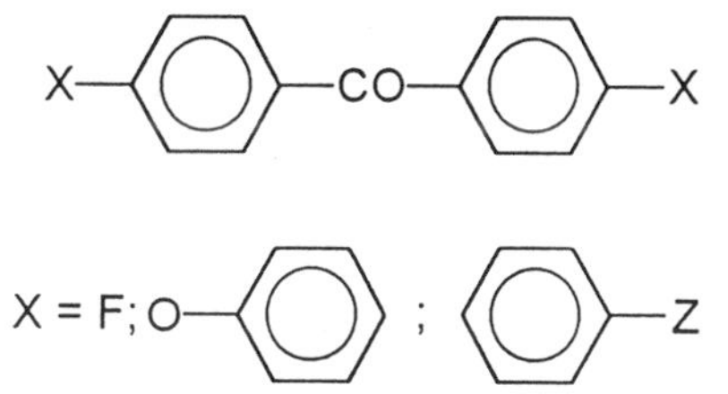

$$X = F; O-\bigcirc \quad ; \quad \bigcirc-Z$$

2. Heterodisubstituted benzophenones (4,4'- and 4,3')

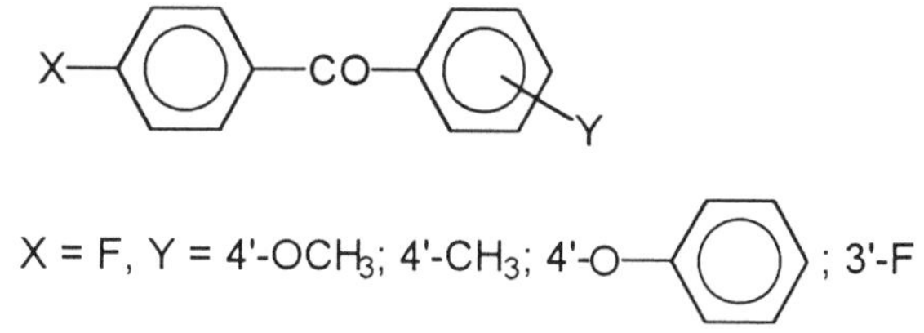

$$X = F, Y = 4'-OCH_3; \ 4'-CH_3; \ 4'-O-\bigcirc \ ; \ 3'-F$$

3. Dibenzosuberone

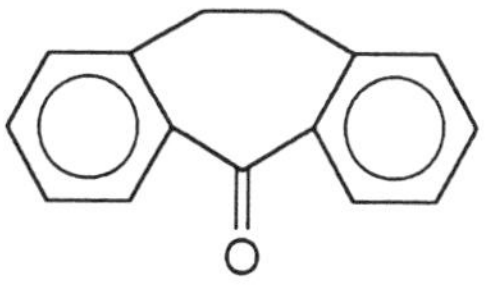

Scheme 5: Preparation of di-heterosubstituted benzophenones

Dibenzosuberone (10,11-Dihydro-5H-dibenzo[a,d]cycloheptene-5-one) is an important intermediate in the preparation of psychotropic agents such as antidepressants (Amittptyline and Nortiptyline), as well as for other drugs (Scheme 6):

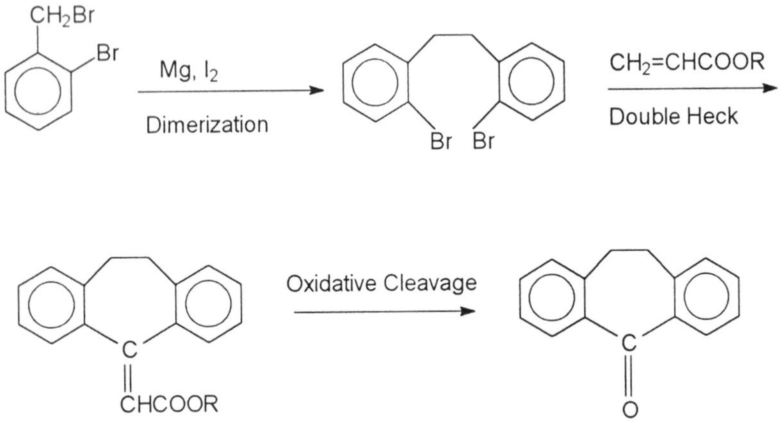

Dibenzosuberone

Scheme 6: Preparation of Dibenzosuberone

Intermediates for Sulindac (an NSAID drug) [12]

The Heck reaction may also be utilized in the preparation of some pharmaceutical intermediates for NSAID drugs, such as Sulindac and DehydroNabumetone, using bromoaromatic precursors and commercially available alkenes. The synthesis of the Sulindac, as developed by Merck, involves in the first stage, the condensation of p-fluorobenzaldehyde with propionic anhydride to give the p-fluoro-α-cinnamic acid. The aldehyde is relatively expensive and unstable, and the Knoevenagel transformation yield is fairly low.

At IMI, the use of the Heck technology enabled the preparation of the same acid [12] in either one step (in aq. medium) or two steps (in a polar nonprotic solvent) according to Schemes 7 and 8:

The original **Merck** *route to Sulindac:*

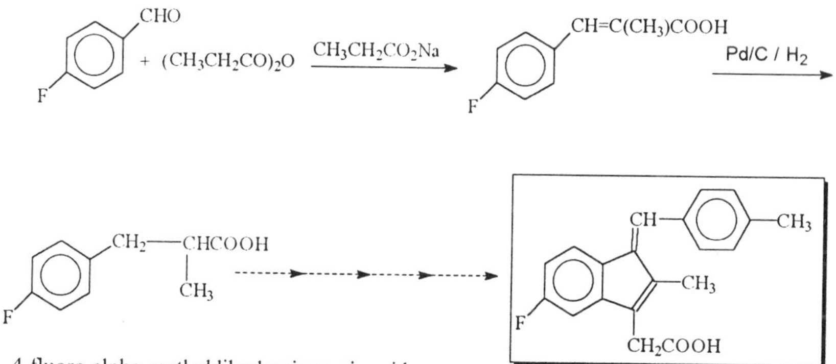

4-fluoro-alpha-methyldihydrocinnamic acid

Sulindac

IMI Route: **Isr. Appl. 108884 (1994)**

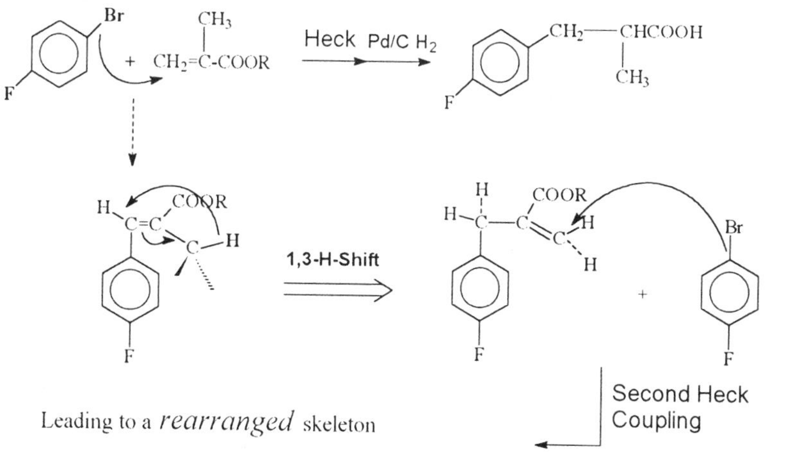

Heck Pd/C H₂

1,3-H-Shift

Leading to a *rearranged* skeleton

Second Heck
Coupling

Scheme 7: Synthesis of intermediates for SULINDAC

The p-fluoroalkylmethacrylate conversion stage involves an unusual 1,3-hydrogen shift and the formation of a rearranged molecular skeleton, in which two p-F-aryl groups are inserted (Scheme 8):

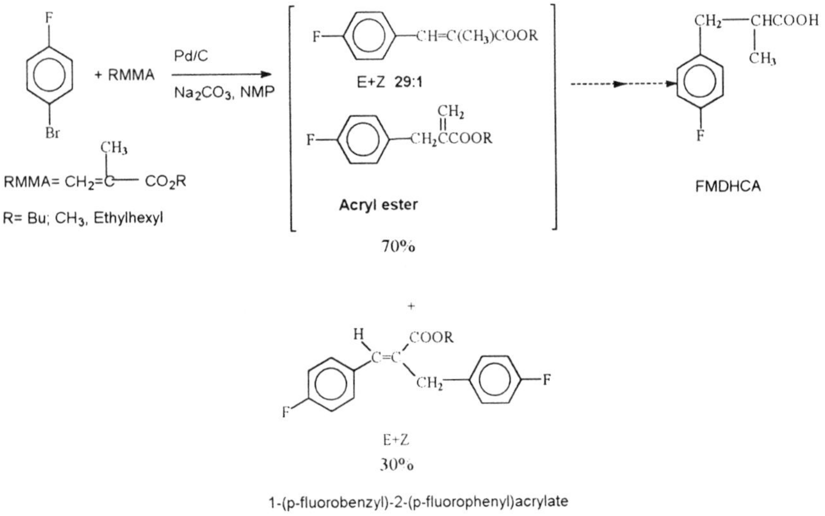

Scheme 8: FMDHCA Production via Heck and hydrogenation technology

Some novel chemical transformations were carried out with the latter skeleton (Scheme 9):

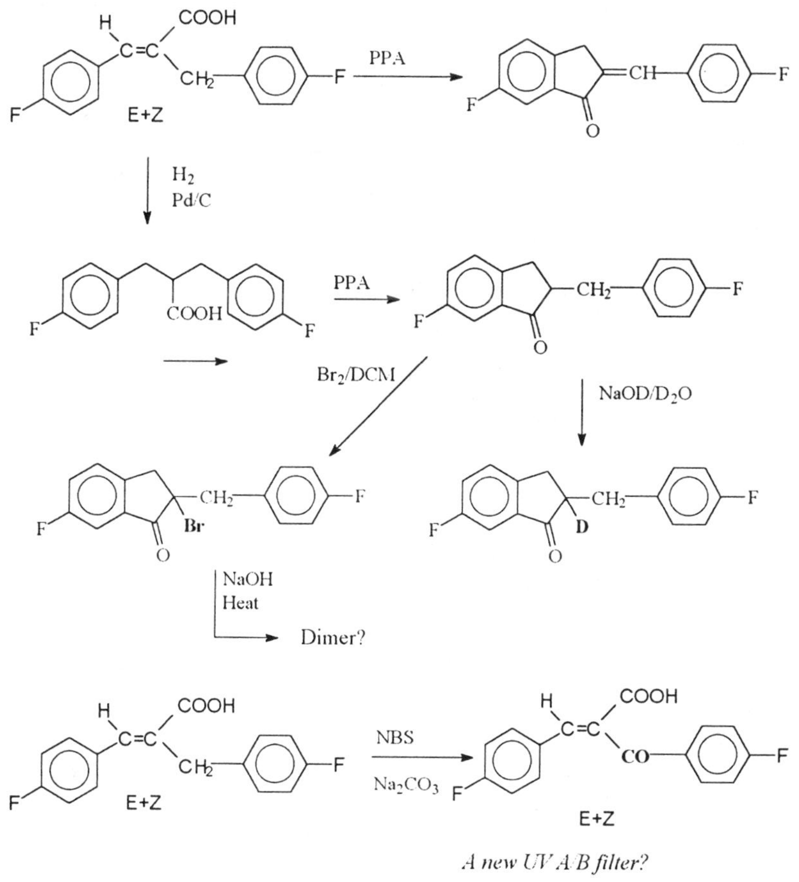

Scheme 9: Transformations starting from FMDHCA

Experimental

Typical example of Heck coupling reaction catalysed by Pd/C for the preparation of OMC:

A 3-litre, 5-necked glass reactor, equipped with a mechanical stirrer, a gas inlet adaptor, a Dean-Stark device for collecting any volatiles or azeotropes evaporating during the coupling reaction (such as water, NMP, p-BA and OA), and a sampling septum, was loaded as follows:

Raw materials	Source	ml	gms	moles	p-BA/Pd
p-BA	BCL Israel	466	374	2	
OA	Union Carbide	466	412	2.24	
Na$_2$CO$_3$			112	1.06	
NMP	GAF	796	775		
Pd/C 37A (4.58%)	JM		.764	3.3×10^{-4}	6057

Loading: $(V_{p-BA}/V_{p-BA} + V_{NMP}) = 47\%$

The reaction mixture was flushed with nitrogen for 15 min, heated to the reaction temperature (180-190°C) then stirred under nitrogen for 2-3 hrs. Samples were taken every half hour. The reactor was then cooled to 70°C and the catalyst and salts were removed by filtration.

The filter cake was washed under vacuum with 2x100 ml fresh NMP, and the washings were combined with the reaction mixture filtrate. The Pd catalyst and the salts (NaBr and Na$_2$CO$_3$) were separated by sieving through an ASTM 2.36 mm mesh. The salt cake was repulped with 600 ml fresh NMP for 1 hr., followed by filtration. The filtrate was combined with the other filtrates.

Purification was by flash distillation of the volatiles (mainly NMP and minor amounts of unreacted p-BA and OA). The distillate was recycled to the next OMC production run. At this stage, the OMC content in the residue (determined by quantitative HPLC and GC) was 83%, together with 3.9% of the dimeric adduct (DAOA).

The residue was subjected to high vacuum fractional distillation (20 theoretical plates) at 181-188°C (0.3-0.5 mBar) yielding 444.8 gms OMC (76.7%) in a purity of 99.1%.

Conclusions

In conclusion, it has been demonstrated that the Heck reaction opens up a great number of new synthetic routes for the preparation of commercial compounds. Numerous products cited in the literature, as well as the above mentioned cinnamates, ketones, aldehydes, and acids, demonstrate the potential of the technology.

This is a relatively new area from an industrial point of view, that demonstrates the prospective growth and advantages of bromine chemistry, specifically over those of the other halides [13]

References

1. R. F. Heck, Accounts of Chemical Research, 12, 146 (1979); Org. Rx., 27, 345 (1982); Palladium Reagents in Organic Synthesis, Acad. Press, 1985
2. J. Tsuji, Expanding Industrial Application of Pd Catalysts, Synthesis, 739 (1990)
3. A. de Meijere, Angew. Chem. Intl. Ed., 33 (23/24) 2379 (1994)
4. W. Cabri, Accounts of Chemical Research, 28, 2, (1995)
5. A. J. Chalk, J. Org. Chem., 41, 1206 (1976); USP 1,548,595 (1976)
6. Halberg et al., Synthetic Comm. 15, 1131 (1985)
7. Hoechst Celanese EP 03765; Beecham PLC 88 03,088; NI 8900,721
8. R. L. Augustine, J. Mol. Cat., 72, 229 (1992)
9. A. Eisenstadt and Y. Keren, "Process for the preparation of OMC", Isr. 97850 (1991), EP 509,426 (1992), US 5,187,303 (1993)
10. A. Eisenstadt, "A process for the preparation of DFBP", Isr. Application 103506 (1992)
11. A. Eisenstadt, "A process for the preparation of disubstituted benzophenones", Isr. Application 115855 (1995)
12. A. Eisenstadt, "Process for the preparation of p-fluoro-α-methylcinnamic acid derivatives", Isr. Application 108884 (1994)
13. W. Bernhagen, Specialty Chemicals, 16(6), 200-206 (1996) and references cited therein

Characterization of Catalytic Reactions by *in situ* Kinetic Probes

C. LeBlond, J. Wang, R. Larsen, C. Orella, and Y.-K. Sun[1]

Merck & Co., Inc.
P.O. Box 2000, RY55-228
Rahway, NJ 07065

Abstract

Several *in situ* probes for continuously monitoring rate of catalytic reactions under reaction conditions are described. They are reaction calorimetry, measurements of hydrogen uptake in the case of hydrogenation, and infrared spectroscopy. In studying catalytic hydrogenation reactions, for example, these *in situ* probes provide kinetic details of the reactions from different perspectives over the entire course of the reaction. The reaction calorimetry and the hydrogen uptake measure directly, continuously, and in a non-invasive manner the rate of reaction, while the *in situ* infrared spectroscopy provides time-resolved compositional information in the liquid phase. A combination of the information thus obtained leads to a clear and coherent kinetic picture of the reaction under study which can greatly facilitate pathway analysis and mechanistic description of the catalytic reaction. In this report, the usefulness of the combination of these *in situ* probes is illustrated with an example of heterogeneously-catalyzed hydrogenation reactions.

1. Introduction

Catalysis for organic synthesis is both technologically important and scientifically challenging. Catalytic processes play particularly important roles in the development of synthetic processes for pharmaceutical applications. Pharmaceutical processes are often characterized by high value-added products, relatively small scale of production, and batch or semi-batch operations. One of the most important criteria in designing a catalytic process is the process yield due to requirements for product purity and high value of the molecules involved in the synthetic process. A major consideration of a catalytic process is how to convert a substrate molecule often containing multiple functional groups with the desired chemo-, regio-, or stereo-, such as enantio- and diastereo-, selectivities.

Development of efficient, high yielding and robust catalytic processes for manufacturing benefits greatly from a thorough understanding of the mechanism by which the catalytic process works. To a great extent, the level of understanding of the mechanism depends critically upon one's ability to observe the reaction under conditions. Among the observables, reaction kinetics are special since they contain some of the most fundamental parameters that describe the catalytic reaction. An

---

[1] Corresponding author. E-mail: yongkui_sun@merck.com

accurate measurement of the reaction kinetics is essential to an understanding of reaction mechanisms, since kinetics are a "reflection" of the mechanism.

Monitoring reactions *in situ* is the most ideal way to observe the progress of the reaction under study. *In situ* monitoring provides information regarding the reaction as it happens under reaction conditions, and makes it possible to observe reactive intermediates, their fates, and reaction kinetics associated with their reactions. These advantages are critical to further our understanding of mechanisms of important chemical and catalytic reactions to a deeper level. They have been nicely demonstrated in a couple of recent applications where *in situ* high-pressure NMR was used for a spectroscopic observation of highly reactive formyl cation formed in a condensed phase from reaction between CO at high pressure and super acids (1) and for the studies of (porphinato)iron-catalyzed isobutane oxidation using dioxygen as the oxidant (2).

The most commonly-used method for monitoring kinetics of synthetic organic reactions are sampling from the reactor followed by chemical analysis using, for example, GC, LC, or NMR. While analysis methods are valuable in providing concentration and conversion information, and in particular, chemical identities of components in the reactor, it has a three-fold deficiency insofar as a determination of kinetics is concerned. First, only integral properties (such as concentration) are measured, *i.e.*, the method measures an integration of the rate from the beginning of the reaction to the point of measurement. To obtain the rate of reaction, one has to differentiate these integral results with respect to time. Second, in reality, only a limited number of samples may be taken per reaction. This factor, when coupled with the first, makes it usually inaccurate to determine instantaneous rate by differentiating the concentration data with respect to time. Finally, chemical analysis is not an *in situ* method. Samples must be taken from the reacting environment for work up and analysis, giving the possibility that the sample may undergo chemical changes which could distort the true profile of components in the reactor as a result.

It is not uncommon to find in the literature reports which limit the kinetic studies to measurements of initial rates for catalytic reactions carried out in batch mode. The initial rates were often obtained by calculating the slope in a plot of the integral properties such as concentration vs. time over a "linear" portion of the plot at the beginning stage of the reaction. In addition to the accuracy issue discussed above, this approach is problematic considering the non-steady state nature of reaction carried out in batch mode. Under the batch mode conditions, the instantaneous rate at the initial stage of the reaction can either increase, decrease, or remain unchanged with increasing time. Consequently, developing a kinetic model and the associated kinetic parameters for a proposed reaction mechanism may be difficult and inaccurate using the initial-rate approach. An example of the initial rate rising over time is the kinetics exhibited by heterogeneously-catalyzed enantioselective hydrogenation of ethyl pyruvate over cinchonidine-modified Pt/Al_2O_3 in which case the rate increased by 30% from the value at the start of the reaction before decreasing with increasing time (3).

Hence, it is desirable to have kinetic probes which can provide rate data accurately over the entire course of the reaction in order to obtain a full picture of the reaction kinetics which should be the basis for development and validity testing of the kinetic models. In this report, three useful methodologies for meeting such a goal are described. These *in situ* kinetic probes provide kinetic data and/or compositional information in real time and in a non-invasive manner. An example is used to illustrate their usefulness in facilitating pathway analysis of catalytic hydrogenation reactions.

2. Three *In situ* Kinetic Probes

Reaction Calorimetry

Calorimetry is hardly a new technique. One often thinks of thermodynamics when thinking of calorimetry since it has long been used to determine thermodynamic properties of materials and chemical/physical processes. Microcalorimetry has been employed to measure under isothermal conditions kinetics of gas-solid interactions, such as the measurements of kinetics of CO chemisorption on nickel single crystal surfaces (4) and kinetics of chemisorption and reactions over supported transition metal catalysts (5, 6). However, only with the advent of *reaction calorimetry*, which carries out an instantaneous enthalpy balance over the course of a reaction, did calorimetry begin to be employed for kinetic studies of pharmaceutical, fine chemical, and biological reactions carried out in the liquid phase (7-10).

The reaction calorimeter employed in this study (Mettler's RC1, MP10 1 L glass reactor), is a highly automated bench scale reactor which allows the reaction to be carried out in a controlled manner. While maintaining a precise control of the temperature of the contents in the reactor, the reaction calorimeter measures the rate of heat flow into or out of reactor during a reaction, information normally lost in an ordinary reactor vessel. In contrast to the sampling technique which measures the integral properties, reaction calorimetry provides directly the differential property, namely, heat flow rate. And the measurements are conducted in an *in situ*, non-invasive, on-line, and continuous fashion.

The heat flow rate measured under isothermal conditions is directly related to a summation of the rates of each individual step, r_i, as scaled by heat of reaction of the corresponding step, ΔH_i, *i.e.,*

$$q_r = V_r \sum_i \Delta H_i r_i, \tag{1}$$

where V_r is the volume of the batch. Hence, the reaction calorimeter provides a mean to map out the overall kinetics over the entire course of the reaction without disturbing the reacting system.

Hydrogen Uptake

In the case of catalytic hydrogenation, hydrogen uptake provides a natural measure of the overall rate of hydrogenation since its rate is related to the rates of each individual step by

$$R_{H_2} = V_r \sum_i \Delta N_i r_i, \tag{2}$$

where ΔN_i is the H_2 stoichiometry for the *ith* hydrogenation step. Similar to reaction calorimetry, the uptake rate measurement also serves as an *in situ* probe that measures kinetics of hydrogenation under reaction conditions, although from a different perspective. The overall kinetics provided by the H_2 uptake have embedded in them H_2 stoichiometry associated with each hydrogenation step, whereas those by reaction calorimetry have embedded in them stepwise heats of reaction. An uncovering of the stepwise H_2 stoichiometry and the stepwise heats of reaction can be helpful to a determination of the reaction pathway.

Different from reaction calorimetry, however, is the much lower cost associated with implementation of the uptake measurements. Fig. 1 shows a schematic of the experimental set up for measuring the rate of hydrogen uptake while keeping the pressure in the reactor constant. The uptake rate may be obtained either directly by measuring the uptake flow rate with a mass flow meter, or indirectly by measuring profile of pressure drop in the H_2 reservoir, followed by differentiating the uptake data with respect to time. The pressure drop is an integral property. But it may be measured with such a high frequency using a fast-response pressure transducer that accurate rate data may be obtained.

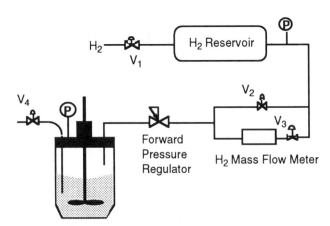

Fig. 1. A schematic of the experimental setup for measuring rate of H_2 uptake during catalytic hydrogenation reactions. The forward regulator keeps the pressure in the reactor constant while the rate of H_2 uptake is measured directly by the mass flow meter or indirectly by measuring pressure drop in the reservoir.

Time-Resolved in situ Infrared Spectroscopy

The infrared spectrometer used in this study is a commercial FTIR unit (ReactIR 1000 from ASI Applied System) operated in ATR (Attenuated Total Reflection) mode. Its DiComp[TM] probe that contains a chemical resistant diamond film as the ATR element is immersed in the reaction media during the reaction under the reaction conditions, allowing the infrared spectrum of the liquid in contact with the diamond film measured for compositional analysis. The infrared spectrum from 600 to 4000 cm^{-1} may be taken and recorded continuously at a time interval down to as short as a few seconds throughout the entire course of the reaction, permitting real-time monitoring and analysis of the reactions under study.

While the kinetics measured by reaction calorimetry and by the hydrogen uptake measurements in the case of hydrogenation reactions are *overall* reaction kinetics that contain contributions from all the steps occurring in the reactor, as a compositional probe, the time-resolved *in situ* infrared spectroscopy yield component-specific and step-specific kinetic information. In contrast to the chemical sampling method, infrared spectrum of the reacting system may be collected at a much faster pace. Consequently, rate data may be derived with good accuracy by differentiating the intensity data with respect to time. The *in situ* FTIR probe is particularly valuable for monitoring and identification of reactive intermediates that are short lived and unstable away from the reacting environments. This makes possible accurate measurements of the reaction kinetics associated with such reactive intermediates, which is difficult, if not impossible, to achieve with the regular sampling technique.

3. An Example of Pathway Analysis Facilitated by the *in situ* Kinetic Probes: *Catalytic Hydrogenation of 1-(4-Nitrobenzyl)-1,2,4-Triazole*

1-(4-Aminobenzyl)-1,2,4-triazole (the aniline) is an intermediate for the synthesis of Merck's drug substance, *Maxalt®*, a selective agonist for the 5-HT$_{1D}$ receptor used for treatment of migraine. It is synthesized via selective hydrogenation of the nitro group in 1-(4-nitrobenzyl)-1,2,4-triazole (4NBT) over Pd/C, *i.e.*,

1-(4-Nitrobenzyl)-1,2,4-Triazole   1-(4-Aminobenzyl)-1,2,4-Triazole

3.1. Kinetics

The catalytic hydrogenation of 4NBT was carried out isothermally at 25 °C and at 40 psig in Mettler's reaction calorimeter as its kinetics were monitored by heat flow, rate of hydrogen uptake, infrared spectroscopy, and analysis of samples from the reactor. When the reaction is complete, each mole of 4NBT is quantitatively converted to the aniline by reactively taking up three molar equivalents of hydrogen. The profiles of rate of hydrogenation during the entire course of the reaction as measured by heat flow and by rate of hydrogen uptake are shown in Fig. 2 which exhibits several distinct features. Firstly, there are two distinct rate regimes with an abrupt transition between them. Secondly, the first regime appears to follow zero-order kinetics, *i.e.*, the rate is invariant with decreasing 4NBT concentration. Furthermore, Fig. 2 shows that the kinetic pictures depicted by both *in situ* probes agree well, particularly in terms of the zero-order kinetics in the first regime and the sharp transition between the two rate regimes.

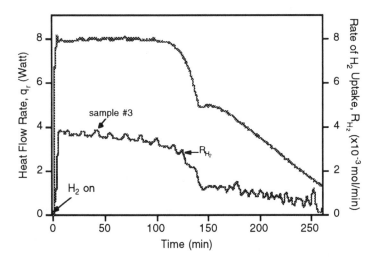

Fig. 2. Rate profiles for isothermal (25 °C) hydrogenation of 1-(4-nitrobenzyl)-1,2,4-triazole to 1-(4-aminobenzyl)-1,2,4-triazole over Pd/C, as measured by heat flow and rate of hydrogen uptake. Mettler's RC1 (1L). Agitation speed: 1200 rpm. Solvent: methanol (400 ml) and aq. NH_3 (28 w/w%, 24 ml).

In terms of the quality of the kinetic data, the hydrogen uptake rate data are relatively noisy, particularly at the later stage of the reaction when the uptake rate is low. In comparison, the heat flow rate is much less noisy. In addition to the inherent differences in the hardware generating these two types of rate data, a factor that contributes to this difference in noise level is the way with which the data were obtained. The heat flow rate data is a direct measurable in reaction calorimetry, whereas the rate of hydrogen uptake in this case was obtained by measuring the uptake followed by differentiating the integral uptake data with respect to time. The hydrogen uptake rate data is sensitive to perturbation in the reactor system. For example, spikes resulted from sampling from the reactor during the reaction due to unavoidable simultaneous removal of a small amount of gaseous hydrogen from the reactor (Fig. 2). On the other hand, the heat flow is far less sensitive to sampling. Notice the absence of perturbation in the heat flow at the points of sampling (sample #3) as indicated by the spikes in the rate of hydrogen uptake curve.

3.2. Reaction Pathways

The heat flow and hydrogen uptake data provide some quick clues to the reaction pathways. The presence of two very different kinetic regimes in Fig. 2 immediately suggests that two different chemical processes are occurring at these two regimes. The sharp transition renders it unlikely that the catalyst is suddenly poisoned at the transition point. The fact that the conversion profiles calculated from the heat flow and from hydrogen uptake do not overlap with each other (Fig. 3) is suggestive of presence of intermediate(s). For hydrogenation reactions requiring more than one molar equivalent of hydrogen for the hydrogenation of each mole of the substrate, the two conversion curves would normally overlap if no hydrogenation intermediate(s) exists in significant amount in the liquid phase except for some special circumstances.

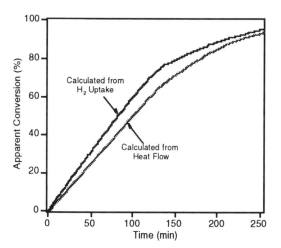

Fig. 3. A comparison of the conversion profiles calculated from the heat flow and from the rate of hydrogen uptake data in Fig 2.

The time-resolved infrared spectra of the liquid phase confirmed the presence of an intermediate in the solution and graphically revealed the sequential nature of the reaction. Fig. 4 shows that upon switching on hydrogen, the concentration of

4NBT as measured by the intensities of the vibrational bands at 1352 cm^{-1} and 1528 cm^{-1} corresponding to the symmetric and asymmetric stretching modes of the ONO group in 4NBT (11) decreases with time. This is accompanied by a corresponding increase in the concentration of an intermediate characterized by a band at 1515 cm^{-1}. The intermediate is further hydrogenated to the aniline characterized by a band at 1520 cm^{-1}. This pathway picture was further confirmed by results from HPLC analysis of the samples taken from the reactor during the course of the hydrogenation displayed in Fig. 5.

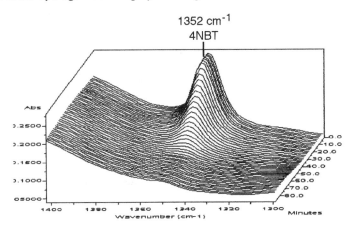

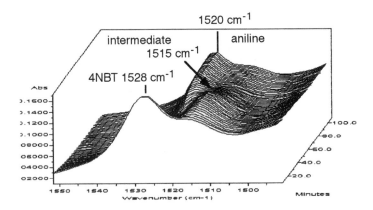

Fig. 4. Compositional evolution of the liquid phase during the 4NBT hydrogenation as revealed by the *in situ* FTIR probe.

A combination of the hydrogen uptake data and the HPLC data can shed light on the possible nature of the intermediate which could either be the nitroso (Ph-NO) or the hydroxylamine (Ph-NHOH) species (12). An analysis of the hydrogen stoichiometry from 4NBT to the intermediate is particularly useful since the hydrogen stoichiometry would be one for the nitroso and two for the hydroxylamine. The results displayed in Fig. 6 show that the hydrogen stoichiometry in the first rate regime in Fig. 2 was approximately two, suggesting that the intermediate is the hydroxylamine, which was verified separately by chemical analysis.

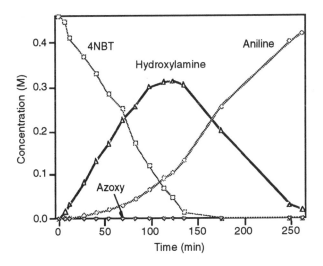

Fig. 5. Profile of components in the reactor corresponding to the heat flow curve in Fig. 2. Determined by sampling from reactor followed by HPLC analysis.

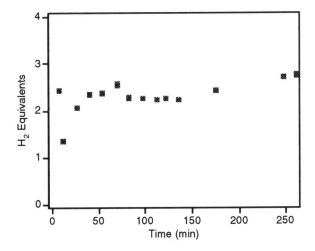

Fig. 6. Hydrogen molar equivalents consumed per mol 4NBT reacted. Derived from hydrogen uptake data in Fig. 2 and the 4NBT concentration data in Fig. 5.

The next most abundant intermediate detectable by HPLC analysis was the azoxy, which may be a condensation product from the hydroxylamine and the nitroso (12). The azoxy, however, accounted for only 0.1% of the intermediates, which is consistent with the fact that the nitroso was not a detectable intermediate in the solution phase, due presumably to the high reactivity of the nitroso to further hydrogenation on the catalytic surface to the hydroxylamine.

Hence, the major hydrogenation pathway may be depicted as follows. 4NBT is hydrogenated on the Pd catalyst following a two-step pathway with the hydroxylamine as the intermediate, *i.e.*,

$$Ph-NO_2 + 2H_2 \longrightarrow Ph-NHOH + H_2O$$

$$Ph-NHOH + H_2 \longrightarrow Ph-NH_2 + H_2O$$

The hydrogenation of 4NBT to the hydroxylamine is mainly responsible for the first kinetic regime in Fig. 2. It proceeds with zero-order kinetics. The subsequent hydrogenation of the hydroxylamine to the aniline is mainly responsible for the second rate regime in Fig. 2.

3.3. Stepwise Thermodynamics

Now that the reaction pathway is delineated, detailed thermodynamic data may be deduced from information provided by these *in situ* probes. Firstly, the overall thermodynamic data can be conveniently extracted from the heat flow data. Integration of the heat flow curve in Fig. 3 with respect to time yields an overall heat of hydrogenation (from 4NBT to the aniline) of -127±4 kcal/mol, in good agreement with the values associated with hydrogenation of nitro benzene compounds to aniline. Secondly, the stepwise heats of hydrogenation of this two-step consecutive hydrogenation reaction may also be determined from a combination of the heat flow data and the hydrogen uptake data.

The instantaneous heat of hydrogenation per mole of hydrogen reacted, ΔH_{H_2}, defined as (using Eqs. 1 and 2)

$$\Delta H_{H_2} = \frac{q_r}{R_{H_2}} = \frac{\sum_i \Delta H_i r_i}{\sum_i \Delta N_i r_i}, \tag{3}$$

can be obtained by dividing a heat flow curve by the corresponding reactive hydrogen uptake curve. Due to the sequential nature, the heat flow at the initial stage of the reaction may be attributed mainly to the hydrogenation of 4NBT to the hydroxylamine, whereas at the final stage mainly to the hydrogenation of the hydroxylamine to the aniline. At these two extremes, Eq. 3 becomes simplified to yield thermodynamic information for each individual step involved in the hydrogenation reaction.

At the early stage of the reaction, since $r_1 \gg r_2$, Eq. 3 becomes

$$\Delta H_{H_2} \approx \frac{\Delta H_1}{\Delta N_1}.$$

At the final stage of the reaction, since $r_2 \gg r_1$, Eq. 3 becomes

$$\Delta H_{H_2} \approx \frac{\Delta H_2}{\Delta N_2}.$$

The instantaneous heat of hydrogenation per mol of hydrogen reacted obtained from Fig. 2 according to Eq. 3 is displayed in Fig. 7. Corresponding to the two distinct kinetic regimes in Fig. 2, there are also two distinct regimes in the heat of hydrogenation, with the transition occurring at the same time.

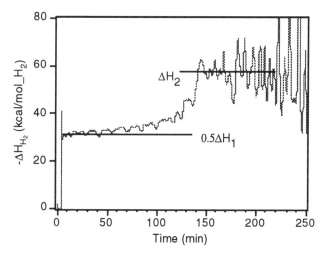

Fig. 7. Instantaneous heat of hydrogenation per mole of hydrogen reacted as a function of time. Derived using Eq. 3 by taking the ratio of the heat flow curve to the reactive hydrogen uptake rate curve in Fig. 2.

At the initial stage of the first regime, ΔH_{H_2} was ~-32.5 kcal/mol_H_2, whereas in the second regime it increased to ~-58 kcal/mol_H_2. Since the hydrogen stoichiometry is two ($\Delta N_1 = 2$) for the first step, the hydrogenation of 4NBT to the hydroxylamine, the heat of hydrogenation shown in Fig. 7 at the beginning of the first regime (-32.5 kcal/mol) corresponds to half of ΔH_1. In other words, $\Delta H_1 = -65$

kcal/mol. Similarly, because the hydrogen stoichiometry is one ($\Delta N_2 = 1$) for the second step from the hydroxylamine to the aniline, the heat of hydrogenation shown in Fig. 7 in the second regime (-58 kcal/mol) equals ΔH_2. Heat of hydrogenation from 4NBT to the aniline is a sum of these two stepwise heats, *i.e.*, -123 kcal/mol. The thermodynamic results are summarized in Table I.

Table I. Stepwise heats of hydrogenation for hydrogenation of nitro group in 1-(4-nitrobenzyl)-1,2,4-triazole

Hydrogenation Reaction	ΔH, (kcal/mol)
Ph−NO$_2$ + 2H$_2$ ⟶ Ph−NHOH + H$_2$O	-65
Ph−NHOH + H$_2$ ⟶ Ph−NH$_2$ + H$_2$O	-58

As a hydrogenation intermediate, the hydroxylamine is rather unstable. It undergoes facile oxidation reaction in the presence of air to dimerize into the azoxy compound. As a result, a direct determination of its heat of formation is difficult. The heat of formation of the unstable hydroxylamine, however, is implicit in the stepwise heats of hydrogenation determined above. The results on the calculated heats of formation are listed in Table II. Note that the analysis of the stepwise thermodynamics described here is possible even without knowledge of the exact concentration of the starting 4NBT.

Table II. Standard heat of formation, ΔH_f (kcal/mol, relative to that of 4NBT), of species associated in hydrogenation of 1-(4-nitrobenzyl)-1,2,4-triazole.

1-(4-nitrobenzyl)-1,2,4-triazole (4NBT)	the hydroxylamine Intermediate	1-(4-aminobenzyl)-1,2,4-triazole
0	-0.7	9.7

4. Conclusions

The advantages of three *in situ* kinetic probes for studying catalytic reactions for organic synthesis have been described and illustrated with two case studies of heterogeneously-catalyzed hydrogenation reactions. Each of these probes is capable of monitoring the kinetics under reaction conditions, over the entire course of the reaction, and in a non-invasive manner. The accuracy in the kinetic measurements rendered and the details in the kinetic profile revealed by these probes make it possible to draw quick conclusions about the reaction pathway. For instance, the number of distinct kinetic regimes is suggestive of the number of different reactions occurring in the reactor. The accuracy and the details may also be used in developing kinetic models, and particularly, in testing the validity of the kinetic models. And each of these probes measures the reaction kinetics from a different perspective. When used simultaneously, these probes generate information that describes different aspects of the same reaction. An analysis of the information thus obtained helps unravel the details of the reaction and facilitates an understanding of the reaction pathways. The accuracy and completeness of the kinetic data provided by the simultaneous application of these probes greatly enhance one's ability in learning more about the reaction under study with each experiment, significantly speed up studies of the reaction and successful development of catalytic processes for organic synthesis.

REFERENCES

1. P.J.F. de Rege, J.A. Gladysz, and I.T. Horvath, *Science* **276**, 777 (1997).

2. K. T. Moore, I.T. Horvath, M.J. Therien, *J. Am. Chem. Soc.* **119**, 1791 (1997).

3. J. Wang, Y.-K. Sun, C. LeBlond, and D. G. Blackmond, *J. Catal.* **161**, 752 (1996).

4. J.T. Stuckless, N. Al-Sarraf, C. Wartnaby, and D.A. King, *J. Chem. Phys.* **99**, 2202 (1993).

5. B.E. Handy, S.B. Sharma, B.E. Spiewak, J.A. Dumesic, *Meas. Sci. Technol.* **4**, 1350 (1993).

6. B.E. Spiewak, P. Levin, R.D. Cortright, and J.A. Dumesic, *J. Phys. Chem.* **100**, 17260(1996).

7. R. Landau, U.S. Singh, F. Gortsema, Y.-K. Sun, S.C. Gomoka, T. Lam, M. Futran, and D.G. Blackmond, *J. Catal.* **157**, 201 (1995).

8. C. LeBlond, J. Wang, R. Larsen, C.J. Orella, A. Forman, R.N. Landau, J. Laquidara, J.R. Sowa, Jr., D.G. Blackmond, and Y.-K. Sun, *Thermochim. Acta* **289**, 189 (1996).

9. R.N. Landau, D.G. Blackmond, and H.H. Tung, *Ind. Eng. Chem. Res.* **33**, 814 (1994).

10. B.A. Williams, and E.J. Toone, *J. Org. Chem.* **58**, 3507, (1993).

11. N.B. Colthup, L.H. Daily, and S.E. Wiberley, "*Introduction to Infrared and Roman Spectroscopy*," 3rd Ed. Academic Press, (1990).

12. (a) F. Haber, *Z. Electrochem.*, **4**, 506 (1898). (b) F. Haber and C. Schmidt, *Z. Physik. Chem.*, **32**, 171 (1900).

Synthesis of Methylpyrazine from Hydroxypropyldiaminoethane. The Mechanistic Study and Catalyst Selection

G.V. Isagulyants and K.M. Gitis.

N.D.Zelinsky Institute of Organic Chemistry (Russian Academy of Sciences).
47, Leninsky prosp., Moscow 117913, Russia.

ABSTRACTS

Synthesis of methylpyrazine from hydroxypropyldiaminoethane over Pt-containing catalysts ($Pt-Al_2O_3$, $Pt-In_2O_3-Al_2O_3$, $Pt-ZnO-Al_2O_3$) was investigated. The catalyst with suppressed acidic function occurred to be most selective. A new mechanistic scheme for the reaction was proposed and substantiated by means of tracer and kinetic studies. The scheme includes dehydrogenation of the initial aminoalcohol and intermediate formation of methyltetrahydropyrazine. The scheme was helpful in the catalyst design. Copper containing catalysts showed high selectivity and stability in methylpyrazine synthesis. Some of them tested in prolonged runs at $350°C$ gave methylpyrazine with yields up to 83%.

INTRODUCTION

Pyrazine derivatives are widely used as drugs, flavouring additives *etc.* (1). Ammoxidation of methylpyrazine allows one to produce pyrazinamide which is known as an antitubercular drug. Several heterogeneous catalytic methods have been reported for the preparation of pyrazine and alkylpyrazines. Among them, dehydrogenation of piperazine (2,3), dehydrocondensation of diamines with diols (4-6) or olefin oxides (7,8) have found application. A number of patents describe the production of pyrazines from diamines and glycoles. Copper-chromia (4,6) and zinc-oxide (5) catalysts with different additives (including $ZnO-ZnCr_2O_4$) were proposed for this process. According to the published data, the yields of pyrazines varies in the range of 55-80%. In a mechanistic study of this process, intermediate formation of a piperazine ring followed by its dehydrogenation was discussed (9):

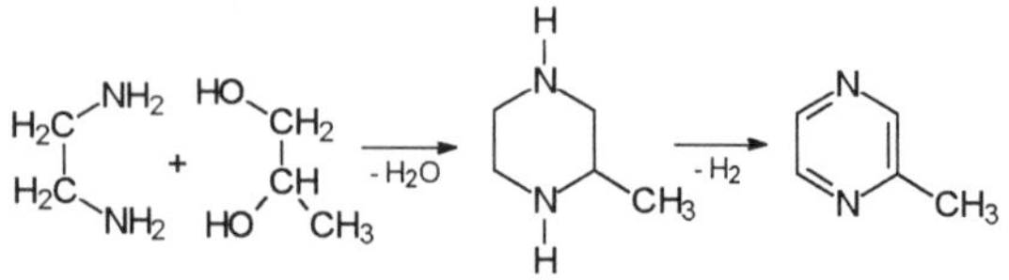

In synthesis of pyrazines from diamines and olefin oxides, the corresponding N-(hydroxyalkyl)diamine should be preliminary obtained:

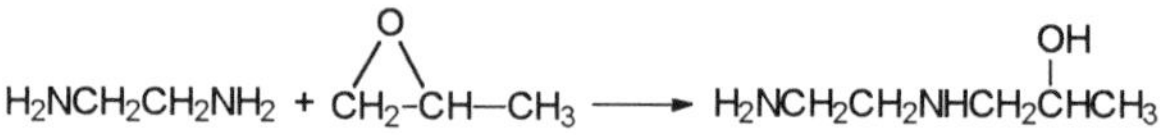

Platinum-alumina (7), chromia-alumina (7) and copper-chromia (7,8) catalysts were proposed for the subsequent transformation of N-(hydroxyalkyl)diamines to the corresponding pyrazines. The yields of the latter obtained in these transformations as a rule are higher than those obtained in the reaction of diamines and glycols and vary in the range of 60-90%. In spite of high efficiency of this process, it has been studied rather scantily and the available information about it is limited to patent data. Nevertheless, the scheme with the intermediate formation of a piperazine ring in the transformation of N-(hydroxyalkyl)diamine to pyrazines seemed to be probable (10):

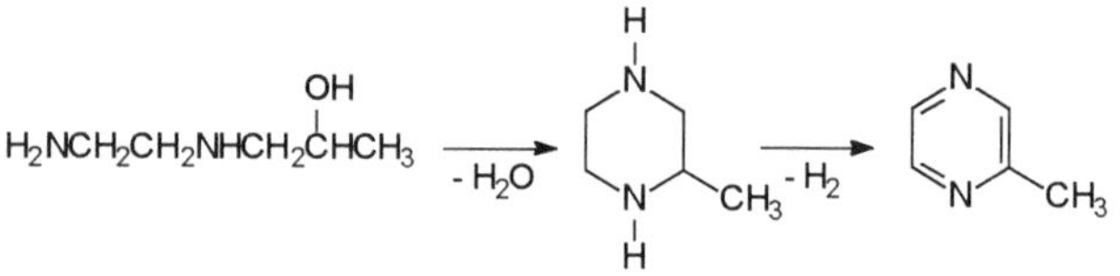

RESULTS AND DISCUSSION

We initiated the study on the synthesis of methylpyrazine from N-(β-hydroxypropyl)diaminoethane (HOPDAE) basing on the scheme mentioned above. Since the cyclization (dehydration) requires acidic sites and dehydrogenation of the cycle requires the redox ones, we believed a dual function catalyst to be suitable for the process.

It has been shown in our previous investigation that modified platinum-alumina catalysts permit pyrazine to be produced with high yields by dehydrogenation of piperazine (3). It seemed to be reasonable to apply catalysts of this type for synthesis of methylpyrazine from HOPDAE obtained by interaction of propylene oxide with diaminoethane. Platinum catalysts supported on carriers of various acidity (Pt-Al$_2$O$_3$, Pt-In$_2$O$_3$-Al$_2$O$_3$, Pt-ZnO-Al$_2$O$_3$) were prepared and tested. Data on their composition, acidity and catalytic properties are presented in Table 1.

Transformation of HOPDAE was carried out under the conditions which were close to those used in piperazine dehydrogenation. One can see in Table 1

TABLE 1
Transformation of HOPDAE over Platinum-containing catalysts (350°C, space velocity of HOPDAE 0.66 h^{-1}, mole ratio hydrogen / HOPDAE = 1.3/1)

Catalysts	Acidity		ESR[a], spin/g x10^{-18}	Liquid product yield, % mass	Composition of liquid product, % mass.				Methyl pyrazine yield, % mole
	Amine titration, mmole/g				Pyrazine	Methyl pyrazine	Ethyl-and dimethyl pyrazines	Other[b] products	
	H$_0$<4.8	H$_0$<-3.0							
Pt(0.6%) -Al$_2$O$_3$	0.3[c](11) 0.3[c](12)	0.2[c](11) 0.3	1.8(13)	64.8	6.1	26.4	6.2	61.3	22.0
Pt(0.6%) -Al$_2$O$_3$-In$_2$O$_3$			0.8(13)	87.6	3.5	44.0	5.4	47.1	43.0
Pt(0.25%) -ZnO-Al$_2$O$_3$[d]	0.1[c](12)	0.0[c](12)		86.1	0.4	73.5	7.2	18.5	65.3
Pt(0.5%) -ZnO-Al$_2$O$_3$[d]	0.1[c](12)	0.0[c](12)		88.5	0.6	86.6	3.4	7.0	78.1

a) Concentration of ion-radicals of adsorbed antracene;
b) Polyalkylpyrazines, products of destruction and condensation
c) Data for the carrier,
d) The ratio ZnO / Al$_2$O$_3$ = 9/1 mole.

that the methylpyrazine yield varied from 22 to 78%. Conversion of HOPDAE in these experiments was 100%. The side-products included ethyl-, dimethyl- and polyalkylpyrazines. In the case of Pt-ZnO-Al$_2$O$_3$ catalyst, methylpiperazine and methyltetrahydropyrazine (MTHP) were detected as well. Alkylpyrazines were also formed in proportional amounts in dehydrogenation of piperazine over corresponding catalysts. Their formation was explained (3) by alkylation of pyrazine with the products formed by hydrogenolysis of the piperazine ring. Deposition of tar products on the catalyst surface was observed as well, and their amount for Pt-ZnO-Al$_2$O$_3$ catalyst was 2-3% calculated on HOPDAE used. One can see in the Table 1 that the yields of methylpyrazine increase essentially as the catalyst acidity diminishes. Thus the Pt-ZnO-Al$_2$O$_3$ catalyst with the lowest acidity was found to be the most selective in the formation of methylpyrazine.

The further investigation of this catalyst in prolonged run revealed that methylpyrazine yields decreased markedly with time on stream (Table 2, Run1).

Table 2. Effect of feed dilution with water on the transformation of HOPDAE over Pt-ZnO-Al$_2$O$_3$ catalyst (350 C, space velocity-0.66 h^{-1}).

Time on stream, h	Liquid products yield, %	Content in liquid products, % mass				Methyl-pyrazine yield, % mole
		Methyl-pyrazine	Dimethyl-pyrazines	MTHP	Other alkyl-pyra-zines	
Run 1. Feed mole ratio HOPDAE/ H$_2$ = 1/1.3						
0-1	88.5	86.6	3.4	1.5	3.8	78.1
1-2	85.5	86.3	2.9	1.7	5.0	76.1
2-3	84.9	85.0	3.9	2.2	5.3	74.3
4-5	87.1	81.0	6.5	4.6	5.4	73.0
5-6	87.7	77.2	6.7	4.6	8.0	70.2
6-7	87.4	72.6	7.9	7.6	7.6	65.7
Run 2. Feed mole ratio HOPDAE /H$_2$/H$_2$O = 1/1.8/1.6						
0-1	90.2	91.9	3.0	0.3	2.1	83.7
1-2	93.8	92.1	2.3	0.7	1.8	89.1
2-3	93.0	92.2	2.9	0.8	1.0	87.9
4-5	92.5	93.8	2.5	0.7	0.8	88.7
5-6	94.9	93.5	2.0	0.8	1.3	92.0
6-7	95.2	89.9	3.4	1.5	1.9	88.8

Simultaneously, the yield of side products increased considerably. This can be explained by the diminishing dehydrogenative function of the catalyst caused by tar deposition on the surface. To decrease the tar formation, water was used as a diluent along with hydrogen. One can see (Table 2, Run 2) that simultaneous use of hydrogen and water enables one to increase and to stabilize to some extent, methylpyrazine yield and to decrease the yield of side products. Thus, the amounts of MTHP, dimethyl- and polyalkylpyrazines at the end of Run 2 were essentially the same as those at the beginning of Run 1.

The effects of feed dilution with water, reaction temperature and space velocity (contact time) were studied to optimise the reaction conditions. Methylpyrazine yield was found to rise with the increase of water/HOPDAE ratio in the feed (Fig. 1).

It should be mentioned, that this effect becomes especially remarkable after the catalyst was on stream for 4-8 hours. The effect of the reaction temperature is depicted in Fig.2. The conversion of HOPDAE at 220° was 70%, though methylpyrazine was formed in minor amounts. The main product of the reaction was MTHP (up to 40%). With the increase of temperature methylpyrazine yield increased. The dependence of MTHP yield had a pronounced maximum, while methylpiperazine formation remained low and depended on temperature insignificantly.

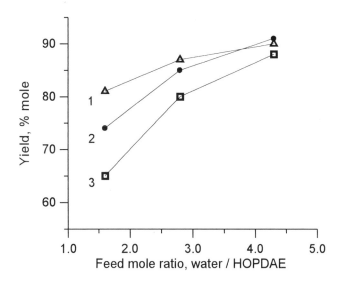

Figure 1. Effect of HOPDAE dilution with water on methylpyrazine yield (350°C, feed space velocity 0.6 h^{-1}). 1, 2, 3 - the second, forth and eighth hour of catalyst operation respectively.

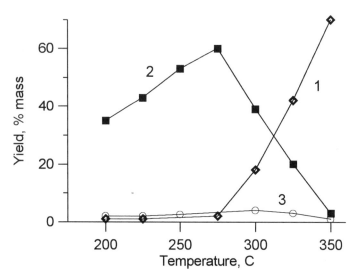

Figure 2. The effect of reaction temperature on the yield of reaction products in HOPDAE transformation over Pt-ZnO-Al$_2$O$_3$ catalyst (feed space velocity 0.66 h^{-1}, feed mole ratio HOPDAE /H$_2$/water 1/1,6/1,8).
1- Methylpyrazine;   2- Methyltetrahydropyrazine; 3- Methylpiperazine.

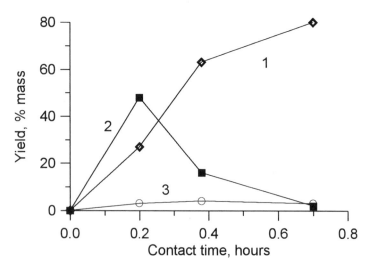

Figure 3. The effect of contact time on the yield of reaction products in HOPDAE transformation over Pt-ZnO-Al$_2$O$_3$ catalyst (350°C, feed mole ratio HOPDAE /H$_2$/water 1/1,6/1,8 mole.).
1- Methylpyrazine; 2- Methyltetrahydropyrazine; 3- Methylpiperazine.

The study of HOPDAE transformation at various contact times showed the yield of the reaction products to change in a similar manner. (Fig 3). One can see that the yield of methylpyrazine increases gradually, that of MTHP changes dramatically and the yield of methylpiperazine is low and changes very little with increasing the contact time.

Among the products accompanying methylpyrazine formation, methylpiperazine and MTHP are of special interest from standpoint of the reaction mechanism. As mentioned above (9), these compounds were supposed to be intermediates in methylpyrazine formation, while methylpiperazine was thought to be a precursor of MTHP. The presented data (Fig. 2, 3) allow one to consider that the intermediate in the reaction is MTHP rather than methylpiperazine.

To verify the reaction scheme, a tracer study was performed using HOPDAE labelled with ^{14}C. One can see in Fig. 4, in the reaction of a mixture of labeled HOPDAE with unlabeled MTHP, the mole radioactivity (MRA) of the latter is higher than that of methylpyrazine in the whole interval of contact times used. According to the kinetic isotope method (14), this result indicates that MTHP formation precedes the formation of methylpyrazine. In the reaction of the mixture of labeled HOPDAE with unlabeled methylpiperazine, MRA of methylpyrazine is close to 1.0 and exceeded MRA of methylpiperazine considerably (Fig. 5). This means that radioactive methylpyrazine is formed from labeled HOPDAE escaping intermediate formation of methylpiperazine. Thus, methylpiperazine does not precede methylpyrazine formation from HOPDAE, but it is rather a side product of the reaction. MRA of methylpiperazine increases slightly with the increase of contact time that seems to be a result of gradual hydrogenation of labelled MTHP. MRA of MTHP remains higher than MRA of methylpyrazine confirming the conclusion that MTHP is the precursor of methylpyrazine in its formation from HOPDAE. At longer contact time, MRA of methylpyrazine decreases to some extent owing to contribution of dehydrogenation of methylpiperazine which the MRA is considerably lower.

To explain the obtained data, we have used the knowledge concerning the mechanism of amination of alcohols and of cycloamination of aminoalcohols (15). Two types of the reaction mechanism are to be considered: the acidic- and redox-type. According to the first one, the reaction of substitution of an OH-group for the NH_2-group with water formation proceeds over catalysts of acid-type (Al_2O_3, Al_2O_3-SiO_2, TiO_2, phosphates). According to the second mechanism alcohol group is dehydrogenated over a metal catalyst (Ni, Cu, Fe) into a carbonyl group. The aldehyde or ketone formed is then subjected to ammonolysis and reduction by hydrogen formed in the dehydrogenation or supplied from outside.

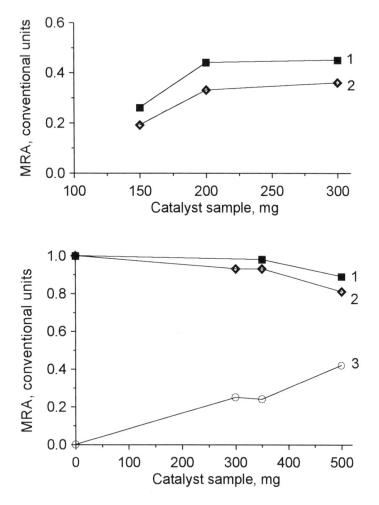

Figures 4. Effect of the mass of the catalyst (contact time) on the mole radioactivity (MRA) of the reaction products in transformation of a mixture of labeled HOPDAE with 5.7% unlabeled MTHP over Pt-ZnO-Al_2O_3 catalyst ($350°C$, dilution HOPDAE /H_2=1/1,3 mole.)

Figure 5. The same in transformation of a mixture of labeled HOPDAE with unlabeled methylpiperazine (9%).

1- MTHP, 2 -Methylpyrazine, 3 -Methylpiperazine

It stands to reason, for aminoalcohol, the processes can proceed intramolecularly with the formation of cyclic amine (the acid-type mechanism) or imine (the redox-type mechanism). Taking into consideration the results obtained in this work, one can obviously conclude that the considered reaction proceeds according to the redox-type mechanism. Consequently, HOPDAE is assumed to be dehydrogenated to the corresponding aminoketone followed by formation of MTHP and then dehydrogenation of the latter to methylpyrazine:

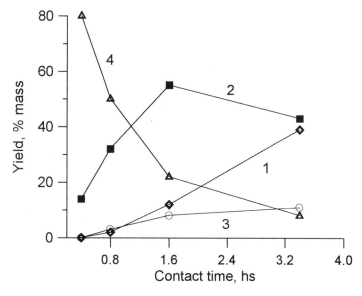

Since the scheme includes dehydrogenation of CH_2OH-group, one can expect high efficiency of catalysts which are used in alcohol dehydrogenation such as ZnO, copper, copper-chromia *etc*. Really, even when metal copper was tested in HOPDAE conversion, methylpyrazine was obtained with the yield up to 40% and the process seems to follow the scheme of consecutive reaction. One can see in Fig.6 that the curve of MTHP yield dependent on contact time has a maximum and that of methylpyrazine has an induction period.

Figure 6. The effect of contact time on the yield of reaction products in HOPDAE transformation over metallic copper catalyst (320°C, feed mole ratio HOPDAE /argon 1/1,3).
1- Methylpyrazine; 2- Methyltetrahydropyrazine; 3- Methylpiperazine, 4 - HOPDAE.

By the way, in the reaction mixture obtained over the copper catalyst, 1-imino-N-cyanomethylpropanone-2 ($CH_3COCH=NCH_2CN$) was detected. Obviously, the latter seems to be formed in dehydrogenation of CH_2OH- and CH_2NH_2-groups of the initial aminoalcohol. In connection with the considered matter, it should be interesting to mention that Cu-containing catalysts were found to be very efficient in cycloamination of 5-aminopentanol (16). The authors pointed out that in the presence of H_2, piperidine was formed, but in the absence of H_2, tetrahydropyridine was obtained additionally. The authors explained formation of the tetrahydropyridine by partial dehydrogenation of piperidine. We consider that under the experimental conditions (210°C), intermediate dehydrogenation of aminoalcohol and formation of tetrahydropyridine are possible. Hydrogenation of the latter results in formation of piperidine. All the more that terminating the dehydrogenation of piperidine after tetrahydropiridine formation seems to be unlikely.

In the light of the data reported above, it becomes clear why platinum and other dehydrogenation catalysts can be used in the process. Therefore, some Cu-containing catalysts were tested and compared with the elaborated Pt-ZnO-Al$_2$O$_3$ catalyst. The results on the testing are presented in Table 3. One can see that all the catalysts are efficient in synthesis of methylpyrazine though they are distinguished by their stability. The Cu-ZnO-Al$_2$O$_3$ catalyst was found to be somewhat more stable as compared to the Pt-ZnO-Al$_2$O$_3$ one, and provides methylpyrazine yield of 83%. Both catalysts were tested in a pilot unit with the reactor having the capacity of 3 dm^3. Methylpyrazine was produced with the yield of 70-80%.

EXPERIMENTAL

The reaction was performed in a flow system at 200-350°C. HOPDAE or its water solution was fed at space velocity of 0,6-8,5 h^{-1} with H_2 dilution. After each run the catalyst was reactivated in the air stream at 550°C.

Pt(0,6%)-Al$_2$O$_3$ and Pt(0,6%)-In$_2$O$_3$(2%)-Al$_2$O$_3$ catalysts were prepared by impregnation γ-Al$_2$O$_3$ with solution of H$_2$PtCl$_6$ and In(NO$_3$)$_3$.4,5H$_2$O. The catalyst Pt(0,5%)-ZnO-Al$_2$O$_3$ was prepared by impregnation with solution of H$_2$PtCl$_6$ the ZnO-Al$_2$O$_3$ carrier (ZnO:Al$_2$O$_3$= 9:1 mole). The latter was prepared by the coprecipitation of the mixture of Al(OH)$_3$ and Zn(OH)$_2$ from the solution of their nitrates using LiOH. The gel obtained was washed, dried and annealed at 500°C. The copper catalysts with the same carrier was prepared in the similar manner.

HOPDAE for synthesis was obtained by mixing 1,2-diaminoethane with propylene oxide at 50°C (1 h) The labeled HOPDAE was obtained using 1,2-diaminoethane-[14]C. Commercial methylpiperazine was used. The MTHP was

Table 3. Effect of time on stream on the conversion of HOPDAE to methylpyrazine over copper-containing catalysts. (350° C, space velocity of HOPDAE 1.12 h^{-1}, feed mole ratio HOPDAE/ H_2O /H_2 = 1/1.6/1.8)

Catalysts	Time on stream, hours								Methylpyrazine mean yield, % mole
	0-1	1-2	2-3	3-4	4-5	5-6	6-7	7-8	
	Methylpyrazine yield, % mole								
CuO-Cr$_2$O$_3$ (C-1) *	76	71	69	67	63	67	64	64	68
CuO-Cr$_2$O$_3$ (C-2) *	81	75	74	67	56	62	61	57	67
CuO-Cr$_2$O$_3$ (C-3)*	84	83	82	79	73	79	77	72	78
CuO-ZnO-Al$_2$O$_3$	88	87	83	86	84	81	77	77	83
Pt- ZnO-Al$_2$O$_3$	79	81	78	74	77	72	65	68	75

* Commercial samples of different mole ratio of CuO and-Cr$_2$O$_3$

separated by preparative GLC-method from the reaction mixture obtained from
HOPDAE on Pt-ZnO-Al$_2$O$_3$ catalyst at 280-300°C.

The reaction products were analyzed by GLC at 110-200°C using a column
packed with 5% PEG-20M on Inerton. Identification of the products was carried
out by means of chromato-mass-spectrometry. Radioactivity of labeled products
was determined using radiochromatograph with flow counter of radioactivity.

REFERENCES

1. G.B. Barlin, in "The Chemistry of Heterocyclic Compounds"
   (A.Weissberger, ed.), Interscience Publ., New-York - London, 1982, **41**, p.8
2. K. Sato, Japan. Patent 76, 56,479, to Tokay Electro-Chem. Co., Ltd. (1976).
3. G.V. Isagulyants, K.M. Gitis, V.A. Myasnikov and G.E. Neumoeva Bull.
   Acad. Sci. of the USSR, Div. of Chem. Sci., **39**, 1340 (1990).
4. S. Yasuda, T. Niva, N. Asegava, Japan. Patent 74, 117,480 to Koei
   Chemical Co., Ltd. (1974).
5. L. Forni, S. Nistori, in "Heterogeneous catalysis and fine chemical" St. Surf.
   Sci. Catal., **41**; M. Guisnet, J. Barrault, C. Bouchoule, D. Duprez, C.
   Montassior and G. Perrot Eds.; Elsevier Science:Amsterdam, p.291 (1988).
6. Japan. Patent 02, 184,679 (90,184,679) to Korea Research Instituteof Chem.
   Technology (1990).
7. S. Yasuda, T. Niva, Japan. Patent 74, 101,391 to Koei Chemical Co., Ltd.
   (1974).
8. Japan. Patent 01, 203,370 (89,203,370) to Korea Research Institute of Chem.
   Technology (1989).
9. L. Forni, P. Pollesel. J. Catal., **130**: 403 (1991).
10. M.V. Shimanskaya, Ya.F. Oshis, L.Ya. Leitis, I.G Iovel, Yu.S. Goldberg,
    L.O. Gollender, A.A. Anderson, A.A. Avots. Advances of heterogeneous
    catalysis in the chemistry of heterocyclic compounds. Ed. by Inst. of organic
    synthesis,"Zinatne" Publ., 1984, p.50 (in Russ.)
11. K. Tanabe. Solid Acids and Bases. Their catalytic properties. (Kodansha,
    Ed.). Academic Press, New York-London, 1970, p.48.
12. K. Shibata, T. Kiyoura, J. Kitagava, T. Sumiyoshi, K. Tanabe. Bull. Chem.
    Soc. Japan, **46**, 2985 (1973).
13. S. B Kogan, A. M. Moroz, O.M. Oranskaya, I.V. Semenskaya, G.I
    Tisowsky, N.R. Bursian. Zhur. Prikl. Chim., **41**, 1884 (1983) (In Russ.).
14. G.V. Isagulyants, Yu.G. Dubinsky and M.I. Rozengart.
    Russian Chemical Reviews, **50**, 805 (1981).
15. M.V.Klyuev and M.L. Khidekel. Russian Chemical Reviews, **49**, 14 (1980).
16. W. Hammerschmidt, A. Baiker, A. Wokaun, W. Fluhr. Applied Catalysis,
    20, 305 (1986).

Non-Linear Effects in Asymmetric Catalysis: Implications for Organic Synthesis Strategies

Donna G. Blackmond

Max-Planck-Institut für Kohlenforschung
Mülheim an der Ruhr   D-45470   Germany

Abstract

The observation of a non-linear relationship between the optical purity of a chiral catalyst and the enantioselectivity obtained in the catalyzed reaction products has led to considerable experimental and theoretical work aimed at developing an understanding of this curious phenomenon. Some of the less-well-recognized kinetic implications of this non-linear behavior are treated here to demonstrate how asymmetric amplification may be exploited for practical strategies in organic synthesis.

Introduction

Asymmetric catalytic synthesis of enantiomerically enriched products has become an important tool in practical organic synthesis. An interesting feature of these reactions which has been highlighted in the recent literature is the observation in certain cases of a non-linear correlation between the optical purity of the chiral catalyst or auxiliary and that of the reaction product (1,2). For example, in some cases the products of an asymmetric catalytic reaction exhibit an enantiomeric excess higher than the enantiopurity of the chiral catalyst, and, as is shown in Figure 1, this departure from proportionality is termed a *positive asymmetric amplification*. The straight line in Figure 1 shows the linear relationship expected in most catalytic reactions. A negative deviation from linearity would mean that the optical yield of the reaction products is lower than would be expected from a simple algebraic sum of the enantioselectivities obtained separately from the individual catalysts in the mixture.

The concept of asymmetric amplification is an appealing one since it addresses the problem of the high cost of preparing extremely enantiopure ligands on a production scale. If high product enantioselectivity is indeed obtainable with less than highly enantiopure catalysts, asymmetric catalysis might find even broader application in the production of pharmaceuticals, agrochemicals, and flavor and aroma chemicals.

455

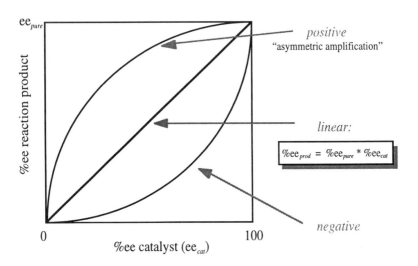

Figure 1. Relationship between the optical purity of the catalyst and the enantioselectivity obtained in the products of the reaction it catalyzes.

Kagan and coworkers (1) reported the first examples of such non-linear effects in asymmetric catalytic reactions more than a decade ago, and many examples of this non-linear behavior have since been observed (2). Kagan's group developed mathematical models to describe this behavior for systems in which an organometallic catalyst contains two or more different chiral ligands, pointing out that an understanding of such seemingly inexplicable behavior may lead to insights about the reaction mechanism and the structure of active catalytic species.

The purpose of this paper is to highlight some less-well-recognized aspects of non-linear behavior in asymmetric catalysis. The mathematical models developed by Kagan and coworkers (1) are used to highlight the potential influence of the kinetic behavior of non-enantiopure catalysts on enantioselectivity in asymmetric reactions. The model predictions of reaction rate may reveal striking consequences for non-linear catalytic behavior. A positive non-linear effect in product enantioselectivity may come at the cost of a severely suppressed reaction rate. In contrast, a system exhibiting a *negative* non-linear effect in product enantioselectivity can provide a significantly *amplified* rate of formation of the desired product. It is also shown that when mixtures of different catalyst species are used to carry out an asymmetric reaction, the enantioselectivity of the reaction products may become a non-linear function not only of the enantiomeric purity of the catalyst but also of the conversion level of the substrate. Consideration of the kinetic behavior of these systems may thus provide valuable mechanistic insights and may help in the development of efficient synthetic strategies using non-enantiopure catalyst mixtures.

Results and Discussion

If partially resolved chiral ligands are used to prepare a chiral organometallic catalyst containing two or more ligands, there exists the possibility of forming achiral and heterochiral complexes in addition to homochiral catalytic species. The models developed by Kagan and coworkers[1] for catalysts of the type ML_n ($n \geq 2$) show how the ultimate enantioselectivity observed in a reaction carried out using such a mixture of catalysts will depend on the relative concentrations and reactivities of each catalytic species in the mixture, as well as on the intrinsic product enantioselectivity obtained with each catalyst in the asymmetric reaction.

Scheme 1 shows the reaction network for the ML_2 model developed by Kagan and coworkers[1] for asymmetric catalysis based on two enantiomeric chiral ligands, L_R and L_S, and a metal center M. The model assumes that at any given level of catalyst optical purity (that is, at fixed amounts of L_R and L_S), a steady-state exists for the concentrations of the three catalyst species in the mixture, two homochiral complexes ML_RL_R, ML_SL_S and a *meso* complex ML_RL_S, with relative concentrations which are set by the equilibrium constant K.

Scheme 1

$$xML_RL_R \quad + \quad yML_SL_S \quad + \quad zML_RL_S$$

$$r_{RR} \downarrow ee_o \qquad r_{SS} \downarrow -ee_o \qquad r_{RS} \downarrow ee = 0$$

$$K = \frac{z^2}{xy}$$

$$\beta = \frac{z}{x+y}$$

reaction   products

$$g = \frac{r_{RS}}{r_{RR}}$$

$$r_{RR} = r_{SS}$$

The catalytic reaction carried out using these complexes is assumed to be pseudo-zero-order in substrate and first order in catalyst concentration. For a reaction carried out with either of the enantiopure catalysts ML_RL_R or ML_SL_S, the reaction rate is identical ($r_{RR}=r_{SS}$) and the product enantioselectivities are opposite (ee_o and $-ee_o$, respectively). Racemic product is formed from the *meso* catalyst which exhibits the relative reactivity equal to the parameter g. With these assumptions, and with the optical purity of the chiral catalyst expressed as ee_{aux}, Kagan and coworkers showed that the enantioselectivity of the reaction products ee_{prod} obtained from this mixture of catalysts will vary with ee_{aux} according to Equation 1:

$$ee_{prod} = ee_o ee_{aux} \frac{1+\beta}{1+g\beta} \qquad (1)$$

Plots of ee_{prod} vs. ee_{aux} will reveal whether a non-linear relationship between the two variables exists in a particular catalyst system. The ML_2 model will be used here to illustrate some aspects of reactions using non-enantiopure catalysts which focus on kinetic implications of non-linear effects in asymmetric catalysis.

Chiral Efficiency vs. Overall Synthetic Efficiency

The values for the parameters K and g determined by the model provide the basis for discussions of the mechanistic implications of the model in a particular application. For example, Kagan and coworkers used this model to fit experimental data in two cases of Sharpless epoxidation of allylic alcohols which exhibited a positive non-linear effect. They found that the model supported experimental evidence indicating the presence of a more abundant but less reactive *meso*-complex.

Large values of K indicate predominance of the *meso* species, and values of g less than one mean that the *meso* species is less reactive than the enantiopure catalyst. Examples of dramatic positive asymmetric amplification in the ML_2 system have been shown in theoretical curves by Kagan and coworkers, corresponding to the case of complete dominance by a very unreactive *meso* species (very high K values, very low g values). By contrast, a strong negative non-linear effect may be observed for a case where high K values are coupled with a highly reactive *meso* complex (g >>1). Figure 2a shows theoretical curves, drawn using Eq. 1, for these two extreme cases of non-linear effects. A product enantioselectivity of greater than 90% is achievable in the case where g = 0.01, while severely suppressed enantioselectivities are observed when g = 100. Thus, for a system with these characteristics, a strategy for high chiral efficiency would be favored by formation of an inactive meso complex.

Both experimental and theoretical studies of non-linear effects have focused almost exclusively on enantioselectivity without considering how the overall reaction rate will vary as a function of catalyst enantiomeric purity. It has recently been shown (3) however, that the theoretical models developed by Kagan and coworkers may also predict reaction rate. When the model is used to calculate reaction rates for the two cases shown in Figure 2a, a striking contrast to the conclusions reached from the enantioselectivity plot is revealed. Figure 2b shows that the system which exhibited a strong *negative* non-linear effect in enantioselectivity shows a strong *positive* amplification in reaction rate. When the catalyst optical purity is ee_{aux} = 50%, the reaction rate is more than two orders of magnitude faster for the case where g = 100 than that for g = 0. This rate difference decreases as ee_{aux} increases further, but a more than twenty-fold difference in rates remains even at ee_{aux} = 85%. The high asymmetric amplification in product chirality is achieved only at a dramatic cost in reaction rate. Chiral efficiency (enantioselectivity) stands in contrast to overall synthetic efficiency (rate of formation of the desired product) in this case.

The result that a catalyst mixture with poorer enantioselectivity gives a higher synthetic efficiency to the desired product may appear to be counterintuitive, but this conclusion is easily rationalized by recalling that selectivity in any reaction is given by a *ratio* of reaction rates. The individual product formation rates may rise or fall as selectivity changes. The overall synthetic yield in a reaction will depend both on how selectively and on how fast a product is formed. Thus a strategy for efficient catalytic synthesis of chiral compounds should take into account both enantioselectivity (chiral efficiency) and reaction rate (overall synthetic efficiency) behavior in the choice of a catalyst system to carry out the reaction. For cases where separation of the two enantiomeric reaction products is very difficult or very costly, a penalty in the form of severely suppressed production rates may ultimately be an acceptable price to pay for high enantioselectivity in the catalytic reaction.

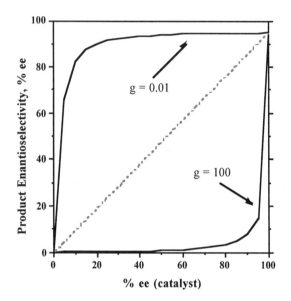

Figure 2a. Product enantioselectivity as a function of catalyst optical purity for two cases of the ML_2 model with K = 2500. The relative reactivity of the meso complex is given by g. $ee_o = 95\%$

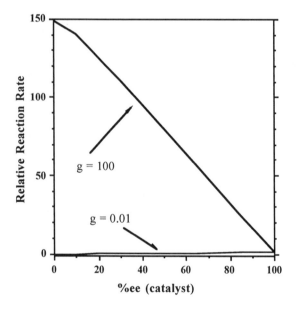

Figure 2b. Overall reaction rate as a function of catalyst optical purity for the two cases of the ML_2 model shown in Figure 2a. K = 2500. The parameter g gives the relative reactivity of the *meso* and enantiopure catalyst species.

Consideration of Reaction Rate Laws

One of the main simplifying assumptions of the ML_2 model presented by Kagan and coworkers (1) is that the reaction pathway for each catalyst obeys zero-order kinetics in substrate concentration. This type of kinetic behavior has also been implicitly or explicitly assumed in many experimental studies in asymmetric catalytic synthesis. In the most general case of asymmetric catalysis using a non-enantiopure catalyst, the reaction mixture will contain different catalyst species, and it is arguably quite likely that the substrate concentration dependence of the reaction rate may *not* be zero-order in substrate and may *not* be the same for each non-enantiomeric catalyst species present in the mixture.

When different catalytic species in a mixture of asymmetric catalysts exhibit different reaction rate laws and different intrinsic product enantioselectivities, the overall observed enantioselectivity for a reaction carried out in such a mixture will vary as a function of the concentration of substrate present in the reacting system. This is a general result and it is not dependent on the particular form of the rate law for each catalyst in the mixture. Most experimental studies of non-linear effects in asymmetric catalysis report only the measurement of enantioselectivity at the reaction endpoint. If the different catalysts in the mixture do indeed exhibit different reaction rate laws, this endpoint enantioselectivity may not be an accurate reflection of the behavior of the catalyst of the course of the reaction.

We consider for the sake of this illustration the general case the ML_2 model as described above, with the exception that now we relax the assumption that all three catalyst species obey zero order kinetics in substrate. Let us assume that the reaction using enantiopure ML_RL_R and ML_SL_S catalysts follows Michaelis-Menten kinetics (4) while the *meso* catalyst ML_RL_S exhibits a rate law which is zero-order in substrate concentration (Eqs. 2 and 3). These forms are chosen because many catalytic reactions are found to be described by one or the other of these two common rate laws.

$$r_{RR} = r_{SS} = \frac{k_{MM}[SUB]}{1 + k'_{MM}[SUB]}$$

[SUB] = substrate concentration
k_{MM}, k'_{MM} = Michaelis-Menten constants   (2)

$$r_{RS} = k_z$$

k_z = zero-order rate constant   (3)

In this case, the product enantioselectivity will not be constant over the course of the reaction, but will be described by the following function of the substrate concentration:

$$ee_{prod} = ee_{aux} ee_o \left\{ \frac{\dfrac{k_{MM}[SUB]}{1 + k'_{MM}[SUB]}}{ee_{aux}\left(\dfrac{k_{MM}[SUB]}{1 + k'_{MM}[SUB]}\right) + \left(1 - ee_{aux}\right)k_z} \right\} \tag{4}$$

Figure 3 gives a graphical example of the relationship in Eq. 4 for three different levels of catalyst optical purity. This illustrates that significant changes in enantioselectivity may be observed over the course of the reaction when different rate laws apply to different catalysts in a non-enantiopure catalyst mixture.

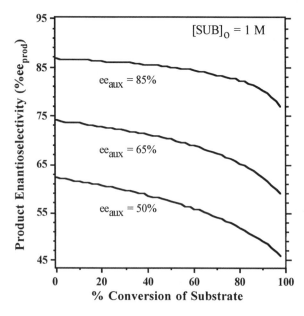

Figure 3. Product enantioselectivity vs. substrate concentration for an ML_2 catalyst system of three different levels of optical purity. The ML_RL_R and ML_SL_S catalysts follow Michaelis-Menten kinetics while the rate of the *meso* catalyst ML_RL_S is zero-order in substrate. $ee_o = 95\%$; $k_{MM} = 1$; $k'_{MM} = 0.5$; $k_z = 0.35$.

In such a case where the product enantioselectivity is a function of the substrate concentration, measurement of enantioselectivity at the reaction endpoint will represent just one point in a family of values describing enantioselectivity in this system. This also means that a different value of the endpoint enantioselectivity will be obtained for different initial concentrations of substrate. Figure 4 shows endpoint enantioselectivity as a function of $[SUB]_o$ for the same example shown in Figure 3. Thus, carrying out the reaction under at least two different starting concentrations provides a simple means of determining whether the complicating kinetic factors discussed here apply to the system under study.

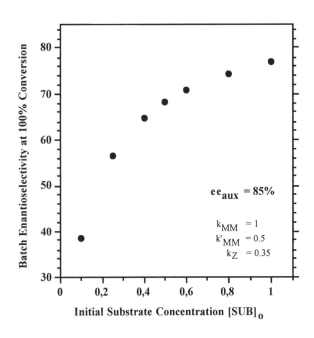

Figure 4. Endpoint enantioselectivity as a function of the initial substrate concentration emloyed in the reaction for one level of catalyst optical purity taken from the example shown in Figure 3.

The observed strong function of initial substrate concentration may be rationalized by considering the competition between the catalysts present in the system. The Michaelis-Menten kinetics exhibited by the enantiopure catalysts are very sensitive to substrate concentration, while the racemic reaction catalyzed by the *meso* complex is impervious to the level of substrate in the system. The use of low initial substrate concentrations therefore allows the racemic route to compete more effectively, resulting in a low overall enantioselectivity for the reaction for the case described here. For such a system, a synthetic strategy might thus be devised where the use of high initial substrate concentrations could considerably improve the enantioselectivity achieved in the reaction mixture as a whole. Thus kinetic information about the reaction may be used to choose reaction conditions to optimize product enantioselectivity for a given non-enantiopure catalyst system.

This illustration emphasizes the importance of understanding the rate behavior of each catalyst in a mixture. The rate laws for the catalysts in this example were chosen to provide an illustration of plausible potential circumstances. The magnitude of the effects as well as the direction of the trends observed will depend on the particular case under study, but the conclusion that enantioselectivity changes with reaction progress is general. One important message of this illustration is that caution is advised in making mechanistic interpretions of catalyst behavior on the basis of reactions carried out under only one set of initial conditions, if the kinetics of each catalyst species present in the system are not well-characterized.

Conclusions

The use of partially resolved or initially racemic chiral catalysts is appealing for larger scale organic synthesis of chiral compounds because of the potential for both increased flexibility and reduced catalyst costs. The examples in this paper show that profound consequences for enantioselectivity may ensue when asymmetric reactions are carried out in a mixture of non-enantiopure catalyst species. A strong amplification in product chirality may come at the cost of a severely suppressed rate of product formation; in comparison, a system exhibiting a *negative* non-linear effect in product enantioselectivity can provide a significantly *amplified* rate of formation of the desired product.

When asymmetric reactions are carried out in a mixture of catalysts exhibiting different reaction rate laws, the product enantioselectivity will show a dependence on substrate concentration. It is also shown how kinetic information may be used to choose reaction conditions to optimize product enantioselectivity for a given non-enantiopure catalyst system. Consideration of the kinetic behavior of these systems can provide valuable mechanistic insights and may help in the development of efficient synthetic strategies using non-enantiopure catalyst mixtures.

References

1. a) Guillaneux, D., Zhao, S.H., Samuel, O., Rainford, D., and Kagan, H.B., *J. Am. Chem. Soc.*, **116**, 9430-39 (1994); b) Puchot, C., Samuel, O., Duñach, E., Zhao, S., Agami, C., and Kagan, H.B., *J.Am. Chem. Soc.*, **108**, 2353-57 (1986); c) Kagan, H.B., Girard, C., Guillaneux, D., Rainford, D., Samuel, O., Zhang, S.Y., and Zhao, S.H., *Acta. Chem. Scand.*, **30**, 345-352 (1996).

2. a) Oguni, N., Matsuda, Y. and Kaneko, T., *J. Am. Chem. Soc.*, **110**, 7877 (1988); b) Noyori, R., and Kitamura, M., *Angew. Chem. Int. Ed. Engl.*, **30**, 49-69 (1991). c) Bolm, C., *Tetr. Asymm.*, **2**, 701-704 (1991); d) Mikami, K., Motoyama, Y., and Terada, M., *J. Am. Chem. Soc.*, **116**, 2812-2820 (1994); e) Mikami, K., and Matsukawa T., *Tetrahedron*, **48**, 5671-80 (1992).

3. Blackmond, D.G., *J. Am. Chem. Soc.*, **119**, 12934, (1997).

4. Michaelis, L., and Menten, M.L., Biochem. Z., **49**, 333 (1913).

Homogeneous Palladium Catalyzed Vinylic Coupling of Aryl Bromides Used to Make Benzocyclobutene Derivatives

Robert A. DeVries, Paul C. Vosejpka, Mezzie L. Ash
The Dow Chemical Company, 677 Bldg, Midland, Michigan 48667

Abstract

Benzocyclobutene (BCB) derivatives are a novel class of thermally polymerized monomers and were first developed at Dow for potential use in a number of electronic, composite, and thermoplastic applications. Cyclotene™ electronic resins based on the monomer divinyltetramethyldisiloxanebis-benzocyclobutene (DVS-bisBCB) were commercialized in 1995. Many of these monomers can be made by a homogeneous palladium-catalyzed vinylic coupling reaction of bromobenzocyclobutene with various olefins. Several modifications of the original Heck arylbromide vinylic coupling chemistry have resulted in faster, more selective, and higher yield reactions. A number of side products and their stereochemistry will be the focus of this paper as well as ligand, temperature, and solvent effects on the product yield and distribution.

Figure 1: BCB Derivatives via Heck Chemistry

Introduction

The palladium-catalyzed coupling of aryl halides and olefins (Heck reaction) is a very efficient and practical method for synthesizing carbon-carbon bonds having a wide variety of functionalities (1-5). This chemistry can be used to prepare fine organics, pharmaceuticals, and specialty monomers. For example, the reaction allows a one-step synthesis of substituted styrenes from aryl bromides (6,7) and is an excellent method for the preparation of a wide variety of styrene derivatives (8,9).

Another example of the utility of this reaction is in the preparation of benzocyclobutene (BCB) derivatives (Figure 1)(10-12). These are a novel class of thermally polymerized monomers, which have recently been commercialized targeting electronic, composite, and thermoplastic applications. Many of the monomers can be made by the reaction of 4-bromobenzocyclobutene (BrBCB) with olefins (13-15). The BCB group when heated to about 180 °C ring opens to form an ortho-quinodimethane group that then undergoes Diels-Alder reactions to form a thermosetting polymer (Figure 2). The Heck reaction is particularly useful in that it attaches an olefin onto the BCB moiety such that the monomer will contain both a diene and a dienophile suitable for efficient Diels-Alder reaction.

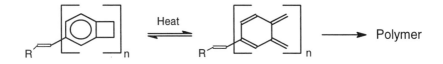

Figure 2. Generalized Concept of Benzocyclobutene Polymerization

The catalyst turnover number, lifetime, and reaction rate were significantly improved over similar arylation reactions by changing the solvent system, catalyst concentration, and ligand to palladium ratio. An inorganic base such as potassium acetate can also be substituted for triethylamine in this reaction and kept soluble by using a mixture of dimethylformamide and water as solvent.(16)

The main coupling reaction to simple olefins usually results in monoarylation of the olefin as shown in Figure 3, however side reactions to form bis-arylation and even tris-arylation products (when R = H) can occur. For example, addition of BrBCB to a large excess of ethylene can give greater than 90% yields of 4-vinyl-BCB, or greater than 90% ethylenebisBCB at lower pressure, and even trisBCB adducts when ethylene is limited (10-12). A mechanism is shown in Figure 3.

Experimental

A 100-mL one-necked flask equipped with a Teflon™ stir bar, condenser (with nitrogen inlet), and thermowell was charged with 10.0 g (0.0546 mol) 4-bromo-benzocyclobutene (BrBCB), 5.08 g (0.0273 mol) tetramethyldivinyldisiloxane (DVS), 25 mL N,N-dimethylformamide, 0.9123 g (0.00055 mol) palladium acetate, and 0.00022 mol (for phosphines containing one phosphorus) or 0.00011 mol (for phosphines containing two phosphorous atoms) of the appropriate ligand. A solution of potassium acetate (16.08 g, 0.162 mol) in water (15 mL) was then added to the flask. A stream of nitrogen was passed through the resulting mixture for approximately 5 minutes, and the reaction was heated by a heating mantle attached to a temperature controller, high limit, and timer.

The set point was 95 °C and the reaction was run for approximately 24 hours. After cooling the reaction to room temperature, the contents of the flask were poured into an 8 oz bottle, the flask rinsed with 50 mL of water and 50 mL of methylene chloride, and the organic layer was analyzed by gas chromatography against known standards.

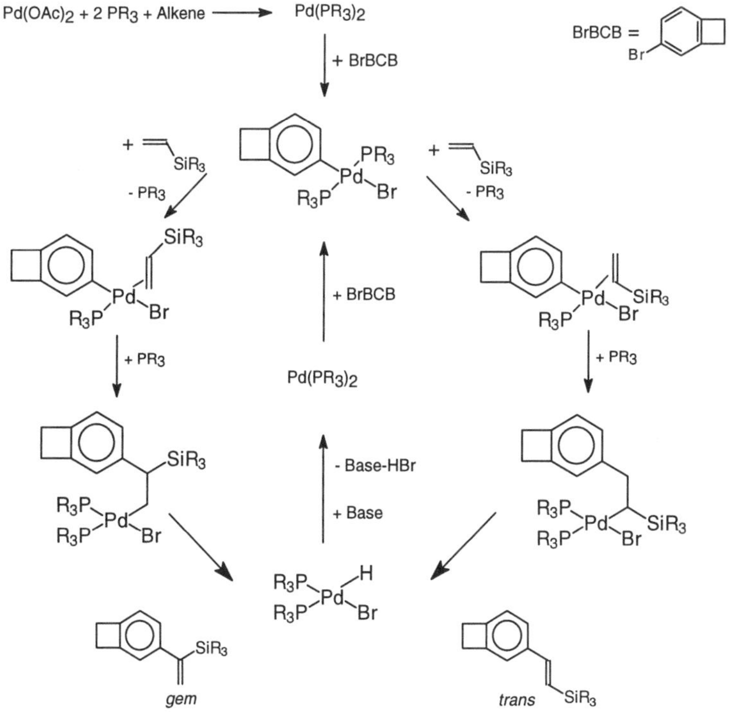

Figure 3. Proposed Coupling Mechanism

Reactions using air-sensitive phosphine ligands were carried out as described above except that the phosphine ligand was weighed out and added to the flask containing the other reagents in a nitrogen-filled glove bag. The flask was removed from the glove box and heated as described.

Reactions of BrBCB with olefins other than DVS were run as above using tri-ortho-tolylphosphine as ligand, DMF as solvent and triethylamine as base. Reactions with low-boiling substrates or olefins, or requiring higher temperatures, were run in an electrically heated 250 mL 316 SS Autoclave Engineer reactor.

Results and Discussion

<u>Stereochemistry of the Vinylic Coupling Reaction with BrBCB</u>
The stereochemistry of vinylic coupling reactions follows a regioselective addition of the palladium to the more highly substituted carbon and a stereospecific cis addition of the aryl-palladium to the olefin (Figure 4)(17). This is followed by a stereospecific cis elimination of hydrido-palladium from the complex to reform an olefin (18,19).

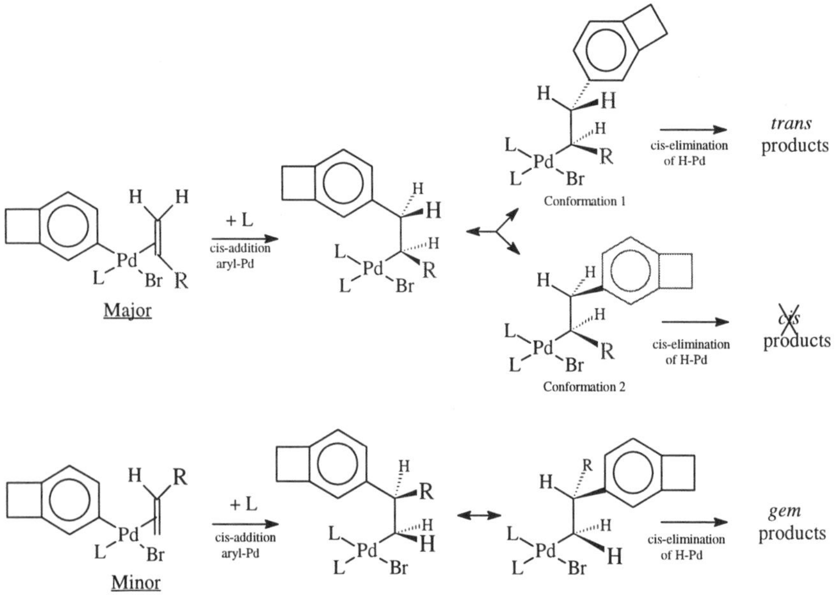

Figure 4. Formation of *trans* and *gem* Products from Vinylic Coupling Reaction

<u>Formation of Mono-benzocyclobutene Adducts - Effect of Olefin Substituents</u>
A series of substituted ethylenes were reacted with BrBCB and the stereochemistry of the mono-arylated products is shown in Table 1. Although addition of the palladium catalyst to the less substituted carbon is not favored, it does occur and is affected by the substituent (Figure 4). In general, a bulky R group on the olefin will give higher levels of gem adducts. The minor product is formed by the addition of palladium to the less substituted carbon and has only one conformation leading to gem products. However, there are two elimination conformations leading to vicinal products following the regiospecific addition of palladium to the olefin, one leading to a trans product and the other to a cis product. The conformation leading to cis products forces the BCB ring and olefin substituent in close proximity, while the conformation leading to trans

products minimizes this steric interaction and was observed in each example. The observed products then are mainly trans and gem mono arylated olefins and the catalytic paths leading to their respective formation are shown in Figure 3.

Table 1. Stereochemistry of Monoarylated Products
BrBCB Conversion >95%, Isomers identified by GC-MS, NMR, and IR

		GC Area %	
Olefin	Trans	Gem	Cis
Ethylene	-	-	-
Styrene	90.3	6.4	0.0
p-Vinyltoluene	91.3	8.7	0.0
m,p-Vinyltoluene*	90.1	9.9	0.0
Vinylbenzocyclobutene	90.8	9.2	0.0
m,p-Divinylbenzene*	87.0	13.0	0.0
Vinyltrimethoxysilane	92.1	2.5	0.0
Vinyltriethoxysilane	96.0	4.0	0.0
Divinyltetramethyldisiloxane	91.6	8.4	0.0

* contains a mixture of meta and para isomers

Formation of Bis-benzocyclobutene Adducts with DVS
The Heck reaction can also be used to made bis-benzocyclobutene derivatives (useful for thermosetting polymer) if divinyl compounds such as divinylbenzene or tetramethyldivinyldisiloxane (DVS) are used. In the case of DVS with two equivalents of BrBCB, a mixture of mono-arylated and bis-arylated products were formed, along with two unexpected ethylene bisbenzocyclobutene products (Figure 5). The formation of these derivatives is thought to originate via acid catalyzed cleavage of carbon-silicon bonds to give vinyl benzocyclobutenes which further react with BrBCB to give ethylene-bisbenzocyclobutene derivatives (Figure 6). The silicon-containing products identified are consistent with the hydrolysis of the silyl-bromide. One advantage of running the Heck reaction with excess potassium acetate dissolved in an aqueous phase is that this reaction is minimized and salts do not precipitate during the reaction (16).

Changes in Ligand - Conversion of BrBCB and Product Selectivity
A diverse collection of phosphines was screened in this reaction including triaryl (with electron-donating and withdrawing substituents), trialkyl, diphosphine (with alkyl tethers), aryl-alkyl, and heteroatom substituted. See Figure 7 for structures and abbreviations. Table 2 summarizes the results of our screening runs for 26 phosphine ligands giving the amount of BrBCB remaining and yields of mono-arylated adducts (DVS-BCB's), ethylene-bisbenzocyclobutenes (Et-bisBCBs), and DVS-bisBCB's.

Of the 26 ligands screened, 15 reacted with palladium acetate to produce catalysts which gave about 90% or greater conversion of BrBCB after 24 hours at 95 °C. The remaining ligands did not produce catalysts which gave high conversion of BrBCB. Some of these ligands did in fact give DVS-bisBCB products but were simply much slower than the 15 "active" ligands. Two of the ligands screened, tri-n-butylphosphine and tris(trimethylsilyl) phosphine, led to virtually no conversion of BrBCB and appeared to give a black precipitate, presumably palladium metal indicating a decomposition of soluble palladium complexes.

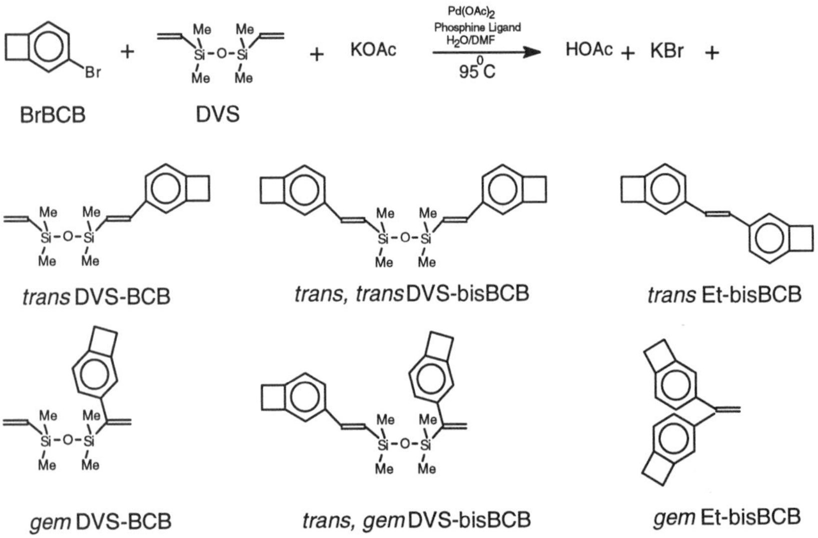

Figure 5. Products of BrBCB and DVS Coupling Reaction

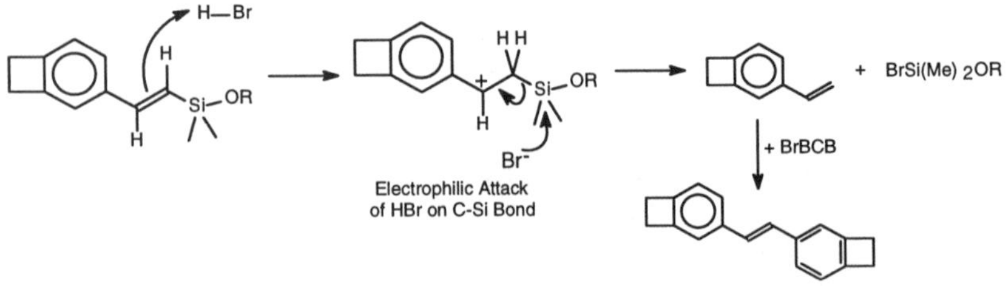

Figure 6. Origin of Ethylene Linked Adducts

For the 15 phosphine ligands that gave high conversion, yields of the desired DVS-bisBCBs ranged from about 60 to 84%. Yields of the major side product ethylene-bisbenzocyclobutene (Et-bisBCBs) ranged from 3 to 11%. Previous studies using only triethylamine as base in DMF are known to give even higher yields of these side products (13-16). The ligand used for production scale, tri-ortho-tolylphosphine, produced the second highest yield of DVS-bisBCB measured during these screening runs, while keeping the amount of Et-bisBCBs below 10%.

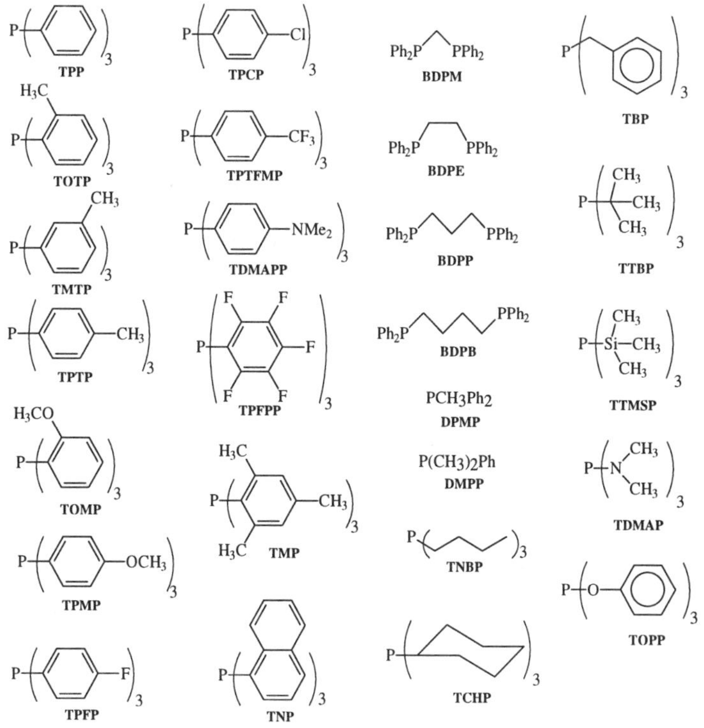

Figure 7. Structure of Phosphine Ligands Screened

Changes in the Phosphine Ligand - Regioselectivity

The palladium-catalyzed coupling of BrBCB with DVS produces two main isomeric products: *trans,trans*-DVS-bisBCB and *gem,trans*-DVS-bisBCB. Since the active catalyst contains Pd-P bonds, the steric and electronic properties of the phosphines should have a large influence on the regiochemistry of the coupling reaction. From the screening of the 26 phosphine ligands, we could determine the effect that a change of ligand has on the regioselectivity of product by the ratio of *trans,trans* to *gem,trans*-DVS-bisBCB isomers.

Table 3 shows the *trans,trans/gem,trans* ratio for various ligands as well as some published cone angles and pK_a values of their conjugate acids. The *trans,trans/gem,trans* ratio varies from about 4.9 for bis(diphenylphosphino)ethane as ligand to about 6.8 for tricyclohexylphosphine as ligand. There does not seem to be any simple correlation between the steric bulk of the phosphine, measured as cone angle (5,20) and conversion of BrBCB, yields of DVS-bisBCB, or regioselectivity. For example, plots of regioselectivity (ratio of *trans,trans* to *gem,trans* isomers) versus cone angel showed no trends. Even with compositionally similar series

Table 2. Results of Ligand Screening Runs

Ligand	Abbrev.	GC Area % BrBCB	GC Area % DVS-BCB	GC Area % Et-bis BCB	GC Area % DVS-bis BCB
tri-t-butylphosphine	TTBP	0.00	2.57	5.88	84.10
tri-ortho-tolylphosphine	TOTP	0.00	2.77	7.42	82.91
tri(1-naphthyl)phosphine	TNP	1.37	10.96	3.85	78.93
triphenylphosphine	TPP	2.28	7.43	6.66	76.55
tri-meta-tolylphosphine	TMTP	3.23	7.84	5.87	75.68
triphenylphosphite	TOPP	3.93	10.13	6.23	73.85
bis(diphenylphosphino)butane	BDPB	3.53	9.98	6.11	72.31
bis(diphenylphosphino)ethane	BDPE	2.81	10.31	6.84	69.34
tribenzylphosphine	TBP	4.60	9.31	7.10	69.16
tri(para-fluorophenyl)phosphine	TPFP	3.83	10.86	5.49	69.09
tri-para-tolylphosphine	TPTP	7.02	15.42	3.38	66.81
bis(diphenylphosphino)methane	BDPM	8.50	16.50	3.71	66.54
tri(ortho-methoxyphenyl)phosphine	TOMP	7.43	14.09	4.50	66.14
dimethylphenylphosphine	DMPP	4.76	12.45	10.66	63.96
tri(para-methoxyphenyl)phosphine	TPMP	9.77	18.24	3.01	63.12
tricyclohexylphosphine	TCHP	11.22	15.66	4.39	60.42
tris(4-dimethylaminophenyl)phosphine	TDMAPP	15.31	20.83	0.75	58.17
tri-1,3,5-mesitylphosphine	TMP	15.27	26.14	1.62	53.73
tri(para-chlorophenyl)phosphine	TPCP	16.63	28.84	1.22	49.85
tris(pentafluorophenyl)phosphine	TPFPP	21.06	32.51	0.00	43.83
tris(dimethylamino)phosphine	TDMAP	38.60	35.57	0.00	24.16
bis(diphenylphosphino)propane	BDPP	46.42	30.39	1.57	12.10
tri(4-trifluoromethylphenyl)phosphine	TPTFMP	58.30	29.76	0.66	9.52
diphenylmethylphosphine	DPMP	62.11	28.92	0.00	5.56
tris(trimethylsilyl)phosphine	TTMSP	96.67	0.00	0.00	0.00
tri-n-butylphosphine	TNBP	97.97	0.00	0.00	0.00

Structures (and abbreviations) of phosphine ligands are shown in figure 7.

of phosphines such as tritolylphosphines (*ortho,meta,para*) there appears to be no clear correlation between cone angle and this ratio (*ortho*/194°/6.1, *meta*/165°/6.3, *para*/145°/6.1). Similarly, there does not appear to be any simple correlation between electronic features of the phosphines, measured as pK_a (21-23) and conversion of BrBCB, yields of DVS-bisBCB, or regioselectivity. Although we could not correlate our data to particular phosphine properties, these studies have allowed us to determine which commercially available ligands react with palladium acetate to give efficient catalysts (those with complete conversion of BrBCB and high yields of DVS-bisBCBs) and to define the range of product regioselectivities available by changing the phosphine ligand (*trans,trans/gem,trans* ratio varies 4.9 to 6.8).

Kinetic and Temperature Effects
The first order rate constants and approximate half-lives for the palladium catalyzed coupling of BrBCB with DVS with three ligands, TOTP, TPP, and BDPE are given in Table 4. The reactions using TOTP as ligand proceed very rapidly at both 120 and 95 °C, about 4 times slower at 80 °C, and quite slow at

Table 3. Product Regioselectivities, Cone Angles, and pK_a's of Various Ligands

Ligand	Abbreviation	*trans,trans/ gem,trans*	Cone Angle	pK_a
tri-t-butylphosphine	TTBP	5.71	182	11.4
tri-ortho-tolylphosphine	TOTP	6.13	194	3.2
tri(1-naphthyl)phosphine	TNP	6.50		
triphenylphosphine	TPP	6.27	145	2.9
tri-meta-tolylphosphine	TMTP	6.28	165	3.4
triphenylphosphite	TOPP	6.43	122	
bis(diphenylphosphino)butane	BDPB	5.93		
bis(diphenylphosphino)ethane	BDPE	4.93	122	3.8
tribenzylphosphine	TBP	6.31	165	
tri(para-fluorophenyl)phosphine	TPFP	5.72	145	2.0
tri-para-tolylphosphine	TPTP	6.10	145	4.0
bis(diphenylphosphino)methane	BDPM	6.48		
tri(ortho-methoxyphenyl)phosphine	TOMP	6.28		
dimethylphenylphosphine	DMPP	6.42	120	6.8
tri(para-methoxyphenyl)phosphine	TPMP	6.22	145	4.8
tricyclohexylphosphine	TCHP	6.84	170	9.6
tris(4-dimethylaminophenyl)phosphine	TDMAPP			
tri-1,3,5-mesitylphosphine	TMP	6.60	212	
tri(para-chlorophenyl)phosphine	TPCP	7.80	145	1.0
tris(pentafluorophenyl)phosphine	TPFPP	6.50	185	
tris(dimethylamino)phosphine	TDMAP			
bis(diphenylphosphino)propane	BDPP			
tri(4-trifluoromethylphenyl)phosphine	TPTFMP	5.60	145	-1.2
diphenylmethylphosphine	DPMP	6.30	135	4.8
tris(trimethylsilyl)phosphine	TTMSP			
tri-n-butylphosphine	TNBP			

Structures (and abbreviations) of phosphine ligands are shown in figure 7.

70 °C. An Arrhenius plot of this data gives approximately 38 kcal/mol for the activation energy. At 95 °C, TPP was about 5 times slower than TOTP, but TPP appears to be more sensitive to impurities and subtle changes in reaction conditions. The reaction with BDPE as ligand at 95 °C was about 10 times slower than TOTP ligand at the same temperature. The temperature using TOTP as ligand also had an effect not only on conversion, but also product yield and product regioselectivity. A higher reaction temperature gave higher yields of DVS-bisBCB and lower yields of Et-bisBCB than reactions at low temperature. At 120 °C the yield of DVS-bisBCB was approximately 87% while the yield of Et-bisBCB was below 4%. At 80 °C, the yield of DVS-bisBCB fell to 74% and the yield of Et-bisBCB rose to over 10%. Temperature also had a considerable effect on the regioselectivity of the DVS-bisBCB products. At 70 °C, the *trans,trans/gem,trans* ratio was about 7.4, while at 120 °C this ratio drops to about 5.75. However, the regioselectivity as measured during the reaction did not significantly change at any given temperature.

Table 4. Effect of Temperature on Kinetics and Regioselectivity of the BrBCB/DVS Coupling Reaction

Ligand	Temp (°C)	k	$t_{1/2}$ (hours)	*trans,trans/gem,trans*
TOTP	120	0.4800	3	5.75
TOTP	95	0.3200	3	6.10
TOTP	80	0.0870	15	6.75
TOTP	70	0.0072	>50	7.40
TPP	95	0.0610	10	6.27
BDPE	95	0.0300	9	4.93

TOTP = tri(ortho-tolyl)phosphine, TPP = triphenylphosphine, BDPE = 1,2-Bis(diphenylphosphino)ethane

Change in the Aryl Bromide Substrate

Since the changes in isomer composition of the products changed little by changing the phosphine ligand and reaction temperatures, we attempted to determine whether minor changes in the aryl bromide reacting with DVS would result in changes in regioselectivity. Figure 8 shows the principle reaction products of DVS with bromobenzene, BrBCB, and 2-bromotoluene. The hindered 2-bromotoluene gave the highest regioselectivity (13.7 *trans,trans* to *gem,trans* ratio), while bromobenzene gave the lowest (5.1 *trans,trans* to *gem,trans* ratio) consistent with adding the bulkiest aryl group to the least hindered carbon of the olefin.

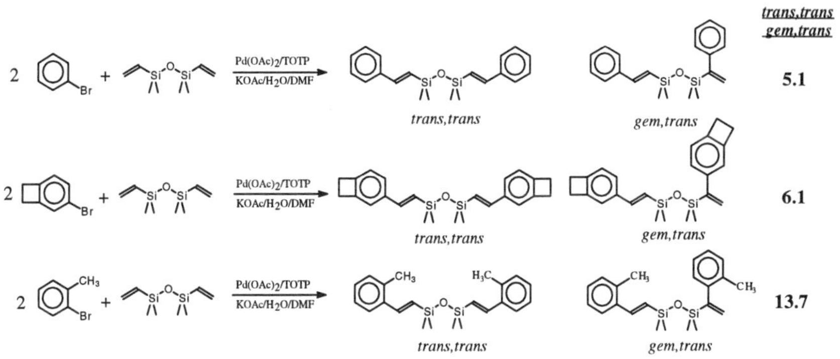

Figure 8. Effect of Substrate Modifications on Regioselectivity

Summary

The palladium-catalyzed coupling of bromobenzocyclcobutene (BrBCB) with olefins such as divinyltetramethyldisiloxane (DVS) is a viable route to produce benzocyclobutene monomers. The effect of changing substituents on the olefin, minor changes of the aryl bromide, ligands used, and temperature were studied to determine the effect on conversion, product yield, and the regioselectivity of the reaction. Of the 26 different phosphines we screened, 15 reacted with palladium acetate to give reasonable conversion of starting material and yields of products. The regioselectivity of product formation is not highly dependent on the choice of ligand. We observed a range of *trans,trans*-DVS-bisBCB to *gem,trans*-DVS-bisBCB ratios of 4.9 to 6.8 depending upon the added ligand. We were unable to correlate the regioselectivity of product formation to either steric or electronic properties of the phosphine ligand. Temperature had some effect on product yields and regioselectivity as well as a large effect on the reaction rate. We observed that at higher temperatures, yields of DVS-bisBCBs increased, while the yield of side product ethylene-bisbenzocyclobutenes (Et-bisBCBs) decreased. Also, higher temperatures decreased the product regioselectivity (lower ratio), while lower temperatures favored a higher *trans,trans*-DVS-bisBCB to *gem,trans*-DVS-bisBCB ratio.

Acknowledgments

The authors would like to thank Ernie Ecker and Hughie Frick for their skillful technical assistance and Tom Wells for his support and encouragement. We also thank the earlier research efforts to develop this chemistry by Ken Bruza, Bob Kirchhoff, Jeff Marra, and Greg Schmidt. Thanks are also due to the Dow Chemical Company for permission to publish this paper.

References

1. R.F. Heck, *Palladium Reagents in Organic Syntheses*; Academic Press: London, 1988.
2. R.F. Heck, *Organic Reactions*, **27**, 345 (1982).
3. R.F. Heck, *Acc. Chem. Res.,* **12**, 146 (1979).
4. R.F. Heck, U.S. Patent 3,922,299, assigned to Univ. of Delaware (1975).
5. J.P. Collman, L.S. Hegedus, J.R. Norton and R.G. Finke, *Principles and Applications of Organotransition Metal Chemistry*, 2nd Ed; University Science Books: Mill Valley, CA, 1987; 69-70.
6. A.J. Spencer, *J. Organomet. Chem.,* **258**, 101 (1983).
7. A.J. Spencer, U.S. Patent 4,564,479, assigned to Ciba-Geigy (1986).
8. W. Heitz, W. Brugging, L. Freund, M. Gailberger, A. Greiner, H. Jung, U. Kampschhulte, N. Niessner, F. Osan, H. Schmidt and M. Wicker, *Makromol. Chem.*, **189**, 119 (1988).
9. R.A. DeVries and A. Mendoza, *Organometallics,* **13**, 2405 (1994).
10. R. Kirchhoff, U.S. Patent 4,540,763, assigned to Dow Chemical (1985).
11. R. Kirchhoff, A. Schrock and J. Gilpin, U.S. Patent 4,642,329, assigned to Dow Chemical (1987).
12. W. A. Gros, U.S. Patent 4,759,874, assigned to Dow Chemical (1988).
13. R.A. DeVries and H.R. Frick, U.S. Patent 5,136,069, assigned to Dow Chemical (1992).
14. R. A. DeVries, M.L. Ash and H.R. Frick, U.S. Patent 5,138,081, assigned to Dow Chemical (1992).
15. R.A. DeVries and H.R. Frick, U.S. Patent 5,243,068, assigned to Dow Chemical (1993).
16. R.A. DeVries, G.F. Schmidt and H. R. Frick, U.S. Patent 5,264,646, assigned to Dow Chemical (1993).
17. R.F. Heck, *J. Am. Chem. Soc.*, **93**, 6896 (1971).
18. R.F. Heck, *J. Am. Chem. Soc.*, **91**, 6707 (1969).
19. H.A. Dieck and R.F. Heck, *J. Am. Chem. Soc.*, **96**, 1133 (1974).
20. C.A. Tolman, *J. Am. Chem. Soc.*, **92**, 2956 (1970).
21. J.R. Soma and R.J. Angelici, *Inorg. Chem.*, **30**, 3534 (1991).
22. R.C. Bush and R.J. Angelici, *Inorg. Chem.*, **27**, 681 (1988).
23. M.M. Rahman, H-Y. Liu, K. Eriks, A. Procks and W.P. Giering, *Organometallics*, **8**, 1 (1989).

Novel Synthesis of 2-(2'-Hydroxyphenyl)Benzotriazole Compounds

V. L. Mylroie, L. Vishwakarma, L. F. Valente, B. Briffa, L. A. Fiorella & M. A. Osypian
Eastman Kodak Company
Kodak Park
Rochester, New York

ABSTRACT

2-(2'-Hydroxyphenyl)benzotriazole compounds are known and used as ultraviolet-absorbing compounds. These compounds are used in paints, plastics, polymers, and coatings and for protection of synthetic fibers against sunlight (1). The dyes are also used in the wool industry and for preventing the sensitivity of organic compounds in products such as cosmetics, fibers, and foods. The compounds are widely used by the photographic industry in silver halide films for protection of image dyes to prevent fading of color photographic prints.

Prior technology included synthetic routes to produce the benzotriazoles involving the reduction of corresponding ortho nitro azo dye using zinc dust in the presence of alkali metal hydroxides. The zinc dust frequently caused dechlorination from the benzotriazole ring. The results were in relatively low yield and utilized a process that is no longer environmentally acceptable.

We have developed a one-step process to make 2-(2'-hydroxyphenyl)benzotriazole compounds of the type shown here;

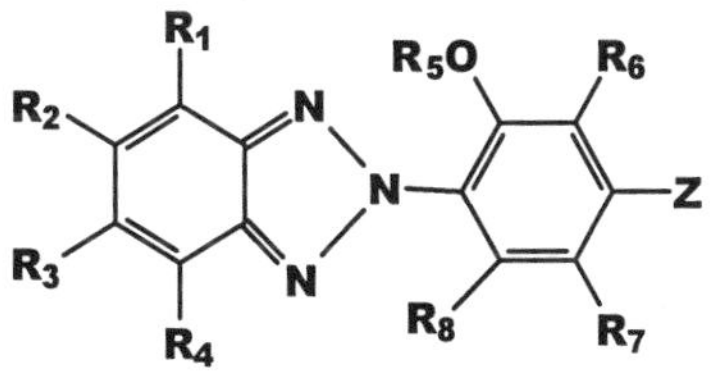

that are useful in stabilizing thermoplastics and coating materials, and also in photographic layers and papers. Such compounds are useful in photographic materials particularly the protection of yellow, magenta, and cyan image dyes from fading in color photographic prints.

We have found that if an acid such as hypophosphorus acid or sulfurous acid is added to the reaction mixture in the presence of a polar solvent, mixed noble metal catalysts, and an organic amine, the hydrogenation and cyclization of the ortho nitro azo dye compound to the corresponding benzotriazole compound occurs in excellent yield. (In addition the azo dye compound can be substituted with a wide variety of substituents.) The optimum ratio of each of the reagents was determined by using Experimental Design Software (JMP) and conducting a designed experiment.

We report the unique preparation of the o-hydroxybenzotriazole from deprotected azo dye precursors in high yield and purity (2). We found that it was required to block Z when Z equals -OH and to use platinum or palladium, or a mixture of platinum and palladium, in the presence of the acids named. Only acids derived from nonmetallic elements such as P and S with lower oxidation states and lower pK_as in the presence of an organic amine are effective. The reaction is simple, clean, and easily completed (3-4). In some of the cases the final product contains the hydroxyl group in position R5 and Z (2).

INTRODUCTION

2-(2'-Hydroxyphenyl)benzotriazole compounds are used in the photographic industry as ultraviolet-absorbing compounds. They are useful for many other applications also, such as ultraviolet-absorbing compounds for paints, coatings, and polymers and for protection of synthetic materials against sunlight. Water-soluble benzotriazole-based UV-absorbing compounds, as disclosed in Japanese patents are used for preventing the sensitivity of organic compounds in various products such as cosmetics, fibers, and foods (5). These compounds are widely used by the photographic industry particularly in silver halide photographic media for protection of yellow, magenta, and cyan image dyes from fading in color photographic prints.

The literature reports the preparation of the benzotriazole type compounds by the use of zinc dust in the presence of alkali metal hydroxides. There are problems associated with preparing these compounds by this route, not the least of which is an environmental issue in the disposal of the large amounts of zinc dust used. The zinc dust reductions give low yields of the benzotriazole and often cause dehalogenation of the compound. Zinc dust reductions frequently fail in the attempted preparation of ortho nitro azo dyes, which are highly substituted with functional groups. Other non-zinc processes give low yields, although some such as thiourea-S,S-dioxide have been used for amino substituted benzotriazoles and do not cause dehalogenation but yields are low (6).

The interest in these compounds is evidenced by the number of patents for the processes in preparation of these compounds. Many begin with the o-nitrophenyl azo dye and utilize reductive cyclization by various methods (7-15). The simple reaction can be represented by the following equation (eq. 1).

$$\text{[structure]} \xrightarrow[\text{Catalyst}]{\text{H2}} \text{[structure]} + 2 H_2O \quad (eq. 1)$$

We investigated the preparation of these compounds using techniques other than those found in the literature or known to us, since most of the cited examples in the literature were prepared by chemical reductions such as the use of large amounts of zinc dust. In most all of the cases the yields were low and quality poor. We tried many of the conditions reported in the patents. Under identical conditions we did not find the same results reported in the patents (12, 16). A particular problem was the complete dehalogenation of compounds with a halogen group on one or both of the aromatic rings. It is desirable in some photographic systems to have a 5-chloro substituent in the benzo ring. Therefore, it was necessary to find a system that was more environmentally acceptable and would not remove the halogens present in the molecule.

DISCUSSION OF DEVELOPMENTAL STRATEGIES

It had been reported that the desired ortho-hydroxy substituted benzotriazole compounds using a mixed metal catalyst such as palladium and platinum in yields of about 90%, could be prepared however, in our hands the products were tarry materials even after numerous attempts (17). The technique of reduction by hydrogen transfer gave yields of only from 60 to 65%. We thought blocking of the hydroxy group while the reduction and cyclization was carried out might give improved results. We could then deblock the hydroxy group to give the desired product. We found that by using a mixed metal catalyst of 4% Pd/C plus 1% Pt/C obtained from Johnson Matthey, the yields were improved as was the quality of the desired product.

The o-nitro azo dye compound of structure (1) can be prepared by the patent procedures (7-9, 12-13, 15-16, 18-27), which detail the preparation of the o-nitroanilines of structure (3) and the subsequent coupling to yield the diazonium salts onto phenols of structure (4), where R_1 through R_8 could be a variety of substituents. Of particular interest in photographic systems are those compounds in which R_5 is H, alkyl, acyl, dialkylcarbamyl, and Z = OH, NH_2, O-alkyl, O-sulfonyl or substituted sulfonyl group, N- alkyl, N-acyl, N-sulfonyl or substituted sulfonyl group, and also when R_8 = H or Z, while Z = OH, or -NH-O-alkyl.

STRUCTURE (1)   STRUCTURE (2)

STRUCTURE (3)   STRUCTURE (4)

DEVELOPMENTAL STRATEGIES

It was necessary to attempt to prepare some of these benzotriazole compounds with a chlorine present on one or both rings. This led us to incorporate into the reduction the use of acids derived from nonmetallic elements with low oxidation states such as hypophosphorous acid, sulfurous acid, or acetic acid, and n-butyric acid in the presence of an organic amine base such as tert-butylamine. The use of acetic acid, however, resulted in complete removal of the chlorines from the aromatic rings. The reaction conditions for the desired reaction were optimized by use of a designed experiment using JMP software from the SAS Institute. The experiment was designed to examine four factors over two levels with center points. This was a 2^{4-1} fractional factorial design. This gave eleven experiments with the center points. The factors examined were catalyst load, amount of the organic base, the amount of the acid, and the temperature. Figure 1 shows the experiment as it was laid out with the measured response factors and values.

Rows	Pattern	Cat. Load Grams	t-Bu--NH$_2$	Hypoph acid eq.	Temp Deg° C	Center Point	Product A % Yield	Dechloro % Yield	De-blocked B	Other
1	--++	0.25	2	3	90	0	10	0	0	90
2	0000	0.375	3	2.5	70	1	75	5	0	20
3	+-+-	0.5	2	3	50	0	5	0	0	95
4	+--+	0.5	2	2	90	0	2	0	0	98
5	-+-+	0.25	4	2	90	0	80	2	5	13
6	-++-	0.25	4	3	50	0	85	0	1	14
7	++++	0.5	4	3	90	0	78	0	5	17
8	0000	0.375	3	2.5	70	1	80	0	0	20
9	0000	0.375	3	2.5	70	1	75	0	0	25
10	++--	0.5	4	2	50	0	75	0	0	25
11	----	0.25	2	2	50	0	80	0	5	15

FIGURE 1. A 2^{4-1} Fractional Factorial Design. In the pattern column a (+) sign indicates the high level and the (-) indicates the low level the acid and base are shown in equivalents.

The catalyst/substrate ratio ranged from 1/50 to 1/10 with the best results at a ratio of 1/20. The organic base and acid were examined in the range of 2 to 4 equivalents. It is known that the desired reaction will take place at room temperature given enough time, but better results were obtained in a range of 50 to 90 degrees C. The response factors were the percentage yield as determined by HPLC in most cases, the percentage of compound with the chlorine removed from the aromatic ring, and the amount of product formed as desired with the exception of the protecting group from the OH being removed. This, of course, was considered a positive benefit since, after the hydrogenation and cyclization, the protecting group was removed in a deblocking step. The reduction can take place at hydrogen pressures of from 5 to 100 psig; however, there is less N-N bond cleavage when the pressure is kept low. This, however, depends on the character of the azo dye compound. There is a tendency for more N-N bond cleavage when the azo dye has more electron withdrawing groups, if the hydrogen pressure is high.

RESULTS AND DISCUSSION

The hydrogenation catalysts used in this work were palladium, platinum, or Pd/Pt on carbon (charcoal) support. The amount of catalyst on each metal can vary, but in this work 4% Pd and 1% Pt on carbon support was the most useful. The catalyst may be recycled by filtration and washing the catalyst with suitable solvent followed by water to give a catalyst of essentially identical activity even after several batch-wise operations. The choice of catalyst loading used in the statistical designed program JMP was in the 1/20 catalyst to substrate ratio.

Organic amines having a NH or NH_2 may be used or even ammonium hydroxide. It is not necessary to use very high boiling amines as described in the patents. The most practical amine appeared to be tert-butylamine, which can be easily removed by a number of techniques. The optimum ratio of amine as predicted from the designed experiment using the JMP software program was 4:1 (mole/mole ratio) amine-to-azo compound. The effective acids used can be acids having one or more ionizable hydrogen atoms. The acid should be of lower oxidation states and for the best results have pKa_1 in the range of 1.8 to 1.9 and a pKa_2 of between 6.0 and 7.0. This is required to retain the chlorine atom on the aromatic ring. When acids such as acetic acid, propionic acid, and butyric acid (with pKa range of 4.2 to 4.85) were used, the halogens were cleanly removed from the aromatic ring. The importance of the ionizable hydrogen atom in the reduction can be demonstrated by using the sodium or potassium salts of the acids, while not changing any other components of the reaction. With alkali salts, 25-40% of an N-oxide was found in the reaction as a result of incomplete reaction and in some cases 5-10% of the overreduced products were observed. We observed "that an anticipated hydrogen-transfer mechanism (i.e. H from acids being transferred to azo compounds of structure (**1**) during the catalytic hydrogenation process) is not operational at least to a significant extent, and is demonstrated by using a well-known hydrogen transferring-reagent, such as 1,4-dihydroquinone, in place of acid(s), but in presence of all remaining reagents under reaction conditions of this work (17). Under this condition, not only dechlorination occurs, but over reduction of the benzo ring predominates by a factor of three. The appropriate acid is essential for the complete conversion to the benzotriazoles.

A prediction profile of the experimental design clearly shows that the ideal load of catalyst is in the 1/20 ratio catalyst to substrate (see Figure 2) to give the maximum yield, while the higher level of amine (base) is also favorable to give the highest yield of product. There was not a significant improvement in the yield with increased amount of the acid; however, as has been indicated earlier, it is essential to have the free protons available in the reaction and increasing the amount of the free protons does not improve the yield. There was a decrease in the yield of desired product as the temperature was increased; however, temperature has very little effect on the amount of deblocked material formed (see Figure 2).

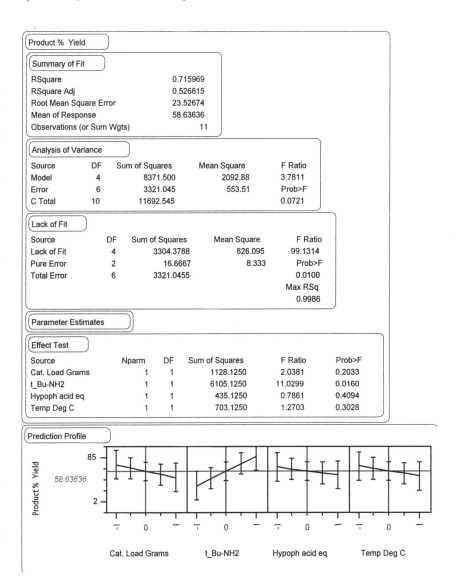

FIGURE 2. Design analysis with respect to product. The slope in the line in the prediction profile indicates a positive or negative effect of the factor shown.

By far the most profound influence on the reaction to increase the amount of product is the amount of the amine (base) that is used. At the same time, with high levels of the amine there is also an increase in the amount of deblocked product formed. Since the reaction product is later treated to remove the blocking groups, the deblocked material is in the positive direction. Thus, the profile shows the amount of A + B (the product plus the deblocked product) is increased and a profile slope in the positive direction (see Figure 3).

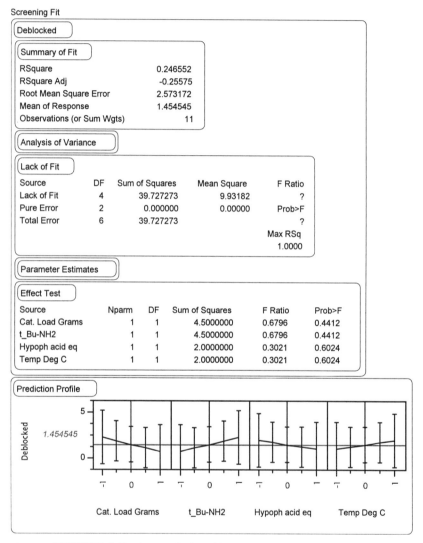

FIGURE 3. Design analysis with respect to the deblocked product. The slope of the line in the prediction profile indicates a positive or negative effect on the factor shown.

The effect of temperature on the reaction is much less significant than is the influence of the amine. With higher temperatures the amount of the product is slightly decreased while the amount of the deblocked material is increased. Since it is helpful to look at the combined yield of the product and the deblocked material, the temperature is not an important factor.

There are interactions as revealed by the design. There is an interaction between the load of catalyst and the acid, which is equal to the interaction of the amine and the temperature relative to the amount of the deblocked product formed (see Figure 4). The other interactions are not significant. Figure 4 shows the statistical summary of the design relative to the yield and the yield plus the deblocked product (A + B) (See Figure 4).

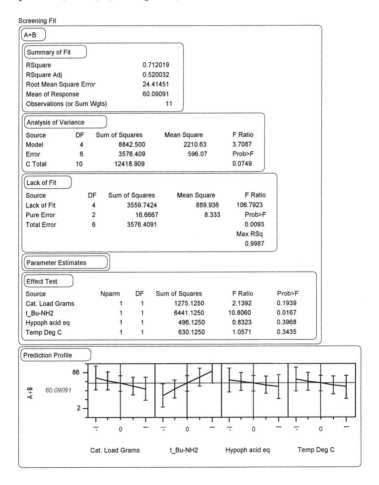

FIGURE 4. Design analysis with respect to the sum of product and yield of deblocked material.

EXPERIMENTAL

The catalysts used were supplied by Johnson Matthey, Engelhard and Degussa and were palladium in 1 to 10% loading on carbon (charcoal), platinum at 1 to 5% loading on carbon and palladium, and platinum mix with 4% palladium and 1% platinum on carbon support.

The mass spectra were recorded by the Analytical Technology Division of Eastman Kodak Company. The proton NMR spectra were recorded in deuterochloroform or in deuterated dimethylsulfoxide (DMSO-d_6) on a General Electric QE-300 MHz Ft-NMR Instrument using tetramethylsilane as an internal reference. The UV-visible absorption spectra in solution were recorded on a Lambda-9 Perkin-Elmer UV-Vis NIR Spectrophotometer and saved as files in the HP1000 System. The purification of the intermediates, or the final UV dye compounds, was carried out either by recrystallization from an appropriate solvent or by silica gel flash column chromatography, eluting with an appropriate mixture of n-heptane and ethyl acetate, or methylene chloride. The routine HPLC analysis in the Dye Research Lab was performed on a Hewlett-Packard 1090 Liquid Chromatograph using the "WORLD" HPLC System World Method.

GENERAL REACTION PROCEDURE

The isolation of the final products from the reaction was completed by conventional methods.

The following example is illustrative of the process described in this work and is the example used in the designed experiment.

Preparation of 2-(2'-4'-Di-NN-dimethylcarbamyloxyphenyl)-5-chlorobenzotriazole (6) CAS # 196790-12-2:

(eq. 2)

(5)   (6)

To a 500 mL stainless steel autoclave equipped with a stirrer, external heater, and internal heating and cooling is added 5.0 g (0.0115 mole) of 4-chloro-2-nitro-2',4'-di-N,N-dimethylcarbamyloxyazobenzene (**5**) , 0.25 g (dry weight) of 4% Pd/C + 1% Pt/C) catalyst, 3.82 g (0.052 mole, 4 equivalents) of tert-butylamine, 5.16 g (0.039 mole, 3 equivalents) of 50% aqueous hypophosphorus acid and 150 mL of methanol. The autoclave was purged with either nitrogen gas or hydrogen gas and then sealed and charged with hydrogen to a pressure of 75 psig. The reaction mixture was stirred at room temperature and 75 psig for one hour. Next the temperature was raised to 50°C and held at that temperature and pressure while stirring for an additional 12 hr. At the end of the reaction time the autoclave and contents were cooled to 45°C and the contents removed and filtered through a Super-cel pad to remove the catalyst. The catalyst and filter-pad were washed with solvent. The organic solvents were removed on a rotary evaporator. The residue was diluted with about 500 mL of brine and treated with hydrochloric acid drop-wise until the Congo Red indicator paper turned blue. The light-brown insoluble material (the desired product) was collected on a sintered glass funnel, washed with cold distilled water to remove any salt, and then air-dried. The crude solid material was recrystallized from isopropanol/water to obtain 4.18 g of 2-(2'-4'-di-N,N-dimethylcarbamyloxyphenyl)-5-chlorobenzotriazole (**6**) as an off-white solid having a melting point 138-139°C, showing a retention time of 15.97 min in HPLC in 100% purity by peak area, and molecular ion at m/e 403 in its FD-mass spectral analysis. It gives a characteristic blue fluorescence under the short wavelength UV light. Its H-NMR in DMSO-d_6 (with tetramethylsilane as an internal reference) showed peaks at 8.2(s, 1H, arom.), 8.1 (t, 2H, arom.), 7.55 (d, 1H, arom.), 3.1 (s, 3H, N-CH$_3$), 2.95 (two singlets, 6H, 2xN-CH$_3$'s).

The above reaction was repeated using either 5% Pd/C or 3-5% Pt/C as the catalyst. The yield with the Pt/C catalyst was 81% and with the Pd/C catalyst was 85% yield. In a similar manner, the tert-butyl amine was replaced by sodium hydroxide in one of the runs using the mixed metal catalyst with similar yields of 83%.

This procedure was scaled to a 20-liter capacity stainless steel autoclave. The crude material obtained from the reaction was recrystallized from aqueous isopropanol to give a 91% yield.

Preparation of 2-(2'-Hydroxy-3',5'-di-tert-amylphenyl)benzotriazole (8) CAS # 025973-55-1:

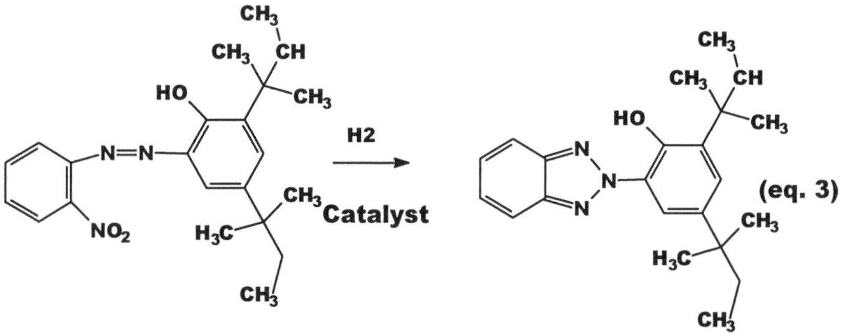

(eq. 3)

(7)

(8)

In a procedure similar to that cited in the preparation of compound (6) 2-nitro-2'-hydroxy-3',5'-di-tert-amylazobenzene (7), yielded compound (8) in a yield of light-brown insoluble material was filtered on a sintered glass funnel, washed with cold distilled water to remove any salt, and air-dried. This crude solid material was recrystallized from isopropanol/ water to obtain 4.56 g of 2-(2'-hydroxy-3',5'-di-tert-amylphenyl)benzotriazole (8). The yield was 99% with melting point 80-81° C showing a retention time of 25.89 min in HPLC analysis.

Preparation of 2-(2',4'-Dihydroxyphenyl)-5-chlorobenzotriazole (10) CAS # 057567-95-0:

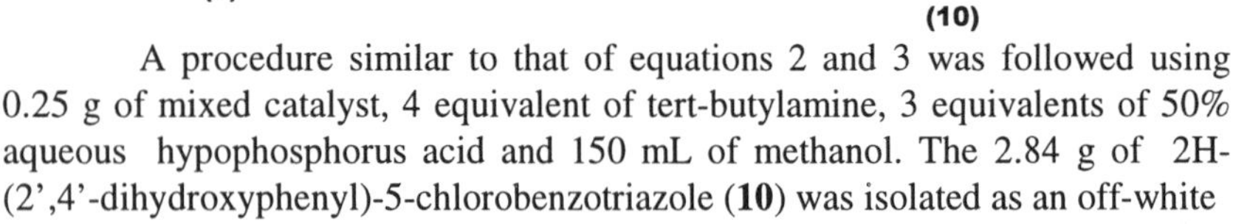

(9)

(10)

A procedure similar to that of equations 2 and 3 was followed using 0.25 g of mixed catalyst, 4 equivalent of tert-butylamine, 3 equivalents of 50% aqueous hypophosphorus acid and 150 mL of methanol. The 2.84 g of 2H-(2',4'-dihydroxyphenyl)-5-chlorobenzotriazole (10) was isolated as an off-white solid (yield 85% of theory). This was characterized by its retention time of 15.03 min in HPLC analysis and its molecular ion at m/e 261 in FD-mass spectral analysis.

This example illustrates the usefulness of this work where *in situ* deprotection of the protecting groups such as acyl or dialkylcarbamyl or alkylsulfonyl or arylsulfonyl or heteroarylsulfonyl is deemed highly desirable.

Benzotriazoles Prepared by Catalytic Hydrogenation

Other compounds prepared by this general procedure are shown in Table 1.

TABLE 1
BENZOTRIAZOLES PREPARED

R1	R2	R3	R4	-OR5 **	R6	Z	R7	R8	mass spec m/e
H	H	Cl	H	OCON(CH3)2	H	H	H	H	316
H	H	H	H	OCON(CH3)2	H	OCON(CH3)2	H	OCON(CH3)2	456
H	H	Cl	H	OH	H	H	(CH2)2OH	H	289
H	H	OCH3	H	OH	t-Bu	H	C9H19	H	563
H	H	OCH3	H	OH	t-Bu	H	OH	H	313
H	H	Cl	H	OH	H	H	(CH2)2COOH	H	317
H	H	Cl	H	OCH3	H	OCH3	H	H	289
H	H	Cl	H	OOCCH3	H	OOCCH3	H	Cl	380
H	H	Cl	H	OOCCH3	H	OOCCH3	H	H	345
H	H	Cl	H	OCON(CH3)2	H	NHC2H5	CH3	H	373
H	H	Cl	H	OH	H	NHC2H5	H	H	288
H	H	OCH3	H	OCON(CH3)2	H	OCON(CH3)2	H	OCON(CH3)2	486
H	H	H	H	OH	t-C5H11	H	H	t-C5H11	351
H	H	Cl	H	OH	H	OH	H	H	261
H	Cl	Cl	H	OH	H	OH	H	H	296
H	H	Cl	H	OH	H	OH	Cl	H	296
H	H	Cl	H	OH	H	NC2H5(COCH3)	H	H	344
H	H	H	H	OH	H	NC2H5(COCH3)	H	H	296
H	H	Cl	H	OCON(CH3)2	H	NC2H5(CON(CH3))2	H	H	430
H	Cl	Cl	H	OCON(CH3)2	H	OCON(CH3)2	H	H	438
H	H	H	H	OH	H	NC2H5(CON(CH3))2	CH3	H	339
H	H	H	H	OH	H	NC2H5(COCH3)	CH3	H	310

See Structure (**2**) for the positions of R_1 - R_8 and Z
** For clarity R_5 is shown with the O such as -OR_5

CONCLUSION

Benzotriazole UV absorbers of the type shown in Structure (**2**) have been successfully prepared by a one-pot synthesis using catalytic hydrogenation in high yield and purity. The process conditions were determined by use of a fractional factorial design experiment, which served as an effective model to develop the process. The design shows the need for four equivalents of the amine and demonstrates that when the acid of the appropriate pK_a is used, the reduction of the nitro group and ring closure takes place, but beyond the required stoichiometric amount, additional acid has little influence on the product.

The synthesis provides a process that has the advantage of being environmentally friendly and yet is effective in providing high-quality UV absorbers used in the photographic industry.

Benzotriazole compounds can be easily and cleanly prepared in one step to give the product where R_5 is H and Z is H or OH when the blocking group used is for example an acetyl in high yield and purity.

ACKNOWLEDGMENT

The work described here is the result of a team effort.We wish to thank the Analytical department from the Kodak Research Laboratories for the analytical work, and Dr. James Kloek and Dr. John Hamer for their encouragement and suggestions.

REFERENCES

1. M. Dexter in Kirk-Othmer *Encyclopedia of Chemical Technology,* 3rd edition, Vol **23**, 615-627 Wiley-Interscience: New York (1983)
2. L.C. Vishwakarma, V.L. Mylroie et. al., U.S. 5,675,015 and U.S. 5,670,654, assigned to Eastman Kodak Company. (1997).
3. Unpublished work by L.C. Vishwakarma and V.L. Mylroie.
4. Unpublished work by L.C. Vishwakarma
5. S. Hotta, Y. Kondo, Kokai JP 50-121178, to Sumitomo Chem. Co. Ltd. (1975).
6. H.S. Freeman and J.C. Posey, Jr. in *Dyes and Pigments*, **20**, 171-195 (1992).
7. E.H. Jancis, U.S. Patent No. 3,978,074, to Uniroyal Inc. (1976).
8. E.H. Jancis, U.S. Patent No. 4,089,839, to Uniroyal Inc. (1996).
9. S. Kintopf, U. Kress, U.S. Patent No. 4,230,867, to Ciba-Geigy (1980).
10. J.W. Long, L. Vacek, U.S. Patent No. 4,363,914, to Sherwin-Williams Co. (1982).
11. A. Davatz, T. Somlo, U.S. Patent No. 4,642,350, to Ciba-Geigy AG (1987).
12. H. Prestel, K. Muller, U.S. Patent No. 4,999,433; E. P. 363318, to Ciba-Geigy (1991).
13. N. Fukuoka et. al., U.S. Patent No. 5,104,992, to Chemipro Kasei Kaisha Ltd. (1992).

14. N. Fukuoka et. al., U.S. Patent No. 5,187,289, to Chemipro Kasei Kaisha, Ltd. (1993).
15. H. Prestel et. al., U.S. Patent No. 5,276,161, to Ciba-Geigy Corp. (1994).
16. J.K. Kaplan et. al., U.S. Patent No. 5,571,924, to Ciba-Geigy Corp. (1996).
17. Private communication.
18. W.B. Hardy, J.S. Milionis, U.S. Patent No. 3,072,585 to American Cyanamid Co. (1963).
19. W.F. Baitinger et. al., U.S. Patent No. 3,159,646 to American Cyanamide Co. (1964).
20. J. Jaeken et. al., U.S. Patent No. 3,813,255, to Agfa-Gevaert (1974).
21. S. Shuichi, U.S. Patent No. 4,780,541, to Chemipro Kasei Kaisha, Ltd. (1988).
22. British Patent No. GB-A 1,494,825, to Ciba-Geigy Corp. (1977).
23. British Patent No. GB-A 1,494,824, to Ciba-Geigy Corp. (1977).
24. British Patent No. GB-A 1,494,823, to Ciba-Geigy Corp. (1977).
25. C.V. Krolewski, H.L. White, U.S. Patent No. 4,219,480, to Ciba-Geigy Corp. (1980).
26. T. Sato, Y. Fukuda, Japanese Patent No. JP-A52/113,973, to Kawaken Fine Chemicals Co., Ltd. (1977).
27. K. Nakajima et. al., JP-A52/113,974, to Kawaken Fine Chemicals Co., Ltd. (1977).

New Data About the Well-Known Dehydrogenation Reaction of Cyclohexanol on Copper-Containing Catalysts

Vladimir Z. Fridman
Reflex, Inc., Brooklyn, New York

Anatoli A. Davydov
Department of Chemical and Material Engineering,
University of Alberta, Edmonton, Canada.

Abstract

Cu-Zn-AL and Cu-Mg catalysts of dehydrogenation reaction of cyclohexanol (CLL) to cyclohexanone (CLN) have been studied by IR-spectroscopy adsorbed probe molecules, X-ray diffraction, X-ray photoelectron spectroscopy and kinetic method. The materials of this investigation show that Cu-containing catalysts: a) have two types of the active surface sites $(Cu^+$ and $Cu^0)$ for CLL dehydrogenation to CLN; b) the partly oxidized copper sites $-Cu^+$ are much more active than the sites of metallic copper; c) the reaction of CLL aromatization to phenol (PHL) takes place on the surface sites- Cu^0 only; d) the CLL dehydrogenation to CLN on Cu^0 sites carries out through the step of the CLL adsorption by the plane of the cyclohexane ring; e) the CLL dehydrogenation to CLN on Cu^+ sites occurs via the adsorbed CLL alcoholate formation, which easily transforms to CLN.

Introduction

Copper-containing catalysts have been used in the industrial process for dehydrogenation of cyclohexanol to cyclohexanone for many years. A number of studies of copper-containing catalysts of CLL dehydrogenation have been reported (1-3), but the following questions remain without answers:

1) What kind of a copper active site is responsible:
 a) for the main reaction of synthesis of CLN from CLL : $C_6H_{11}OH{=\!=\!=}C_6H_{10}O+H_2$ (Reaction 1)?
 b) for the concurrent unwanted reaction of CLL aromatization with the formation of phenol: $C_6H_{11}OH{=\!=\!=}C_6H_5OH+3H_2$ (Reaction 2)?
2) How does the degree of oxidation of copper sites influence the level of activity in the Reaction 1?
3) What are the mechanisms of the CLL dehydrogenation on the sites in different

degrees of the copper oxidation? To answer these questions, we are reporting an investigation of the reaction of CLL dehydrogenation on the Cu-Zn-Al and Cu-Mg catalysts with different copper loading and the degree of copper oxidation on the catalyst surface.

EXPERIMENTAL

Cu-Zn-Al and Cu-Mg catalysts were prepared by a co-deposition procedure from an aqueous solution of $Cu(NO_3)_2$, $Al(NO_3)_3$, $Zn(NO_3)_2$ and $Mg(NO_3)_2$ using a 10% solution of Na_2CO_3 followed by washing of the Na^+ ions, drying at 120 °C and calcining at 350° C. The Cu-Mg catalyst was calcined at 500 °C. The activity of catalysts was studied by the kinetic method in flow type reactor with a fluidized catalyst layer in the temperature range at 220 ‑ 300 °C. Initially, the samples were reduced by H_2 at 250 °C. The activity of the catalysts was estimated for the Reaction 1 by the rate of the CLL dehydrogenation at 250 °C and also for the Reaction 2 by the rate of the formation of PHL at 300 °C. The Cu -Zn-Al samples were studied by X-ray diffraction. X-ray photoelectron spectroscopy (XPS) was used to estimate the distribution of copper and zinc atoms on the catalyst surface. The measurements were recorded before and after use of catalysts in the reaction of dehydrogenation. IR-spectroscopy of adsorbed probe molecules (4,5) was used to determine the oxidation degree of the copper sites and the forms of the adsorption of CLL and CLN on the surface of catalysts. Specific surface values were determined for the fresh samples of catalysts.

RESULTS AND DISCUSSION

1. Catalyst activity as a function of the degree of copper oxidation.
The Cu-Zn-Al catalysts were prepared with the different atom loading of Cu, Zn, and at a constant Al loading equal to 15% at. Cu-Zn-Al catalysts copper loading 5, 15, 20, 40, 60 and 80% at. were prepared. The XRD data (phase compositions), the XPS data (the ratio of the copper and zinc atoms on the catalyst surface) and values of the specific surface of catalysts are presented in Table 1. In this work we could not determine the presence of aluminum atoms on the catalysts surface because of the low level of the section of XPS lines of Al2p and Al2s. However, this shows the insignificant amount of the aluminum atoms on the surface of Cu-Zn-Al catalysts. The IR- spectra of CO adsorbed on the Cu-Zn-Al catalysts are presented on Fig.1. To assign the absorbed bands (AB) two criteria were used, namely: the value of CO vibration and thermal stability of the CO complexes with the copper cation. It is known (6,7) that spectra of CO adsorbed on the copper cations with the different oxidation degrees have different values of vCO (Cu^0- CO < 2100 cm^{-1}, Cu^+ -CO -2110-2160 cm^{-1} , Cu^{2+}-CO > 2160 cm^{-1}) and thermal stabilities of the complexes are different (Cu -CO is easily removed at temperature around 100 °C and while the Cu^+- CO is a stable

Table 1.
PROPERTIES OF THE Cu-Zn-Al CATALYSTS BEFORE AND AFTER USE IN THE REACTION OF DEHYDROGENATION OF CLL

Concentration of components Cu-Zn-Al in catalysts (at.%)	Phase Composition XRD		Surface ratio of the atoms Cu/Zn XPS	
	Fresh	After use	Fresh	After use
1. 5.0:80.0:15.0	ZnO	ZnO	0.068	0.096
2. 15.0:70.0:15.0	ZnO	ZnO	0.170	0.160
3. 20.0:65.0:15.0	ZnO	ZnO	0.290	0.320
4. 40.0:45.0:15.0	ZnO, Cuo traces	ZnO,Cu	0.625	0.523
5. 60.0:25.0:15.0	ZnO, CuO	ZnO,Cu	3.545	1.150
6. 80.0: 5.0:15.0	CuO	Cu	10.700	1.983

complex and it cannot be removed at 150 °C. The spectral data were obtained for the catalysts with the copper loading 5 , 15 and 20 % at. (Fig.1).

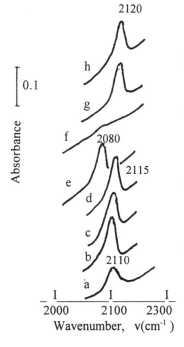

Fig.1 IR-spectra of adsorbed CO: a-on the Cu-Zn-Al (Cu-5%at.) reduced by H_2 at 523 K and evacuated (condition 1); b-on the Cu-Zn-Al (Cu-15%at.)-condition 1; c-on the Cu-Zn-Al (Cu-20%at.) -condition 1; d-on the Cu-Zn-Al(Cu-15%at.)-condition 1 and evacuated at 150 °C after adsorption of CO; e-on the Cu-Mg (Cu-52%at.)-condition 1; f-on the Cu-Mg (Cu-52%at.) - condition 1 and evacuated at 100 °C after adsorption of CO and C; g-on the Cu-Mg (Cu-52%at.) -reduced by CO at 100 °C and evacuated (condition 2); h- on the Cu-Mg (Cu-52%)- condition 2 and evacuated at 150 °C after adsorption of CO.

Prior to the measurements have been taken, the catalysts were reduced by H_2 at

250 °C. The IR spectra of CO adsorbed on the reduced surface of these samples display one band at the range 2110-2115 cm^{-1} (Fig.1a,b,c). The following evacuation at 150 °C does not change the spectral picture (Fig. 1d). These data demonstrate the high level of thermal stability of these complexes and allow us to attribute this AB to v(CO) in Cu$^+$ -CO. The reduced samples at 250 °C with the Cu loading equal and more than 40% at. have not passed IR-radiation. This phenomenon has been observed earlier for Cu-Zn-containing catalysts with a high Cu- loading (8), where it was shown that the samples with bulk and surface metallic copper in Zn-containing catalysts have not passed the IR-radiation. This fact allowed us to assume the lowest oxidation degree of copper on the surface of the Cu-Zn-Al catalysts with copper loading more than 40% at.. This conclusion is in agreement with the data (9-11).Thus, the IR data indicates that in the catalysts with low copper loading (Cu<20%), even after reduction by H$_2$ at 250 °C, copper stabilizes on the surface in partly oxidized state(Cu$^+$). At the same time, in the catalysts with copper loading more than 40% at. is easily reduced to the metallic state. The dependence of Reaction 1 rate to the loading of copper is presented in Fig.2, which demonstrates that the initial Zn-Al system does not work in CLL dehydrogenation. The addition of copper to Zn-Al system leads to the significant growth of the activity of catalysts in the Reaction 1 (Fig.2).The maximum rate of Reaction 1 is observed on the catalyst with Cu-20% at., but the subsequent increase of copper loading gradually decreases the rate of Reaction 1. A different dependence is observed in the reaction of the PHL formation (Reaction 2) (Fig.2). PHL is not formed on the catalysts with copper loading less than 20% at.. A very small amount of PHL was determined on the catalyst with 20% at. of copper. The maximum rate of Reaction 2 is observed on the catalyst with copper loading of 40% at.. The subsequent increase of copper loading leads to a small decrease of Reaction 2 rate (Fig.2).

The following conclusions regarding the influence of copper oxidation degree on the kinetic behavior can be derived from the above - mentioned results.

Reaction 1

After reduction the copper atoms are in two oxidation states on the surface of Cu-Zn-Al catalysts, which are the active sites of Reaction 1. Apparently, the catalysts with copper loading less than 20%at. have only one kind of active sites of copper- Cu$^+$. The catalysts, which contain 20-40% at. of copper, have both Cu$^+$ and Cu^0, and the catalysts with high copper loading more than 40%at. have only one kind of active sites Cu^0. To compare the activities of these sites (Cu$^+$ and Cu^0) of copper the catalysts with copper loading of 5% and 100 % at. were used, since there is only one type of the active copper sites Cu$^+$ and Cu^0 on these samples respectively. The values of specific area and XPS data were used to determine the copper surface in the catalysts with 5% at. and 100% at. of copper, and they are equal to 5.7 and 15.0 m^2 /g respectively. The activities of one square meter of copper surface are equal to 6.1*10^{-5} mol/min for Cu$^+$ and to 1.2*10^{-5} mol/min for Cu^0. This data indicates that the sites Cu$^+$ are much more active sites in Reaction 1 than the sites of Cu^0.

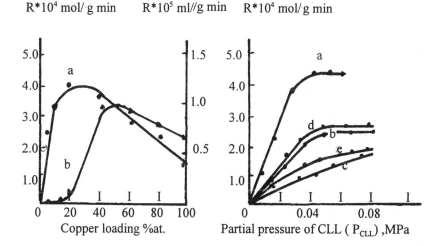

R*10⁴ mol/ g min   R*10⁵ ml//g min   R*10⁴ mol/ g min

Fig.2 The rates of Reaction 1(a) (250 °C) and Reactions 2 (b) (300 °C) of copper as function loading in the Cu-Zn-Al catalyst.

Fig.3 The rate Reaction 1 as function of partial pressure of CLL under different conditions on Cu-Mg catalyst with metallic state of copper (Cu⁰) on the surface a-t-280 °C, Pi_{CLL}-0.080 MPa; b-t-250 °C, Pi_{CLL}- 0.08 ; c-t-220 °C, Pi_{CLL}-0.080 MPa; d-t-250 °C, Pi_{CLL}-0.080; Pi_{H_2}-0.0128 MPa; e-t-250 °C, Pi_{CLL}-0.080, Pi_{CLN}-0.018 MPa. Pi-initial partial pressure of CLL,CLN,H_2

Reaction-2

The comparison of dependence of Reaction 2 rate to the copper oxidation degree suggests that the catalysts which have only the Cu^+ sites on their surface do not form PHL. At the same time, the appearance of metallic copper sites in the Cu-Zn-Al catalysts leads to formation of PHL from CLL. It means that the partly oxidized copper sites are not active in Reaction 2 and the formation of PHL carries out only on the Cu^0 sites.

Apparently, the different mechanism of Reaction 1 on Cu^+ and Cu^0 sites is the main reason the activities of these sites in the CLL dehydrogenation differ. That is why the next part of the present work is dedicated to investigate the mechanism of CLL dehydrogenation on the catalysts which contain only one kind of active sites.

2. CLL dehydrogenation on the sites of metallic copper.

The investigation of CLL dehydrogenation on the sites of metallic copper was carried out on the Cu-Mg catalyst with Cu-52% at. Kinetics of Reaction 1 was studied in the temperature range at 220-280 °C at different initial partial pressures

CLL, CLN and H2 in the range of the flow rate at 0.5-4.0 h^{-1}. Prior to dehydrogenation the catalyst was reduced by hydrogen at 250 ^{0}C. To estimate the copper oxidation degree on the surface of Cu-Mg catalyst this sample was explored by IR-spectroscopy of adsorbed CO. The IR- spectrum of CO adsorbed on the reduced surface of the Cu-Mg catalyst displays one band at 2080 cm^{-1} (Fig.1e). The following evacuation of the samples at 100 ^{0}C changed the spectral pictures, and the 2080 cm^{-1} band disappeared (Fig.1,f) . These data show the low level of thermal stability of CO complexes and suggest to assign the spectrum to v(CO) in Cu^0-CO. Therefore there is only one type of copper sites (Cu^0) on the surface.

The kinetic data of CLL dehydrogenation on the Cu-Mg catalyst are presented in Fig.3 and show that the rate of Reaction 1 grows with increasing the partial pressure of CLL. It should be noted, that the latter is observed only in the low range of CLL coverage of the surface (Fig.3,a.b,c). The Reaction 1 rate attains a constant value after subsequent increase of the CLL partial pressure. These results can be explained by two possible reasons of a negative influence of CLL adsorption the rate of the main reaction . First one is that CLL has the strong adsorption form on the surface of copper which blocks the subsequent adsorption of CLL. Second one is that the step of the CLL conversion into CLN is very slow and, probably, this step is the limiting one in Reaction 1.The addition of H$_2$ to the reaction atmosphere does not influence the rate of Reaction 1 (Fig.3,d). Contrary to the influence of hydrogen, CLN significantly decreases the rate of Reaction 1 (Fig.3,e). Probably, there is a very strong adsorption of CLN on the surface of metallic copper. The strongly adsorbed CLN blocks new adsorption of CLL on the surface and decreases the rate of the main reaction. To determine the influence of the opposing reaction, CLN hydrogenation was studied at 220 and 250 ^{0}C. It was shown that CLN does not transform into CLL under these conditions at all. Therefore, the step of CLL conversion into CLN is not reversible.

To understand the mechanism of Reaction 1 on the sites of metallic copper, the adsorption and both the CLL and CNL were studied by IR -spectroscopy.

CLL and CLN adsorption on the metallic copper surface of the Cu-Mg catalyst.
Prior to the CLL adsorption, the Cu-Mg catalyst was reduced by H$_2$ to form the metallic state of copper on the surface of the catalyst. The IR -spectrum after exposure of CLL at 20 ^{0}C on the reduced surface of the Cu-Mg catalyst does not display any AB. After exposure of CLL at 50 ^{0}C in the wavenumber region 1000-1800 cm^{-1} the spectrum displays three AB: at 1590,1490,1300 cm^{-1} (Fig 4, c). According to (12,13) the band at 1300 cm^{-1} can be attributed to the stretching of the C-O bond in PHL. The AB at 1490 and 1590 can be assigned to the stretching of the aromatic -C=C- bond. This spectrum does not have either the band of the deformation of the C-O-H bond at 1370cm^{-1} or the the band of the stretching of O-H at 3500cm^{-1}. Both of them exist in the initial CLL spectrum (13). This result indicates that the CLL adsorption is accompanied by a breakage of the O-H bond. Moreover, this spectrum does not have the band at 1070-1100 cm^{-1} which belongs

to the stretching of the C-O bond of CLL. However, in the wavenumber region more than 2500 cm^{-1} the spectrum has AB at 2880 and 2995 cm^{-1} which obviously can be attributed to the stretching of the C-H bond in CH_2. The subsequent heating of CLL on the surface of the Cu-Mg catalyst up to 200 °C leads to the appearance of the weak band at 1690 cm^{-1}. The increase of the temperature up to 250 °C enlarges the intensity of the band at 1690cm^{-1} (Fig.4d,e).

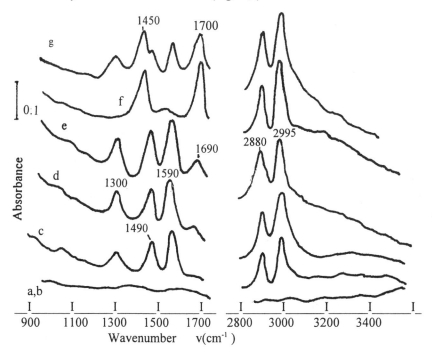

Fig.4 IR-spectra of adsorbed CLL and CLN on the Cu-Mg catalyst after formation on surface metallic state of copper: a-initial Cu-Mg catalyst after reduction in H_2 and vacuum treatment at 250 °C; b-sample a after exposure of CLL at 20 °C; c-sample a after adsorption of CLL at 50 °C; d- sample a after adsorption of CLL at 50 °C and a treatment at 200 °C; e-sample a after adsorption of CLL and a treatment at 250 °C; f-sample a after adsorption of CLN at 20 °C; g-sample a after adsorption of CLN and a treatment at 150 °C.

Obviously, this band can be assigned to stretching of C=O bond in CLN (12), which was formed from the adsorbed CLL. The observed results of CLL adsorption are unusual. The spectrum of CLL adsorbed has the AB which are very close to AB in the spectrum of PHL, and this spectrum does not have the band of the stretching C-O bond of initial CLL. In another side, this spectrum has the AB (2880 and 2995 cm^{-1}), which prove the existence of cycloalkane species on the surface of the catalyst. It is possible to assume that after adsorption on the surface of metallic copper CLL forms two different kind of species. The first step of the adsorption of CLL is the breakage of O-H bond. The spectrum of the first kind of species has the AB which correspond to the AB of the C-H stretching in the spectrum of initial CLL, but it does not have the band of the C- O stretching. In our opinion, it is possible only in case if this bond becomes inactive to IR-radiation, probably, because of a very strong bonding with the surface of metallic copper. Apparently, these species form CLN. The spectrum of the second kind of species is very close to spectrum of PHL, that is why we called it phenol-like species, and these species can be a source of the PHL formation. We have considered two versions of the formation of these species. According to the first version, these species are not linked to each other, because they have been formed after adsorption of CLL on different sites of the surface. The second version is that CLL after adsorption on the surface of metalic copper forms the cycloalkane species of CLL. These species of CLL adsorption transform to the phenol-like species and to CLN by parallel reactions. In this case the CLN and the phenol-like species have to be connected to each other. To answer to this question the adsorption of CLN was studied on mettalic sites of copper. The spectra of CLN adsorbed on the surface of the Cu-Mg catalyst at 20 ^{0}C are presented in Figure 4 and display the AB, which belong to the initial CLN, where the most intensive band is the band at 1700 cm^{-1} (Fig.4,f). The subsequent heating of adsorbed CLN up to 150 ^{0}C leads to decrease of the intensity of the band at 1700 cm^{-1} and to the appearance of the AB at 1590, 1490 and 1290 cm^{-1} (Fig.4,g) which belong to the phenol-like species of CLN adsorption. These data prove our assumption about connection between CLN formation and the phenol like species of adsorbed CLL, and that CLN and the phenol-like species are formed from the same cycloalkane species of CLL. It should be noted that the formation of the phenol-like species from cycloalkane species of CLL occurs even at 50 ^{0}C. In our opinion, this transformation, which does not request a lot of energy, is possible only in case of the adsorption of CLL by a plane of a cyclohexane ring. This assumption is in agreement with work of Balandin (14). These results suggest the following scheme of Reaction 1 on metallic copper. The first step of the Reaction 1 is the CLL adsorption by the plane of the cyclohexane ring. The CLL adsorption is accompanied by the breakage of the O-H bond and the formation of cycloalkane and phenol-like species of CLL. Apparently, both kinds of species can transform to each other. The next step is the transformation of the CLL species into CLN. This step is reversible.

A number of equations has been obtained on the basis of this scheme and assumptions about different limiting steps of Reaction 1. The results of calculations of the rates on basis of the equation 1 $R=K_0 P_{CLL}/ (1+b_1P_{CLL}+b_2P_{CLN})$ correspond well to the experimental data at the values of the constants:

$$K_0=9.6* 10^3 *exp^{-51000/RT}; b_1=2.1* 10^5*exp^{-63000/RT}; b_2=7.0 *10^3 *exp^{-63000/RT}$$

This equation is based on the assumption that the step of the transformation of the CLL species to CLN is a limiting step of Reaction 1.

Obviously, the phenol-like species of CLL can be the source of PHL formation. The proposed mechanism of Reaction 1 makes understandable the reason of a low selectivity of industrial catalysts with metallic copper sites at high temperature, because of the same initial adsorbed complex as a source for the formation of CLN and PHL.

3. CLL dehydrogenation on partly oxidized copper sites

The investigation of kinetics of the CLL dehydrogenation on the sites of the partly oxidized copper (Cu^+) was carried out on the Cu-Zn-Al catalyst with copper loading 15% at.. There is only one type of the copper sites (partly oxidized copper) on the surface of such catalyst after the reduction by hydrogen (Fig.1,b). The kinetic data of CLL dehydrogenation on the Cu-Zn-Al catalyst with copper loading of 15 % at. are presented on Figure 5. They demonstrate that the rate of Reaction 1 grows with increasing of the partial pressure of CLL in all ranges of the coverage of the catalyst surface by CLL.

$R*10^4$ mol/g min

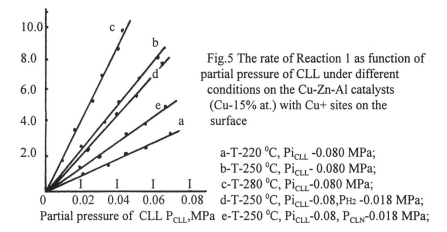

Fig.5 The rate of Reaction 1 as function of partial pressure of CLL under different conditions on the Cu-Zn-Al catalysts (Cu-15% at.) with Cu+ sites on the surface

a-T-220 ^{0}C, Pi_{CLL} -0.080 MPa;
b-T-250 ^{0}C, Pi_{CLL}- 0.080 MPa;
c-T-280 ^{0}C, Pi_{CLL}-0.080 MPa;
d-T-250 ^{0}C, Pi_{CLL}-0.08,P_{H2} -0.018 MPa;
e-T-250 ^{0}C, Pi_{CLL}-0.08, P_{CLN}-0.018 MPa;

Pi_{CLL} -initial partial pressure of CLL,CLN,H_2

We presume the reasons of observed dependence are, either the weak form of CLL adsorption on the Cu^+ sites or the high rate of the CLL conversion to CLN. The dependence to the reaction products on the rate of Reaction 1(Fig.5) shows that H_2 does not influence the reaction rate and that CLN decreases the rate of ketone formation. This result can be explained by the strong adsorption of CLN on the Cu^+ sites and, probably, by the fact that the step of desorption of CLN is the limiting stage of Reaction 1.

Adsorption of CLL on partly oxidized sites of copper.
For investigation of adsorption of CLL on partly oxidized sites of copper we used

the Cu-Mg catalyst with Cu-52% at..   To form the Cu⁺ sites the Cu-Mg catalyst
was reduced by CO at 100 °C followed by evacuation of CO. The spectrum of CO
adsorbed on the surface of such catalyst (Fig.1.g) displays one band at 2120 cm⁻¹
which is not removed by evacuation at 150 °C (Fig.1,h). This band can be assigned
to vCO in Cu⁺ species. The subsequent adsorption of CLL on this surface at 20 °C
leads to the appearance in the spectrum of the whole collection of the AB which
correspond to AB of initial CLL (the AB at 890, 970, 1025, 1070, 1360, 1450 cm⁻¹)
(Fig.6b,f). These species were not removed by vacuum at 20 °C (Fig.6 c).

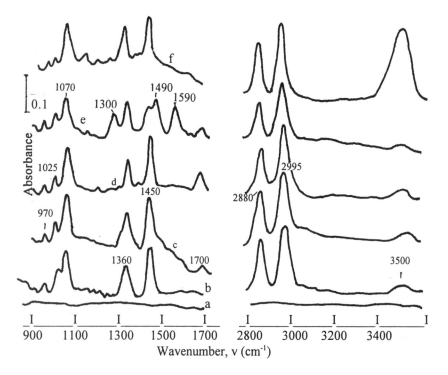

Fig.6 IR-spectra of CLL adsorbed on the Cu-Mg catalyst after forming
on the surface of a partly oxidized state of copper: a-initial Cu-Mg catalyst after
reduction by CO at 100 °C and vacuum treatment at 250 °C; b-sample a after
adsorption of CLL at 20 °C; c- sample a after adsorption of CLL at 20 °C and
evacuating at 20 °C; d-sample after adsorption of CLL and a treatment at 100 °C; e-
sample a after adsorption of CLL and an evacuation at 200 °C; f-initial CLL.

This fact indicates the strong connection of CLL to the surface of the catalyst.
The comparison of the AB in the spectrum of initial CLL (12) to the AB in the
spectrum of adsorbed CLL (Fig.6,b,f) shows that both spectra have the same AB
with the exception of very intense band v(OH) at 3500 cm⁻¹ Hence, the
disappearance of the intense AB at 3500cm⁻¹ and the stability of the CLL species
suggest to assign these species to surface alcoholates of Cu⁺ (6,15,16). Moreover,

the spectrum of CLL on the surface (Fig.6,b) displays weak band in the range at 3500 cm^{-1} which does not change even after an evacuation at 200 ^{0}C. These results can be explained, either by the additional bonding of the hydrogen atoms of CLL to surface oxygen atoms, or by a formation of a surface O-H group. Therefore, we can conclude, that CLL on the surface of partly positive charged copper forms adsorbed structure of CLL alcoholate, which is, probably, double-connected. These CLL species have high reaction ability and form CLN and even after desorption at 20 ^{0}C in a vacuum (the appearance of the AB at 1700 cm^{-1}) (Fig.6c). They form CLN at the high rate.

A number of equations have been obtained on the basis of this scheme and assumptions about different limiting steps of Reaction 1. The results of the calculation on basis of equation 2 $R=K_0 (P_{CLL} / \sqrt{P_{H2}} - P_{CLN}\sqrt{P_{H2}})/ (1 +b_1 P_{CLN})$ of Reaction 2 rates correspond well to experimental data at the values of the constants:

$$K_0 = 7.4*10^{3}* exp^{-59000/RT} \qquad b_1 = 6.1*10^{5}*exp^{-4200/RT}$$

This equation 2 is based on the assumption that the step of the desorption of the CLN species is a limiting step of Reaction 2.

CONCLUSIONS

1) The copper-containing catalysts have two types of the active surface sites (Cu$^+$ and Cu^0) for CLL dehydrogenation to CLN. These sites have the different activities;

2) The partly oxidized copper sites-Cu$^+$ are much more active than the sites of metallic copper;

3)The reaction of CLL aromatization to PHL takes place on the surface sites-Cu^0 only;

4) The CLL dehydrogenation to CLN on Cu^0 sites carries out through the step of CLL adsorption by the plane of the cyclohexane ring, this CLL species are also the intermediate compounds for the PHL synthesis;

5) The CLL dehydrogenation to CLN on Cu$^+$ sites occurs via adsorbed CLL alcoholate formation which easily transforms to CLN.

References

1. V.Z.Fridman, E.D.Mikalchenko, L.M.Plasova,A.N.Ziborov, B.G.Traysunov. *Kinetics and Catalysis(in Russia)*.32:922 (1991).
2. O.N.Medvedeva,A.S.Badrian,S.L.Kiperman. *Kinetics and Catalysis (in Russia)*.17:1530 (1976).
3. C.Sivarai,R.B.Manipal,R.P.Konta. Applied Catalysis.45:11 (1988).
4. A.A.Davydov, IR-spectroscopy in chemistry of a surface of oxide, (Science),Novosibirsk,1984, pp.23-47.
5. U.M. Shekochikin, A.A. Davydov. The method's questions of applying

of IR- spectroscopy for the investigation of physics-chemistry of a surface of semi-conductors, (Vinity), Moscow,1983, N 5014-83 ,pp. 1-37.

6.   A.A.Davydov. Infrared spectroscopy of adsorbed species on the surface of transition metal oxides, (Chichester),New York, 1990, pp.64-112,215-231.

7.   J.M.Gallardo Amores,V. Sanchez Estribano, G. Busca, and V.J. Loreazelli. *J. Mater. Chem.* 4:965(1994).

8.   J.Jhotty and F.Boccuzzi. *Catalysis Review, Science and Engineering.*29: 151(1987).

9.   D.J. Elliott and F.Penella *J. of Catalysis.* 102 :464 (1986).

10. Y.Okamoto, K.Fukino, T.Jmanara and S.Teranishi. *J Phys.Chem.* 87:3747 (1983).

11. T.M.Yurieva. *Reaction Kinetics and Catalysis Letters.*23:267 (1983).

12. G.Socrates. Infrared characteristic group frequencies: table and charts, (Chichester),New York, 1994, pp.35-67,122-125.

13. C.J.Pouchert. The Aldrich library of infrared spectra, (Milwaukee), Aldrich Chemical Co.,1981,pp.97,544.

14. A.A.Balandin. Multiplet theory,(Chemia), Moscow, 1969,pp.66-98.

15. N.A.Osipova,A.A.Davydov andL.N. Kyrina. *Russian J Phys.Chem.*112: 1296 (1988).

16. N.A.Osipova,A.A. Davydov,L.N.Kyrina and B.E.Loyko. *Russian J Phys. Chem.* 109:1479 (1985).

The Prins Reaction Catalyzed by Heteropoly Acids

Á. Molnár[1], Cs. Keresszegi[1], T. Beregszászi[1], B. Török[2], M. Bartók[1,2]

[1] Department of Organic Chemistry and
[2] Organic Catalysis Research Group of the Hungarian Academy of Sciences,
József Attila University, H-6720 Szeged, Dóm tér 8, Hungary

Abstract

Heteropoly acids and acidic cesium salts are found to be excellent catalysts for the Prins reaction to produce 1,3-dioxanes from styrenes and formaldehyde in dioxane as solvent in good yield under mild conditions. Less satisfactory results are achieved with acetaldehyde. Substituted 3,6-dihydro-2H-pyran derivatives are always formed from α-methylstyrene. Both thermal activation (with Cs salts) and microwave irradiation (with the use of heteropoly acids) are efficient, synthetically valuable methods to produce 1,3-dioxanes from styrenes.

Introduction

Wide-ranging studies have provided substantial experimental information in recent years which prove the outstanding properties of various heteropoly acids (HPAs) as catalysts in electrophilic transformations (1). According to our own experiences in this field, HPAs exhibit excellent catalytic properties in the dehydration of diols (2), the rearrangements of oxygen-containing compounds (3), the tetrahydropyranylation of alcohols (4) and Friedel–Crafts alkylation (5). To further widen the application of these interesting and important catalytic materials detailed studies of the Prins reaction (6) have been carried out. Since only a single paper is available on the transformation of styrene in the presence of various HPAs (7), it was our intention to acquire important new information including the possibility to develop synthetically useful reactions.

A model study, therefore, has been undertaken on the Prins reaction of styrene and α-methylstyrene with formaldehyde and acetaldehyde. HPAs with the general formula $H_n[XM_{12}O_{40}]$ (X= Si, P; M= Mo, W) and acidic cesium salts were used as catalysts. The activity of the various catalysts will be evaluated, and a comparison of thermal activation and microwave irradiation will also be discussed.

Experimental

Materials. Four commercially available HPAs, $H_4[SiMo_{12}O_{40}]$ (denoted as SiMo), $H_3[PMo_{12}O_{40}]$ (PMo), $H_4[SiW_{12}O_{40}]$ (SiW) and $H_3[PW_{12}O_{40}]$ (PW) were used. $Cs_{2.5}H_{0.5}[PW_{12}O_{40}]$ ($Cs_{2.5}PW$) and $Cs_{2.5}H_{0.5}[PMo_{12}O_{40}]$ ($Cs_{2.5}PMo$) were synthesized according to the literature (8). Paraformaldehyde (95% purity), styrene and α-methylstyrene (99% purity) were purchased from Aldrich and used as received. Paraldehyde was a BDH product. Solvents (toluene and 1,4-dioxane) were analytical grade Fluka products. *Methods*. Thermal reactions were carried out in a 25 ml Pyrex flask with magnetic stirring at 74 °C. The reaction mixture consisted of 0.573 ml (5 mmol) styrene, 0.3 g (10 mmol) paraformaldehyde, 0.05 mmol HPA, 10 ml solvent (toluene or 1,4-dioxane) and 0.732 ml hexadecane (internal standard). After reaction the products are diluted with CH_2Cl_2, washed with water then $NaHCO_3$ solution and dried over $MgSO_4$. Cesium salts, in turn, were simple filtered off. A Samsung M6148 domestic microwave oven was used for reactions under microwave irradiation (5 ml Pyrex vial, 400 Watt, two 2-min or 1-min irradiation periods, solvent-free conditions). Product formation was negligible (about 1%) when reactions were carried out by either method in the absence of catalyst. *Analysis*. Product distributions were determined by gas chromatography (Carlo Erba Fractovap Mod G GC, 1.2 m SE-52 column, or CHROM 5 GC, 2.4 m OV-17 column, thermal conductivity detector). Authentic samples and the GC-MS method (HP-5890 GC, 50 m HP-1 capillary column, HP-5970 mass selective detector) were used for product identification.

Results

All four heteropoly acids and the cesium salts were found to catalyze the Prins reaction of styrene and α-methylstyrene to yield the corresponding 4-phenyl-1,3-dioxanes as indicated in Scheme 1 under both thermal activation and microwave irradiation. Yields and selectivites, however, vary widely depending on structure and experimental conditions.

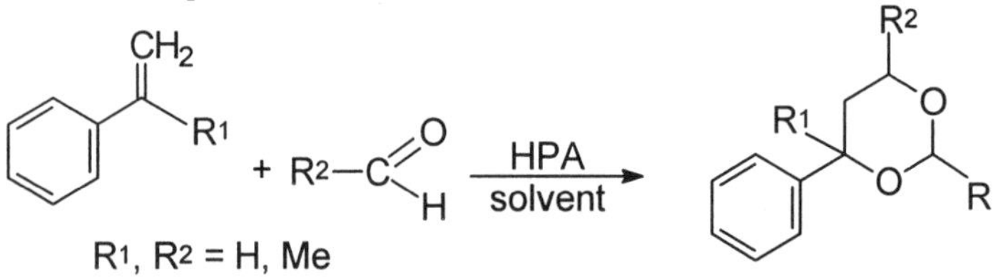

Scheme 1

Transformation of styrenes with formaldehyde

Styrene was chosen to study the effect of the reaction parameters on product yields and selectivities. At first, toluene as solvent and a 6-h reaction time was selected on the basis of earlier studies (7, 9). It is seen (Table 1) that all HPAs and the cesium salts catalyze the reaction to give 4-phenyl-1,3-dioxane as the main product. In addition, we always observed the formation of 1-phenyl-1-*p*-tolylethane in low yields (usually about 25%) as a result of Friedel–Crafts alkylation of toluene by protonated styrene. The yields, however, are not satisfactory. Since GC analysis showed that unreacted styrene was not present in the reaction mixture, the reaction was stopped at a shorter reaction time (2 h). Yields were very similar to those determined after 6 h.

Table 1. Yield of 4-phenyl-1,3-dioxane in the reaction of styrene with paraformaldehyde catalyzed by HPAs and cesium salts in various solvents at 74 °C[a]

Solvent/reaction time	SiW	PW	PMo	SiMo	$Cs_{2.5}PMo$	$Cs_{2.5}PW$
Toluene/6 h	35	41	54	54	32	44
Toluene/2 h	41			55		
1,4-Dioxane/2 h	81	11[b]	57	83	63	82(81)[c]

[a] Conversion = 100%. [b] 45% of styrene was recovered.
[c] Yield in a second experiment with recovered catalyst.

This observation indicates that side-reactions – alkylation and styrene polymerization – is the possible reason for the relatively low yields. Since the change of the solvent was already shown to have beneficial effects on the yield of the Prins reaction (6b), further experiments were carried out in 1,4-dioxane. Indeed, in most cases, significant increases in yields were observed (Table 1).

In contrast to the transformations discussed above the reaction of α-methylstyrene is not selective. In addition to 4-methyl-4-phenyl-1,3-dioxane the expected product, a second compound 4-phenyl-3,6-dihydro-2*H*-pyran is always formed, which can even become the main product (Table 2). In most cases, however, increased amount of formaldehyde improved selectivites.

Table 2. Yield in the transformation of α-methylstyrene with paraformaldehyde catalyzed by HPAs and acidic cesium salts at 74 °C[a,b]

Solvent/ reaction time	SiW	PW	PMo	SiMo	$Cs_{2.5}PMo$	$Cs_{2.5}PW$
Toluene/6 h	10/0	44/5	17/0	5/0	20/0	12/4
1,4-Dioxane/2 h	3/17	57/4	67/7	20/21	55/8	31/21
1,4-Dioxane/2[c]	37/5	49/6	69/3	44/11	59/14	26/6

[a] Conversion = 100%. [b] 4-Me-4-Ph-1,3-dioxane/4-Ph-3,6-dihydro-2*H*-pyran.
[c] 50% excess of paraformaldehyde was used.

Transformation of styrenes with acetaldehyde
The Prins reaction with homologous aldehydes is known to result in lower yields. This is exactly what we observed in the present study by reacting styrenes with acetaldehyde in the form of paraldehyde (Table 3). The selectivity pattern is similar to that observed in the transformation with formaldehyde.

Transformations under microwave irradiation
As being observed in other organic transformations (10), microwave heating is proved to be a convenient and fast method to carry out the Prins reaction (Table 4). Yields with formaldehyde are slightly lower than in the thermal process, whereas the results with acetaldehyde are surprisingly high. A single byproduct (a few percent of 2,4-diphenyltetrahydropyran) was also formed.

Table 3. Yield of substituted 1,3-dioxanes in the transformation of styrenes with paraldehyde catalyzed by HPAs and acidic cesium salts at 74 °C[a,b]

Compound	SiW	PW	PMo	SiMo	$Cs_{2.5}PMo$	$Cs_{2.5}PW$
Styrene	21	1	6	9	3	4
α-Methylstyrene	59/18	18/1	41/8	34/36	21/31	55/3

[a] Reaction time = 2 h, solvent = 1,4-dioxane. Starting materials (5-90%) were always recovered. [b] Substituted 1,3-dioxane/substituted 3,6-dihydro-2*H*-pyran.

Table 4. Yield of 4-phenyl-1,3-dioxane in the transformation of styrene catalyzed by HPAs and acidic cesium salts under microwave irradiation[a]

Irradiation time	SiW	PW	PMo	SiMo	$Cs_{2.5}PMo$	$Cs_{2.5}PW$
$(CH_2O)_x$, 2x2 min	10	36	49	52	1[b]	15[b]
$(CH_2O)_x$, 2x1 min		2[b]		52		
paraldehyde, 2x1 min	24[b]	59	26	47		

[a] Reaction time = 2 h, solvent = 1,4-dioxane. [b] Starting material was recovered.

Discussion

Alkenes reacting with formaldehyde in acidic medium may yield allylic alcohols, 1,3-diols and 1,3-dioxanes as the characteristic major products of the Prins reaction. Due to the water content of HPAs, the actual catalyst is H_3O^+ generated in solution. The key intermediate is carbocation **1** formed in the electrophilic attack by protonated formaldehyde on the double bond (Scheme 2). Further reaction of **1** with water (HPAs are known to contain water) as a nucleophile leads to the main product 1,3-dioxanes (**2**) through the 1,3-diol intermediate. Non-protonated styrene can also serve as the nucleophile leading to tetrahydropyran derivative **3**. This compound was formed only when microwave heating was applied. Formation of a third product (**5**) is characteristic of the Prins reaction of homoallylic alcohols. Since, in our case, it is only observed in the transformation of α-methylstyrene, the involvement of unsaturated alcohol **4** can account for the formation of compound **5**. It was argued (11) that deprotonation of **1** to give the homoallylic alcohol **4** over the corresponding allylic alcohol is greatly facilitated via a cyclic process.

Concerning the activities of the various HPAs it is surprising that PW is often exhibits the lowest activity though it has the highest acidity of the four HPAs used in the present study (12). The reason, however, is simple: we always observed that a substantial part of the catalyst adhered to the wall of the reaction flask. No wonder, that in all these reactions unreacted starting material was also recovered.

With respect to the acidic cesium salts their activity is usually comparable with that of the acids. For a more accurate comparison turnover numbers were calculated by supposing that only a single proton dissociates from the acids in

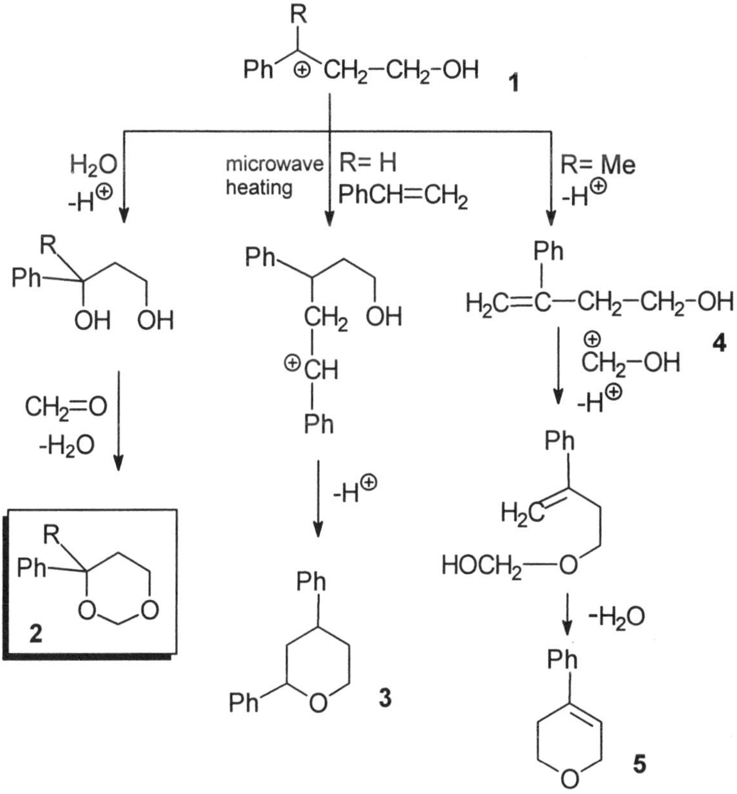

Scheme 2

solvents of low basicity as indicated in a recent paper (13). Surface acidity values of the salts were calculated according to the literature (14). When the highest conversions are compared cesium salts exhibit about one order of magnitude higher activity than HPAs (5-6.5 min^{-1} vs. 0.6-0.7 min^{-1}).

The Prins reaction can be efficiently carried out under microwave activation in the presence of HPAs, whereas the cesium salts are not effective. The reason for this is that all acids but not the salts contain water. In the lack of other material capable of absorbing microwaves water is essential for microwave heating. The advantages of the process (very short reaction time, solvent-free conditions, selective, clean reaction) make microwave activation to be a useful and attractive alternative to thermal activation in the Prins reaction.

Conclusions

Heteropoly acids catalyze the Prins reaction to yield 1,3-dioxanes under both thermal activation and microwave irradiation. The best results are achieved with paraformaldehyde in dioxane solvent with thermal activation. The mild reaction conditions, the easy work-up (simple filtration), and the possibility of repeated use make acidic cesium salts the catalyst of choice for carrying out the Prins reaction under the thermal activation. Microwave activation, in turn, works best with HPAs as the catalysts. Both processes are useful and efficient, synthetically valuable method to produce 1,3-dioxanes from styrenes.

Acknowledgment
Financial supports by the Hungarian National Science Foundation (OTKA grants T016941and F023674) and the US-Hungary Science and Technology Program (JF No 553) are gratefully acknowledged.

References
1. T. Okuhara, N. Mizuno and M. Misono, *Advances in Catalysis*, 41, 113 (1996).
2. B. Török, I. Bucsi, T. Beregszászi, I. Kapocsi and Á. Molnár, *J. Mol. Catal. A: Chem.*, 107, 305 (1996).
3. B. Török, I. Bucsi, T. Beregszászi and Á. Molnár, in: *Catalysis of Organic Reactions* (R.E. Malz, Jr., ed.) Marcel Dekker, New York, 1996, p. 393.
4. Á. Molnár and T. Beregszászi, *Tetrahedron Lett.*, 37, 8597 (1996).
5. T. Beregszászi, B. Török, Á. Molnár, G.A. Olah and G.K.S. Prakash, *Catal. Lett.*, 48, 83 (1997).
6. a) V.I. Isagulyants, T.G. Khaimova, V.R. Mleikyan and S.V. Pokrovskaya, *Russ. Chem. Rev.* (Engl. Transl.)., 37, 17 (1968).
b) D.R. Adams and S.P. Bhatnagar, *Synthesis*, 661 (1977).
c) B.B. Snider, in : *Comprehensive Organic Synthesis* (B.M. Trost and I. Fleming, eds.), Vol. 2: *Additions to C–X π-bonds*, Part 2 (C.H. Heatchock, ed.), Pergamon Press, Oxford, 1991, ch 2.1, p. 527.
7. K. Urabe, K. Fujita and Y. Izumi, *Shokubai*, 22, 223 (1980).
8. T. Okuhara, T. Nishimura, H. Watanabe and M. Misono, *J. Mol. Catal.*, 74, 247 (1992).
9. J. Tateiwa, K. Hashimoto, T. Yamauchi and S. Uemura, Bull. *Chem. Soc. Jpn.*, 69, 2361 (1996).
10. a) D.M.P. Mingos and D.R. Baghurst, *Chem. Soc. Rev.*, 20, 1 (1991).
b) R.A. Abramovitch, *Org. Prep. Proced. Int.*, 23, 683 (1991).
c) S . Caddick, Tetrahedon, 51, 10403 (1995).
11. P.R. Stapp, *J. Org. Chem.*, 34, 479 (1969).
12. a) Y. Izumi, K. Matsuo and K. Urabe, *J. Mol. Catal.*, 18 (1983) 299.
b) Y. Izumi, R. Hasebe and K. Urabe, *J. Catal.*, 84 (1983) 409.
13. R.S. Drago, J.A. Dias and T.O. Maier, *J. Am. Chem. Soc.*, 119, 7702 (1997).
14. T. Okuhara, T. Nishimura and M. Misono, in: *11th Int. Congr. Catal. - 40th Anniversary* (J.W. Hightower, W.N. Delgass, E. Iglesia and A.T. Bell), *Stud. Surf. Sci. Catal.*, Vol. 101, Elsevier, Amsterdam, 1997, p. 581.

α-Pinene Oxide Isomerization Promoted by Mixed Cogels

Nicoletta Ravasio*, Michela Finiguerra, Michele Gargano

*Centro C.N.R. SSSCMTBO, via Venezian 21, I-20133 MILANO, Italy
Centro C.N.R. MISO, via Amendola 173, I-70126 BARI, Italy

Abstract

The use of commercial mixed cogels for the acid catalyzed isomerization of α-pinene oxide is proposed. SiO_2-Al_2O_3 and SiO_2-ZrO_2 reach 72% yield of the desired product, α-campholenic aldehyde, under mild conditions.

Introduction

The need to reduce the environmental impact of many processes operated by the fine and speciality chemical industries push towards the implementation of heterogeneous catalytic technologies even for small-scale processes. Particularly challenging is the search for solid alternatives to homogeneous Lewis and Bronsted acids. Mesoporous materials are particularly suitable for use in liquid phase systems for the fine chemical industry, as the mesopores provide improved access of the big substrate molecules to the active sites.

We recently reported the results obtained in the cyclization of citronellal in the presence of different mixed cogels under mild conditions (1). These oxides proved to be an effective alternative to homogeneous Lewis acids, particularly to $ZnBr_2$ which is used in the Takasago process to carry out the same reaction (2). This prompted us to investigate another reaction catalyzed by $ZnBr_2$, namely the isomerization of α-pinene oxide 1. This reaction is used industrially to produce α-campholenic aldehyde 2, a valuable intermediate for the synthesis of α- and β-santalol, but the acidic rearrangement gives several other products, some of which are shown in the Scheme. The homogeneous catalyzed reaction is quite selective, giving 85% selectivity of the desired product (3) whereas alternative systems have so far been quite unsuccessfull (4, 5). Only recently the use of dealuminated zeolites giving up to 75% selectivity of 2 has been proposed (6).

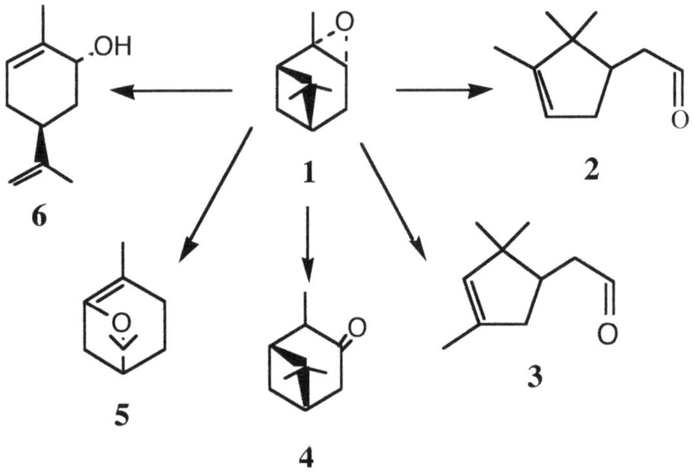

Scheme 1

Results

Our results are summed up in Table 1. All the cogels examined were found to be active under mild conditions, oligomeric products were never detected, and the product distribution found is in agreement with the Lewis acid character of these materials (3). Thus, although we recently found evidence for the presence of protons when SiO_2-TiO_2 and SiO_2-ZrO_2 supported copper catalysts were used under hydrogenation conditions (7), preliminar results on surface characterization rule out the presence of Bronsted acidic sites on the pure supports. FT-IR spectra of SiO_2-TiO_2 2,3% , SiO_2-Al_2O_3 0,6% and SiO_2-ZrO_2 4.7% after absorption of pyridine showed the 8a and 19b vibration modes at 1596 and 1445 cm^{-1}, diagnostic of the presence of Lewis acid sites, but no evidence of pyridinium ion formation (8).

As already observed in the cyclization of citronellal, SiO_2 and SiO_2-TiO_2 showed similar behaviour and low activity. Best activity and selectivity were obtained in the presence of SiO_2-Al_2O_3 and SiO_2-ZrO_2 . In particular SiO_2-Al_2O_3 with a 1.2 mass % content in Al_2O_3 (Si/Al = 70) was found to be more active than the 13% one (Si/Al = 6), suggesting that this reaction requires well dispersed Lewis acid sites instead of Bronsted ones, as already proposed by Hölderich et al. (6). However, a high number of active sites is required. When the amount of catalyst was reduced, both activity and selectivity dropped down (entry 8). Table 2 shows that selectivity to campholenic aldehyde was constant during the reaction.

Table 1[a]

Isomerization of α-pinene oxide in the presence of different solid acids

Entry	cat	solvent	T (°C)	t (min)	Yields: 2 (%)	3	4	5	6
1	SiO$_2$	tol	90	60	55	8	6	5	9
2	Si-Ti 2.3	tol	90	60	57	9	7	2	14
3	Si-Al 0,6	tol	90	15	62	7	10	4	3
4	Si-Al 1.2	tol	90	5	63	6	13	2	9
5	Si-Al 1.2[b]	tol	90	5	64	7	11	4	2
6	Si-Al 1.2[b]	tol	25	5	72	7	14	2	3
7	Si-Al 1.2[b,c]	tol	25	5	75	6	13	2	2
8	Si-Al 1.2[b,d]	tol	25	3 h	60	10	8	4	6
9	Si-Al 1.2[b]	dce	25	15	69	5	6	3	7
10	Si-Al 1.2[b]	diox	25	7 h	56	5	5	6	8
11	Si-Al 1.2[b]	tol	0	15	72	7	12	2	5
12	Si-Al 13	tol	90	10	65	6	17	1	4
13	Si-Zr 4.7	tol	90	15	64	6	11	1	5
14	Si-Zr 4.7	tol	60	40	66	6	11	1	7
15	Si-Zr 4.7[b]	tol	60	20	72	6	12	-	4

[a] = 100% conversion, catalyst dehydrated at 270°C, [b] = catalyst dehydrated at 450°C
[c] = 0,3 g of catalyst; [d] = 0,05 g of catalyst

Table 2

Isomerization of α-pinene oxide in the presence of SiO_2-Al_2O_3 1,2% at low catalyst loading (entry 8 in Table 1)

t (min)	conv. (%)	sel (%)
5	77	56
30	93	59
60	97	60
180	100	60

A significant improvement was introduced by dehydrating the catalysts at 450°C, probably owing to better dipersion of the active sites. Lowering of the reaction temperature was found to have a beneficial effect on selectivity, best yields in aldehyde 2 being obtained at room T. However, lowering to 0°C did not introduce any further improvement.

As far as solvent effects are concerned we observed a trend similar to that reported in the presence of $ZnCl_2$ (3) Thus, both activity and selectivity decreased when the solvent polarity increased, particularly when dioxane was used (entries 6, 9, 10), showing poisoning of the active sites by means of the solvent oxygen atoms.

In conclusion, we can easily reach a 72% yield by using SiO_2-Al_2O_3 1.2% or SiO_2-ZrO_2 4.7%, dehydratedat 450°C, in toluene, at 25 or 60°C (entry 6 and 15).

Work is in progress to improve this system and to extend it to the isomerization of limonene epoxide to carvenone .

Experimental

SiO_2-Al_2O_3 13% (Grade 135), listed in the Table as Si-Al 13, was purchased from Aldrich, all the other cogels were obtained from Grace Davison, Worms (Germany). The catalysts were treated at 270°C or 450°C for 20 minutes in air and for 20 minutes under reduced pressure at the same temperature.

α-pinene oxide (Aldrich, 97%, 0,2 g) was dissolved in toluene (8 ml) and the solution transferred into a glass reaction vessel where the catalyst (0,2 g) had been previously dehydrated. Reaction mixtures were analyzed by GC using a polyethylene glycol (Supelcowax 10) capillary column (60 m), injection T = 140°C. Aldehyde **2** was characterized through its 500 MHz ^{1}H NMR spectrum (9): δ 0.8 (3H, s), 1.0 (3H, s), 1.6 (3H, dt), 1.85 (1H, m), 2.25 (1H, m), 2.35 (2H, m), 2.49 (1H, m), 5.20 (1H, m), 9.77 (1H, t). Minor reaction products were identified by GLC-MS and/or by comparison of their retention times with those of standard samples.

References

1. N.Ravasio, M.Antenori, F.Babudri, M.Gargano, *Stud. Surf. Sci. Catal.*, **108** (Heterog. Catal. Fine Chem., IV), 625 (1997)
2. B.Arbusow, *Chem. Ber.* , **68**, 1430 (1935)
3. S.Akutagawa, *Chirality in Industry* (A.N.Collins, G.N.Sheldrake and J.Crosby eds.), John Wiley and Sons, Chichester, 1992, p. 313
4. J.Kaminska, M.A.Schwegler, A.J.Hoefnagel, H. van Bekkum, *Recl. Trav. Chim. Pays-Bas*, **111**, 432 (1992)
5. K.Tanabe, A.Harada, *Chem. Letters*, 1017 (1979)
6. A.T.Liebens, C.Mahaim, W.F.Hölderich, *Stud. Surf. Sci. Catal.*, **108** (Heterog. Catal. Fine Chem., IV), 587 (1997); W.F.Hölderich, J. Röseler, G.Haitmann, A.T.Liebens, *Catalysis Today* , **37**, 353 (1997)
7. N.Ravasio, V.Leo, F.Babudri, M.Gargano, *Tetrahedron Lett.*, **38**, 7103 (1997)
8. G.Ramis, L.Yi, G.Busca, N.Ravasio, II Conv. Scient. Cons. Chim. Mater., Firenze (I), 13-15 feb. 1995, Abstracts: C23.
9. L.Lopez, G.Mele, V.Fiandanese, C.Cardellicchio, A.Nacci, *Tetrahedron*, **50**, 9097 (1994)

Multi-Iron Polyoxanions. Synthesis, Characterization and Catalysis of H_2O_2-Based Hydrocarbon Oxidations

Xuan Zhang[#] and Craig L. Hill[*]

Department of Chemistry, Emory University, Atlanta, GA 30322.

Abstract

Two novel multi-iron polyoxoanions $[Fe^{III}_4(H_2O)_2(P_2W_{15}O_{56})_2]^{12-}$ (1) and $[Fe^{III}_2(NaH_2O)_2(P_2W_{15}O_{56})_2]^{16-}$ (2) have been synthesized by the reaction of Fe^{2+} or Fe^{3+} with the trivacant laucunary $[P_2W_{15}O_{56}]^{12-}$. The structures have been characterized by single crystal X-ray diffraction analyses. The compounds have been further characterized by UV-Visible, infrared, ^{31}P NMR spectroscopy and elemental analyses. The compounds have been tested for the ability to catalyze the oxidation of alkenes by H_2O_2. Despite the similarity in overall structure, the two compounds have very different catalytic reactivities. Compound 1 has very low catalytic activity for the oxidation of alkenes by H_2O_2. Alkene oxidation is characterized by dominant allylic attack on aliphatic substrates and oxidative cleavage of stilbenes. In contrast, 2 catalyzes the epoxidation of a variety of alkenes with H_2O_2 including primary alkenes with >90% selectivities and ca. 80% yield based on H_2O_2.

Introduction

Environmental concerns have forced the chemical industry to turn their efforts to the development of new and effective catalytic systems for selective oxidation of hydrocarbons by H_2O_2. One of several investigated general processes is catalytic epoxidation by H_2O_2 (eq. 1). There are two basic requirements for such a targeted process to be viable: first, the reaction has to be highly selective, and the oxidant has to be consumed in high efficiency during the product formation; second, the catalytic system has to be stable

$$\text{（alkene）} + H_2O_2 \xrightarrow{\text{catalyst}} \text{（epoxide）} + H_2O \qquad (1)$$

with respect to oxidation. These conditions are not fulfilled by many of the catalytic systems involving transition metals coordinated by organic ligands.[1] One of the problems is that transition-metal-catalyzed hydrogen peroxide

[*] To whom correspondence should be addressed.
[#] Current address: Department of Chemistry, Carnegie Mellon University, Pittsburgh, PA 15213.

activation often leads to homolytic cleavage of the O-O and/or O-H bonds.[2] Hydroxy and hydroperoxy radicals formed tend to react with each other or with H_2O_2. This results in the net catalyzed dismutation of H_2O_2. The other problems are the irreversible deactivation of the catalytic species resulting from oxidative degradation of organic ligands.[3, 4] As a partial solution to those problems, many systems involving incorporation of low-valent transition metals into lacunary polyoxometalates have been investigated.[5, 6, 7, 8]

This paper reports the preparation and characterization of polytungstophosphates of formula $[Fe^{III}_4(H_2O)_2(P_2W_{15}O_{56})_2]^{12-}$ (1) and $[Fe^{III}_2(NaH_2O)_2(P_2W_{15}O_{56})_2]^{16-}$ (2). The catalytic reactivities for the epoxidation of alkenes of these two catalysts have been evaluated and compared.

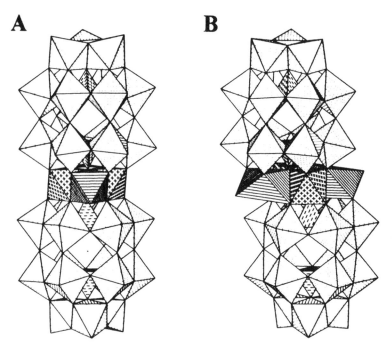

Figure 1. A polyhedral representation of **1** (A) and **2** (B), respectively.

Experimental

Synthesis of 1 and 2. The compounds were first prepared as sodium salts by crystallizing from H_2O. The compounds were later converted to tetra-*n*-butylammonium (TBA) salts by metathesis reactions. The detailed synthesis and characterization are reported elsewhere.[9]

Catalytic Oxidation of Alkenes. See the Table footnotes for detailed reaction conditions. The organic products were identified and quantified by GC and GC/MS using trimethylacetonitrile or decane as an internal standard.

The final H_2O_2 concentration was measured using standard iodometric analysis. The reactions were followed for 36 h.

Results and Discussion

Synthesis and Physical Properties of Tetra-Iron Sandwich Polyoxanion, $[Fe^{III}_4(H_2O)_2(P_2W_{15}O_{56})_2]^{12-}$ (1). Na1 was prepared by stoichiometric (2:1) reaction between Fe^{3+} and the trivacant polyoxoanion $[P_2W_{15}O_{56}]^{12-}$ at 25 °C. The complex was readily prepared in ~ 30% isolated yield and in high purity. Figure 1A shows polyhedral representation of the structure. The polyanion consists of an oxo and aqua tetranuclear iron core, $[Fe^{III}_4O_4(H_2O)_2]$, sandwiched by two trivacant Wells-Dawson structural moieties, $[P_2W_{15}O_{56}]^{12-}$. The polyanion has C_{2h} symmetry and approximate axial (P1···P2 direction) and equatorial dimensions of 21.88 Å and 10.64 Å, respectively. The paramagnetic ground electronic state renders only one clearly observable in ^{31}P NMR spectrum, that for the P atoms distal to the Fe_4 unit (-11.1 ppm, $\Delta\upsilon_{1/2} = 70$ Hz). The cyclic voltammogram of Na1 exhibits three principal unresolved reduction peaks and three corresponding oxidation peaks for the Fe(III/II) redox processes.

Synthesis and Physical Properties of Di-Iron Sandwich Polyoxanion, $[Fe^{III}_2(NaH_2O)_2(P_2W_{15}O_{56})_2]^{16-}$ (2). Na2 was prepared by the reaction between 2:1 ratio of Fe^{2+} and $[P_2W_{15}O_{56}]^{12-}$, followed by O_2 oxidation in aqueous solution at 25 °C. As shown in Figure 1B, the structure of **2** consists of a mixed alkaline and mid-transition metal cluster unit, Fe_2Na_2 sandwiched by two defect Wells-Dawson moieties, $[P_2W_{15}O_{56}]^{12-}$. Each sodium atom of the Fe_2Na_2 unit is loosely bound to seven oxygen atoms, forming a very distorted, pro-spherical coordination geometry. The two sodium atoms are easily replaced by other transition metal ion of similar size, such as, Fe^{3+}, Mn^{2+}, etc. Only one signal is observable in ^{31}P NMR spectrum (-12.9 ppm, $\Delta\upsilon_{1/2} = 200$ Hz) due to the paramagnetic ground electronic state and the C_{2h} symmetry of the polyanion. The cyclic voltammogram of Na2 exhibits no reduction or oxidation peaks corresponding to the Fe(III/II) redox processes.

Alkene Oxidation by H_2O_2 Catalyzed by TBA1 and TBA2. Table 1 reports the oxidation of representative alkenes by H_2O_2 catalyzed by TBA1 in CH_3CN solution and the distribution of alkene-derived oxidation products. The data clearly show that H_2O_2-based oxidations catalyzed by **1** exhibit minimal selectivity; allylic oxidation predominates for most of the aliphatic alkenes and oxidative cleavage predominates for the aromatic alkene, *cis*-stilbene. These selectivities are not compatible with electrophilically activated peroxide and heterolytic oxidation mechanism(s). They are compatible with homolytic oxidation mechanism(s). The likely substrate attacking specie(s) in the mechanism are oxy radicals and/or high-valent oxo-iron intermediates. In contrast, TBA2 catalyzes the facile oxidation of alkenes at 25 °C with extremely high selectivity for epoxide (~95%). The yields of oxidative

Table 1. Organic Products from Oxidation of Representative Alkenes with H_2O_2 Catalyzed by TBA1. [a]

Substrate	Products Selectivity [initial rate, mM/h] [b]		

11% [0.133]   40% [0.469]   48% [0.566]

25% [0.125]   0[c]   74% [0.374]

88% [0.570]   0[c]   11% [0.077]

0[c]   0[c]   0[c]

[a] Reaction conditions: A solution of TBA1 (4 mM) and alkene (0.9 M) in 1 mL of CH_3CN was degassed and stirred under Ar at 25 °C. The reaction was initiated by injection of a solution of 30% aqueous H_2O_2 (25 µL). Aliquots (1.5 µL) of reaction mixtures were removed at calibrated time intervals and analyzed by GC. [b] Selectivity = moles of indicated product / moles of all organic products derived from substrate [initial rate = the slope of the indicated product vs. time plot at < 1% conversion]. [c] Detection limit < 0.2%.

Table 2. Organic Products from Oxidation of Representative Alkenes with H_2O_2 Catalyzed by TBA2. [a]

Substrate	Products
	Selectivity [average rate, mM/h] [b]

94% [6.2]   4% [0.25]   2% [0.13]

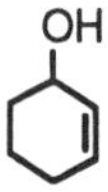

93% [5.5]   7% [0.40]   0 [c]

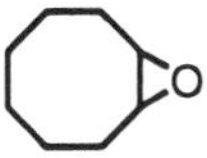

98% [12]   0 [c]   2% [0.23]

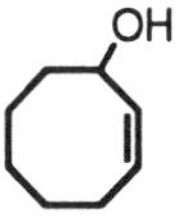

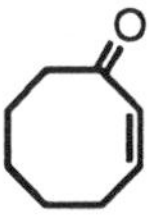

99% [0.71]   0 [c]   0 [c]

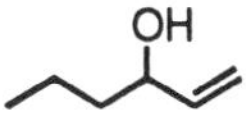

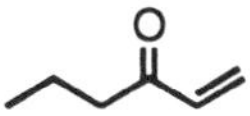

[a] Reaction conditions: A solution of TBA1 (4 mM) and alkene (0.9 M) in 1 mL of CH_3CN was degassed and stirred under Ar at 25 °C. The reaction was initiated by injection of a solution of 30% aqueous H_2O_2 (25 µL). Aliquots (1.5 µL) of reaction mixtures were taken out at certain time interval and analyzed by GC. [b] Selectivity = moles of indicated product / moles of all organic products derived from substrate [average rate = the amount of the indicated product produced divided by the reaction time of 30 h]. [c] Detection limit < 0.2%.

cleavage products and allylic oxidation products were very low, even for substrates vulnerable to these processes. Alkene-derived product distributions for the oxidation of representative alkenes are summarized in Table 2. When compared with the literature Mn-substituted Zn Tourne system,[7, 8] this Fe-based Wells-Dawson sandwich complex has the advantage of being reactive and selective for alkene epoxidation in a nonhalogenating solvent. By measuring the remaining H_2O_2 after the catalytic oxidation reaction using iodometric titration, the product yield based on H_2O_2 can be determined. The typical product yield is around 20% and 80% for TBA1 and TBA2, respectively. Comparison of the rates and product distributions for TBA1 and TBA2 indicates that the oxidation reactions proceed via quite different mechanisms. While the results from TBA1 are indicative of an Fe initiated radical chain process (Fenton type chemistry), the results from TBA2 suggest a peroxometal intermediate or high-valent oxo intermediate. The difference in reactivity and selectivity of the two compounds might be attributed to the presence of the two labile sodium ions in TBA2. However, more evidence including full kinetic analysis must be garnered before a detailed reaction mechanism can be put forward. Unlike most previously documented transition-metal-substituted polyoxometalates (TMSPs), both TBA1 and TBA2 are relatively stable at high H_2O_2 concentration. There is no significant change in IR and ^{31}P NMR spectra after 48 h of incubation with ca. 0.25 M aqueous H_2O_2.

Acknowledgments. This research was supported by the National Science Foundation (grant CHE-9412465). We thank Dr. Qin Chen and Dr. Dean C. Duncan for their contribution to this project.

Reference
1. R. A. Sheldon, *Chemtech* **21**, 566-576 (1991).
2. G. Strukl, *Catalytic Oxidations with Hydrogen Peroxide as Oxidant*. G. Strukul, Ed. (Kluwer Academic Publishers, Boston, 1992).
3. J. Renaud, P. Battioni, J. F. Bartoli, D. Mansuy, *J. Chem. Soc. Chem. Commun.* , 888-889 (1985).
4. P. Battioni, J. Renaud, J. F. Bartoli, D. Mansuy, *J. Chem. Soc. Chem. Commun.* , 341-343 (1986).
5. C. L. Hill, R. B. J. Brown, *J. Am. Chem. Soc.* **108**, 536-538 (1986).
6. D. E. Katsoulis, M. T. Pope, *J. Am. Chem. Soc.* **106**, 2737-2738 (1984).
7. R. Neumann, M. Gara, *J. Am. Chem. Soc.* **116**, 5509-5510 (1994).
8. R. Neumann, M. Gara, *J. Am. Chem. Soc.* **117**, 5066-5074 (1995).
9. X. Zhang, Q. Chen, D. C. Duncan, C. Campana, C. L. Hill, *Inorg. Chem.* **36**, 4208-4215 (1997).

The Preparation of Amino Acids via Rh(DIPAMP)-Catalyzed Asymmetric Hydrogenation

Scott A. Laneman,* Diane E. Froen, and David J. Ager

NSC Technologies; 601 E. Kensington Rd.; Mt. Prospect, IL 60056

Abstract: A continued study of the asymmetric hydrogenation of enamides catalyzed by [Rh(COD)(*R,R*-DIPAMP)]$^+$BF$_4^-$ to prepare several substituted-phenylalanine derivatives is presented.

Introduction

[Rh(COD)(*R,R*-DIPAMP)]$^+$BF$_4^-$ (**1**) has shown considerable versatility in the asymmetric reduction of various enamides to form protected amino acids with high catalyst activities and stereoselectivities.(1-6) Since the development of the L-DOPA process by Monsanto, [Rh(COD)(*R,R*-DIPAMP)]$^+$BF$_4^-$ has been the benchmark of other new asymmetric hydrogenation catalysts that contain novel chiral bisphosphines, such as DuPHOS (7), PennPHOS (8),and CarboPhos (9)

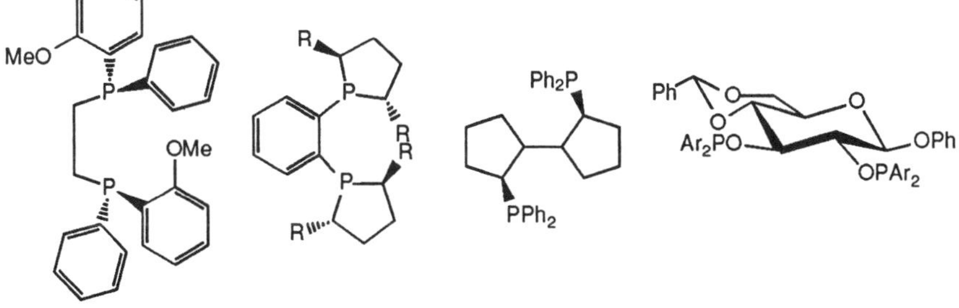

2	**3**	**4**	**5**
R,R-DIPAMP	(S,S)-DuPhOS	S,S-PennPHOS	CarboPHOS
	R = Me, Et, i-Pr, Cy		Ar = 3,5-(CH₃)₂C₆H₃-

Figure 1 Various asymmetric bisphosphines emplemented in rhodium catalyzed asymmetric hydrogenations of prochiral enamides.

shown in Figure 1, for the preparation of amino acids *via* asymmetric hydrogenation. Although these novel catalyst systems can provide enantioselectivities higher than **1**, asymmetric catalysis with **1** continues to prove useful for the large scale preparation of unnatural amino acid in high enantioselectivities. Our goal is to contributed additional examples of asymmetric reduction of enamides, but with the intention of applications towards industrial manufacturing.

Results and Discussions

L-Unnatural amino acids are prepared by a four step sequence shown in Scheme 1. Z-Enamides are selectively formed by an Erlenmeyer condensation of an aldehyde and *N*-acetylglycine in moderate yields followed by ring opening with HOAc/H$_2$O/NaOAc in moderate to high yields. Asymmetric hydrogenation of the enamides with **1** followed by acid hydrolysis produces amino acids with enantioselectivities >95%ee with the exception of 3-pyridinylalanine (77%ee). Enantiomeric enrichments are observed upon isolation of the amino acids after hydrolysis.

The key step in the preparation of amino acids is the generation of the chiral center by asymmetric hydrogenation. [Rh(COD)(R,R-DIPAMP)]$^+$BF$_4^-$ (**1**) has demonstrated high catalyst activities for the majority of enamides, with the exception of the *p*-nitrophenylalanine analog (**9d**). Substrate to catalyst ratios (S/C) of 10,000 are achieved in the reduction of many enamides with **1** while the reduction of **9d** occurs with a ratio of 200. Reduction of the nitro group is not observed. High catalyst activities are observed in the asymmetric hydrogenation of phenylalanine derivatives that contain chloro (**9b**) and fluoro (**10c**) groups in the *para* position.

The enantioselectivities did not change upon decrease of the catalyst loading (S/C) from 5,000 to 20,000 in the reduction of the 2-naphthyl enamide (**9a**), but the catalyst performance is susceptible to trace amounts of oxygen, which can significantly decrease the catalyst activity. High catalytic turnovers are achieved in the asymmetric reduction of enamides that contain methoxy and acetoxy groups in the *meta* position. Hydrolysis of **10e** and **10f** affords L-*m*-tyrosine and L-*O*-methyl-*m*-tyrosine in 99.3%ee and 98.7%ee, respectively. A slightly higher catalyst loading (S/C = 5,000) is required for complete conversion in the presence of a cyano functionality. L-3-Cyanophenylalanine (**11g**) can be prepared in 98.0%ee without reduction or hydrolysis of the cyano group.

[Rh(COD)(R,R-DIPAMP)]$^+$BF$_4^-$ slowly catalyzes the reduction of 3-pyridinyl enamide due to the pyridinyl unit. The asymmetric reduction of the methyl ester analog with [Rh(COD)(R,R-DIPAMP)]$^+$BF$_4^-$ has been reported to be 99%ee with a substrate to catalyst ratio of 28.(11) A similar decrease in catalyst activity has been observed with other catalysts.(12) Protonation of the

Scheme 1

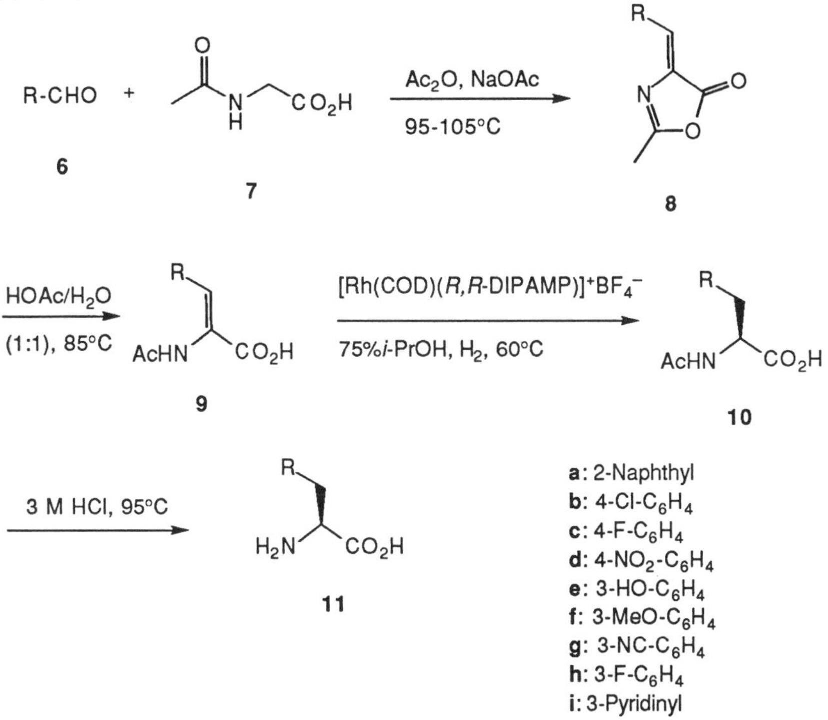
R-CHO
6
7
Ac_2O, NaOAc
95-105°C
8
$HOAc/H_2O$
(1:1), 85°C
9
$[Rh(COD)(R,R\text{-DIPAMP})]^+BF_4^-$
75% *i*-PrOH, H_2, 60°C
10
3 M HCl, 95°C
11
a: 2-Naphthyl
b: 4-Cl-C_6H_4
c: 4-F-C_6H_4
d: 4-NO_2-C_6H_4
e: 3-HO-C_6H_4
f: 3-MeO-C_6H_4
g: 3-NC-C_6H_4
h: 3-F-C_6H_4
i: 3-Pyridinyl

Table 1: Isolation yields and enantioselectivities on Several Intermediates and Amino Acids.

R	8 (%yield)	9 (%yield)	10 (%ee)[a]	S/C	11 (%yield)	11 (%ee)[a]
a	72%	90%	92%	5,000-20,000	89%	94.7%
b	58%	95%	–	10,000	89%	94.6%
c	67%	94%	91%	10,000	80%	95.4%
d	94%	96%	–	200	90%	96.0%
e	69%[b]	43%[b]	--[b]	10,000	87%	99.3%
f	46%	84%	–	10,000	88%	98.7%
g	74%	87%	–	5,000	69%	98.7%
h	72%	81%	94%	10,000	86%	97.2%
i	45%[c]	70%[d]	--[d]	1,000	33%	77%

[a] Products are of the *S*-configuration. Enantioselectivities determined by chiral HPLC. [b] Contains *O*-acetate. [c] Prepared by method literature method.(10) [d] Isolated as the HBF_4 salt.

pyridinyl group with a non-coordinating strong acid, HBF$_4$, improves the S/C to 1,000 (12), but 50% esterification occurs during hydrogenation. The enamide 9i•HBF$_4$ is prepared by ring opening of 8i with HBF$_4$ instead of HOAc at room temperature. Hydrolysis of the acid/ester hydrogenation mixture with HCl in water produces 3-pyridinylalanine in 77%ee.

Conclusions

In summary, [Rh(COD)(R,R-DIPAMP)]$^+$BF$_4^-$ catalyzed asymmetric hydrogenation of enamides still provides an excellent route for the preparation of unnatural amino acids in high enantioselectivities. In most cases, low catalyst concentrations are attained due to the high catalytic activity of 1 and is industrially advantageous. Also, the air-stability of [Rh(COD)(R,R-DIPAMP)]$^+$BF$_4^-$ provides added convenience for industrial operations.

Experimental

Each procedure is representative of the preparation of other derivatives unless stated otherwise.

(Z)-2-Methyl-(2'-Naphthyl)methylene-5H-oxazolone (8a): A 2L round-bottom flask is charged with 600.0 g (3.85 mole) of 2-naphthylaldehyde, 420.5 g of NaOAc (5.13 mole), 600.0 g N-acetylglycine (5.13 mole), and 1.27 L of acetic anhydride (13.46 mole) and heated to 95-105°C for 2 hours. The reaction mixture is cooled to 75°C, diluted with 240 mL of HOAc, and cooled to 45°C. H$_2$O (80 mL) is added to the slurry, and the reaction mixture is cooled to room temperature. The solids are filtered, rinsed with 250 mL of HOAc (2 times) and 600 mL of H$_2$O (3 times), and dried in a vacuum oven (25°C, 28 in. Hg) overnight to give 658 g (72% yield) as a yellow solid.

Typical Procedure for Enamide Preparation (9a): A 2L round-bottom flask is charged with 640 g of 8a (2.76 mole), 1.16 g of NaOAc (14.2 mmole), 1.1 L of HOAc and 1.1 L of H$_2$O and the mixture heated to 85°C for 2 hours. The reaction mixture is cooled to room temperature, and the solids are filtered, rinsed with 600 mL of H$_2$O (2 times), and dried in a vacuum oven (50°C, 28 in. Hg) overnight to give 657 g of a yellow solid (95% yield).

Typical Procedure for the Asymmetric Hydrogenation of Enamides (10a): A 2L Parr bomb is charged with 150.0 g of 9a (0.588 mole), 0.089 g of [Rh(COD)(R,R-DIPAMP)]$^+$BF$_4^-$ (0.118 mmole), and 1.65 L of 75% i-PrOH in H$_2$O. The vessel is sealed, sparged with N$_2$ for 15 minutes, heated to 60°C, purged with H_2 (4 times) and pressurized to 50 psig H_2. Hydrogen pressure is maintained at 50 psig with a regulated reservoir. The reaction mixture is cooled to room temperature and depressurized after 18 hours and the volatiles are removed *in vacuo* to give 148.6 g of **10a** as a solid (98% yield). The enantioselectivity of the product has been determined by chiral HPLC to be 92%ee.

Typical Procedure for the Hydrolysis of N-Acetyl Amino Acids (**11a**): A round bottom flask is charge with 578 g of **10a** (2.23 mole) and 3.7 mL of 3M HCl, and heated to 95°C for 6 hours. The reaction mixture is cooled to room temperature and pH is adjusted to ~5 with 50% NaOH. The solids are filtered, rinsed with 0.5 L of H_2O (2 times), and dried overnight under vacuum (50°C) to give 427 g of **11a** (89% yield). Chiral analysis of **11a** by chiral HPLC indicates 94.7%ee.

Acknowledgements

The authors are grateful for the analytical support by Jeff Webb, Eric Sodeman, Winfred Ivy, Renata Schaad, and Marsha Little of Nutrasweet Kelco Company.

References

1. K. E. Koenig, in *Asymmetric Synthesis*, J. D. Morrison Ed., Academic Press, Inc.: Orlando, FL, Vol., p. 31 (1985).
2. W. S. Knowles *Acc. Chem. Res.*, **16**, 106 (1983).
3. H.-J. Zeiss *J. Org. Chem.*, **56**, 1783 (1991).
4. A. Hammadi, J. M. Nuzillard, J. C. Poulin, and H. B. Kagan *Tetrahedron: Asymmetry*, **3**, 1247 (1992).
5. I. Ojima, C.-Y. Tsai, and Z. Zhang *Tetrahedron Lett.*, **35**, 5785 (1994).
6. T. Masquelin, E. Broger, K. Muller, R. Schmid, and D. Obrecht *Helv. Chim. Acta*, **77**, 1395 (1994).
7. M. J. Burk, J. E. Feaster, W. A. Nugent, and R. L. Harlow *J. Am. Chem. Soc.*, **115**, 10125 (1993).
8. G. Zhu, P. Coa, Q. Jiang, and X. Zhang *J. Am. Chem. Soc.*, **119**, 1799 (1997).
9. T. V. RajanBabu, T. A. Ayers, and A. L. Casalnuovo *J. Am. Chem. Soc.*, **116**, 4101 (1994).
10. C. Cativiela, M. D. Diaz de Villegas, J. I. Garcia, J. A. Mayoral, and E. Melendez *An. Quim., Ser. C.* **81**, 56 (1985).
11. J. J. Bozell, C. E. Vogt, and J. Gozum *J. Org. Chem.*, **56**, 2584 (1991).
12. C. Dobler, H.-J. Kreuzfeld, M. Michalik, and H. W. Krause *Tetrahedron: Asymmetry*, **1**, 117 (1996).

Enantioselective Hydrogenation of α-Keto Esters Over Cinchona-Pt/Al₂O₃ Catalyst. Kinetic and Molecular Modelling

J. L. Margitfalvi, E. Tfirst, M. Hegedüs, and E. Tálas

Chemical Research Center, Institute for Chemistry, Hungarian Academy of Sciences, 1525 Budapest POB 17, Hungary.

Abstract

In this work a new kinetic model is given for the enantioselective hydrogenation of ethyl pyruvate over cinchonidine-Pt/Al₂O₃ catalysts. This is the first attempt to give a kinetic model, which describes the kinetics up to high conversion levels. It is known that the existing kinetic models (1, 2) cannot describe either the conversion-time or the enantioselectivity-conversion dependencies. The kinetic modelling is based on results obtained in (i) reaction kinetic studies, and (ii) computer modelling of interactions involved in the control of the asymmetric induction. Further important information was provided by conformational analysis of the modifier. The kinetic model reflects all of the characteristic kinetic behaviour of the enantioselective hydrogenation of ethyl pyruvate (EtPy).

Introduction

The enantioselective hydrogenation of ethyl pyruvate over Cinchona-Pt catalyst has been investigated by different research groups (1-7). Characteristic feature of the above reaction is the strong rate acceleration and the high enantioselectivity in the presence of the chiral modifier. In the presence of derivatives of cinchonidine (CD) enantioselectivities higher than 90 % were obtained.

The asymmetric induction can be attributed either to surface phenomena (1,2,4-6) or to modifier - substrate interactions taking place in the liquid phase (7-11). This work is considered as a continuation of our recent model, the "shielding effect" model (8). The "shielding effect" model rationalizes both the rate acceleration and the asymmetric induction using terms generally accepted in organic chemistry. The aim of this work is to give further insight into the interactions involved in the enantioselective hydrogenation of α-keto esters in the presence of cinchona-Pt/alumina catalysts. In this work the following tools were used (i) reaction kinetics, (ii) molecular modelling, and (iii) kinetic modelling. The final goal of this work is to demonstrate that the reaction scheme derived from the "shielding effect" model can provide a kinetic model, which can describe all of main kinetic patterns of the enantioselective hydrogenation of ethyl pyruvate in different solvents.

Experimental

Hydrogenation of ethyl pyruvate. The hydrogenation of EtPy was carried out in a 300 cm^3 SS autoclave equipped with an injection chamber for separate introduction of cinchonidine (CD) as described earlier (7-9). The transient experiments were done by injecting CD during racemic hydrogenation. The optical yield was calculated as e.e. = ([R]-[S])/([R]+[S]) or e.e., % = e.e. x 100. Corrections were always done for the amount of ethyl lactate (EtLa) in pure EtPy. In this series of experiments a new batch of EtPy was used. It is the reason that the e.e. values obtained in this series were slightly higher than in earlier studies (7-9).

Product analysis. GC analysis was used for product analysis. The modified cyclodextrine coated capillary column resulted in complete separation of (R)- and (S)-ethyl lactate and EtPy. The analysis of the CD and its derivatives in the reaction mixture was done as described earlier (12).

Molecular modelling. The molecular modelling was done as described in (8,9). Computer programs provided by MSI (Insight II, Discover etc.), and Hypercube (Hyperchem) were used.

Kinetic modelling. The material balance was calculated for EtPy, ethyl lactates (EtLa) and CD by solving the set of differential equation derived from the reaction scheme. More details are given in Ref. 11. The hydrogen pressure was constant during the kinetic run. Some of the rate constants were determined or approximated from independent kinetic measurements. The initial ratio of the open and closed forms of the modifier ([CD$_{open}$]$_o$/[CD$_{close}$]$_o$) was estimated based on NOESY NMR spectra.

Results and discussion

Reaction kinetic data.

Hydrogenation experiments. Figure 1. shows typical kinetic curves of the hydrogenation of EtPy in ln[1-X] vs. time coordinates (X = conversion). The kinetic curves of the enantioselective hydrogenation can be described by two pseudo-first order rate equations. The first equation describes the kinetics in the first 6-10 minutes (up to 20-40 % conversion), while the second equation is valid after 10 - 20 minutes of reaction time. It is very important to note that in ethanol $k_1 > k_2$, while in toluene under standard experimental condition $k_1 < k_2$. At low concentration of CD in toluene $k_1 > k_2$. Similar results were observed in our earlier studies, where upon increasing the modifier concentration k_1 decreased, while k_2 increased (11).

Figure 2. shows the dependence of the enantioselectivity of the conversion. The above series of experiments were obtained in toluene at different modifier concentration. The monotonic increase type behaviour of the e.e. vs. conversion dependencies is well demonstrated. These data strongly indicate that the earlier model (1,2) should be revised. Please note that according to the earlier model the optical yield should be constant during the whole reaction.

Transient experiments. The results of transient experiments are shown in Figure 3A and B. Figure 3A shows the conversion vs. time, while Figure 3B the enantioselectivity vs. conversion dependencies. As emerges from Figure 3A upon introduction of CD the rate acceleration is instantaneous, while the e.e. vs. conversion dependencies show a pronounced monotonic increase character. In transient experiments the monotonic increase character of the e.e. vs. conversion dependencies is even more pronounced than in standard experiments

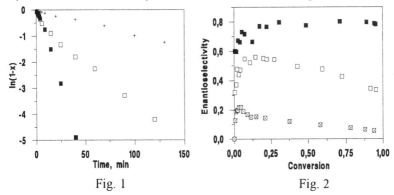

Fig. 1   Fig. 2

Figure 1. Kinetic curves of the hydrogenation of ethyl pyruvate in $\ln[1-X]$ - time coordinates. + - racemic hydrogenation in toluene, $\square$, ■ - enantioselective hydrogenation in ethanol ($[CD]_o = 3.4 \times 10^{-4}$ M) and toluene ($[CD]_o = 8.5 \times 10^{-4}$ M), respectively. T = 22 °C, $P_{H2}=50$ bar, $[EtPy]_o = 1.0$ M.

Figure 2. Enantioselectivity vs. conversion dependencies obtained at different modifier concentration in toluene. T = 23 °C, $P_{H2} = 50$ bar, $[EtPy]_o =1.0$ M. $[CD]_o$: ⊠ - 1.7×10^{-6} M, $\square$ - 6.8×10^{-6} M, ■ - 1.7×10^{-5} M.

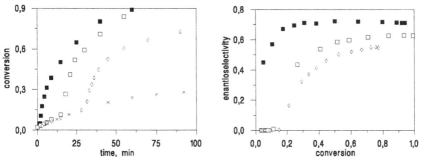

Figure 3. Transient kinetic results obtained by injection of CD during racemic hydrogenation. **A** - conversion vs. time, **B** - e.e. vs. conversion dependencies. $[Etpy]_o = 1.0$ M, $[CD]_o = 8.5 \times 10^{-4}$; injection time, in minutes: ■ - 0, $\square$ - 15, ◊ - 30. Reaction in ethanol.

Transformation of cinchonidine during the hydrogenation reaction.
During the hydrogenation reaction both the vinyl and the quinoline group of CD can be hydrogenated. The hydrogenation of quinoline ring leads to the formation of CD derivatives (hexa- and decahydro CD) which have no ability to induce

enantio-differentiation (13). The loss of active forms of CD during the enantioselective hydrogenation is shown in Figure 4. The lower the concentration of CD in the liquid phase the higher the extent of loss of CD. The above loss of cinchonidine is responsible for the decrease of the optical yield at high conversion observed at low concentration of CD (see Figure 2).

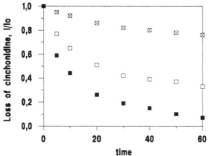

Figure 4. Kinetic curves of the relative loss of cinchonidine under condition of enantioselective hydrogenation of EtPy. Reaction in ethanol, $[EtPy]_o = 1.0$ M, T $=23\ ^\circ$C, $P_{H2} = 50$ bar, [CD] : ⊠ - 1.7×10^{-3}, ☐ - 1.7×10^{-4}, ■ - 9×10^{-5}.

Computer modelling

Conformational analysis. The molecular modelling of the [substrate - modifier] interaction required detailed conformational analysis of cinchonidine (CD). The energy map of CD was calculated by changing the torsion angles (C3')-(C4')-(C9)-(C8) (phi) and (C4')-(C9)-(C8)-(C7) (psi) using the MM+ forcefield. The numeration of the cinchonidine molecule is given in Scheme 1.

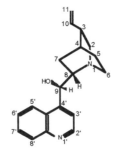

Scheme 1.

In our previous study (7-9) the above analysis was done by using rigid quinoline and quinuclidine parts. As a result of earlier investigations four stable conformations have been found. In the present study all of these calculations were repeated in such a way that only the phi and psi torsion angles were forced to be constant for all gridpoints of the map, and all the other degrees of freedom of the molecule were left to relax. The conformational analysis and full geometry optimization calculations (using other forcefields, too) indicate that CD can exist at least in nine different forms: A1, A2, A3, B1, B2, B3, C1, C2, and C3. (The notation of the conformations is the following: phi is near to -90°, 90°, and 20° for forms numbered 1, 2, and 3, respectively, and psi is near to -80°, 60°, and 180° for A, B, and C forms, respectively).

The calculated CD conformers are shown in Figure 5. The main difference between the CD conformers is the direction of the lone pair orbital of the quinuclidine nitrogen. For the C1, C2, and C3 forms this orbital is directed towards the quinoline ring, while in other conformers the nitrogen atom of the quinuclidine moiety and the quinoline ring are far away form each other. C1, C2, and C3 forms will be denoted as closed, while all the other forms as open conformations. The A2 conformer corresponds to the crystallographic form of CD.

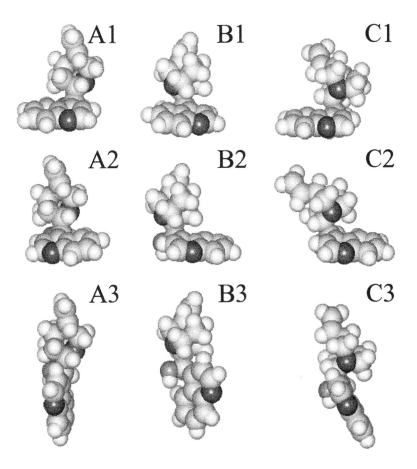

Figure 5. Calculated conformers of cinchonidine.

The rotational barriers between the optimized structures are shown in Figure 6A and 6B. These calculations were carried out in such a way that only one torsion angle was restrained to maintain a certain value and all the other degrees of freedom were left to relax. This way the other torsion angle sets the value which minimizes the total energy with the other relaxed degrees of freedom. Figure 6A shows the molecular mechanics energy as a function of psi along the path C1-A1-B1-C1 (solid line) and C2-A2-B2-C2 (dotted line). Along these paths while psi changes its value from -180° to +180° the value of phi changes from -120° to -70° for C1-A1-B1-C1 path, and from 50° to 110° for C2-A2-B2-C2 path. Similar functions are shown in Figure 6B, however in this case the total energy is the function of phi, and psi can change in order to minimize the total energy. Solid line shows the function along the A1-A3-A2 path, while dotted line demonstrates the C1-C3-C2 path. It should be mentioned that there are no "natural" C3-A3-B3-C3 and B1-B3-B2 paths along which the molecule can stable rotate, hence these paths do not appear in Figures 6A and 6B.

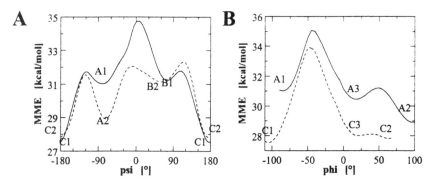

Figure 6. Rotational barriers between different conformers. A - rotation around (C4')-(C9) axis, B - rotation around (C9)-(C8) axis.

The conformational analysis indicates that there are only three stable conformers of cinchonidine, A2, C1 and C2. All of the other conformers are very unstable hence around one kcal/mol energy is sufficient to transform A1, A3, B1, B2, and C3 conformers into another most stable conformer. These results indicate how easy it is to rotate both the quinoline and quinuclidine moiety around the (C4')-(C9) and (C9)-(C8) axes, respectively. Upon increasing the temperature the probability of free rotation increases, thus at high temperature no preferential stabilization of CD can be expected. We consider that this fact is responsible for the loss of enantio-differentiation if the hydrogenation reaction was carried out above 35-40 °C (4). Please note that none of the earlier models can explain the loss of enantioselectivity observed at high temperature.

In this study we shall use only the open A2 and the closed C2 forms for modelling. The conformational change of CD from open A2 form to closed C2 one requires the rotation of the quinuclidine ring around the (C9)-(C8) axis, the energy needed for this change is less than 4 kcal/mole (see Figure 6A).

Similarly, the rotation of the molecule around the (C4')-(C9) axis from the closed C2 conformation to the C1 one is about 6 kcal/mole (see Figure 6B). Thus, the conformational analysis strongly indicates that CD can exist both in open and closed forms and both forms of CD can be involved in the formation of [substrate - modifier] complex. It should also be mentioned that in all of the previous modelling (19) only the open conformer of CD was used.

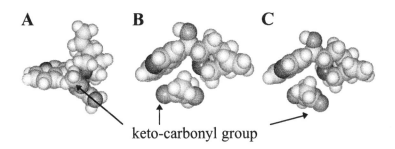

keto-carbonyl group

Figure 7. Modelling the [substrate-modifier] complex. **A** - [substrate - CD$_{open}$], **B** - [substrate - CD$_{closed}$] favourable alignment, **C** - [substrate - CD$_{closed}$] unfavourable alignment.

Modelling substrate modifier interactions. Figure 7A shows the model of the [substrate - modifier] complex for the CD$_{open}$, while Figure 7B and 7C represent the same model for the CD$_{closed}$. The difference between complexes shown in Figure 7B and 7C is that upon subsequent hydrogenation the former results in (R)-lactate, while the latter (S)-lactate.

In the open [substrate - modifier] complex shown in Figure 7A the substrate molecule is not rigidly fixed. In this complex the modifier behaves as a monodentate ligand. Consequently, this complex cannot be involved in asymmetric induction. Contrary to that the modifier in the closed form behaves like a bidentate ligand. The above results show that in closed conformation cinchonidine can provide a specific shielding effect for the substrate molecule (see Fig. 7B and C). Due to the above shielding the substrate can only adsorb onto the platinum surface by its unshielded side. The specificity in the above shielding lies in the right directionality between the quinuclidine nitrogen and the keto-carbonyl group (compare complexes shown in Fig. 7B and C). In complex B the above directionality is favourable, while in complex C it is unfavourable (8). We consider that in complex B the reactivity of the keto group strongly increases due to the favourable directionality resulting in the observed rate acceleration and preferential formation of (R) -lactate.

Kinetic modelling

Basic kinetic information. The most important information obtained in kinetic experiments are as follows: (i) the form of kinetic curves in ln(1-X) vs. time coordinates (X = conversion) (see Fig. 1), (ii) the form of enantioselectivity vs. conversion dependencies (see Fig. 2), the loss of CD in the hydrogenation experiments (see Fig. 4), and (iv) the reaction rate - time dependencies. The latter very interesting observation was recently found (15). These results are shown in Figure 8.

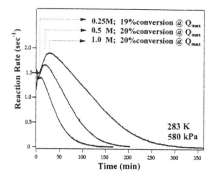

The simplest model. In the simplest model there are two catalytic cycles: (i) racemic hydrogenation (reactions (1a) and (1b)), (ii) and the enantioselective hydrogenation of the [substrate - modifier] complex [X], formed in an equilibrium reaction, onto (R) - lactate. In the simplest model no differentiation was done between the open and closed forms of CD.

Figure 8. The reaction rate - time dependencies [15].

Reactions included into the simpliest model are as follows:

Etpy	+	H_2	-----> (R) - lactate	(1a)
Etpy	+	H_2	----> (S) - lactate	(1a)
Etpy	+ CD		<----> [X]	(2)
[X]	+	H_2	----> (R) - lactate + CD	(3)

Kinetic results (i.e. first order rate constants) obtained in the racemic hydrogenation were directly used in the kinetic model. A brief mathematical analysis of the above model has been recently given (10). The above analysis showed that even this simple model should give the monotonic increase type e.e. -conversion dependencies observed in kinetic experiments (see Fig. 1A).

Models with the involvement of two forms of modifier. The computer modelling indicated that the modifier can form at least two stable conformers, i.e. CD_{closed} and CD_{open}. Both forms of modifier can form [substrate - modifier] complexes, i.e. $[X]_{closed}$ and $[X]_{open}$. The two conformers of CD are in equilibrium. Based on NOESY NMR data obtained in benzene solvent the ratio of CD_{closed}/CD_{open} was estimated as 5:95 (16). The above ratio shows a strong solvent dependence, the amount of CD_{closed} increases in the following way: $CD_3OD < C_6D_6 < CD_3COOD$ (16).

We consider that $[X]_{open}$ can be transformed into $[X]_{closed}$. This assumption is based on the change of the NMR pattern of the C(9) proton in the presence of substrate (8,9). In the presence of methyl pyruvate the C(9) proton was shifted from 5.95 to 6.3 ppm and its doublet form was changed to singulet.

These changes in the NMR pattern were attributed to the substrate induced conformational changes of CD.

The hydrogenation of $[X]_{closed}$ results in (R) - lactate, while the hydrogenation of $[X]_{open}$ results in racemate. Both $[X]_{closed}$ and $[X]_{open}$ can be hydrogenated if the carbon atom of the keto group and the quinuclidine nitrogen are in favourable alignment, similar to that shown in Figure 7A and 7B. The involvement of $[X]_{closed}$ and $[X]_{open}$ in the hydrogenation is shown in Scheme 2.

The hydrogenation via $[X]_{open}$ was modelled based on experiments, where quinuclidine (QNU) was used as a modifier. The addition of QNU resulted in threefold rate acceleration with respect to the pure racemic hydrogenation reaction (17).

$$[CD]_{open} \longleftrightarrow [CD]_{closed}$$

(R) and (S) -lactate

↑

$$\text{EtPy} + [CD]_{open} \longleftrightarrow [X]_{open} \longleftrightarrow [X]_{closed} \longleftrightarrow \text{EtPy} + [CD]_{open}$$

↓

(R) -lactate

Scheme 2.

Further improvements of the model. The improvement of the model required to include into the model the reaction responsible for the loss of CD in the hydrogenation reaction.

$$CD_{open} + H_2 \longrightarrow [CD_{open}]^{inact.}$$
$$CD_{closed} + H_2 \longrightarrow [CD_{closed}]^{inact}$$

The lower the concentration of CD the higher the rate of the formation of inactivated forms of CD.

Based on the homogeneous catalytic analogs of the enantioselective hydrogenation reactions (18) it was also suggested that the formation of the product can take place via two step hydrogenation, and an equilibrium was suggested in the first step:

$$[X]_{closed} + H_a \longleftrightarrow [XH]_{closed} \longrightarrow (R) \text{-lactate}$$

$$[X]_{open} + H_a \longleftrightarrow [XH]_{open} \longrightarrow (S) \text{-lactate}$$

$$[X]_{open} + H_a \longleftrightarrow [XH]_{open} \longrightarrow (R) \text{-lactate}$$

It has also been suggested that in the enantioselective hydrogenation of ethyl pyruvate the product formation is a two-step hydrogenation reaction (4,14).

Results obtained in kinetic modelling. Results obtained in kinetic modelling are shown in Figure 9-10. The kinetic constants used for the calculations are given in Table 1. Upon increasing $[CD]_o$ the following rate constants were altered: (i) addition of adsorbed hydrogen to $[X]_{open}$ and $[X]_{closed}$, and (ii) the hydrogenation of $[CD]$ to $[CD]^{ina}$. These alterations are based on the decrease of k_1 values when $[CD]_o$ increased (11) and on the results presented in Figure 4 (loss of CD during the hydrogenation reaction).

Figures 9A-D show the four main dependencies: (i) conversion vs. time; (ii) $\ln[1-x]$ vs. time, (iii) e.e. - vs. conversion, (iv) rate vs. time. In these figures results obtained at different modifier concentration are presented. Figure 10 shows the transient behaviour of the hydrogenation, when CD is injected in the 30th minute.

As emerges form Figures 9-10 the new kinetic model developed in this work describes all of the main kinetic feature of the enantioselective hydrogenation of ethyl pyruvate. The most important issue is that both the monotonic increase type optical yield - conversion dependencies and the decrease of the optical yield with conversion at low CD concentration could be modelled both under normal and transient experimental condition (compare results shown in Figure 9 and Figure 1B or Figure 10 and Figure 2).

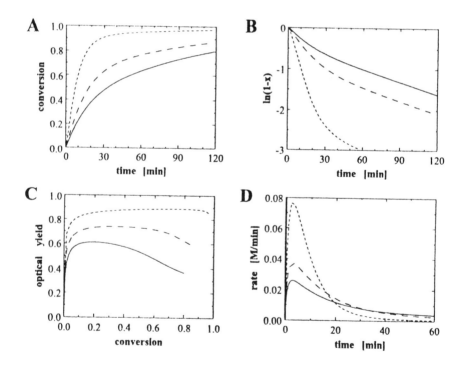

Figure 9. Modelling the kinetic behaviour of the enantioselective hydrogenation of ethyl pyruvate at different modifier concentration. $[EtPy]$ = 1.0 M, $[CD]$, M: —— 6.8×10^{-6}, - - - - 1.2×10^{-5}, ... 3.4×10^{-4}.

Table 1. Reactions and rate constants used for modelling. (Rate constants are given in 1/min for first order and hydrogenation reactions, and in M/min for second order reactions.)

Reaction	$[CD]_0 = 6.8$ x 10^{-6} M	$[CD]_0 = 1.2$ x 10^{-5} M	$[CD]_0 = 1$ x 10^{-4} M
$[CD]_{closed} \rightarrow [CD]_{open}$	0.475	0.475	0.475
$[CD]_{open} \rightarrow [CD]_{closed}$	0.025	0.025	0.025
$EtPy + [CD]_{open} \rightarrow [X]_{open}$	500	500	500
$[X]_{open} \rightarrow EtPy + [CD]_{open}$	50	50	50
$EtPy + [CD]_{closed} \rightarrow [X]_{closed}$	12000	12000	12000
$[X]_{closed} \rightarrow EtPy + [CD]_{closed}$	1200	1200	1200
$[X]_{open} \rightarrow [X]_{closed}$	1	1	1
$[X]_{closed} \rightarrow [X]_{open}$	0.05	0.05	0.05
$EtPy + H_2 \rightarrow (R) / (S)$ prod.	0.006	0.005	0.004
$[X]_{open} + H_a \rightarrow [XH]_{open}$	25	20	15
$[XH]_{open} \rightarrow [X]_{open} + H_a$	1	1	1
$[XH]_{open} + H_a \rightarrow (R)/(S)$ prod.	250	250	250
$[X]_{closed} + H_a \rightarrow [XH]_{closed}$	12000	10000	7000
$[XH]_{closed} \rightarrow [X]_{closed} + H_a$	1	1	1
$[XH]_{closed} + H_a \rightarrow (R)$ prod.	12000	12000	12000
$[CD]_{op/cl} + H_2 \rightarrow [CD]^{ina}_{op/cl}$	0.15	0.12	0.10

The model also shows that the reaction rate passes through a maximum (compare Figure 8 and 9). However, the results also indicate that the above maximum can be observed if the reaction rate is continuously monitored, similar to that described in Ref. 15.

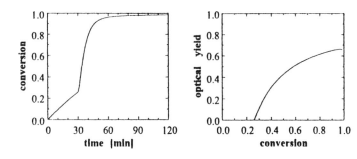

Figure 10. Modelling the transient behaviour of the enantioselective hydrogenation. Injection of CD in the 30th minute. [EpPy] = 1.0 M, [CD] = 3.4 x 10^{-5}.

Summary

Results obtained in molecular and kinetic modelling provided further prove that the enantioselective hydrogenation of α-keto esters over cinchona-Pt catalyst is more complex than it was suggested in the earlier models (1-2,4). The result obtained in this study provided additional proofs with respect to the importance of interactions in the liquid phase. Further proofs are in progress in our laboratories to demonstrate the role of interactions in the liquid phase both in the rate acceleration and enantio-differentiation.

Acknowledgment

Partial financial help from OTKA grant (T 023317) is greatly acknowledged.

References

1. H. U. Blaser and M. Müller, *Stud. Surf. Sci. Catal.*, **59,** 73 (1991).
2. H. U. Blaser, M. Garland and H. P. Jallett, *J.Catal.*, **144,** 569 (1993).
3. Y. Orito, S. Imai, and S. Niwa, *J. Chem. Soc. Jpn.*, 1118 (1979).
4. G.Webb and P.B. Wells, *Catal. Today*, **12,** 319 (1992).
5. U. K. Singh, R.N. Landau, Y. Sun, C. LeBond.,D. G. Blackmond, S.K. Tanielyan, and R. L. Augustine, *J.Catal.* **154,** 91 (1995).
6 T.Mallat, Z.Bodnar, B.Minder, K.Borszeki and A. Baiker, *J. Catal.*, **168,** 183 (1997).
7. J. L. Margitfalvi, in M.G. Scaros, M. L. Prunier (Editors), Chem. Ind., **62,** *Catal. Org. React.*, Dekker, 1995, p. 189.
8. J. L. Margitfalvi, and M. Hegedüs, *J. Mol. Catal. A*, **107,** 281 (1996).
9. J. L. Margitfalvi, M. Hegedüs and E. Tfirst, *Stud. Surf. Sci. Catal.* (11th International Congress on Catalysis) **101,** 241 (1996).
10. J. L. Margitfalvi, M. Hegedüs and E. Tfirst, *Tetrahedron: Asymmetry*, **7,** 571 (1996).
11. J. L. Margitfalvi, B. Minder, E. Tálas, L. Botz and A. Baiker, in L. Guczi et al. (Editors), "New Frontiers in Catalysis", (Proc. 10th Int. Cong. Catal. Budapest, July 1992), Elsevier, Amsterdam, 1993, p. 2471.
12. E. Talas, L. Botz, J. Margitfalvi, O. Sticher, and A. Baiker, *J. Planar Chrom.*, **5,** 28 (1992).
13. J. L. Margitfalvi, P. Marti, A. Baiker, L. Botz and O. Sticher, *Catal. Lett.*, **6,** 281 (1990).
14. A. Baiker, *Stud. Surf. Sci. Catal.* (11th International Congress on Catalysis) **101,** 51 (1996).
15. J. Wang, J. Sun, C. LeBlond, R. N. Landau and D. G. Blackmond, *J. Catal.*, **161,** 752 (1996).
16. J.L. Margitfalvi, L. Radics, unpublished results.
17. J.L. Margitfalvi, and M. Hegedüs, *J. Catal.*, **156,** 175 (1995).
18. J. Halpern, in *"Asymmetric Synthesis"*, ed. by J. D. Morrison, (Academic Press, New York, 1985) p.41.

New Rhodium Catalysts for the Regioselective and Enantioselective Hydroboration of Styrene

John A. Brinkman and John R. Sowa, Jr.*

Department of Chemistry, Seton Hall University, South Orange, NJ 07079

Abstract

New rhodium catalysts that give varying regioselectivity for the addition of catecholborane to substituted styrenes depending on the co-ligands added to the reaction have been developed. The catalysts were prepared *in situ* from $[(\eta^5\text{-}C_9H_7)Rh(COD)]$ and CO, monodentate, and bidentate phosphine ligands. With phosphine ligands the reaction is regiospecific (>99%) for Markovnikov addition. When performed in the presence of chiral ligands the hydroboration reaction was mildly enantioselective (ee = 62%). The enantiomeric excess for this reaction was determined by formation of the ester using commercially available (S)-naproxen.

Introduction

The catalytic hydroboration of alkenes is a reaction that has recently received considerable attention. In 1985, Manning and Noth demonstrated that Wilkinson's catalyst, $[RhCl(PPh_3)_3]$, catalyzed the hydroboration of alkenes with catecholborane under mild conditions (1). However, the chemo-, regio- and stereoselectivity was different from the uncatalyzed reaction (1). For example, metal catalyzed hydroboration of styrene favors the Markovnikov product **1** rather than the anti-Markovnikov product **2** formed by the uncatalyzed hydroboration reaction (eqn.1).

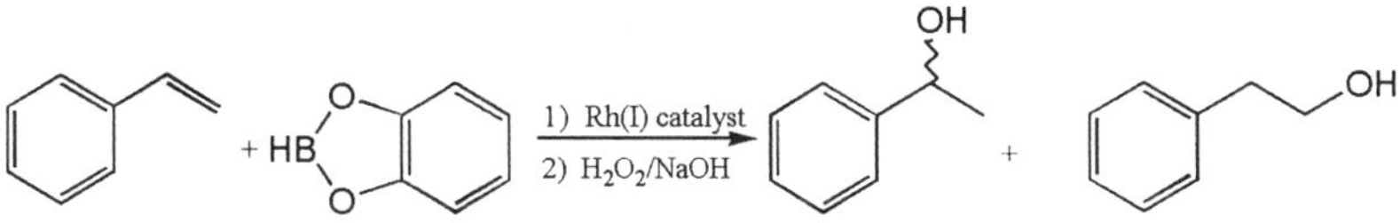

$$(1)$$

The catalytic asymmetric synthesis of optically active 1-arylalkanols with up to 96% ee has also been demonstrated with this reaction using a catalyst prepared *in situ* by addition of catecholborane to a solution of $[Rh(COD)_2]^+BF_4^-$ and (R)-(+)-2,2'-bis(diphenylphosphino)-1,1'-binapthyl [(R)-binap] (2). Subsequently Marder et al. reported Markovnikov selectivities for catalyzed hydroboration of substituted styrenes using [Rh(η^3-2-Me-allyl)(DiPPE)] (DiPPE = 1,2-bis(diisopropylphosphino)ethane) as a catalyst (Fiqure 1).

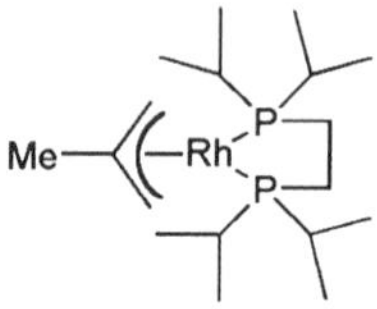

Fiqure 1. [Rh(η^3-2-Me-allyl)(DiPPE)]

The same selectivity was observed by generating the catalyst *in situ* by treatment of [Rh(η^3-2-Me-allyl)(COD)] with DiPPE in THF. Excellent Markovnikov selectivity was also reported with the bidentate phosphine analogue [Rh(η^3-2-Me-allyl)(DPPB)] (DPPB = 1,4-bis(diphenylphosphino)-butane) (3).

Our work has focused on the development of a new rhodium catalyst system based on (η^5-C$_9$H$_7$)Rh(COD) (Fiqure 2) for the addition of catecholborane to substituted styrenes. Our catalyst system employs the indenyl ligand as we felt that this ligand would have similar properties to the allyl ligand used in the work of Marder and the pre-catalyst is conveniently prepared. In addition, the COD ligand can be easily removed to allow *in situ* addition of co-ligands. In our study, we examined the effect of CO, PR$_3$, cyclooctadiene, chiral and achiral bidentate phosphine ligands attached to rhodium. The catalysts were generated *in situ* by addition of different ligands to a solution of the (η^5-C$_9$H$_7$)Rh(COD) complex to enable convenient exploration of electronic and steric effects on the hydroboration stereoselectivity.

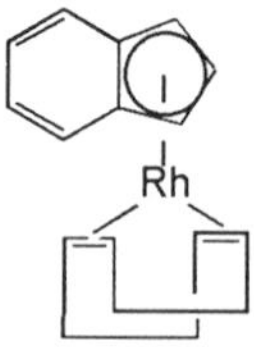

Figure 2. (η^5-C$_9$H$_7$)Rh(COD)

Experimental

The following complexes were prepared using previously reported procedures: bis(1,5-cyclooctadiene)dirhodium(I) dichloride, [Rh(COD)Cl]$_2$ (4), indenyl rhodium(I) (1,5-cyclooctadiene), (η^5-C$_9$H$_7$)Rh(COD) (3) (5a). The complex, $(\eta^5\text{-}1\text{-}CF_3\text{-}C_9H_6)Rh(COD)$ (**4**) was prepared by reaction of $1\text{-}CF_3\text{-}C_9H_7$ and $[Rh(COD)Cl]_2$ in the presence of KH in THF solvent (5b). All other reagents were purchased from the commercial sources of Aldrich and Strem and used without further purification.

In situ preparation of $(\eta^5\text{-}C_9H_7)Rh(CO)_2$: A 50 mL flame dried Schlenk flask was charged with 13 mg (0.04 mmol) of $(\eta^5\text{-}C_9H_7)Rh(COD)$ and 5 mL of THF. A stream of carbon monoxide was bubbled into the reaction solution for 10 minutes. This crude solution was then used in the hydroboration reaction without further purification. IR: (hexane) υ_{CO} 2062 cm^{-1}, υ_{CO} 2008 cm^{-1}.

In situ preparation of $(\eta^5\text{-}C_9H_7)Rh(CO)(PPh_3)$: A solution of $Rh(\eta^5\text{-}C_9H_7)(CO)_2$ in THF was prepared as described above, then 11.5 mg of PPh_3 (0.044mmol, 1.1 equiv) was added and stirring continued for 20 min. The crude solution was then used without isolation or further purification. IR: (THF) υ_{CO} 1965 cm^{-1}.

Example procedure for the hydroboration reaction with styrene: A 50 mL flame dried Schlenk flask under an argon atmosphere was charged with 13 mg of $(\eta^5\text{-}C_9H_7)Rh(COD)$ (0.04 mmol) and 5.0 equiv/Rh of PPh_3 ligand or 2.5 equiv/Rh for bidentate phosphine ligands. THF (5 mL) was added and stirring was continued for 30 min. To the flask was added 208 mg of styrene (2 mmol) followed by 2.2 mL of catecholborane (1.1 equiv, 1M in THF). After stirring for 30 min the reaction was quenched with 4 mL of methanol. The boronate ester was converted to the alcohol by addition of 4.8 mL of 3N NaOH and 0.5 mL of 30% H_2O_2, and stirring at room temperature for 3h. Extraction with ether followed by flash chromatography and eluting with hexane/ethyl acetate (4/1) gave 212 mg (87%) of 1-phenylethanol. The identity of the 1-phenylethanol was established by comparison with an authentic sample which was purchased from Aldrich.

Example procedure for the esterification of 1-phenylethanol with (S)-naproxen (eqn. 2): To 20 mL of ether was added 95 mg of racemic 1-phenylethanol (0.78 mmol) and 270 mg of (S)-naproxen (1.17 mmol, 1.5 equiv). To this stirred mixture was added dropwise 147 mg of 1,3-diisopropylcarbodiimide (1.17 mmol, 1.5 equiv) and 5 mg of 4-dimethylaminopyridine dissolved in 5 mL of ether. After stirring for 3h the mixture was filtered and then concentrated *in vacuo*. The crude product was purified by flash chromotography eluting with hexane/ethyl acetate (9/1) to yield 198 mg (76%) of ester as a white solid. ^1H NMR: 7.80-7.50 (m, 3H), 7.40-7.00 (m, 8H), 5.85 (q, 1H), 3.95 (s, 1H), 3.90 (q, 1H), 1.56 (dd, 3H), 1.49 (d, 3H), 1.41 (d, 3H). The ee analysis was performed by ^1H NMR integration of the signals at 1.49 and 1.41 and by gas chromatography (GC) using a HP 5890 instrument with a SPB-1 or SPB-5 column.

(2)

Results and Discussion

Our initial work focused on the hydroboration of styrene. We found the regiochemistry of this reaction to be dependent on the type of ligands attached to rhodium. In the absence of catalyst we found the major product of the hydroboration reaction to be 2-phenylethanol (89% after oxidative workup) resulting from anti-Markovnikov selectivity (Table 1). When the reaction was performed with 2 mol % (η^5-C$_9$H$_7$)Rh(COD) the major product was still the primary alcohol, however, the percentage of the Markovnikov product, 1-phenylethanol increased slightly to 25%. The slight increase in Markovnikov selectivity was attributed to involvement of the rhodium catalyst in the reaction.

We then explored ways to modify the catalyst to increase selectivity. The strategy we employed was to explore the effects of electron withdrawing ligands such as CO and electron donating ligands such as phosphines. Thus, the next catalyst we examined was (η^5-C$_9$H$_7$)Rh(CO)$_2$ which was prepared *in situ* by bubbling CO gas through a solution of (η^5-C$_9$H$_7$)Rh(COD) in THF. The hydroboration selectivity for this catalyst was similar to the uncatalyzed system indicating no catalytic reaction occurred with electron withdrawing CO ligands. However, it is well known that addition of phosphine ligands to (η^5-C$_9$H$_7$)Rh(CO)$_2$ produces (η^5-C$_9$H$_7$)Rh(CO)(PR$_3$) complexes (6). We felt that the phosphines would increase the electron richness of the metal center and effect catalysis (7). Indeed, adding 1.1 equiv. of PPh$_3$ or PPh$_2$Me to (η^5-C$_9$H$_7$)Rh(CO)$_2$ to give (η^5-C$_9$H$_7$)Rh(CO)(PPh$_3$) or (η^5-C$_9$H$_7$)Rh(CO)(PPh$_2$Me), respectively, increased the selectivity of the hydroboration reaction to favor 1-phenylethanol (67%). Encouraged by this result we returned to the (η^5-C$_9$H$_7$)Rh(COD) catalyst and added 5.0 equiv of triphenylphosphine *in situ*. The hydroboration reaction now proceeded with regiochemistry indicating only 1-phenylethanol. High Markovnikov selectivity (>99%) was also observed for the chelating (S)-(-)-2,2'-bis(diphenylphosphino)-1,1'binapthyl (S-binap) ligand.

These results indicate the strong dependence of the ligand coordinated to the rhodium catalyst. Markovnikov selectivities result when the electron donating phosphine ligands are present. The anti-Markovnikov selectivities that result when the electron withdrawing carbonyl group is bound to rhodium are likely a result of poor catalytic activity and the dominating noncatalytic reaction. Although the structure of the catalyst is not known, it is likely that the catecholborane removes the indenyl and COD ligands from rhodium via hydroboration. That $(\eta^5\text{-}C_9H_7)Rh(COD)$ and $(\eta^5\text{-}1\text{-}CF_3\text{-}C_9H_6)Rh(COD)$ give the similar regioselectivities also suggests that the indenyl ligands are removed during the catalytic cycle (Table 1). The rhodium species is then trapped by the phosphine ligand in solution to provide a catalytic species.

Table 1. Effect of ligand on regioselectivity.

entry	Ind`Rh(COD)[a] (3, 4) complex + added ligand[b]	time (min)	% 1-phenyl- ethanol	% 2-phenyl- ethanol
1	no catalyst	30	11	89
2	3 + CO	30	12	88
3	4 + CO	30	12	88
4	3	30	25	75
5	4	30	20	80
6	3 + CO + PPh₃	30	67	33
7	4 + CO + PPh₃	30	57	43
8	3 + CO + PMePh₂	30	67	33
9	3 + 2.5equiv dppb	30	89	11
10	3 + 5.0equiv PPh₃	30	>99	-
11	4 + 5.0 equiv PPh₃	30	>99	-
12	2.5equiv (S)-binap	30	>99	-

[a]Complex **3** corresponds to $(\eta^5\text{-}C_9H_7)Rh(COD)$ (see Figure 2) and complex **4** corresponds to $(\eta^5\text{-}1\text{-}CF_3\text{-}C_9H_6)(Rh(COD))$.
[b]See experimental section for the *in situ* preparation of catalysts and reaction conditions.

The catalytic hydroboration of substituted styrenes also proceeds with high Markovnikov selectivity in the presence of 2 mol % of $(\eta^5\text{-}C_9H_7)Rh(COD)$ + 5.0 equiv/Rh of PPh₃. As shown in Table 2 the hydroboration reaction is highly selective (>99%) with electron donating or electron withdrawing groups on the aromatic ring.

Table 2. Effect of substrate on regioselectivity.[a]

entry	substrate	% 1-phenylethanol	% 2-phenylethanol	% yield
1	styrene	>99	-	87
2	2-vinylnapthalene	>99	-	71
3	4-fluorostyrene	>99	-	87
4	4-vinylanisole	>99	-	72

[a] 1) $(\eta^5\text{-}C_7H_9)Rh(COD)$, PPh_3 (5.0 equiv/Rh), catecholborane, 2) H_2O_2/NaOH

We next determined the enantioselectivity of the hydroboration reaction of styrene using $(\eta^5\text{-}C_9H_7)Rh(COD)$ + 2.5 equiv/Rh of (S)-binap as the catalyst. The enantioselectivity was strongly dependent on the reaction temperature and (S)-binap gave predominately (S)-1-phenylethanol as shown in Table 3. At 25°C, 0°C and -22°C the ee's were 20%, 62% and 22%, respectively. Since our catalyst system gives lower ee's than the previously reported $[Rh(COD)_2]^+BF_4^-$ /binap (2) system, this indicates that the active form of the two of the catalysts is different.

Table 3. Enantioselectivity of the hydroboration reaction.[a]

entry	temp (°C)[b]	time (h)	% ee[c]
1	25	0.5	20±1
2	0	2	62±4
3	-22	4	22±2

[a] 1) $(\eta^5\text{-}C_9H_7)Rh(COD)$, (S)-Binap, catecholborane, 2) H_2O_2/NaOH
[b] No reaction occurred at -78 °C.
[c] S-1-phenylethanol is the major enantiomer. Average and std. dev. over three runs.

The enantiomeric purity was determined by coupling the 1-phenylethanol with commercially available (S)-naproxen to give the ester as a diastereomeric mixture of products. The ratio of products were determined by ^1H NMR integration of the signals at δ 1.49d and 1.41d which are the methyl groups at the benzylic positions of the styrene moiety of each diastereomer. In addition, integration was done by GC analysis as the diastereomers have retention times that differ by over 1 minute. This novel approach takes advantage of the ready availability of (S)-naproxen which is a widely used non-steroid anti-inflammatory drug that is sold in the enantiomerically pure form (8).

Conclusion

We have developed a new catalyst system for *in situ* preparation of hydroboration catalysts. We have shown that the regiochemistry of the hydroboration reaction is strongly dependent on the ligands added to the reaction. However, with phosphine and bidentate phosphine ligands the reaction is specific (>99%) for Markovnikov addition. In addition, the substituents on styrene do not affect the ratio of Markovnikov to anti-Markovnikov products. We have also shown using (S)-binap that the reaction is mildly enantioselective and the enantioselectivity of the hydroboration reaction is temperature dependent. The enantiomeric purity was determined by formation of the diastereomeric ester using a readily available chiral auxillary, (S)-naproxen.

References

1. D. Manning and H. Noth, *Angew. Chem., Int. Ed. Engl.*, **24**, 878 (1985).
2. T. Hayashi, Y. Matsumoto and Y. Ito, *J. Am. Chem. Soc.*, **111**, 3426 (1989).
3. S. A. Westcott, H. P. Blom, T.B. Marder and R. T. Baker, *J. Am. Chem. Soc.*, **114**, 8863 (1992).
4. G. Giordano and R. H. Crabtree, *Inorg. Synth.*, **28**, 88 (1990).
5. a) H. Eshtiagh-Hosseni and J. F. Nixon, *J.Less Common Metals*, **61**, 107 (1978), b) J. W. Mickelson, Ph.D. Thesis, University of Minnesota, 1992.
6. F. Basolo, L. N. Ji and M.E. Rerek, *J. Chem. Soc., Chem.Commun.*, 1208 (1983).
7. J. R. Sowa, Jr., V. Zanotti, G. Facchin, and R. J. Angelici, *J. Am. Chem. Soc.*, **113**, 9185 (1991).
8. For related uses of naproxen as a chiral derivatizing agent see: R. Büschges, H. Linde, E. Mutschler, and H. Spahn-Langguth, *J. Chromatogr. A* , **725**, 323 (1996).

New Catalytic Systems for the Selective Hydrogenation of Halogenated Aromatic Nitro Compounds

E. Auer, A. Freund, M. Groß, R. Hartung and P. Panster
*Degussa AG, Silicas and Chemical Catalyst Division
Research and Applied Technology Chemical Catalysts and Zeolites,
P. O. Box 1345, 63403 Hanau, Germany*

Abstract

For the selective reduction of halogenated aromatic nitro compounds a new modified supported iridium based catalyst system has been developed. These catalysts are characterized by a high activity combined with an extreme low rate of dehalogenation. Moreover, the formation of other undesired side products such as azo or azoxy compounds can be kept under a very low level. The new modified iridium system is an interesting alternative to existing supported platinum and modified platinum catalysts for this application.

Introduction

Halogen substituted aromatic amines are widely used for the production of pharmaceutical and agrochemical substances. For the synthesis of haloaniline derivatives the reduction of the corresponding nitro compound is applied as one of the most important technical routes. Reaction is preferably performed with group VIII metal catalysts. Most of the time the reduction of the halogenated aromatic nitro compound to the corresponding amine is accompanied by a simultaneous dehalogenation which decreases the yield of desired product and causes problems due to the formation of corrosive halogen acids. [1,2,3,4] The amount of dehalogenated side-products depends strongly on the reactivity of the aromatic halogen.

Various types of catalysts that keep the elimination of halogen as low as possible are described in the literature, especially in several patents, and are applied in industrial processes. Hydrogenation is performed as a batch process in liquid phase with a slurry catalyst, in a stirred tank reactor or in a Buss loop reactor. Selectivity of supported group VIII metal catalyst, preferably platinum on activated carbon, can be affected by certain compounds, e.g. partial poisoning by sulfur [5,6], triphenylphosphite [7] or catalyst treatment with phosphoric acid [8].

Additionally cocatalytically active metals, e.g. cobalt, nickel [9], copper [10] or silver, lead and bismuth [11] largely avoid the formation of undesired dehalogenated side-products. Addition of morpholine [12] to the reaction mixture also leads to a significant improvement of the selectivity.

Although some of these modifiers and inhibitors can keep the elimination of halogen below 0.1 wt%, the purity of the corresponding amine is not as high as expected due to the formation of other undesired side products, e.g. azo- or azoxy compounds. Triphenylphosphite modified Pt/C catalysts are the most effective systems minimizing dehalogenation (< 0.01 wt%) , but also yield the lowest purity of final product (approx. 98.5 %). Sulfided supported platinum catalysts or morpholine/platinum systems are characterized by slightly higher dehalogenation (0.2 wt%), but a higher quality of the corresponding amine [5, 6].

Not only dehalogenation, but also the formation of azo- or azoxy compounds is a severe problem especially with dihalogenated nitro compounds. In the presence of morpholine modified platinum catalysts hydrogenation of 3,4-dichloronitrobenzene is accompanied by the formation of 3,3′,4,4′-tetrachloroazobenzene (TCAB) or 3,3′,4,4′-tetrachloroazoxybenzene (TCAOB) which should be completely avoided due to their toxicity similar to dioxin compounds [13].

Experimental

Catalyst Preparation and Characteristic Data

For the selective reduction of halogenated aromatic nitro compounds a new modified supported iridium-based catalyst system has been developed. Various types of suitable supports, such as aluminum oxide, titanium oxide, silicon dioxide or activated carbon can be used for the catalyst preparation in aqueous media applying standard impregnation methods [14,15,16]. Peat or wood based activated carbons are preferably used as supports and have shown the best results. The new multimetallic catalyst contains iridium and promoters such as manganese, iron, cobalt, nickel, copper and ruthenium. These promoters can be fixed onto the support either by simultaneous or sequential precipitation in the presence of hexachloroiridium(IV) acid. The amount of iridium on the catalyst is kept at 5 wt%, typically the total loading of promoters varies in the range of 0.1 - 0.3 wt%.

This new catalytic system is characterized by reduced precious metal particles which are typically smaller than 5 nm. The dispersion of iridium depends on the preparation method as well as on the reducing agent. TEM analysis confirms a shell character of the catalyst. Platinum or modified platinum catalysts used as comparable examples have been prepared according to standard methods reported in the literature [5,10].

Catalytic Hydrogenation

Catalytic hydrogenation of 2-chloronitrobenzene dissolved in an inert solvent, e.g. toluene or methanol was performed in a stirred batch stainless steel autoclave (total volume 0.5 l) using the recently developed catalyst in comparison to standard platinum, sulfided platinum or other modified platinum catalysts.

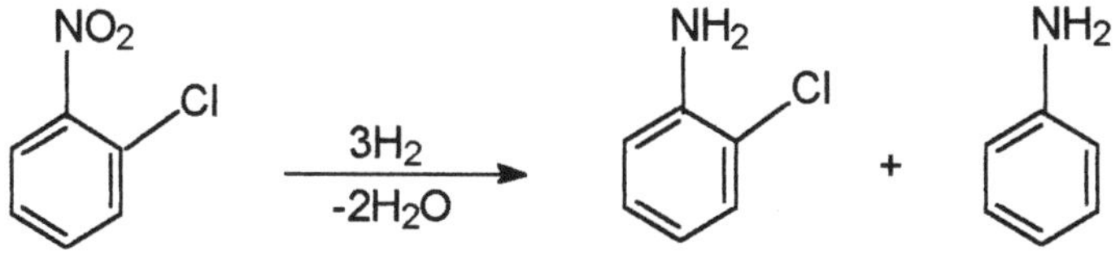

fig. 1: Hydrogenation of 2-chloronitrobenzene

Following representative industrial processes, typical reaction conditions for the hydrogenation of 2-chloronitrobenzene have been applied as described subsequently. Conversion, yield and formation of side-products have been precisely detected by GC. A sample was taken from the autoclave for GC-analysis 15 minutes after the apparent 100% conversion of reactant.

Modified iridium catalysts have also been applied for the selective hydrogenation of 2,4,5-trichloronitrobenzene (TCNB) to 2,4,5-trichloroaniline (TCA), which typically undergoes dehalogenation reactions very easily.

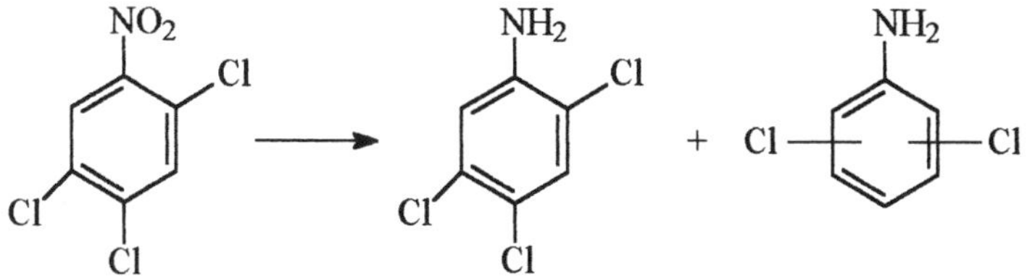

fig. 2: Hydrogenation of 2,4,5-trichloronitrobenzene

The conversion of reactant, yield of 2,4,5-trichloroaniline and the level of undesired side-products, such as 2,4-, 2,5- or 3,4-isomers of dichloroaniline, dimers of TCA as azo or azoxy derivatives and other oligomeric compounds can be determined from the analysis of the reaction mixture by GC.

Results and Discussion

Aim of the catalyst development was to find a new precious metal based catalyst system which is distinguished in particular by improved selectivity compared with known catalysts and which does not require the addition of modifiers or treatment with promoters during the hydrogenation reaction. Due to environmental aspects the catalyst system should also avoid the formation of undesired side products, especially highly toxic chlorinated dimers such as azo or azoxy derivatives.

Hydrogenation of 2-chloronitrobenzene

Test results obtained with the new iridium catalysts show excellent selectivities and activities in the hydrogenation of 2-chloronitrobenzene. In comparison to modified and unmodified platinum or palladium catalysts dehalogenation can be usually kept below 0.1 % if manganese, iron, cobalt, nickel and ruthenium have been used as suitable doping elements. Depending on the stirring rate, formation of undesired dehalogenated products could be kept below 0.05 % in some cases.

Figure 3 summarizes that iridium-based catalysts are not only characterized by an excellent selectivity but also by a significantly higher activity compared to sulfided platinum catalysts.

Modified supported iridium catalysts lead to reliable and convincing selectivities that are independent of the chemical properties of the halogenated aromatic reactant. However, a variation in catalyst composition, e.g. kind of promoter and its amount, is recommended to select the best iridium catalyst system for a certain reactant. This precision work helps to guarantee reproducible results and opens a broad variety of different applications.

Figure 4 illustrates the influence of different promoters on the selective hydrogenation of 2-chloronitrobenzene. It is recognizable that iron considerably hinders the dehalogenation. Addition of manganese or cobalt increases the activity by shortening the overall reaction time. However, selectivities of these promoters are worse than that of iron.

fig 3: Activities and selectivities of various catalysts in the hydrogenation of 2-chloronitrobenzene; temperature 90 °C, pressure 10 bar, solvent toluene (160 g), reactant 0.5 mol, catalyst concentration 0.5 wt%, stirring rate 700 rpm; the conversion of 2-chloronitrobenzene was 100%

catalyst type	precious metal loading [wt%]	activity: reaction time [min]	selectivity: amount of aniline [wt%]
Ir-Fe-Mn/C	5	45	< 0.05
Ir-Fe-Co/C	5	55	< 0.07
Ir-Fe/C	5	60	< 0.1
Pt-S/C	3	80	< 0.2
Ir/C	5	45	< 0.2
Pt-Cu/C	1	30	0.9
Pt/C	1	25	1.5

Additionally, catalyst development work dealt with the interaction between iridium and more than one doping element. Trimetallic iridum-based catalysts not only significantly improve the selectivity of the hydrogenation, but also shorten the reaction time. Especially iron combined with cobalt or manganese increases the performance of supported iridium as shown in figure 5.

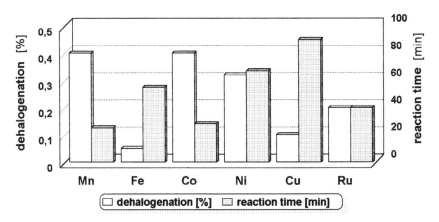

fig. 4: Activities and selectivities of various doping elements on Ir/C in the hydrogenation of 2-chloronitrobenzene

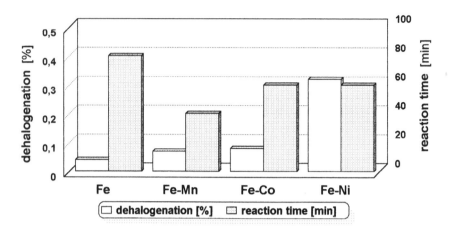

fig.5: *Activities and selectivities of trimetallic iridium-based catalysts in the hydrogenation of 2-chloronitrobenzene*

Hydrogenation of 2,4,5-trichloronitrobenzene (TCNB)

In this section autoclave test results for the hydrogenation of 2,4,5,-trichloronitrobenzene are summarized as an example of the selective hydrogenation of a polychlorinated aromatic nitro compound. Compared to monochlorinated systems, especially the reaction of polyhalogenated aromatic compounds to their corresponding amines requires catalytic systems exhibiting high performances as the above mentioned type of molecules are known for undergoing dehalogenation reaction very easily.

Figure 6 gives an overview of the test results obtained with multimetallic iridium catalysts in comparison to conventional platinum and sulfided platinum catalysts.

Modification of the iridium-based catalyst by iron and copper leads to an extremely pure product which contains only trace amounts of dechlorinated side-products. Even the formation of toxic dimers and other products could be reduced significantly compared to platinum systems. These results clearly underline the approach to develop a well-designed catalyst for each type of halogenated aromatic nitro compound.

fig.6: Activities and selectivities of various catalysts in the hydrogenation of 2,4,5-trichloronitrobenzene; temperature 90 °C, pressure 10 bar, solvent: toluene (160 g), reactand 0.5 mol, catalyst concentration 0.5 wt%, stirring rate 700 rpm; conversion of 2,4,5-TCNB was 100 %

catalyst type	yield of 2,4,5-TCA [wt%]	activity: reaction time [min]	selectivity: amount of dichloroaniline derivatives and side-products [wt%]		
			2,4-;2,5-DCA	3,4-DCA	others
Ir-Fe-Cu/C	> 99.3	40	< 0.04	< 0.10	< 0.5
Ir-Fe/C	> 99.0	35	< 0.05	< 0.15	< 0.5
Ir/C	> 98.5	30	< 0.09	0.5	0.5
Pt-S/C	98.0	30	< 0.1	0.4	1.2
Pt/C	96.0	25	0.2	0.5	3.0

Summary

Using this commercially available supported iridium based catalyst system, hydrogenation of various halogenated aromatic nitro compounds can be carried out with excellent selectivities. The formation of dehalogenated products remains below 0.1 wt%.
The new multimetallic iridum catalyst could be an interesting alternative to the broad variety of conventional platinum catalysts due to the improvement in productivity of the hydrogenation of substituted aromatic nitro compounds. Moreover, the formation of undesired toxic side-products can be kept as low as possible applying this new type of heterogeneous catalysts. In many cases, an additional post-treatment of the halogenated amine derivatives to remove the undesired side-products is unnecessary. This is an important contribution to our efforts to strengthen the responsibility for environmental awareness.
Moreover, refining of spent activated carbon supported iridium catalyst can be carried out with sufficient results.

References

1 Ullmann, Enzyklopädie der technischen Chemie, 5th edition, vol. A2, page 46 (1985)
2 J.R. Kosak, *Catalytic Hydrogenation of Aromatic Halonitro Compounds, Ann. New York Acad. Sci.*, **172**, 174-185, (1979)
3 R.K Rains, E.A. Lambers, R.A: Genetti, *Chem. Ind.*, **8** (Catalysis of Organic Reactions), 43-52 (1996)
4 F. Figueras, B. Coq, *Nato Asi. Ser. Ser. C*, **474**, 163-184 (1996)
5 US 4 059 627 (Bayer AG, 1977)
6 G. Cordier, J.-M. Grosselin, R.-M. Ferrero, *Ind. Chem. Libr.*, **8** (Roots of Organic Development), 336-342, (1996)
7 J.R. Kosak, Catalysis in Organic Synthesis, Academic Press Inc., (1980), 107-117,
8 DE-OS 30 030748 (Johnson Matthey, 1980)
9 DE 39 28 329 (Bayer AG, 1991); DE-OS 42 36 203 (Bayer AG, 1992),
10 EP 573 910 A1 (Degussa AG, 1993)
11 US 3 253 039 (Engelhard Corporation, 1966)
12 US 4 760 187 (Du Pont, 1988); DE 38 21 013 (Hoechst, 1989)
13 Pergamon Series on Environmental Science, Vol 5, 543-544; J. Soc. Occup. Med, 31, (1981), 158-163
14 Edelmetalltaschenbuch, Degussa AG, 2. Auflage, 1995
15 A.B. Stiles, Catalyst Manufacture, Marcel Dekker Inc, New York, 1983
16 A.B. Stiles, Catalyst Supports and Supported Catalysts, Butterworths Publishers, Stoneham, 1987

Catalytic Reactions of Heterocycles: The Hydrogenation of Furan Over Palladium Catalysts

S. David Jackson, Ian J. Huntingdon, and Naseem A. Hussain.

ICI Katalco, R T & E Group, P O Box 1, Billingham, Cleveland TS23 1LB, U.K.

Abstract

The hydrogenation of furan was studied over Pd/zirconia and Pd/alumina catalysts. A relationship between conversion and selectivity was observed. The Pd/alumina catalyst initially had no selectivity to THF. The surface carbonaceous deposit was found to unique to each catalyst.

Introduction

Furan hydrogenation can be performed over most of the group VIII metals (1 - 3). However nickel (3 - 5) and palladium (1, 6) systems have been used the most often. Even so the research on the hydrogenation of furan is not extensive. Therefore we set out to investigate the fundamentals of furan hydrogenation over palladium catalysts. From the limited information in the literature (1) it appeared that there may be a problem of catalyst deactivation as unsupported palladium deactivated rapidly. Given this problem we decided to study the system using a pulse-flow reactor and to examine the surface deposit by infra-red spectroscopy.

Experimental

Pulsed reaction studies were performed in a dynamic mode using a pulse-flow microreactor system with on-line GC (6), in which the catalyst sample was placed in a glass-lined stainless steel tube (8 mm od), in a vertical position, inside a furnace. Using this system the catalysts (typically 0.27 g) could be reduced in situ in flowing dihydrogen (90 cm^3 min^{-1}) by heating to 523 K at 10 K min^{-1} and then holding at this temperature for 1 h. After reduction had ceased the catalyst was maintained, at the desired temperature, in flowing 5% dihydrogen in dinitrogen (90 cm^3 min^{-1}). The liquid furan was admitted by injecting pulses of known size (typically 2.5 μl) into the 5% dihydrogen/dinitrogen carrier-gas stream with resulting vaporisation and hence to the catalyst. After passage through the catalyst bed the total contents of the pulse were analysed by GC. The amount of gas reacted, from any pulse, was determined from the difference between calibration peak areas and the peak areas obtained following the injection of pulses of comparable size onto the catalyst. Adsorption, desorption, and reaction were followed using a gas chromatograph, fitted with a thermal conductivity detector and Porapak Q-S column, coupled to a mass spectrometer (Spectramass Selectorr).

The infra-red spectra were obtained using a commercial FTIR spectrometer (Nicolet 5DXC). The studies were performed in transmission mode, with the catalyst in the form of a pressed disc, using an environmental cell. All spectra were recorded at 4 cm^{-1} resolution with the co-addition of 100 scans using an TGS detector. Using the environmental cell the catalysts could be reduced in situ, and furan admitted with or without dihydrogen.

The two catalysts used in this study were prepared by impregnation. For both catalysts palladium chloride (PGP Industries) was dissolved in sufficient dilute hydrochloric acid to fully wet the supports, zirconia (Degussa, S.A. 50 m^{2}g^{-1}) and alumina (Degussa Aluminium Oxid C, S.A. 102 m^{2}g^{-1}). The resulting mixtures were evaporated to dryness at 353 K. The weight loadings obtained were 0.81% w/w Pd/alumina and 0.99% w/w Pd/zirconia. The dispersion of the catalysts was determined by carbon monoxide chemisorption, and assuming a ratio of 1:2 CO:Pd, were calculated to be 29% for the Pd/alumina and 96% for the Pd/zirconia.

Results

The support materials were subjected to a reduction procedure followed by pulses of furan at 373 K. No reaction to tetrahydrofuran (THF) was observed over any of the supports. However a chromatographic effect was noted on the GC peak shape indicating that the furan was chromatographed, but not retained, by the supports.

Ten pulses of furan were passed over freshly reduced catalysts and the output monitored. The results are shown in Figure 1 and Table 1. From Figure 1 it can be seen that there is a linkage between conversion and selectivity, such that as conversion increases selectivity decreases. No other gas phase products were detected. Carbon deposition however was significant. Over the first ten pulses the average deposition on the Pd/alumina catalyst was 61%, while the deposition on the Pd/zirconia catalyst averaged 37%. In Table 1 it is apparent that with both catalysts there is an increase in the rate of production of THF over the first few pulses. Indeed no THF is formed over the Pd/alumina catalyst until pulse 3. In terms of rate of production of THF the Pd/zirconia catalyst gives a higher yield after the first few pulses, but we can see that in terms of Turnover frequency (TOF) the surface palladium atoms on the alumina supported catalyst are more effective at catalysing the hydrogenation of furan than those on the zirconia supported catalyst.

Table 1. Activity of palladium catalysts at 373 K and a GHSV of 9000 h⁻¹.

Catalyst	Pulse No.	% Conversion	Rate (μmol.g⁻¹.s⁻¹)ᵃ	TOF (s⁻¹)
Pd/zirconia	1	92	4.26	0.05
	2	29	53.55	0.61
	3	59	43.21	0.49
	4	52	42.11	0.48
	5	45	53.08	0.6
	10	33	56.37	0.64
Pd/alumina	1	94	0	0
	2	82	0	0
	3	75	14.82	0.67
	4	62	51.35	2.32
	5	71	46.14	2.08
	10	68	27.13	1.23

a) Rate of formation of THF.

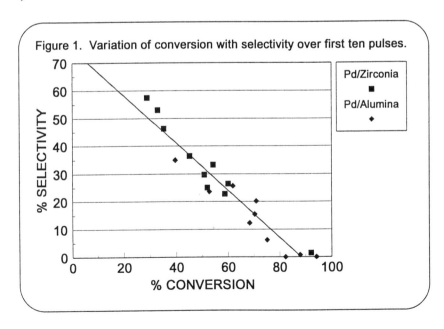

Figure 1. Variation of conversion with selectivity over first ten pulses.

Table 2. Infra-red analysis of surface residue (cm^{-1}).

Pd/zirconia	Pd/alumina	Furan	THF	DEK[a]	Butyraldeh.[a]
		3142w			3125mw
3000mw			2985s	2992s	3068mw
2945mw	2950vw		2966s	2949ms	2987ms
2925mw				2914ms	2956ms
2875mw	2889vw		2871ms		2900m
2800w					
	1863mw				
1738w	1750m	1714m	1742w	1735s	1769ms
1688mw	1688m				1712s
1558w		1578ms			
1461w	1461mw	1478ms	1464mw	1464ms	1475mw
1361w	1370w	1371m	1364mw	1357ms	
1250w	1243mw				1281mw
1187w	1194vw	1193ms	1185mw	1164m	1175w
1124w				1121ms	1150m
1086w			1085s		1088mw

a) DEK, diethyl ketone; Butyraldeh., butyraldehyde.

Infra-red spectroscopic analysis of the hydrogenation reaction over the two
catalysts confirmed that the gas phase product was THF. Analysis of the surface
deposit however revealed that the deposit was unique to each catalyst (Table 2).
When deuterium was added to the gas phase over each catalyst after they had
been used for furan hydrogenation changes in the spectra were noted. The most
significant change was the absence of exchange into the surface deposit on the
Pd/alumina catalyst. On this catalyst all H-D exchange was confined to the
alumina surface hydroxyls. No change could be detected in the hydrocarbon
region of the spectrum either in the loss of C-H band intensity or growth of C-D
band intensity. Whereas with the Pd/zirconia catalyst the CH_3, CH_2 bands reduced
in intensity by approximately 15% while there was a corresponding increase in the
C-D band intensity.

Discussion.

Both catalysts are active and selective for the hydrogenation of furan to THF. However there are significant differences in their behaviour patterns. From Table 1 it is apparent that a freshly reduced Pd/alumina catalyst initially has no selectivity to THF. Indeed all the furan reacted is retained by the catalyst. Once a quantity of carbon has been deposited there is then formation of THF. A similar pattern is observed with the Pd/zirconia catalyst although to a lesser extent. This type of behaviour is not uncommon with hydrogenation/dehydrogenation reactions (7 - 9). Once selectivity has been generated it is clear that the alumina supported catalyst has a far higher intrinsic activity, as measured by TOF, for furan hydrogenation than the zirconia supported catalyst. However we see here the difficulty of using TOF as a measure of intrinsic activity. The fresh Pd/alumina catalyst has no intrinsic activity for furan hydrogenation, therefore the catalytic site is generated in situ. The Pd/zirconia catalyst appears to have initial activity for hydrogenation but given the extent of carbon deposition from the first pulse and the low rate of hydrogenation, it is more likely that again the hydrogenation site is generated in situ but required a lesser quantity of carbon deposit. Why a smaller amount of deposit is required may be related to the palladium crystallite size and hence morphology. The Pd/alumina catalyst has an average particle size of 3.8 nm, whereas the Pd/zirconia has an average crystallite size of 1.1 nm. This behaviour agrees with the literature (1) as when unsupported palladium was used the catalyst rapidly deactivated, whereas supported palladium had a longer life. Taken in conjunction with our own results, this suggests that there may be spillover of the hydrocarbonaceous species to the support. It is worth noting that bi-layer adsorption of furan (i.e. adsorption on top of an existing adsorbed layer) has been observed by Sexton (10) in a study of furan adsorption on copper.

From Figure 1 it is clear that the reaction system follows the often seen relationship between activity and selectivity, i.e. as activity increases selectivity decreases. Therefore we can calculate that the typical single-pass yield of THF under any of the conditions will be limited to approximately 16%.

The infra-red analysis of the system confirmed that furan and THF could be unambiguously assigned to account for all bands in the gas phase spectra. Analysis of the deposit on the Pd/zirconia system suggested that the main species was adsorbed THF with a non-cyclic ketonic/aldehydic species also being present. At first sight it does not appear that the species present on the Pd/alumina catalyst is significantly different from that on the Pd/zirconia. However the band shapes were considerably different and there is the absence some C-H and C-O bands, but the most telling difference was observed when deuterium was added to the system. When deuterium is added to a catalyst that has a hydrocarbonaceous residue it is

When deuterium is added to a catalyst that has a hydrocarbonaceous residue it is usual for there to be exchange (11). As long as there are free metal sites the dihydrogen can adsorb/dissociate. We can be sure that this process is occurring on the Pd/alumina catalyst because the support hydroxyls exchange. Therefore we may speculate that the initial contact between the surface and the residue on the Pd/alumina catalyst is through at least one quaternary carbon that cannot take up hydrogen.

The formation of aldehydic/ketonic species was not unexpected as it is known that in some hydrogenation systems butanol has been detected as a minor product (12), while under more forcing conditions (13) the furan molecule will fracture to carbon monoxide and C-3. However it is interesting to note that there was no evidence of adsorbed carbon monoxide. Hence it would appear that a ring-opening has taken place rather than a complete fracture.

References

1. G. Godawa, A. Gastet, P. Kalck, and Y. Maire, *J. Molec. Catal.*, **34**, 199 (1986).

2. US Patent 3828077, assigned to Phillips Petroleum Co. (1971).

3. Y. Ikushima, M. Arai, and Y. Nishiyama, *Applied Catal.*, **11**, 305 (1984).

4. Polish Patent 104070, assigned to D. Zdzislaw and M. Gasiorek (1979).

5. Y. Ikushima, M. Arai, and Y. Nishiyama, *Bull. Chem. Soc. Jpn.*, **59**, 347 (1986).

6. SU Patent 417150, assigned to T. M. Beloslyudova (1974).

7. S. D. Jackson, R. B. Moyes, P. B. Wells, and R. Whyman, *J.Catal.*, **86**, 342 (1984).

8. B. J. Brandreth and S. D. Jackson, *J.C.S.Faraday 1*, **85**, 3579 (1989).

9. S. D. Jackson and G. J. Kelly, *J.Mol.Catal.*, **87**, 275 (1994).

10. B. A. Sexton, *Surf. Sci.*, **163**, 99 (1985).

11. see e.g. S. D. Jackson and N. J. Casey, *J. Chem. Soc. Faraday Trans. 1*, **91**, 3269 (1995).

12. SU Patent 438648, assigned to D. Z. Zavelskii (1975).

13. K. C. Pratt and V. Christoverson, *Fuel Processing Tech.*, **8**, 43 (1983).

Synthesis of Nitroarylhydroxylamines by Selective Reduction of 2,4-Dinitrotoluene Over Pd/C Catalysts

M. G. Musolino, A. Pistone, S. Galvagno
Department of Industrial Chemistry, University of Messina, Italy

G. Neri and A. Donato
Faculty of Engineering, University of Reggio Calabria, Italy

Abstract

The selective liquid phase hydrogenation of 2,4-dinitrotoluene (2,4-DNT) to the corresponding 2,4-nitroarylhydroxylamines has been studied over carbon supported Pd catalysts. The influence of the nature of metal precursor and of the addition of a second element (Fe, Sn, Ca) on the catalytic activity and products distribution has been investigated. The highest yield to nitroarylhydroxylamine intermediates (up to 88%) was obtained on Ca-promoted palladium catalysts.

1. Introduction

Selective catalytic hydrogenation of aromatic nitro-compounds into corresponding arylhydroxylamines finds a great practical application because these intermediates can undergo rearrangement to a variety of important chemicals (1,2). Aromatic hydroxylamines are generally prepared by chemical reduction or selective hydrogenation of aromatic nitrocompounds by using metal catalysts promoted with dimethyl sulfoxide. However, such methods of synthesis are characterized by difficult products purification and low yields.

Nitroarylhydroxylamines coming from the selective hydrogenation of dinitrocompounds have received little attention and only few studies have been reported in the literature. In previous papers on the hydrogenation of 2,4-dinitrotoluene (2,4-DNT) over supported metals we obtained the highest yields to the 2,4-nitroarylhydroxylamine isomers on Pd/C catalysts (3,4).

In this work we have investigated the selective hydrogenation of 2,4-DNT over Pd/C samples with the aim of studying the influence of some parameters (nature of metal precursor and addition of a second element such as Sn, Fe, Ca) on the activity and yield to 2,4-nitrohydroxyaminotoluenes.

2. Experimental

Monometallic Pd catalysts supported on carbon were prepared by incipient wetness impregnation of the support (Chemviron SCXII, 100-200 mesh, surface area 1100 m^2/g) with aqueous solutions of $PdCl_2$ and $Pd(acac)_2$, or $Pd(CH_3COO)_2$ in benzene. After impregnation, the catalysts were dried at 120 °C and reduced at 200 °C under flowing hydrogen for 1h.

Promoted catalysts were prepared by impregnation of the reduced monometallic Pd3C sample with aqueous solutions of $Fe(NO_3)_3 \cdot 9H_2O$, $SnCl_2$ and $CaCl_2$, containing the suitable amount of promoter. The Me/Pd molar ratio was varied between 0.01 and 0.5. The catalysts were dried at 120 °C and reduced as above.

Chemisorption of CO was measured at room temperature by using a pulse flow technique. The metal dispersion was calculated assuming a stoichiometry CO/Pd = 1. Table 1 shows the main characteristics of the catalysts used.

Table 1. Main characteristics of the Pd supported catalysts

Catalysts code	Pd loading (wt %)	Me loading (wt %)	Me/Pd	CO/Pd
Pd3C[1]	3	-	-	0.17
Pd2A[2]	2	-	-	0.29
Pd3Ac[3]	3	-	-	0.46
PdFe1	3	0.16	0.1	0.12
PdFe2	3	0.78	0.5	0.10
PdSn1	3	0.03	0.01	0.16
PdSn2	3	0.10	0.03	0.07
PdSn3	3	0.33	0.1	0.01
PdCa1	3	0.11	0.1	0.17
PdCa2	3	0.34	0.3	0.08
PdCa3	3	0.56	0.5	0.11

[1] = from $PdCl_2$; [2] = from $Pd(acac)_2$; [3] = from $Pd(CH_3COO)_2$

The hydrogenation of 2,4-DNT was carried in liquid-phase (ethanol as solvent) in a batch reactor at 50°C. The progress of the reaction was followed by gas-chromatography and liquid chromatography.

3. Results and discussion

3.1. Pd/C catalysts

On Pd/C catalysts the hydrogenation of 2,4- DNT was found to proceed through a complex consecutive/parallel reaction pathway involving the formation of hydroxyaminonitrotoluene isomers (HANT), aminonitrotoluene isomers (ANT) as reaction intermediates and 2,4-diaminotoluene (2,4-DAT) as the final product (4,5).

Typical conversion-time plots showing the course of the reaction over monometallic Pd catalyst prepared from $Pd(CH_3COO)_2$ and $PdCl_2$ are reported in Figure 1.

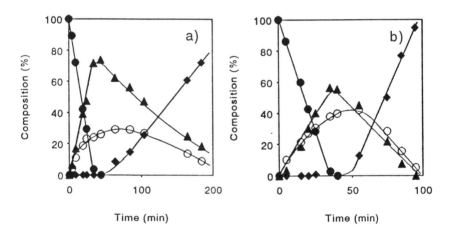

Figure 1. Hydrogenation of 2,4-DNT on monometallic Pd samples: a) Pd3C; b) Pd3Ac. (●) 2,4-DNT; (▲) HANT; (○) ANT; (♦) 2,4-DAT.

The specific rate of disappearance of 2,4-DNT, r_1 (expressed per atom of palladium on the surface), measured on the unpromoted catalysts prepared from different palladium precursors is reported in Table 2. In the same table the overall yields to nitroarylhydroxylamines is also shown.

Table 2. Specific activity and yield to 2,4-nitroarylhydroxylamines on Pd/C catalysts prepared from different precursors.

Catalysts	CO/Pd	r_1 (molecules DNT sec^{-1} Pd$_{(s)}$$^{-1}$)	Yield to HANT isomers (%)
Pd3C	0.17	$3.04 \cdot 10^{-1}$	74.0
Pd2A	0.29	$2.47 \cdot 10^{-1}$	76.2
Pd3Ac	0.46	$1.07 \cdot 10^{-1}$	55.5

Catalyst prepared from impregnation of precursors coming from aqueous solutions have shown higher yields to nitroarylhydroxylamines. Over these samples, the HANT isomers were infact the main intermediates obtained with an overall yield of about 75 %. It should be noted that Rylander et al. have shown that a small amount of dimethylsulfoxide (DMSO) had to be added to the reaction mixture in order to obtain high selectivity (1).

A correlation between the catalytic behaviour and the CO/Pd ratio can be observed (Tab. 1). The specific activity was found to increase as the CO/Pd ratio decreases. Moreover, the larger metal particles were the most selective towards the formation of arylhydroxylamine intermediates. This is in agreement with previous results which have shown that the activity and selectivity of this reaction is sensitive to variation in the metal particle size (4).

3.2. Promoted Pd/C catalysts

It is well known that the addition of suitable promoters is a way to improve the performance of monometallic catalysts. In order to evaluate the possibility to achieve higher yields to arylhydroxylamines, the influence of adding of Fe, Sn and Ca to Pd catalysts has been investigated. The performance of Fe- and Sn-doped catalysts on the 2,4-DNT reduction has been previously reported, but no detailed information on the products distribution has been given (5). No data on this reaction, to the best of our knowledge, are available on Ca-doped catalysts.

Fig. 2 reports typical conversion-time plots obtained on promoted catalysts. Ca and Sn-promoted catalysts were the most selective catalysts.

Addition of Fe causes instead a strong decrease in the yield to desired intermediates.

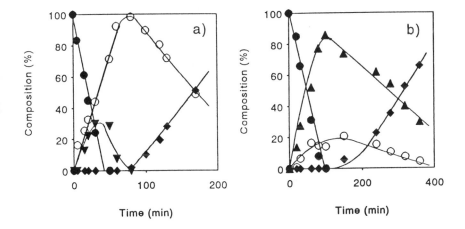

Figure 2. Hydrogenation of 2,4-DNT over promoted samples. a) PdFe; b) PdCa. (●) 2,4-DNT; (▲) HANT; (○) ANT; (◆) 2,4-DAT.

The influence of the Me/Pd ratio on the rate of hydrogenation of 2,4-DNT and yield to 2,4- nitroarylhydroxylamines is reported in Fig. 3. It can be observed that addition of either Sn or Ca to Pd/C increases the yield to the nitroarylhydroxylamine intermediates. However, small amounts of tin poison the palladium catalyst due to a strong decrease of the palladium surface active sites. On addition of Ca instead the activity increase reaches a maximum at a Me/Pd = 0.1 then decreases slowly.

The unique behaviour of Ca-doped catalysts can be attributed to its acid-base character. It has been previously suggested that the selective hydrogenation to hydroxylamines is promoted on less acid support (4). It can be therefore suggested that the effect of Ca is to favour an increase in the basicity of support. However it cannot be excluded that the alkaline promoter causes a change in the electronic density of the metal particles (6). Studies are in progress to characterize doped catalysts in order to better understand the mechanism through which the promoters modify the behaviour of the Pd active sites.

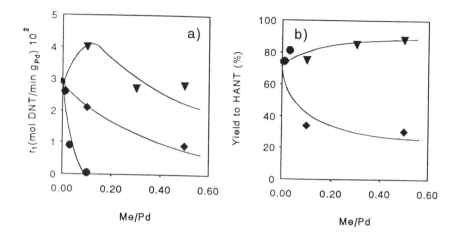

Figure 3. Catalytic properties of doped catalysts as a function of Me/Pd ratio. a) rate of DNT disappearance; b) yield to HANT. (♦) PdFe; (●) PdSn; (▼) PdCa.

References

1. P. N. Rylander, Catalytic Hydrogenation over Platinum Metals, Academic Press, New York, 1967.
2. S. L. Karwa and R. A. Rajadhyaksha, *Ind. Eng. Chem. Res.*, **26**, 1746 (1987).
3. G. Neri, M. G. Musolino, C. Milone, A. M. Visco and A. Di Mario, *J. Mol. Cat. A:Chemical*, **95**, 235 (1995).
4. M. G. Musolino, C. Milone, G. Neri, L. Bonaccorsi, R. Pietropaolo and S. Galvagno, *Stud. Surf. Sci. Catal.*, H. U. Blaser et al. (eds.), Elsevier Science, Amsterdam, **108**, 239 (1997).
5. A. Benedetti, G. Fagherazzi, F. Pinna, G. Rampazzo, M. Selva and G. Strukul, *Catal. Lett.*, **10**, 215 (1991).
6. W. D. Mross, *Catal. Rev. - Sci. Eng.*, **25**, 591 (1983).

Selective Hydrogenation of Citronellal into Menthols on Ru Supported Catalysts

C. Milone, C. Gangemi, S. Minicò and S. Galvagno

Department of Industrial Chemistry, University of Messina, Italy.

G. Neri

Faculty of Engineering, University of Reggio Calabria, Italy.

Abstract

Hydrogenation of citronellal has been performed on Ru/SiO$_2$ catalysts using cyclohexane as solvent. The influence of metal loading and ruthenium precursor on the products distribution has been investigated. On the catalysts prepared from RuCl$_3$ the highest yield to menthols is observed at low metal loadings. It is suggested that the cyclization reaction occurs mainly on the support and it depends on its acid-base properties.

Introduction

Citronellal (3,7-dimethyl-6-octen-1-al) represents an interesting raw substrate for fine chemicals production. Till now, the attention has been mainly paid to its selective hydrogenation to the corresponding unsaturated alcohol, citronellol (3,7-dimethyl-6-octen-1-ol) (1).

Another important synthetic route starting from citronellal is the one leading to menthol isomers. The synthesis of menthol stereoisomers (2-isopropyl-5-methyl-cyclohexanol) is carried out through the isomerisation of citronellal to isopulegols (2-isopropenyl-5-methyl-cyclohexanol) on acid catalysts (2-4) followed by their hydrogenation.

Wismeijer et al. (1) in their study on the hydrogenation of citronellal have observed a yield to isopulegol up to 50% on Ru/SiO$_2$ catalysts prepared from RuCl$_3$ (ex-RuCl$_3$/SiO$_2$). SiO$_2$ itself was reported to show a low activity. An influence of the nature of the ruthenium precursor has also been pointed out. The formation of isopulegol was attributed to the presence of partially reduced ruthenium species (RuCl$_x$, and/or RuO$_x$). G. Neri et al. (5) have also observed that the formation of isopulegol from citronellal is strongly suppressed when ethanol is used as solvent due to the formation of acetals of citronellal which block the C=O group.

A goal of this work is to synthesize menthols from citronellal in one single step by carrying out the reaction in the presence of isomerisation and hydrogenation sites. The selective formation of menthols can be obtained by minimizing the formation of products deriving from the hydrogenation of the double bond and /or the carbonyl group of citronellal. The influence of parameters such as metal loading and nature of the precursor on the product distribution has been investigated.

Experimental

Silica gel (surface area= 360 m^2/g, grain size 100 -200 mesh) was used as support. Catalysts derived from RuCl$_3$ and Ru(NO)(NO$_3$)$_3$ were prepared by incipient wetness impregnation, then dried for 2 h at 393 K in air and reduced in H$_2$ at 623 K for 2 h. For the ex-RuCl$_3$ catalysts the metal loading was varied between 1 and 10 wt%. The ex-Ru(NO)(NO$_3$)$_3$ catalyst was prepared with a 2 wt.% ruthenium loading.

Citronellal hydrogenation was carried out at 333 K, at atmospheric pressure under H$_2$ flow in a batch reactor, with 25 ml of cyclohexane as solvent. Citronellal (ASSAY > 97 %) and cyclohexane (ASSAY 99.8 %) were supplied by Fluka. Before catalytic activity measurements the catalysts were reduced "in situ" at 343 K for 1 h. After cooling at reaction temperature, the substrate (0.1 ml) was injected through one arm of the flask. The reaction mixture was stirred at 500 rpm.

The progress of the reaction was followed by sampling a sufficient number of microsamples.

Chemical analysis was performed with a gas cromatograph equipped with a flame ionization detector. Preliminary runs carried out at different catalyst loadings and grain size demonstrated the absence of external and internal diffusion limitations.

Results and discussion.

The hydrogenation of citronellal on the ex-RuCl$_3$ catalysts leads to a wide product distribution. Typical results obtained on the 5% Ru/SiO$_2$ catalyst (5RuClS) are reported in Fig.1.

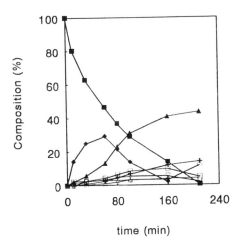

Figure 1. - Hydrogenation of citronellal over 5% Ru/SiO₂ catalyst.
(■ citronellal;◆ Isopulegols; ▲ Menthols; ☐ Citronellol; △ 3,7-dimethyloctanal; ▽ 3,7 dimethyloctanol)

Up to 20 % conversion, the main products are: isopulegols obtained from the isomerisation of citronellal, citronellol and 3,7 dimethyloctanal obtained from the hydrogenation of the C=O and C=C bonds of citronellal. Menthols, which are produced from the hydrogenation of isopulegol, are also formed. At longer reaction time citronellol and 3,7-dimethyloctanal are hydrogenated to 3,7-dimethyloctanol, whereas the isopulegols are converted to menthols. The scheme of the reaction is reported on Fig. 2

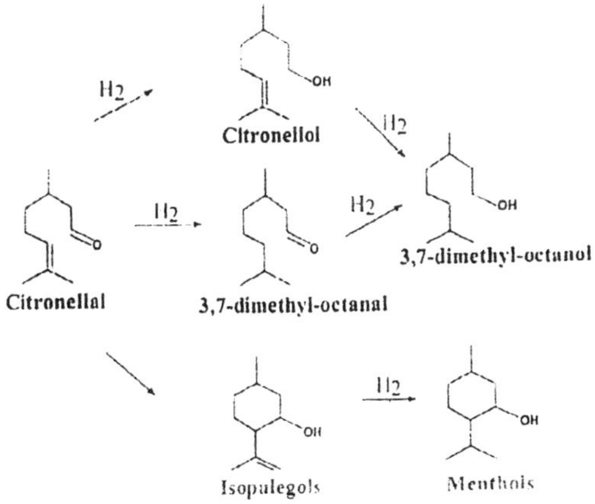

H₂

Citronellol

H₂

Citronellal

H₂

3,7-dimethyl-octanal

H₂

3,7-dimethyl-octanol

H₂

Isopulegols

Menthols

Figure 2

Fig. 3 shows the selectivity to cyclisation products (isopulegols and menthols) measured at 80 % conversion of citronellal as a function of the metal loading. It should be noted that after the total disappearance of citronellal, the isopulegols formed are quantitatively converted to menthols.

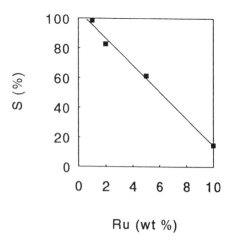

Figure 3. Selectivity to cyclisation products, S (%), measured at 80% conversion as a function of the Ru loading.

On increasing the metal loading, the selectivity to cyclisation products decreases from 98 to 14 %. The 1RuClS catalyst (1% Ru/SiO$_2$) was found to be the best catalyst for menthols production. On this sample a yield to menthols higher than 95% was obtained.

Fig. 4 shows the initial rate of isomerisation measured on all the investigated Ru/SiO$_2$ catalysts and it is compared with the activity of pure SiO$_2$ and SiO$_2$ impregnated with an aqueous solution of HCl. On this latter sample the nominal content of HCl was 10%. Both, pure and acidified SiO$_2$ were treated at 623 K in H$_2$ for two hours before reaction.

SiO$_2$ itself shows a low isomerisation activity which increases by more than 5 times when the SiO$_2$ is acidified with HCl. In the absence of ruthenium the isomerisation products are essentially isopulegols. Menthols cannot be formed due to the lack of hydrogenation active sites.

When ruthenium is added to the support the isomerisation activity decreases

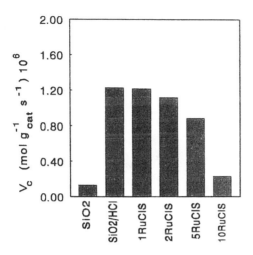

Figure 4. - Initial rate of isomerisation of citronellal

with increasing metal loading. The catalyst with 1% Ru (1RuClS) shows an isomerisation activity similar to that measured on the SiO$_2$/HCl sample. When all the citronellal was converted, the isopulegols are hydrogenated to menthols.

These results suggest that the noble metal has a little, if any, isomerisation activity. The discrepancy with the conclusions of Wijsmeier et al. (1) which attributed the isomerisation activity of Ru/SiO$_2$ catalysts to unreduced ruthenium species could be related to the fact that the authors have neglected the effect of chlorine introduced by impregnating the support with RuCl$_3$.

Hydrogenation of citronellal was also performed on a 2% Ru/SiO$_2$ sample prepared from Ru(NO)(NO$_3$)$_3$ (2RuNIS). Cyclisation products were not observed during this run. Citronellol, 3,7-dimethyloctanal and 3,7-dimethyloctanol were the main products. The specific activity towards hydrogenation reactions (expressed per mole of Ru on the surface) is the same as that measured on the ex-RuCl$_3$ catalysts (1.1×10^{-2} s^{-1}). The absence of cyclisation products is due to the low activity of SiO$_2$ when chlorine is not present on the surface.

Wijsmeir et al (1) have reported on chlorine free catalyst a yield to isopulegol equal of 11%. They claim that the formation of isopulegol occurs on the RuO_x sites generated from a metal - support interaction. This conclusion is not supported by our results because, even with a highly dispersed catalyst (2RuNIS D =0.70), where a metal - support interaction is likely to occur, no formation of isopulegol is observed. Further investigation is in progress to clarify this point.

Conclusions

The hydrogenation of citronellal on $exRuCl_3/SiO_2$ catalysts has shown that:

i.   cyclisation of citronellal occurs on the surface of SiO_2 and it depends on the acid properties of the support.

ii. in the presence of Ru, isopulegols are quantitatively converted to menthols.

iii. increasing the Ru loading the selectivity to cyclisation products decreases probably due to a blocking of the acid sites.

iv. On the sample with low Ru loading the yield to menthols is higher than 95 %.

References

1.   A.A. Wijsmeier, A.P.J. Kieboom, and H. Van Bekkum, Applied Catalysis, **25**, 181 (1986).
2.   K. Arata and C. Matsuura, Chemistry Letters, 1797, (1989).
3.   M. Fuentes, J. Magraner, C. De Las Pozas and R. Roque-Malherbe, Applied Catalysis, **47**, 367, (1989).
4.   K. Kogami, J. Kumanotani, Bull. Chem. Soc. Jpn., **41**, 2530, (1986).
5.   G. Neri, L. Mercadante, A. Donato, A.M. Visco and S. Galvagno, Catalysis Letters, **29**, 379 (1994).

Shape Selective Hydrogenation of Benzene to Cyclohexene

G. Srinivas, D.B. MacQueen, M. Dubovik, and R.K. Mariwala

TDA Research, Inc.
12345 W. 52nd Ave.
Wheat Ridge, CO 80033

Abstract

Although cyclohexene is a potentially valuable chemical intermediate, it is difficult to manufacture. Essentially, any catalyst active enough to hydrogenate benzene to cyclohexene is more than active enough to hydrogenate the cyclohexene to cyclohexane. One way to increase the yield of cyclohexene is to use a shape selective carbon molecular sieve (CMS)-based catalyst to shift the product distribution by exploiting the differences in size and shape of benzene, cyclohexene and cyclohexane molecules. CMS catalysts were synthesized, and characterized using multipoint BET and XRD, and their pore size was determined using adsorption isotherms. The catalysts were tested in a laboratory reactor to determine their benzene conversion and cyclohexene selectivity. The conversion and selectivity were determined as a function of pore size, reaction temperature, water, benzene, and hydrogen concentration, and as a function of time-on-stream. The selectivity of the catalyst to cyclohexene increased with decreasing pore diameters of the CMS, increasing water concentrations and increasing benzene/hydrogen ratios in the feed.

Introduction

Cyclohexene is a very useful but difficult to manufacture chemical intermediate that can be used in the environmentally benign manufacture of important chemicals such as adipic acid. Adipic acid, (AA) (HOOC-$[CH_2]_4$-COOH) is one of the top 50 chemicals produced in the U.S. The manufacture of AA involves a number of steps: benzene is hydrogenated to cyclohexane in the first step. Next, cyclohexane is catalytically oxidized to cyclohexanone, and cyclohexanol, known as KA oil (ketone and alcohol), at 100-170°C and 0.9-1.8 MPa. The typical conversion of cyclohexane and yield of KA oil in this step are 3-12 mol % and 76-87 %, respectively. Because of the low conversion of cyclohexane, it is separated by distillation from KA oil and recycled. In the third step, KA oil is oxidized with HNO_3 at 60-90°C and 0.1-0.3 MPa to produce AA (1). Since HNO_3 is used to oxidize KA oil to AA, 1 mole of nitrous oxide (N_2O) is produced as a by-product per mole of AA. Although AA plants use downstream NO_x abatement systems, only 32% of the total N_2O generated is either recycled or destroyed. The AA industry worldwide accounts for 5-8% of anthropogenic

NO_x emission. The oxidation of cyclohexene to AA could offer potential solutions to this problem.

Unfortunately, a catalyst to hydrogenate benzene to cyclohexene with high selectivity at high conversion is not readily available; the reason for this problem can be understood from the following reaction:

$$C_6H_6 \xrightarrow{3H_2} C_6H_{10} \xrightarrow{H_2} C_6H_{12}$$

Benzene is hydrogenated step wise to form cyclohexene, which is further hydrogenated to form cyclohexane (2). To stop the reaction at the cyclohexene step is difficult because the formation of cyclohexane is thermodynamically more favorable. Cyclohexene formation can be favored by shifting the cyclohexene-cyclohexane equilibrium to the left. One way to achieve the shift in equilibrium is to exploit the differences in size and shape of benzene, cyclohexene and cyclohexane molecules. Benzene is a planar molecule (0.69 nm x 0.35 nm x 0.63 nm), whereas the energetically most favorable form for cyclohexane is the non-planar chair form (0.69 nm, x 0.57 nm x 0.68 nm). Cyclohexene is slightly skewed (0.72 nm x 0.53 nm x 0.64 nm). A catalyst which prevents the transport of cyclohexane based on its larger size but allows the transport of cyclohexene could shift the cyclohexene-cyclohexane production in favor of cyclohexene.

CMS are a class of amorphous microporous (pore size < 2 nm) molecular sieves that can differentiate small molecules based on their size and shape. Commercially, CMS are used for the recovery of nitrogen from air, hydrogen from coke oven gas and separation of methane, and carbon dioxide by the Pressure Swing Adsorption (PSA) technology. The CMS pore is made of aromatic microdomains loosely oriented in a slit-like shape. The pore dimension in the center is larger than the pore mouth dimension. Fitzer et al. (3) have reported that for polyfurfuryl alcohol (PFA)-derived CMS, the central pore dimension can be as large as 1.5 to 2 nm. The pore mouth dimensions are in the range of 0.4 to 0.8 nm and can be controlled by the time and temperature of CMS synthesis (4). If the transport of cyclohexane is stopped in the pore, and only cyclohexene can get out of the pores, the formation of cyclohexene will be favored by equilibrium, increasing its selectivity.

Experimental

Synthesis of Ru/CMS Polyfurfuryl alcohol (PFA) was used as a precursor for the synthesis of CMS. The pyrolysis of the PFA under controlled conditions (temperature and time) results in the formation of CMS. The PFA was first synthesized from furfuryl alcohol (Aldrich) by polymerization with dilute nitric acid. Following the polymerization of furfuryl alcohol to PFA, the PFA was washed with water to eliminate residual acid in the PFA. The water was then distilled out of the mixture to obtain the PFA.

The PFA was then pyrolyzed in a quartz reactor apparatus under nitrogen. This apparatus consists of a quartz reactor (7.5 cm o.d., 100 cm long) heated by a Thermcraft tube furnace (7.5 cm diameter, 30 cm heated zone, max. temp. 1200°C) fitted with an Omega temperature controller. The PFA was placed in a quartz sample boat inside the reactor. A thermocouple placed above the quartz boat was used to measure and control the temperature. The sample temperature was ramped gradually in the presence of N_2 from room temperature to the pyrolysis temperature at 2°C/min, and held at the pyrolysis temperature for 2 - 6 hours. This heat treatment of the PFA forms a CMS with a large fraction of the pores with diameters of 5-15 Å. Following the synthesis of the CMS, the incipient wetness impregnation method was used to load the catalyst pores with Ru using a solution of ruthenium chloride in water and alcohol. The catalyst was reduced in a stream of hydrogen prior to reaction at temperatures (T_r) between 250 - 300°C. The procedure resulted in the formation of Ru crystallites in the pores of the CMS. Dispersion of the Ru in the catalyst was less than 5%.

Catalyst Characterization
Figure 1 shows the pore size distribution of the CMS samples as a function of the CMS synthesis temperature. The pore size distribution of the of the CMS catalysts was measured using n-butane adsorption isotherms on a Cahn 2000 microbalance apparatus using the Horvath-Kawazoe equation (5). The X-axis (D_a-l) is the pore

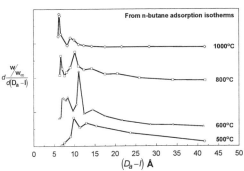

Figure 1. Pore size distribution of CMS samples.

diameter of the carbon and the Y-axis is the differential of the normalized uptake of n-butane with respect to the pore diameter (5). All samples showed multiple peaks in the 5 - 15 Å range. In all the samples, there was a peak with pore sizes greater than 10 Å, and a peak at or below 10 Å. For the sample synthesized at 500°C, the peaks were not sharp, and two peaks were found at 10 and 12.5 Å. As the synthesis temperature was increased to 600°C, the peaks grew more pronounced, and sharp. In addition, the peak positions also shifted to lower diameters, and were located at 7 and 11 Å. A third peak was also found, located at about 14.5 Å. When the CMS synthesis temperature was further increased, to 800°C, the features on the pore size distribution remained the same; however, the peaks were shifted to lower pore diameters and the peaks were found at 6.5, 11, and 14 Å. The BET surface area of the catalyst powders was determined using a Gemini multipoint N_2 adsorption apparatus to be 100 - 150 m^2/g. XRD measurements were undertaken on a Philips 1729 X-ray diffraction spectrometer and showed that the carbon was amorphous. The gas-phase benzene hydrogenation reaction was carried out in a stainless steel flow reactor. The benzene was introduced into the apparatus through a saturator maintained at 85°C. The temperature of the reaction (T) was maintained between 80 and 110°C. The reactants and the products of the reaction were analyzed for benzene, cyclohexene and cyclohexane content using an SRI gas chromatograph equipped with an automated sampling system and thermal conductivity and flame ionization detectors.

Results and Discussion

Effect of CMS Synthesis Temperature The benzene hydrogenation reaction was carried out in a differential reactor at a high gas hourly space velocity high enough to maintain the conversion at less than 20% most of the time. Figure 2 summarizes the benzene conversion and cyclohexene selectivity (ratio of moles of cyclohexene

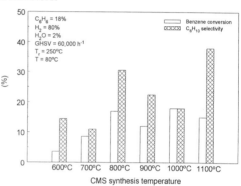

Figure 2. Activity of Ru/CMS catalysts with different synthesis temperatures.

formed to moles of benzene reacted) for the Ru/CMS catalysts prepared with different CMS carbonization temperatures. The weight loading of Ru on all the catalysts was 3%. The ratio of H_2 to benzene was about 4.5, and the catalyst reduction temperature was 250°C in all cases. As Figure 2 indicates, the conversion of benzene on the Ru/CMS catalyst synthesized at 600°C was less than 4%, and the selectivity to cyclohexene was about 15%. The conversion increased as the CMS synthesis temperature increased and appeared to maximize at about 20% while, the cyclohexene selectivity also increased. The highest selectivity to cyclohexene was about 40%, obtained at a benzene conversion of about 15%, at a CMS synthesis temperature of 1100°C. In order to compare the activity of the Ru/CMS catalysts with those of Ru/Al$_2$O$_3$ catalysts, tests were conducted with a 3%Ru/Al$_2$O$_3$ catalyst at 80°C. The Ru/Al$_2$O$_3$ catalyst showed close to 100% conversion of benzene, but the selectivity to cyclohexene was less than 2%.

Effect of Water Concentration The concentration of water in the feed to the reactor played a significant role in affecting the benzene conversion and the cyclohexene selectivity. Figure 3 shows the effect of water on the catalyst activity and selectivity. The concentration of water was increased stepwise from 0 to 17%. The conversion of benzene decreased with increasing water

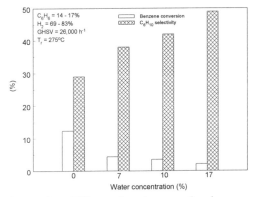

Figure 3. Effect of water on the benzene conversion and cyclohexene selectivity.

concentration; however, the selectivity to cyclohexene increased gradually with increasing water concentration and the cyclohexene selectivity reached 50% with 17% water in the feed.

The effect of water vapor on the selectivity of cyclohexene in the vapor phase hydrogenation of benzene on Ru powder catalyst has been studied before (2). The researchers found that upon addition of water to the feed stream of a pure Ru powder catalyst, the selectivity to cyclohexene increased gradually over a 10 hour period. Pre-covering the catalyst with water resulted in the immediate increase in the selectivity to cyclohexene. The authors ascribe the effect of water addition and increase in cyclohexene selectivity to repulsive forces on the catalyst surface. They suggest that the water molecules may repel the adsorbed cyclohexene molecules by physical repulsive forces, reducing the bond-strength of the cyclohexene molecule with the Ru surface. This leads to the desorption of the cyclohexene before it can be hydrogenated, increasing its selectivity.

The catalyst reduction temperature also affected the cyclohexene selectivity. Increasing the catalyst reduction temperature from 250 to 300°C increased the selectivity of the catalyst to cyclohexene from 30 to 47%. The conversion of benzene decreased slightly when the reduction temperature was increased. Ru loading (between 3 and 5%) did not seem to change the catalyst activity or the selectivity significantly. As expected from the proposed reaction mechanism, decreasing the hydrogen/benzene ratio during the experiment resulted in increasing the selectivity to cyclohexene. However, it decreased the conversion of benzene. We also undertook a catalyst stability for about 25 hours. During the stability test, the benzene conversion and the cyclohexene selectivity were monitored continuously. The catalyst selectivity increased from about 10% to about 60% over 25 hours; however, the benzene conversion simultaneously decreased from about 25% to less than 5%. There was no loss in the Ru content on the catalyst during the run.

The data obtained with different pore sizes, while not conclusive, suggest that smaller pore diameters result in higher selectivities of cyclohexene. The catalyst stability data showed that the selectivity increased, as the conversion of benzene decreased over time. This could be due to a poisoning, or coking effect on the catalyst. Further research into the exact nature of the catalysts is needed to explain the effect of pore size and metal support interaction effects, if any.

Process Analysis

A U.S. patent relating to the hydrogenation of benzene to cyclohexene has been recently issued (6). The patent is assigned to Asahi Kasei Kogyo Kabushiki Kaisha, Japan, where one plant is in operation using the technology. The patent discloses a method for partially hydrogenating a monocyclic aromatic hydrocarbon to produce a cycloolefin in a three phase slurry reactor with water, benzene, catalyst, and hydrogen phases. The catalyst used is powdered Ru. Crystallites cyclohexene yields of 25-40% and cyclohexene selectivities up to 80% were reported. The Ru/CMS catalysts described in this work have a higher selectivity than conventional powdered Ru crystallites or Ru/Al$_2$O$_3$.

Conclusions

A series of Ru/CMS catalysts were synthesized, characterized, and tested for their benzene conversion and cyclohexene selectivity during the hydrogenation of benzene. The cyclohexene selectivity of the catalyst increases with increasing CMS synthesis temperature and increasing water concentration in the feed. The Ru/CMS catalyst exhibits a higher selectivity to cyclohexene than Ru powder and Ru/Al$_2$O$_3$.

Acknowledgment

The authors wish to thank the National Science Foundation (Grant No. DMI-9560625) for funding this project.

References

1. D. Davis and D. Kemp, *Kirk-Othmer Encycl. of Chem. Tech.* **1**, 894 (1993).
2. P.J. Van der Steen and J.J.F. Scholten *Appl. Catal.* **58**, 281 (1990).
3. E. Fitzer and W. Schaefer, *Carbon* **8**, 353 (1970).
4. R.K. Mariwala and H.C. Foley, *Ind. Eng. Chem. Res.* **33**, 607 (1994).
5. G. Horvath and K. Kawazoe *J. Chem. Eng. of Japan*, **16 (6)**, 470 (1983).
6. K. Yamashita, H. Obana and I. Katsuta *U.S. Pat. 5,457,251* (1995).

Selective Hydrogenation of α,β-Unsaturated Aldehydes Using Pt and Ru Catalysts in Three Phase Reactors

S.Sharma, W. Koo-amornpattana, L. Zhang, S.Raymahasay and
J.M.Winterbottom.*

*School of Chemical Engineering, University of Birmingham, Edgbaston,
Birmingham, B15 2TT, United Kingdom.*

Abstract - The selective hydrogenation of α,β-unsaturated aldehydes using platinum and ruthenium catalysts in three phase systems has been examined. The importance of deriving optimised process conditions, which can enhance both catalyst activity and selectivity towards the desired unsaturated alcohol is discussed. Enhanced selectivity values were obtained by using the catalyst promoters of iron and cobalt for crotonaldehyde and cinnamaldehyde hydrogenation respectively and the catalyst poisons; thiophene, quinoline and triethyl phosphite for crotonaldehyde hydrogenation.

1. INTRODUCTION

The selective hydrogenation of an α,β-unsaturated aldehyde to its corresponding unsaturated alcohol, for example crotonaldehyde to crotyl alcohol or cinnamaldehyde to cinnamyl alcohol, is a reaction which interests both industrial and academic researchers. These unsaturated alcohols can be synthesised by using classical stoichiometric reactions utilising dissolved metals or metal hydrides (1,2). Although these reactions are highly selective they unfortunately produce copious amounts of waste product. Therefore to replace these wasteful stoichiometric reactions, highly selective and clean catalytic processes need to be developed. This research therefore aims to optimise both catalyst and process conditions for three phase catalytic reactions in conjunction with high selectivity towards the unsaturated alcohol.

---

* To whom correspondence should be sent.

2. CATALYTIC REACTION STUDIES

Hydrogenation of crotonaldehyde and cinnamaldehyde as shown in Reaction Schemes 1 and 2, exemplify the selectivity problem that is encountered when catalytic hydrogenation of α,β-unsaturated aldehydes are undertaken.

Reaction Scheme 1. Hydrogenation of Crotonaldehyde.

Crotonaldehyde   ➔   Crotyl Alcohol*

⬇   ⬇

n-Butyraldehyde   ➔   n-Butanol

*Isomerisation of crotyl alcohol to n-butyraldehyde can occur (3).

Hydrogenation of crotonaldehyde to the intermediate n-butyraldehyde is readily achieved using palladium based catalysts, as the reaction stops after one mole of hydrogen uptake with selectivities approaching to unity (4). Obtaining equivalent high yields of crotyl alcohol has, however, shown to be a more difficult process to achieve. Considerable research using gas-solid phase reactions have been performed utilising various platinum (5), cobalt (6) and ruthenium (7) catalysts. However, investigations using three phase systems for crotonaldehyde hydrogenation using water as the reaction solvent and platinum based catalysts have seldom been investigated.

Reaction Scheme 2. Hydrogenation of Cinnamaldehyde.

Cinnamaldehyde   ➔   Hydrocinnamaldehyde

⬇   ⬇

Cinnamyl Alcohol   ➔   Phenyl Propanol

⬇   ⬇

Methyl Styrene   ➔   Propyl Benzene

Hydrogenation of cinnamaldehyde has been investigated more frequently in three phase systems using alcoholic (8) and hydrocarbon (9) solvents. However, use of the former has shown to lead to the formation of acetal and diacetal components as the result of homogeneous side reactions (8). High selectivities to the unsaturated alcohol using various platinum (10) and ruthenium (11) catalysts have been reported, with values as high as 92% at 60% conversion cited for a biphasic reaction system utilising a 5 wt.% Pt/G catalyst (12).

3. EXPERIMENTAL

For both crotonaldehyde and cinnamaldehyde hydrogenation, a high pressure stainless steel baffled autoclave incorporating a four blade pitch stirrer was used. The rate of hydrogen uptake for both reactions were monitored using an Engelhard hydrogenation unit. The range of process conditions investigated using unmodified catalysts are shown in Table 1.

Table 1.   Process Parameters Investigated Using Unmodified Platinum and Ruthenium Catalysts for Crotonaldehyde and Cinnamaldehyde Hydrogenation.

Process Conditions	Crotonaldehyde	Cinnamaldehyde
Catalyst	5 wt.% Pt/G (J/M, Type 286)	5 wt.% Pt/G and 5 wt.% Ru/G
Catalyst Loading (kg/m^3)	0.5 - 10	3.3
Substrate Concentration (kg/m^3)	50	10
Solvent	Water, Methanol, Ethanol, Propan-2-ol, Decalin	Toluene
Hydrogen Partial Pressure (bar absolute)	1 - 10.0	10.0
Temperature (°C)	25 - 95	25
Reactant solution pH	3.4 - 10	--
Reactor Volume (m^3)	2 x 10^{-4}	3 x 10^{-4}
Stirring Speed (rpm)	300 - 1200	1000

Further modifications of the platinum and ruthenium catalysts described in Table 1, were undertaken using catalyst promoters and poisons as described in Table 2. The preparation method for these catalysts has been described previously (13).

Table 2.   Platinum and Ruthenium Catalyst Promoter Variables.

Process Conditions	Crotonaldehyde	Cinnamaldehyde
Catalyst	5 wt.% Pt/G	5 wt.% Pt/G and 5 wt.% Ru/G
Catalyst Promoter[1]	Fe, Sn, Co, Ni, Cu, Au	Co
Catalyst Poison[2]	Thiophene, quinoline and triethyl phosphite	--

1 Platinum: Promoter mole ratio, (1 : 0.5).
2 Volume of poison added in pre-treatment of the catalyst: 1 - 10 mls.

Product analysis for crotonaldehyde and cinnamaldehyde hydrogenation were undertaken using gas chromatography techniques utilising a FID detector. The products of crotonaldehyde and cinnamaldehyde hydrogenation were analysed on a DB-Wax capillary column (0.25mm i.d x 30m) and DB-1 capillary column (0.25mm i.d. x 60m) respectively.

4. RESULTS AND DISCUSSION

4.1 CROTONALDEHYDE HYDROGENATION

Resistances to gas-liquid mass transfer were eliminated at stirring speeds of 1000 rpm and above as indicated by the independent relationship between reaction rate and stirring speed. Therefore subsequent process evaluation experiments and their effects on crotyl alcohol selectivity were undertaken using a stirring speed of 1000 rpm.

Investigations for the optimisation of process conditions showed that variation of the conditions described in Table 1, played a significant role in increasing selectivity towards crotyl alcohol. Using a reaction temperature of 25°C and catalyst loading of 0.5 g/l, crotyl alcohol selectivity values were typically less than 1.6% at 40 % conversion. However, this selectivity was enhanced up to 15% by increasing both reaction temperature and catalyst loading to 95°C and 10g/l respectively. This selectivity increase could be attributed to the positive effects obtained due to an increase in liquid-solid diffusional resistances at higher temperatures, as indicated by the shift in activation energy from 8.3 kcal/mol (between 25 - 60°C) to 3.6 kcal/mol (between 85 - 115°C). This controlling reaction mechanism also increases significantly on increasing catalyst loading as suggested by Doraiswamy and Sharma (14). A marginal increase in crotyl alcohol selectivity from 16% to 23% (at 40% conversion) was obtained by adjusting the pH of the initial aqueous solution of crotonaldehyde from 3.4 - 10, by addition of aqueous solutions of sodium hydroxide. However, subsequent reactions were performed using unadjusted aqueous solutions of crotonaldehyde (pH : 3.4) as the selectivity enhancement at high pH was considered negligible.

No significant increase in crotyl alcohol selectivity was observed with increasing hydrogen partial pressure, substrate concentration or stirring speed. Therefore the optimised process conditions derived for liquid phase hydrogenation of crotonaldehyde within the autoclave are described in Table 3.

Table 3. Optimised Process Conditions for Crotonaldehyde Hydrogenation
in the Stainless Steel Autoclave.

Process Conditions	Crotonaldehyde
Catalyst	5 wt.% Pt/G
Catalyst Loading (kg/m^3)	10.0
Substrate Concentration (kg/m^3)	50
Solvent	Water
Hydrogen Partial Pressure (bar absolute)	10.0
Temperature (°C)	95
Reactant Solution pH	3.4
Stirring Speed (rpm)	1000
Typical Reaction Rate (mol/lmin)	0.178
Typical Crotyl Alcohol Selectivity Using Above Mentioned Conditions (%)	15.0

Investigations into solvent effects indicated that water was a better solvent in comparison to the alcoholic solvents, in terms of eliminating the formation of acetal compounds as obtained with methanol, ethanol and propan-2-ol, and in allowing higher rates of reaction to be obtained. Lower reaction rates (0.119 mol/lmin) and crotyl alcohol selectivity (8.1%) were also obtained in decalin as compared to reactions performed in water.

Modification of the 5 wt.% Pt/G catalyst with iron using the conditions described in Table 3, showed the most significant shift towards crotyl alcohol, exhibiting selectivities of 54% at 35% conversion. The positive promoting effects of iron have been well documented (15), whereby electron acceptor $Fe^{\delta+}$ species decorating the surface of the platinum crystallites, act as relatively strong electron acceptor sites for the lone pair of electrons free on the oxygen of the carbonyl group. The C=O bond, thus activated, is hydrogenated by dissociated hydrogen on nearby platinum atoms, which also maintains the iron in a low oxidation state.

Enhancements in crotyl alcohol selectivity were also achieved by partial poisoning with thiophene, quinoline and triethyl phosphite as shown in Figure 1, with an enhancement from 15% (using 5 wt.% Pt/G) to the maximum selectivity value of 36% with thiophene modification. However, selectivity enhancements were achieved at the expense of reducing the reaction rate from 0.178 mol/lmin to 0.012 mol/lmin, as a consequence of blocking the active sites of platinum by the deposited poison.

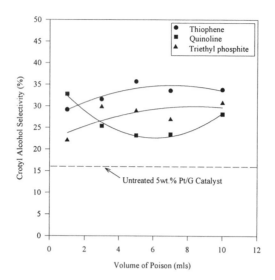

Figure 1. Comparison of Crotyl Alcohol Selectivity Using Poison Modified
Platinum Catalysts. Process Conditions Described in Table 3.

4.2 CINNAMALDEHYDE HYDROGENATION

A comparison of cinnamyl alcohol selectivities using the unmodified
5wt.% platinum and ruthenium catalysts, showed that higher selectivities were
obtained with the former catalyst, with values typically in the range of 55% as
compared to 30% for the ruthenium catalyst. Cinnamyl alcohol selectivity was
further enhanced by promoting the platinum catalyst with increasing amounts of
cobalt additives as shown in Figure 2. A maximum cinnamyl alcohol
selectivity of 75% at 45% cinnamaldehyde conversion was obtained using a
80mol% Pt - 20mol% Co/G catalyst, using toluene was the solvent.

Experiments have also been performed using a standard 5wt.% Pd/C
catalyst (J/M, Type 37) within the mass transfer efficient cocurrent downflow
contactor (C.D.C.) reactor (16). Product selectivity in this case was 85% to
predominately hydrocinnamaldehyde. The C.D.C. reactor system has a reactor
volume of 12×10^{-3} m^3 and has shown higher rates of hydrogenation as
compared with equivalent experiments undertaken within the smaller autoclave
of reactor volume 5×10^{-4} m^3. This enhancement in catalyst activity can be
attributed to the greater turbulence and shear created within the C.D.C. reactor
leading to greater mass transfer efficiency and hence better utilisation of the
catalyst. Initial studies using the stirred autoclave with biphasic solvent
systems e.g. toluene/water have indicated that very high cinnamyl alcohol

selectivities are attainable, in the range 80 - 95%, which are similar to those of Satagopan and Chandalia (12).

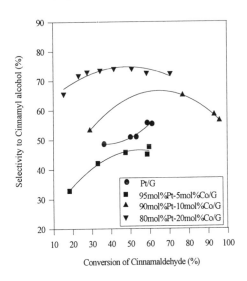

Figure 2.   Selectivity of Cinnamyl Alcohol vs. Cinnamaldehyde Conversion. Process Conditions Described in Table 1.

5 CONCLUSIONS

1. Determination of optimum process conditions is imperative for effective utilisation of the catalyst in three phase reactors to enhance both reaction rates and selectivity.
2. Higher selectivities to crotyl alcohol were obtained under liquid-solid diffusional control conditions.
3. Promotion of a 5 wt.% Pt/G catalyst with iron increased selectivity to crotyl alcohol from 15% to 54% at similar conversion levels.
4. The use of catalyst poisons increased selectivity from 15% to 36%, but at the expense of reducing significantly the reaction rate.
5. Promotion of platinum catalysts with cobalt enhanced selectivity to cinnamyl alcohol from 55% to 75%.

ACKNOWLEDGEMENTS

The authors wish to thank Johnson Matthey (UK) for loan of the precious metal catalysts and salts. I.C.I. (Katalco), E.P.S.R.C, The Chinese Government, The Thai Royal Government and the British Council for financial support.

REFERENCES

1. A. Vogel, in *Textbook of Practical Organic Chemistry-Including Qualitative Organic Analysis*, Revised by B.S. Furniss, A.J. Hannaford, V. Rogers, P.W.G Smith and A.R. Tatchell, 4th Edition, Longman Scientific & Technical, UK. p. 357 (1978).
2. S.W. Chaikin and W.G. Brown, J. Am. Chem. Soc., **71**, 122 (1949).
3. J. Šimoník and L. Beránek, J. Catal., **24**, 348 (1972).
4. P.N. Rylander, N. Himelstein and M. Kilroy, Engelhard Ind. Tech. Bull., **4**, 49 (1963).
5. M.A. Vannice and B. Sen, J. Catal., **115**, 65 (1989).
6. Y. Nitta, K. Ueno and T. Imanaka, Appl. Catal., **56**, 9 (1989).
7. B. Coq, F. Figuèras, C. Moreau, P. Moreau and M. Warawdekar, Catal. Lett., **22**, 189 (1993).
8. S. Galvagno, G. Capannelli, G. Neri, A. Donato and R. Pietropaolo, J. Mol. Catal., **64**, 237 (1991).
9. A. Chambers, S.D. Jackson, D. Stirling and G. Webb, J. Catal., **168**, 301 (1997).
10. S. Galvagno, Z. Poltarzewski, A. Donato, G. Neri and R. Pietropaolo, J. Chem. Soc. Chem. Commun., 1729 (1986).
11. B. Coq, P.S. Kumbhar, C. Moreau, P. Moreau and M.G. Warawdekar, J. Mol. Catal., **85**, 215 (1993)
12. V. Satagopan and S.B. Chandalia, J. Chem. Tech. Biotechnol., **60**, 17 (1994).
13. S.Sharma, PhD Thesis, University of Birmingham (1998).
14. L.K. Doraiswamy and M.M. Sharma in *Heterogeneous Reactions: Analysis, Examples and Reactor Design*, Volume 2, John Wiley and Sons, New York, p. 282 (1984).
15. V. Ponec, Appl. Catal., **149**, 27 (1997).
16. A.P. Boyes, A. Chughtai, X.X. Lu, S. Raymahasay, S. Sarmento, M.W. Tilston and J.M. Winterbottom, Chem. Eng. Sci., **47**, 3729 (1992).

Catalysts for Selective Oxidation of Alcohols to Carbonyl Compounds

Isaguliants G.V., and Belomestnykh I.P.

N.D.Zelinsky Institute of Organic Chemistry, Russian Academy of Sciences.
47, Leninsky prospect, Moscow, 117913, Russia

Abstract

The results of the complex investigation of the process and the catalysts for synthesis of carbonyl compounds are presented. The efficiency of zinc-chromium ferrite and vanadium-magnesium oxide catalysts in alcohol dehydrogenation has been established. The optimum reaction conditions have been determined. The problems are discussed concerning formation of optimum structure of the catalysts that are highly active and selective in production of aldehydes and ketones.

Introduction

Carbonyl compounds (aldehydes and ketones) are used widely in organic synthesis. The main method for their production is dehydrogenation (or oxidative dehydrogenation) of alcohols (1-3) and catalysts for this process draw increasing attention (4-6).

Recently we have elaborated zinc, chromium iron spinel and vanadium-magnesium oxide systems which proved to be rather efficient in production of compounds with double C=C bonds by dehydrogenation of olefins and oxidative dehydrogenation of alkylaromatic hydrocarbons (7,8). The nature of the active components has been established. The scientific principles for preparing of the catalysts have been developed.

It has been found, that catalysts of these types are also efficient in dehydrogenation of butanediol and ethanol leading to formation of compounds with double C=O bonds (9). Later on, V-Mg oxide catalysts were applied in oxidative ethanol dehydrogenation (6), and it was reaffirmed that the system can be an alternative to other catalysts applied in producing of carbonyl compounds.

This communication presents data on application of these catalysts in dehydrogenation and oxidative dehydrogenation (respectively) of alcohols of different structure. Efforts have been made to obtain new data on the correlation of the nature of the catalysts activity in dehydrogenation of hydrocarbons and of alcohols, especially for V-Mg oxide catalysts.

Experimental

Conversion of alcohols was carried out in a fixed-bed flow quartz reactor under atmospheric pressure. A catalyst (20-40 cm^3) was sandwiched with quartz chips (20-40 cm^3) (7,9). Parameters of the reaction were varied in a broad range: temperature 200-390°C, LHSV 1.0-1.7 h^{-1}. Oxidative dehydrogenation was performed in the presence of air; the mole ratio of alcohol/O$_2$ was kept equal to 1/1. The experiment duration was 1-10 hours. The reaction products were analyzed by GLC using 1.5 m column packed with Porapak Q. The amount of products deposited on the catalyst surface was determined by the derivatographic technique.

Zinc, chromium and iron oxide catalysts were prepared by co-precipitation from aqueous solutions of corresponding nitrates with aqueous ammonia. The precipitated hydroxo-compounds were mixed with supports (ZnO and/or Fe$_2$O$_3$) and the mixtures were extruded, then dried at 120°C and calcined. In course of the calcination, basic nitrates interact with one another and the supports yielding active catalyst component - zinc-chromium spinel (390-470°C) or zinc-chromium-iron spinel (650°C). It should be stressed that only preparation of basic nitrates of definite composition and the appropriate mode of their calcination ensure obtaining of an active catalyst for the process. The spinel catalysts preparation procedure has been described in detail elsewhere (7). The ZnFe$_2$O$_4$ catalysts were prepared in the similar manner.

Vanadium oxide catalysts were prepared by impregnation of a MgO-support with aqueous solutions of ammonium vanadate. Samples were dried at 120°C, then calcined in air stream by gradual temperature elevation (8,9).

Results and Discussion

1. Isopropanol dehydrogenation over Zn-Cr-Fe oxide catalyst.

The catalysts such as α-Fe$_2$O$_3$, ZnCrFeO$_4$, ZnFe$_2$O$_4$, ZnCr$_2$O$_4$, have been used for isopropanol dehydrogenation at 340°C (Table 1).

ZnCrFeO$_4$ spinel was found to be the most active and selective for isopropanol dehydrogenation, that is why the process using this catalyst has been studied in a broader temperature range, 270-370°C (Table 2). More than 88% yield of acetone has been obtained with 96.7% selectivity at 370°C. No dehydration product was formed, only minor amount (2-3%) of carbon deposited on the catalyst was detected by DTG.

Table 1. Isopropanol dehydrogenation. (340°C, space velocity of 1.0 h⁻¹)

Catalyst	Acetone yield, mole %	Selectivity, %
ZnCrFeO$_4$	79.0	97.0
ZnFe$_2$O$_4$	35.0	72.0
ZnCr$_2$O$_4$	65.0	87.0

Table 2. Effect of temperature on isopropanol dehydrogenation over ZnCrFeO$_4$ catalyst. (Isopropanol space velocity of 1.0 h⁻¹)

Temperature, °C	Acetone yield, mole %	Selectivity, %
270	10.0	97.0
295	39.5	97.5
300	42.6	98.0
320	68.9	98.0
340	79.0	97.6
370	88.6	96.7

Table 3. Oxidative dehydrogenation of isopropanol over V-Mg oxide catalyst. (Isopropanol space velocity of 1.2 h⁻¹, the mole ratio of isopropanol/O$_2$ = 1/1)

Temperature, °C	Acetone yield, mole %	Selectivity, %
260	20.0	99.0
285	45.6	98.5
300	60.0	98.3
320	68.0	98.3
341	86.8	98.0
368	95.7	98.0

2. Oxidative dehydrogenation over V-Mg oxide catalysts.

Table 3 presents data on oxidative dehydrogenation of isopropanol over V-MgO catalyst described earlier for oxidative dehydrogenation of 2,3-butanediol and ethanol (9). Varying the process temperature in the range of 200-370°C revealed the optimum process temperature to be equal to 370°C. The product contained acetone, hydrogen, minor amounts of CO$_2$ and negligible amount of propene. The CO$_2$ selectivity was around 1-3%. The catalyst became stable after 0.5 hour on stream producing acetone with more than 98% selectivity.

Data given in Table 4 show that the catalyst operates in oxidative dehydrogenation steadily without reducing the activity and selectivity towards

Table 4. Oxidative dehydrogenation of isopropanol over V-Mg oxide catalyst. (370°C, isopropanol space velocity of 1.0 h^{-1}, isopropanol/O$_2$ mole ratio of 1/1)

Acetone production	Time on stream, hours				
	1	3	5	8	10
Yield, mole %	96.0	95.0	95.5	94.9	95.0
Selectivity, %	97.8	98.0	98.0	98.4	98.2

acetone in prolonged runs. After 10 hours on stream, the catalyst kept the initial activity. Thus, comparing isopropanol dehydrogenation over Zn-Cr-Fe-spinel system (Table 2) and oxidative dehydrogenation of isopropanol over V-Mg oxide catalyst (Tables 3, 4), one can see that the latter process is more selective (98%) providing higher (96%) acetone yields.

Since V-Mg oxide catalyst is much more efficient in oxidative dehydrogenation of isopropanol to acetone, it has been studied in oxidative dehydrogenation of alcohols of different structure. Corresponding aldehydes and ketones, minor amount of CO$_2$ and traces of olefins were obtained in oxidative dehydrogenation of the alcohols. That proves alcohol dehydrogenation to be predominant in the process while dehydration is negligible. The yield of the carbonyl compounds (Table 5) reached 68-77% at the selectivity of 88-98%. According to the reactivity in oxidative dehydrogenation, the alcohols can be arranged in the following row: isopropanol > *di*-isopropanol > ethanol > *n*-propanol > *sec*-butanol. For oxidative dehydrogenation of the alcohols, as it has been established previously for hydrocarbons and ethanol (8, 9), the activity and selectivity of the catalyst were found to depend on vanadium content in the sample and conditions of heat treatment. The content of V$_2$O$_5$ of 12% and calcination at 550°C were found to be optimum. The results for *sec*-butanol dehydrogenation are presented as an example in Table 6. One can see that over the optimum catalyst, methylethylketone yield amounted 71% at the selectivity of 97%.

The agreement of the results with those obtained recently allows one to conclude that dehydrogenation of alcohol groups and of ethylbenzene requires one and the same active phase which provides high efficiency of the catalyst. It has been suggested to be Mg$_2$V$_2$O$_7$ with irregular structure and V-ions in octahedral coordination. It undergoes facile redox transformations. This very phase is predominantly formed when catalysts with 12% V$_2$O$_5$ are treated at 550°C in contrast to Mg$_3$(VO$_4$)$_2$ or Mg(VO$_3$)$_2$ (8, 9). Thus, the activity of the catalysts in dehydrogenation of hydrocarbons and alcohol groups occurs to be of the same nature and both reactions necessitate the same species.

Table 5. Oxidative dehydrogenation of alcohols over V-Mg oxide catalyst. (Alcohol/O_2 mole ratio of 1/1)

Temperature, °C	Feed space velocity, h⁻¹	Product yield, mole %	Selectivity, %
feed - ethanol		product - acetaldehyde	
325	1.2	14.5	95.0
338		19.1	94.8
351		25.7	91.2
360		29.0	95.0
366		33.5	94.9
375		41.0	94.8
377		42.1	95.0
398		62.7	95.0
			95.0
318	1.7	10.9	96.0
340		19.1	96.0
360		30.5	95.8
390		49.2	95.6
feed -*n*-propanol		product - propylaldehyde	
300	1.2	42.0	88.0
320		50.2	86.7
340		60.7	86.5
feed - *di*-isopropanol		product - *di*-isopropylketone	
280	1.2	37.8	
300		58.2	
320		64.4	
358		77.3	
feed - *sec*-butanol		product - methylethylketone	
280	1.2	30.2	98.0
305		47.0	98.2
320		55.5	97.6
330		62.9	97.0
340		70.6	96.8
280	1.7	25.0	99.0
312		44.8	98.5
320		49.5	98.0
340		68.1	98.0

Table 6. Oxidative dehydrogenation of *sec*-butanol over V-Mg oxide catalysts. (340°C, space velocity of 1.5 h^{-1}, mole ratio of *sec*-butanol/O_2 = 1/1)

V_2O_5 content, %	Heat treatment, °C	Catalyst area, m^2/g	Methylethylketone, % Yield	Methylethylketone, % Selectivity
5.0	550	98	42.0	98.0
12.0	550	105	71.0	96.8
29.0	550	70	50.0	90.0
12	120		30.0	94.0
12	550	100	71.0	96.8
12	750	60	35.0	79.0
12	850	50	30.0	68.0

Conclusion

The efficiency of dehydrogenation of isopropanol to acetone over Zn-Cr-Fe spinel and oxidative dehydrogenation of alcohols of different structure over V-Mg oxide catalysts have been established.

The optimum reaction conditions has been found for acetone and methylethylketone production. The processes proved to proceed with high yields and selectivity to carbonyl compounds at long life-time of the catalysts.

References

1. M. Rusek, *Stud.Surf.Sci.Catal.*, **59**, 359 (1991).
2. C. Gremmelmaier, *German Pat* N 2801496 (1978).
3. Ch. Sivaraj, B.M. Reddy and P.K. Rao, *Appl.Catal.*, **45**, L11 (1988).
4. Y. Matsumura, K Hashimoto and Y. Yoshida, *J.Catal.*, **122**, 352 (1990).
5. T. Mallat, A. Baiker and J. Patscheider, *Appl.Catal.*, **79**, 59 (1991).
6. M.F. Gomez, L.A. Arrua and M.C. Abello, Ind.Eng.Chem.Res. **36**, 3468, (1997).
7. I.P. Belomestnykh, G.V. Isaguliants and V.P. Danilov, *Kinetika i Kataliz*, **28**, 691 (1987).
8. I.P. Belomestnykh, G.V. Isaguliants and N.N. Rozhdestvenskaja, *Appl.Surf.Sci.*, **72**, 40 (1992).
9. G.V. Isaguliants and I.P. Belomestnykh, *Stud.Surf.Sci.Catal.*, **108**, 415 (1996).

One-Step Production of Low Salt Vinylamine Polymers

M. E. Ford and J. N. Armor

Air Products and Chemicals, Inc., 7201 Hamilton Boulevard, Allentown, Pa. 18195

Abstract

Treatment of aqueous poly(N-vinylformamide) with a supported transition metal catalyst provides partially hydrolyzed poly(vinylamine) in a single step. Formate is catalytically decomposed in situ; only low concentrations of it remain. With the preferred conditions, good conversions are obtained.

Introduction

Poly(vinylamine) (pVAm; 1) is the nitrogen analog of poly(vinyl alcohol). It

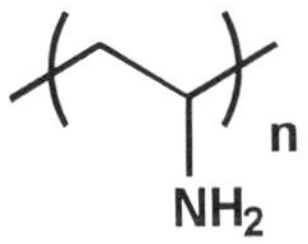

1

has significant potential applications, which include wet strength additives for nonwovens, water treatment chemicals, and as a curing agent for epoxies and polyurethanes. Owing to the tautomeric instability of vinylamine, pVAm is made indirectly by polymerization of a derivative of vinylamine, such as N-vinylformamide (NVF), and subsequent removal of the derivatizing group (1). Previous methods for conversion of poly(N-vinylformamide) (pNVF) to pVAm entail hydrolysis with either strong acid (2) or base (3,4). In either case, inorganic coproducts are formed in conjunction with pVAm: base hydrolysis leads to alkali metal formates, while acid hydrolysis gives the corresponding salt of pVAm and formic acid. Neutralization provides pVAm, accompanied by a salt of the acid used for hydrolysis and (unless formic acid was removed) a formate salt. Although some applications of pVAm are insensitive to the presence of inorganics, many, including those in adhesives and coatings, require essentially salt-free pVAm. Separation of these coproducts from pVAm has been accomplished by traditional routes such as precipitation or selective extraction. In all instances, however, preparation of salt-free pVAm entails tedious removal and disposal of stoichiometric quantities of an inorganic coproduct. This adds needless separation costs. An alternate route to pVAm, which would avoid separation and disposal of coproduct formate, would facilitate development and large scale production of pVAm. Previously, we discovered a route to essentially salt-free pVAm from poly(vinylammonium formate) (pVAF) by catalytic decomposition of formate (5). However, application of this process required initial hydrolysis of pNVF to pVAF.

Results and Discussion

We now report that partially hydrolyzed pVAm which contains a low concentration of formate is obtained directly from pNVF. Carbon dioxide and hydrogen are produced; no other salts or ionic coproducts are formed. The process consists of heating aqueous pNVF in the presence of a supported transition metal; initial hydrolysis of pNVF is most likely followed by in situ catalytic formate decomposition (Table 1, cf runs 2 and 5 with run 6). As suggested by literature on the related topic of transfer hydrogenations (6,7) and our experience with catalytic formate decomposition (5), supported palladium catalysts showed good activity (Table 1, runs 1-5, 7,8). However, high reaction temperatures are required, owing to the absence of added acid or base. Possibly as a result of these high reaction temperatures, moderate to good results were also obtained with other Group VIII catalysts (Table 1, cf runs 9-11 with, eg, runs 1, 3, 4). Although good conversions of formate were obtained, evolution of carbon dioxide was poor. This reflects the facile formation of amine (bi)carbonates/ carbamates, particularly in a closed system with an aqueous medium. However, virtually complete removal of carbon dioxide was achieved in a second step by heating the reaction product to reflux, and sparging gently with nitrogen.

Reactions with medium (409,000) and high molecular weight (900,000) pNVF proceeded similarly (Table 1, runs 12-14). Good conversions and removal of formate were obtained with both substrates.

In conclusion, treatment of aqueous pNVF with supported Group VIII metal catalysts, and especially supported palladium catalysts, provides partially

Table 1 Preparation of Poly(vinylamine) from Poly(N-vinylformamide)

Run	pNVF MW (x 10^3)	Catalyst (loading)[a]	T(°C)[b]	t(hr)	Conv(%)[c]	Selectivity[d] pNVF	pVA	Amidine	Formate Dec (%)[e]
1	1.3	5% Pd/C (1)	165	3	67	21	56	23	61
2	1.3	5% Pd/C (1)	180	2	73	15	62	23	72
3	53	5% Pd/C (1)	165	3	64	19	48	33	62
4	53	5% Pd/C (1)	165	6	66	21	54	25	78
5	53	5% Pd/C (1)	180	2	79	13	71	16	79
6	53	--	180	2	38	28	5	67	0
7	53	5% Pd/alumina (1)	180	2	65	23	53	24	68
8	53	5% Pd/CaCO$_3$ (1)	180	2	64	23	51	26	62
9	53	5% Ru/C (1)	180	2	67	17	51	32	62
10	53	5% Pt/C (1)	180	2	73	14	61	25	74
11	53	5% Rh/C (1)	180	2	60	18	39	43	51
12	409	5% Pd/C (1)	165	6	72	18	63	19	88
13	409	5% Pd/C (1)	180	2	71	18	61	21	84
14	900	5% Pd/C (1)	165	6	68	23	59	18	83

[a] Wt% catalyst (dry basis), based on volume of hydrolyzate, expressed as [gm catalyst (dry basis)/mL hydrolyzate] x 100.

[b] Reactions carried out under autogeneous pressure (200 - 350 psig).

[c] Mole percent of formamide reacted; balance remains either as pendant formamide or incorporated in cyclic amidine. Since each amidine is a 1:1 adduct of pendant amine and formamide groups, conversion equals selectivity to pVA plus 1/2(selectivity to amidine).

[d] Mole percent of each functionality in the isolated terpolymer.

[e] Mole % formate decomposed.

hydrolyzed pVAm which contains a low concentration of formate. High conversions are obtained; the resulting carbon dioxide is quantitatively removed by subsequent sparging with nitrogen. Formate conversion and carbon dioxide evolution are not inherently dependent on polymer molecular weight.

Experimental

Reactions were carried out with aqueous solutions of pNVF (50 mL; 7 wt% polymer) in a 100 mL stirred stainless steel autoclave. The appropriate amount of the desired catalyst was added, the reactor was purged and pressure checked with nitrogen, and then the reaction was heated with stirring to the desired temperature for the desired time (see Table 1) under autogeneous pressure (typically 200 - 350 psig). Products were analyzed by ^{1}H and ^{13}C NMR.

To remove entrained carbon dioxide, the hydrolyzate (25 mL) was added to a 100 mL three-necked round bottom flask equipped with a magnetic stirrer, reflux condenser, and gas dispersion tube. Nitrogen flow was introduced, and the reaction heated under reflux for 1 hr.

Acknowledgments

We thank F. M. Mullen for technical assistance, and Air Products and Chemicals, Inc. for permission to publish this work.

References

1.   R. K. Pinschmidt and D. J. Sagl, The Polymeric Materials Encyclopedia, Synthesis, Properties and Applications, J. C. Salamone, ed., CRC Press, Boca Raton, in press.
2.   T. Itagaki, M. Siraga, S. Sawayama, and K. Satoh, US Pat. 4,808,683, to Mitsubishi Chemical Industries Limited (1989).
3.   D. J. Dawson and P. J. Brock, US Pat. 4,393,174, to Dynapol (1983).
4.   F. Brunmueller, M. Kröner, and F. Linhart, US Pat. 4,421,602, to BASF (1983).
5.   M. E. Ford and J. N. Armor, Chem. Ind. (Dekker), **68**, (*Catal. Org. React.*), 367 (1996).
6.   H. Weiner, J. Blum, and Y. Sasson, *J. Org. Chem.*, **56**, 6145 (1991), and references therein.
7.   R. A. W. Johnstone, A. H. Wilby, and I. D. Entwhistle, *Chem. Rev.*, **85**, 129 (1985).

The Selective Dehydrogenation and Dimerization of Amines Over Cu Containing MFI Zeolites

T. F. Guidry and G. L. Price

Department of Chemical Engineering, Louisiana State University, Baton Rouge, Louisiana 70803

ABSTRACT

A copper exchanged MFI zeolite prepared by reaction of the acidic zeolite with gaseous CuCl is demonstrated to be active for the catalytic dehydrodimerization of 1-propanamine to a 1-propanamine, N-(1-propylidene). It is proposed that the active catalyst is a bifunctional material consisting of acid sites and dispersed copper metal.

1. INTRODUCTION

Zeolites have been shown to be capable of catalytically converting amines into other nitrogen containing compounds. In the acidic form, H-MFI zeolite, also referred to as H-ZSM-5, has been shown to be capable of catalyzing the dimerization of propanamines to dipropanamines and ammonia (1). Addition of transition metals, such as gallium, indium, or copper, into MFI zeolite (2-3) has been shown to significantly increase the selectivity to dehydrogenated products, such as imines and nitriles (2-3).

In this paper, we highlight the reaction of 1-propanamine over a copper containing MFI zeolite prepared by the reaction of H-MFI with gaseous CuCl.

2. EXPERIMENTAL

MFI provided by PQ (Lot CBV-3020) zeolite with SiO_2/Al_2O_3 mole ratio of 40 was used as the base material in a fully protonated form, which will be referred to as H-MFI. Another material, which will be referred to as Cu-MFI, was prepared by reaction of 5 g of the dry H-MFI material with 1 g of CuCl vapor in Ar. Physically adsorbed CuCl was removed by heating the material to 973 K for 6 hours in Ar and the sample was then oxidized in pure O_2 at 373 K for 5 hours.

A Perkin-Elmer TGA-7 microbalance interfaced to a PC was used to detect weight change during the temperature programmed desorption (TPD) of 1-propanamine. H_2 temperature programmed reduction (TPR) experiments were performed in a Carle Analytical Gas Chromatograph 111 H Series S which was reconfigured to measure hydrogen consumption. Catalytic reactor experiments were performed in a gradientless recirculating batch reactor system utilizing 100 mg of catalyst. In a typical experiment, the samples were pretreated by heating them in vacuum from 303 K to 773 K at 10 K/min and held at 773 K for 1.8 ks. Most of the samples were then reduced by cooling the sample to 573 K or 673 K and then circulating 127 kPa He and 6.6 kPa H_2 in the reactor system for 1.8 ks. The samples were then evacuated, cooled to the reaction temperature, the reactor filled with 133 kPa of He, and then isolated from the rest of the system. For 1-propanamine reactions, the circulation loop was filled with about 6.6 kPa of reactant and 127 kPa He. Dipropanamine reactions were carried out using 1.3 kPa of DPA and 131.7 kPa He. The reactions were started by diverting the circulating mixture through the reactor. Samples were withdrawn periodically through a traced line to the evacuated loop on a gas sampling valve of an HP 5890 II GC with a 50 m PONA capillary column and a HP 5972 Mass Selective Detector. A detailed description of the reactor system is published elsewhere (1).

3. RESULTS AND DISCUSSION

The catalytic conversion of 1-propanamine, C_3H_9N, was used as a model test reaction to investigate the mechanism of the dehydrogenation and dimerization reactions of light amines over the Cu-MFI zeolite. The product distribution as a function of time for the catalytic conversion of 1-propanamine over four different materials is shown in Figures 1-4. The ordinate of these figures is the area percent of the total ion chromatogram signal for the respective compound.

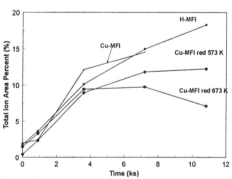

Figure 1: Comparison of the dipropanamine concentration from the conversion of 1-propanamine over several catalysts.

The material labeled H-MFI is the copper free base zeolite heated to 773 K for 1.8 ks. The material labeled Cu-MFI is the Cu-MFI material heated to 773 K for 1.8 ks. The other two catalysts are Cu-MFI zeolites dried to 773 K for 1.8 ks and then reduced at either 573 K or 673 K prior to contact with the reactant.

In agreement with previous studies, the H-MFI material catalyzes the conversion of 1-propanamine into dipropanamine, $C_6H_{15}N$, and ammonia (1). Negligible amounts of dehydrogenated products are observed during this reaction. The unreduced Cu-MFI material is also active for the conversion of 1-propanamine. In addition to dipropanamine, this material also catalyzes the conversion of 1-propanamine to 1-propanamine, N-(1-propylidene), $C_6H_{13}N$. Prereduction of the Cu-MFI material at 573 K increases both the rate of 1-propanamine conversion and the yield of dehydrogenated products. The amount of 1-propanamine, N-(1-propylidene) produced is greater than that of the unreduced Cu-MFI sample and small amounts of propananitrile, C_3H_9N, and 4-methylpentanitrile, $C_6H_{11}N$, are also detected. The amount of dipropanamine formed is lower than that of either the unreduced Cu-MFI or the H-MFI material, especially at longer reaction times. Reduction of the Cu-MFI material at 673 K further increases the rate of 1-propanamine conversion and the yield of the dehydrogenated products. Significant quantities of both propananitrile and 4-methylpentanitrile are observed. The initial yields of 1-propanamine, N-(1-propylidene) produced are greater over the Cu-MFI material reduced at 673 K than any of the other materials. At longer times, however, the concentration of the 1-propanamine, N-(1-propylidene) falls below that of the Cu-MFI material reduced at 573 K. The Cu-MFI reduced at 673 K also shows a lower selectivity to dipropanamine than any of the other materials.

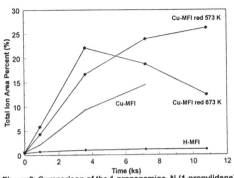

Figure 2: Comparison of the 1-propanamine, N-(1-propylidene) concentration from the conversion of 1-propanamine.

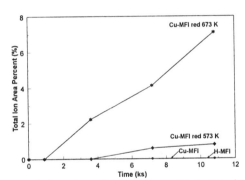

Figure 3: Comparison of the 4-methylpentanitrile concentration from the conversion 1-propanamine of over several catalysts.

As shown in Figure 5, the initial product of dipropanamine conversion over Cu-MFI reduced at 573 K is the 1-propanamine, N-(1-propylidene). At longer reaction times, its concentration decreases and falls below that of propananitrile.

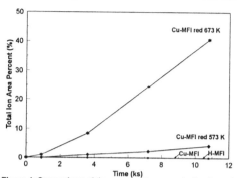

Small amounts of 4-methyl-pentanitrile are also observed.

The interaction of 1-propan-amine with the catalytic mater-ials was also investigated by TPD of 1-propanamine in the microbalance. These results are shown in Figure 6. The broad plateau in the desorption spec-trum for the H- MFI material between 550 K and 650 K has been shown to correspond to a 1/1 proton/ amine complex

Figure 4: Comparison of the propanitrile concentration from the conversion of 1-propanamine over several catalysts.

(4-6). This complex decomposes at higher temperatures and produces the large peak in the H-MFI derivative curve. The 1-propanamine TPD curves for the Cu-MFI samples reduced at 573 K and 673 K are very similar. Both of them have a plateau region in the weight curve between 550 K and 650 K and a high temperature desorption peak in the weight derivative curve, suggesting the presence of zeolitic protons. The unreduced Cu- MFI material does not possess either a plateau region in the weight curve or a major high temperature desorption peak in the weight derivative curve.

The consumption of hydrogen by the samples during hydrogen TPR is shown in Figure 7. The Cu-MFI sample dried to 373 K for 1.8 ks shows both

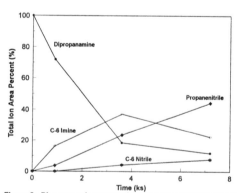

low and high temperature hydrogen consumption peaks. Previous studies of the hy-drogen temperature program-med reduction of copper con-taining MFI zeolite suggested that reduction peaks in the 415 K to 550 K temperature region were due to the reduction of Cu^{2+} to Cu^+ ions and that the higher temperature reduction peaks were due to reduction of Cu^+ ions to copper metal (7,8).

Figure 5: Dipropanamine conversion catalyzed by reduced Cu-MFI

The total amount of hydrogen consumed is equivalent to approximately 1.04 Cu^{2+} per framework Al. The sample heated to 773 K for 1.8 ks, however, only possesses one hydrogen con-sumption peak. This result suggests that thermal reduction of the Cu^{2+} to Cu^+

occurred. Similar observations have been reported previously for other copper containing MFI zeolites (9,10). Neither the H-MFI sample nor the Cu-MFI sample reduced at 573 K show significant hydrogen consumption peaks during the hydrogen temperature programmed reduction. The absence of any reduction peaks for the prereduced Cu-MFI material suggests that the hydrogen reduction at 573 K reduced all of the copper ions to the metal.

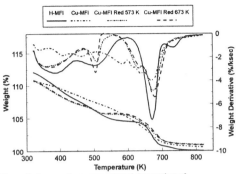

Figure 6: Temperature programmed desorption of 1-propanamine.

The reaction data coupled with hydrogen TPR data suggest that the reduced Cu-MFI material consists of dispersed copper metal on the acidic MFI zeolite base. The observed dimerization of the amines proceeds over the acidic sites and the dehydrogenation reactions occur over the dispersed copper metal. The data in Figures 1-4 for the H-MFI material indicate that the major product of 1-propanamine conversion over acid sites is dipropanamine. The results in Figure 5 indicate that the primary initial product of the reaction of dipropanamine with the reduced Cu-MFI material is the 1-propanamine, N-(1-propylidene). Both propanenitrile and 4-methylpentanitrile appear at longer reaction times. The data suggest that the reaction pathway for the conversion of 1-propanamine to the 1-propanamine, N-(1-propylidene) occurs by the dimerization of 1-propanamine to dipropanamine and ammonia. The dipropanamine is then dehydrogenated to the 1-propanamine, N-(1-propylidene) by the dispersed copper sites. Nitriles are then formed by subsequent reaction of the 1-propanamine, N-(1-propylidene). The increased yield of dehydrogenated products over the Cu-MFI material reduced at 673 K relative to the Cu-MFI material reduced at 573 K is likely due to sintering of the copper metal. The larger copper crystallite size may a be more effective dehydrogenation site. A previous study of a Cu-containing MFI prepared by solid-state ion-exchange of

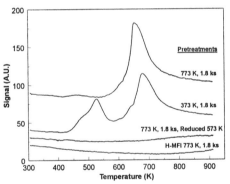

Figure 7: TPR Curves for Pretreated Cu-MFI.

CuO with H-MFI (11) reported that a 4-methylpentanitrile was the major product of 1-propanamine conversion under similar reaction conditions, which contrasts with the results published here. Future studies will investigate the reasons for these differences.

4. CONCLUSIONS

A copper containing MFI zeolite can be prepared by reaction of the acidic zeolite with CuCl. Heating this material to 773 K and then reducing it in hydrogen at 573 K produces a material capable of selectively dimerizing and dehydrogenating 1-propanamine to imines and nitriles. The active material is a bifunctional catalyst consisting of dispersed copper metal on the acidic base zeolite. It is proposed that the acid sites catalyze the dimerization reactions and the dehydrogenation reactions occur over the dispersed copper metal.

5. ACKNOWLEDGMENTS

The authors gratefully acknowledge the financial support of the National Science Foundation (Grant # CTS-9634754).

6. REFERENCES

1. P. A. Jacobs and J. B. Uytterhoeven, *J. Catal.*, **26**, 175 (1972).
2. V. Kanazirev, K. M. Dooley and G. L. Price, *J. Catal.*, **146**, 228 (1994).
3. V. I. Kanazirev, G. L. Price, and K. M. Dooley, *J. Catal.*, **148**, 164 (1994).
4. J. D. Parrillo, A. T. Adamo, G. T. Kokotailo, and R. J. Gorte, *Appl. Catal.*, **67**, 195 (1990).
5. T. J. Gricus Kofke, R. J. Gorte, and G. T. Kokotailo, *Appl. Catal.*, **54**, 177 (1989).
6. A. I. Baiglow, A. T. Adamo, G. T. Kokotailo, and R. J. Gorte, *J. Catal.*, **138**, 377 (1992).
7. J. Y. Yan, G. D. Lei, W. M. H. Sachtler, and H. H. Kung, *J. Catal.*, **161**,43 (1996).
8. T. Beutel, J. Sarkany, G. D. Lei, J. Y. Yan, and W. M. H. Sachtler, *J. Phys. Chem.*, **100**, 845 (1996).
9. J. O. Petunchi, G. Marcelin, and W. K. Hall, *J. Phys. Chem.*, **96**, 9967 (1992).
10. W. Grunert, N. W. Hayes, R. W. Joyner, E. S. Shpiro, M. R. H. Siddiqui, and G. N. Baeva, *J. Phys. Chem.*, **98**, 832 (1994).
11. V. Kanazirev and G. L. Price, *Stud. Surf. Sci. Catal.*, **84**, 1935 (1994).

Alkane Conversion and Arene Hydrogenation by Platinum Based Catalysts Prepared from Supported Molecular Precursors: Effects of Phosphorus and Gold

B. D. Chandler, A. B. Schabel, L. I. Rubinstein, and L. H. Pignolet,
Department of Chemistry, University of Minnesota, Minneapolis, MN 55455

Abstract: Silica supported mono- and bimetallic Pt and Pt-Au molecular precursors were thermolized and evaluated as alkane conversion catalysts. Whenever triphenylphosphine is a ligand in the precursor, the resulting catalyst is remarkably selective for hydrogenation and dehydrogenation reactions and is much less active for cyclization, isomerization, and cracking reactions relative to non-phosphorus containing wetness impregnated Pt- and Pt-Au on silica catalysts. Propane dehydrogenation experiments show 90% selectivity at 35% conversion for phosphine containing catalysts, as well as greatly enhanced stability over a conventional Pt/SiO_2 catalyst. DRIFTS of adsorbed CO implicates an electronic effect of phosphorus on Pt. The effects of Au on catalysis by Pt are much less dramatic; however, significant differences in hexane conversion selectivity are observed.

Introduction: Supported mono- and bimetallic catalysts are widely used in a multitude of industrial and laboratory scale reactions, but rarely have bimetallic cluster compounds been used as catalyst precursors. Phosphine ligated precursors are almost never used in this regard due to suspected catalyst poisoning by residual phosphorus; however, suspected "poisons" such as sulfur are frequently added to industrial catalysts. In this paper we examine a series of silica supported Pt and Pt-Au catalysts prepared from heterogenized supported mono- and hetero-metallic precursors and examine their performance as hexane conversion, propane dehydrogenation, and arene hydrogenation catalysts. We evaluate the effects of Au and triphenylphosphine (PPh_3) on the Pt based catalysts and compare these systems to conventionally prepared Pt- and Pt-Au catalysts. Some of this work has been published (1).

Experimental: The mono- and bimetallic precursors used in this study are: H_2PtCl_6 (**Pt**), $HAuCl_4$ (**Au**), $Pt(PPh_3)_2O_2$ (**PtP2**), $[PtAu_2(PPh_3)_4](NO_3)_2$ (**PtAu2P4**), $[PtAu_8(PPh_3)_8](NO_3)_2$ (**PtAu8P8**), and $Pt_2Au_4(C{\equiv}CBu^t)_8$ (**Pt2Au4**). All catalysts were 1% Pt (excpt **PtAu8P8**, which was 1% adsorbed cluster) supported on Davisil' silica (1, 2); the **P** in the abbreviation indicates the presence of PPh_3 in the precursor. The Pt and Au chloride salts were supported by wetness impregnation and coimpregnation, respectively, from aqueous solution with a 1:2 Pt:Au ratio for the **Pt+2Au** catalyst. Phosphine ligated precursors were spontaneously adsorbed from organic solution. The catalysts were calcined at 300 °C for 2 hr followed by H_2 reduction at 200 °C for 1 hr. Catalytic tests for alkane conversion and toluene hydrogenation were carried out using a fixed-bed microreactor system operated in a continuous flow mode. Pt availability was determined by CO chemisorption and catalysts were further characterized by DRIFTS, TEM, and XPS. A more detailed experimental description can be found in (1).

Results and Discussion: Hexane Conversion. Catalysis experiments were carried out using a hydrogen to hexane ratio of 16:1 at 400 °C with hexane conversions between 2 and 45%. The major product classifications used hereafter are: *cracking* (C_1 - C_5 hydrocarbons), *isomerization* (2-methylpentane & 3-methylpentane), *hexenes*, *cyclization* (methylcyclopentane (MCP) & cyclohexane), and *benzene*. It is clear that the primary (and most dramatic) factor governing the differences in these catalysts is the presence or absence of phosphorus in the precursor. For illustrative purposes, however, we will begin our discussion with the effects of gold.

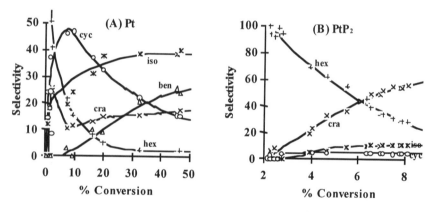

Figure 1. Product selectivity dependence on conversion of hexane for the catalysts **Pt** (A) and **PtP2** (B). The products are designated as follows: hex = hexenes, iso = isomerization, cyc = cyclization, cra = cracking, and ben = benzene. Lines are drawn only to help see the trends. Reprinted from Ref. (1) with kind permission of Elsevier Science - NL, Sara Burgerhartstraat 25, 1055 KV Amsterdam, The Netherlands.

Effects of Gold. For the wetness impregnated **Pt** and **Pt+2Au** catalysts, the effects of Au on catalysis by Pt are subtle relative to the effects of phosphorus. In general, for non-P containing catalysts, hexenes are produced at very low conversion (< 2%) while MCP and methylpentanes are the major products of catalysis at conversions above ca. 5% (Figure 1A). These observations are consistent with literature data (3) and with the accepted model for metal reforming catalysis (4). The coimpregnation of Au with Pt does, however, retard the production of benzene and increase selectivity towards isomerization and cracking products. The Au containing catalyst also has a somewhat slower rate of hexane conversion (0.50 vs. 0.96 min^{-1} for **Pt**).

Using the $Pt_2Au_4(C≡CBu^t)_8$ cluster as a catalyst precursor yields a catalyst (**Pt₂Au₄**) with different properties than either of the conventionally prepared catalysts. Catalyst selectivity follows the same general trends as **Pt**; however, at 5-20% conversion selectivity towards cracking products is increased and selectivity towards isomerization products is decreased relative to both the **Pt** and **Pt+2Au** catalysts. At conversions between ca. 25-45%, cracking and benzene selectivites are comparable to Pt, isomerization selectivity is lower, and selectivity towards hexenes and cyclization products is higher. The rate of hexane conversion for **Pt₂Au₄** is greater than that for **Pt+2Au**, but less than that of **Pt**. The cluster derived catalyst also has greatly enhanced longevity under constant flow conditions (Figure 2).

Effects of Phosphorus. Some of this work has been published elsewhere (1). Selectivities of different catalysts at low conversion (ca. 2%) are shown in Table 1. When phosphorus is present in the catalyst precursor, there is a dramatic shift in observable activity and in catalysis product selectivities (Table 1, Figure 1). For all three phosphorus containing catalysts, (**PtP₂**, **PtAu₂P₄**, and **PtAu₈P₈**), we were unable to achieve conversions above 8%, even when contact times exceeded 10 seconds. At very short contact times (ca. 0.3 seconds), however, the catalysts maintained conversions of ca. 2%. In this lower conversion regime, selectivity towards dehydrogenation products is nearly 100%. As the contact time increased, greater cracking activity was observed, as shown in Figure 1B.

Table 1. Selectivities (%) for hexane conversion at low conversion (ca. 2%).[a]

Catalyst	Cracking	Isomerization	Cyclization	Hexenes
Pt	14	12	37	35
Pt+2Au	8	15	46	31
PtP₂	2	0	0	98
PtAu₂P₄	6	3	0	91
PtAu₈P₈	9	3	0	89
Pt+2P[b]	7	6	8	79

[a]See text for product classifications and catalyst abbreviations.
[b]Prepared from sequential deposition of H_2PtCl_6 and 2 equivalents of PPh_3.

The above data indicate that the hexane dehydrogenation reaction runs at equilibrium, a conclusion supported by thermodynamic data. In addition, this conclusion about the thermodynamic control of dehydrogenation is consistent with accepted models and with the observations that this reaction is faster than other processes that occur in reforming (4). The dependence of selectivity on percent conversion for the phosphorus containing catalyst **PtP₂** is shown in Figure 1B. The data for the other phosphorus containing catalysts are similar. The absolute production of olefins remains stable while production of cracking products increases at higher conversion. It is clear that the presence of phosphorus in the catalyst severely inhibits the formation of cyclization and isomerization products. Cracking is also significantly inhibited with the phosphorus containing catalysts relative to **Pt**. The cracking selectivity increases at higher conversion for **PtP₂** (Figure 1B) because it is the dominant secondary reaction available for the rapidly produced olefin to undergo.

The phosphorus containing catalysts show much greater stability on stream under the hexane conversion reaction conditions than the non-phosphorus containing catalysts. Figure 2 shows a comparison of stability on stream for **PtP₂** and **Pt**. Because the phosphine derived systems have very low rates of cracking and dehydrocyclization, coking for these systems is much slower and the catalyst is more stable.

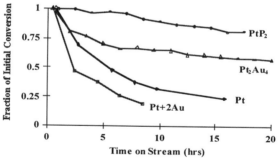

Figure 2. Catalyst Stability for Hexane Conversion.

Toluene Hydrogenation. Because the dehydrogenation reaction is at equilibrium, the rate of this process cannot be determined from the hexane conversion data. Since hydrogenation and dehydrogenation reactions are considered to occur at the same site, rates and activation energies for toluene hydrogenation to methylcyclohexane were measured for several catalysts to investigate the effects of P on these processes. Results are summarized in Table 2. The rates for the phosphorus containing catalysts at 60 °C are slightly lower than for **Pt**, but not dramatically so. The measured rates and activation energies are similar to those reported elsewhere for **Pt** on SiO_2 (5). This data indicates that the presence of P does not greatly affect hydrogenation/dehydrogenation rates, despite its dramatic effect on hexane conversion selectivity.

Table 2. Effects of Phosphorus on Toluene Hydrogenation.[a]

Catalyst	Activity[b] $(x10^3)$	Activity corrected[c] for CO/Pt $(x10^3)$	Ea (kcal/mol)
Pt	53	120	13.0
PtP$_2$	41	85	11.2
Pt+2P	25	72	11.6

[a]continuous flow catalysis with H_2:tol = 37:1 at ambient pressure
[b]mol methylcyclohexane (mch) produced per (mol Pt sec) at 60 °C
[c]mol mch produced per mol (surface Pt sec) at 60 °C

Propane Dehydrogenation. Since the phosphorus containing catalysts showed high activity and selectivity for dehydrogenation of hexane, some preliminary experiments were run with propane in order to evaluate a more practical reaction. At 550 °C with a pure propane flow, the phosphorus containing catalysts show excellent selectivity for propylene production, typically around 90% at 35% conversion. They also have greatly enhanced stability relative to the **Pt** catalyst, which became inactive due to coking in 30 min. The results for the phosphorus containing catalysts are encouraging and are approaching better patented industrial catalysts reported for this reaction (6-8). Work in progress is directed at optimizing the catalyst and conditions of reaction.

Catalyst Characterization. Pt availability was determined with CO chemisorption at ambient temperature. All of the samples showed similar CO uptakes (~.25-.60 mol CO/mol Pt) except Au which did not irreversibly bind CO and was inactive in all catalysis reactions studied. These results agree qualitatively with DRIFTS results and, for the **Pt** catalyst, CO uptake (44%) agrees very well with TEM data indicating average particle sizes to be 3 nm (corresponding CO uptake = 45%). These values are in good agreement with the literature data for Pt on silica. Chemisorption data obtained for bimetallic compounds are more difficult to interpret because they do not necessarily provide direct information about particle size and because restructuring is known to occur in Pt-Au alloys. Chemisorption data for the phosphorus containing catalyst **PtP2** (48%) may also be difficult to relate to Pt particle size since bound phosphorus may be blocking some surface Pt sites. Preliminary TEM results on **PtP2** show very small Pt particles between 0.5 and 1 nm suggesting that this may indeed be the case.

Electronic effects of Au and P were probed with DRIFTS experiments of adsorbed CO. The **Pt** catalyst has a CO stretch at 2066 cm^{-1}, in agreement with literature values (8). Surprisingly, addition of gold does not greatly alter the CO stretching frequency. A lowering (red shift) of only 4 cm^{-1} was observed with **Pt+2Au**. This shift is somewhat smaller than reports in the literature (10-

15 cm^{-1} upon Au alloying) (9-12) and is attributed to differences in supports, pretreatment and sample preparation conditions, and spectroscopic technique. Preliminary DRIFTS results for the **Pt$_2$Au$_4$** catalyst also shows a greater lowering of the CO stretch, indicating that this catalyst differs from the coimpregnated **Pt+2Au** catalyst.

The presence of phosphorus in the catalysts has a significant effect on the CO stretching frequency and causes an upward (blue) shift of about 10 cm^{-1}. Two explanations have been forwarded for CO stretching frequency shifts in alloyed and doped Pt catalysts: geometric and electronic effects (13). We believe the blue shift results from an electronic effect caused by direct interaction between Pt and the phosphine residues. Preliminary XPS results indicate that the phosphorus remains in some oxidized form after catalysis. An electron poor oxide of phosphorus bound to Pt would be expected to cause such a blue shift. Evidence for direct Pt-P bonding was recently reported in an EXAFS study on a similar silica supported triphenylphosphine Pt-Au cluster derived system (14).

In addition, TPD experiments of chemisorbed CO indicate significant differences in the temperature of maximum desorption between the **Pt** and **PtP$_2$** catalysts. Desorption of CO from **PtP$_2$** had a maximum at ca. 140 °C while CO desorption from **Pt** peaked between 240-270 °C. Moreover, desorption from **PtP$_2$** was very clean, giving only CO, while the desorption products from **Pt** contained a significant amount of CO$_2$. This supports the argument for an electronic effect of phosphorus on Pt; however, a geometric compression of the CO adlayers also remains a possibility. One also cannot rule out the possibility that the use of triphenylphosphine ligated precursors results in Pt particles with a different morphology, thus causing a shift in the adsorbed CO stretch.

Conclusions: A primary objective of this study is to examine the utility of bimetallic Pt-Au clusters as precursors to supported heterogeneous catalysts. In this examination, it is apparent the presence or absence of phosphine ligands in the catalysis precursor is key in determining the properties of the resulting catalyst. Phosphine residues on the catalyst direct selectivity towards nearly exclusive production of hexenes at low conversion by inhibiting secondary cracking, cyclization, and isomerization reactions. Toluene hydrogenation data indicates that phosphorus is not a general poison; rather, it selectively poisons sites responsible for these secondary reactions. One possible explanation may be that phosphorus inhibits the production of the key intermediates that have been implicated in reforming catalysis by Pt (15).

Both the phosphorus containing catalysts and cluster derived **Pt$_2$Au$_4$** catalyst have greatly increased stability on stream over the conventionally prepared **Pt** and **Pt+2Au** catalysts. For the former, this stability is attributed to the suppression of cracking reactions by the phosphine residues. It is interesting that the **Pt$_2$Au$_4$** catalyst produces similar quantities of cracking products to **Pt** and **Pt+2Au**, yet is less susceptible to deactivation processes. The enhanced stability appears to be an effect of decomposing the preformed cluster, perhaps generating catalyst particles with a unique morphology.

Acknowledgements: We would like to thank Dr. Larry Ito of Dow Chemical and Dr. Bruce Alexandar of Amoco for helpful discussions and the University of Minnesota Graduate School for monetary support.

References:
(1) B.D. Chandler, L.I. Rubinstein, L.H. Pignolet, J. Mol. Catal., in press (1998).
(2) I.V.G. Graf, J.W. Bacon, M.E. Curley, L.N. Ito, L.H. Pignolet, *Inorg. Chem.*, **35**, 689 (1996).
(3) A. Sachdev, J. Schwank, *J. Catal.*, **120**, 353 (1989).
(4) B.C. Gates, J.R. Katzer, G.C.A. Schuit, *Chemistry of Catalytic Processes*, McGraw-Hill: New York (1979).
(5) S.D. Lin, M.A. Vannice, *J. Catal.* **143**, 539, 554, 563 (1993).
(6) R.D. Cortright, J.A. Dumesic, *J. Catal.*, **148**, 771 (1994).
(7) B.D. Alexander, G.A. Huff, U. S. Patent 5,453,558, Assigned to Amoco (1995).
(8) B.V. Vora, P.R. Pujado, R.F. Anderson, *Energy Progress*, **6**, 171 (1986).
(9) Y. Soma-Noto, W.M.H. Sachtler, *J. Catal.*, **34**, 162 (1974).
(10) K. Balakrishnan, A. Sachdev, J. Schwank, *J. Catal.*, **121**, 441 (1990).
(11) J.W.A. Sachtler, G.A. Somorjai, *J. Catal.*, **81**, 77 (1983).
(12) F. Stoop, F.J.C.M. Toolenaar, V. Ponec, *J. Catal.*, **73**, 50 (1982).
(13) F.J.C.M. Toolenaar, F. Stoop, V. Ponec, *J. Catal.*, **82**, 1 (1983).
(14) Y. Yuan, K. Asakura, H. Wan, Y. Iwasawa, *Chem. Lett.*, 129 (1996).
(15) F. G. Gault, *Adv. Catal.*, **30**, 1 (1981).

Cyclohexane Functionalization Catalyzed by Octahedral Molecular Sieve (OMS-2) Materials

Guan-Guang Xia[1], Jin-Yun Wang[2], Yuan-Gen Yin[2], Steven L. Suib[1,2,3*]

[1]Institute of Materials Science, [2]Department of Chemistry, [3]Department of Chemical Engineering, University of Connecticut, Storrs, CT 06268

Abstract

Cyclohexane Functionalization under mild conditions is a challenge to chemists due to the low reactivity of cyclohexane, the large economic potential of products, and environmentally friendly conditions. Manganese oxides with the cryptomelane structure (OMS-2) were used to catalyze the oxidation of cyclohexane with tert-butyl hydroperoxide (TBHP) as an oxidant. Tert-butyl alcohol (TBA), the reduced product of tert-butyl hydroperoxide, was used as a solvent to ensure the solubility of the reaction mixture and to avoid introducing solvent impurities. The effects of reaction temperatures (30°C, 45°C, 60 °C) and solvents were tested. Preliminary studies of the oxidation mechanism were carried out. OMS-2 materials showed very good catalytic properties for cyclohexane oxidation. For example, the total yield is 16.5% (60°C, 36 hours). The yields of cyclohexanone (K), cyclohexanol (A), cyclohexyl hydroperoxide (CHP) and cyclohexyl tert-butyl peroxide (P) are 9.65%, 5.15%, 0.72% and 1.02%, respectively. The catalysts still retain their crystallinity after several reaction cycles without losing activity and can be easily separated from the system.

I. Introduction.

Although the oxidation of cyclohexane to cyclohexanol/one has been commercialized for many years, this is still a very challenging area for chemists to improve efficiencies for major industrial chemical processes.[1] A lot of effort has been made to try to find more economic, more effective, and highly selective catalytic systems. The work involved in liquid phase reactions can be classified into two categories: homogeneous and heterogeneous catalytic systems. The former focuses on imitating biological systems and using transition metal complexes as catalysts such as iron enzymes,[2] μ-oxo-bridged diiron and binuclear manganese complexes,[3] and Gif-systems.[4] These homogeneous systems possess good activity and high selectivity under mild reaction conditions. The main problems, however, are the low stability of catalysts and difficulty of separation of the catalysts from reactants and products. Heterogeneous catalysts such as transition metal doped zeolites[5] and Mn^{2+} exchanged clays[6] have been found to have some advantages over metal complexes due to their reusability and ease of separation, but their activities are still not high.

Manganese oxide molecular sieves (OMS materials)[1] have shown promising activities towards hydrocarbon Functionalization. In our previous work, transition metal doped OMS-1 materials (with 3x3 tunnel structure) have been used as catalysts for cyclohexane oxidation with tert-butyl hydroperoxide (TBHP) as an oxidant[8]. The results are similar to homogeneous catalysts in terms of yield and conversion. Tert-butylalcohol (TBA) was used as a solvent in this system. Since TBA is a reduced product of TBHP, there will be no additional impurities introduced into the system. In this work, OMS-2 materials were used as catalysts. OMS-2 has a composition of $KMn_8O_{16} \cdot nH_2O$ with the cryptomelane structure. This structure is composed of 2x2 edge and corner shared octahedral MnO_6 units with an opening of 4.6 Å. OMS-2 materials are mixed valent manganese oxides with an average oxidation state (AOS) of manganese of ca. 3.9.

II. Experimental Section.

The methods for synthesizing OMS-2 have been reported elsewhere[7]. The reaction mixture consists of 1.94 grams of cyclohexane (CyH), 3.41 grams of solvent, 2.37 grams of TBHP (70% in water) and 0.28 grams of acetophenone (AP) as an internal standard. A 40 mg sample of OMS-2 was added to the mixture. All reactions were carried out in stirred autoclaves with a Teflon liner, unless otherwise specified. The temperature was controlled with a paraffin oil bath and a temperature controller. The stirring speed for all the experiments was set at 600 r.p.m. The reaction system is under autogenous pressure. The resultant mixtures were filtered to remove the catalysts and quantitatively analyzed with a GC (HP5890 series II with a HP-5 capillary column).

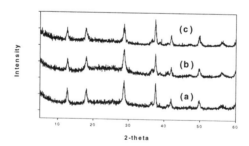

Fig.1. XRD Paterns of OMS-2 Catalysts (a) fresh, (b) used once, (c) used four times

Qualitative analyses of reaction products were performed with a gas chromatograph-mass spectrometer (GC-MS) and a gas chromatograph-infrared spectrometer (GC-IR). The crystallinity of catalysts was examined using X-ray diffraction methods (see Fig 1.).

III. Results.
3.1. Effects of Reaction Temperature.

The results of cyclohexane oxidation with TBHP catalyzed by OMS-2 under different conditions are shown in Table 1. Entries 1, 2, 3 represent the reactions carried out at 30°C, 45°C and 60°C over a 36 hour period of time, respectively. The main products were cyclohexanol (A), cyclohexanone (K), cyclohexyl tert-butyl peroxide (P), and cyclohexyl hydroperoxide (CHP). As the temperature was elevated from 30°C to 60°C, the overall conversion of cyclohexane increased significantly. The ratio of ketone (K) to alcohol (A) was increased from 0.93 to 1.87. Ketone became a dominating product over alcohol at higher temperatures (60°C). In terms of yield, the formation of P at temperature above 45°C did not change much, but CHP increased at first, then went down as the

Table 1. Cyclohexane Oxidation with TBHP*

Entries	Conditions[a]	Conv.% CyH[b]	Eff., % TBHP[c]	S. % A	S.% K	S.% CHP	S.% P	K/A
1	30°C, TBA	4.7	26.2	37.4	34.6	14.8	13.1	0.93
2	45°C, TBA	10.7	26.9	34.5	45.8	10.2	9.5	1.33
3	60°C, TBA	16.5	27.2	31.1	58.3	4.4	6.2	1.87
4	60°C, AC	15.6	28.8	30.4	60.8	6.3	2.4	2.00
5	60°C, B	8.1	18.0	24.3	50.8	14.0	3.4	2.30
6	60°C, TBA[d]	11.2	17.8	21.7	59.2	0.0	7.1	2.73

* The reaction mixture consists of 1.94 g of cyclohexane, 3.41 g of solvent, 2.37 g of TBHP (70% in water), 0.28 g of AP and 40 g of OMS-2.

a. Sample was sealed in an autoclave with a Teflon liner, stirred at 60 r.p.m. for 36 hours.

b. Conversions of cyclohexane are based on total yield of products.

c. Efficiency of TBHP is based on the number of moles of cyclohexane converted per number of moles of TBHP consumed.

d. 0.13 g of TEMPO was added. Selectivity of other products (derivatives of cyclohexane and TEMPO) was 11.9%. The recovery of TEMPO was 35%.

temperature increased. The efficiency of TBHP did not change significantly with change of temperature.

3.2. Effects of Solvents.

To compare the effects of solvents on the reaction, TBA, acetonitrile (AC), and benzene were chosen as solvents, respectively. The results are shown in Table 1 (entries 3, 4, 5). The reaction mixture in AC or in benzene was in two phases before undergoing reaction. After 36 hours of reaction, the resultant mixture in AC became homogenous and the results were similar to those obtained by using TBA. However, benzene showed significantly different results. The reaction system was still in two phases. The catalyst particles were found only in the water phase. The conversion value was half that obtained in TBA solution. The efficiency of TBHP towards oxidizing cyclohexane was low. The reaction under these conditions produced a large amount of CHP but a small amount of P.

3.3. Effects of Reaction Time.

Experiments were carried out in sealed glass vials and stirred with a magnetic stirring bar. The degrees of reaction as a function of time at 60°C are shown in Figure 2. The reaction rate was fast in the first few hours. During the reaction, CHP yield first increased, then decreased. Selectivity of alcohol was higher than that of ketone in the beginning, but after four hours, this relationship was reversed.

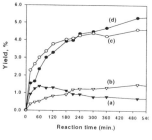

Fig. 2. Effects of Reaction Time

- ● K vs Time
- ○ A vs Time
- ▼ CHP vs Time
- ▽ P vs Time

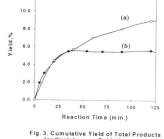

Fig. 3. Cumulative Yield of Total Products for Cyclohexane Oxidation

- ● removal of OMS-2 after 35 minutes reaction
- ○ OMS-2

3.4. Effects of Removal of Catalysts

The same amount of reactant mixture and OMS-2 catalyst shown as entry 3 in Table 1 was sealed in a glass vial with a magnetic stirring bar. The reaction was carried out at 60°C with a stirring speed of 600 r.p.m. After starting the reaction for 35 minutes, the solid catalysts were filtered. Following this the reaction system was run at 60°C for 2 additional hours. These catalytic results are shown in Figure 3. No free manganese cations were found in the solution phase.

In addition, the crystallinities of all used catalysts were checked by X-ray diffraction. The patterns show almost no change after the catalyst was reacted for 36 hours (Figure 1b), and even after several reaction cycles (Figure 1c). The activity of the used catalysts did not decrease.

3.5. Effects of Radical Competitive Agents (TEMPO).

When a small amount of 2,2,6,6-tetramethyl piperidinooxy (TEMPO), a radical trapping or competitive agent,[4] was introduced to the oxidation system, the conversion of cyclohexane decreased by ca. 32% as compared to entry 3 in

Table 1. The data are shown in Table 1 (entry 6). The ratio of K/A is almost 1.5 times higher than the ratio obtained without TEMPO. The selectivity to alcohol dropped by ca. 30% while the selectivity to ketone and P remained unchanged. CHP was unable to be detected after 36 hours of reaction.

IV. Discussion.
4.1. Effects of Reaction Conditions.
(a) Temperature: The results for different temperatures indicate that the activity of cyclohexane with TBHP was largely affected by temperature. The ratio of K/A changes with temperature, which was a result of further oxidation of A and the decomposition of CHP to form K at higher temperatures. These trends are similar to those in our previous work with OMS-1 catalysts.[8] The fact that the efficiency of TBHP did not change significantly with temperature implies that the decomposition speed of TBHP may be mainly governed by the catalyst instead of temperature over the temperature range studied here.

(b) Solvents: TBA was as an excellent solvent for the reaction system but benzene was not. This is probably due to the polarity of solvents and miscibility of reactants in the solvents. The mixture in AC became miscible during reaction because TBA, which was generated from decomposition of TBHP, enhanced the miscibility.

(c) Time: The conversions of cyclohexane and the distributions of products changed with time. Over the first few hours, the yields of products had an order of A>K>CHP>P. But after four hours reaction, K became a dominant product and the yield of CHP decreased. The results indicate that CHP acted as an intermediate in the oxidative dehydrogenation of cyclohexane. The further oxidation of A and the decomposition of CHP increased the selectivity of K. Additionally, ketone may result from alkane oxidation without going through the alcohol stage.[2]

4.2. Mechanistic Studies.
The mechanism for OMS-2 acting as a catalyst for cyclohexane oxidation has not been investigated in detail. Preliminary studies provide some evidence that this reaction is a heterogeneous catalytic process and radicals may be involved in the oxidation. The data of Fig. 3 clearly show that the oxidation could not be carried out without catalysts; the X-ray patterns (Fig. 1) show that the structure of OMS-2 remained unchanged; and no free manganese cations were found in the reaction solution. These results indicate that this oxidative process is a heterogeneous catalytic process. The facts that TEMPO largely affected the conversion of cyclohexane and the formation of A may imply that radical chain reactions may dominate these oxidation processes.

The efficiencies of TBHP are not high in these systems as compared to homogeneous systems. This may be due to the rapid decomposition of TBHP into dioxygen. The dioxygen may not be effectively utilized to oxidize hydrocarbons under these reaction conditions. If an oxidant decomposed into dioxygen at a very fast speed, the conversion of cyclohexane would be very low. In fact, hydrogen peroxide can be decomposed by OMS-2 at an extremely fast speed even at room temperature. When using hydrogen peroxide as an oxidant under reaction conditions similar to TBHP, a low conversion (<1%) of cyclohexane was obtained.

V. Summary

OMS-2 materials, like their OMS-1 analogues, exhibit very promising activities in catalytic oxidation of cyclohexane even at room temperature. The conversion of cyclohexane in the liquid phase oxidation system is as high as 16.5% at 60°C. Cyclohexanone and cyclohexanol are the two major products. The ratio of K to A changes with temperature. The reduced product of TBHP, tert-butyl alcohol, is the best solvent among those tested. Preliminary mechanistic studies indicate that radical reaction processes might be involved and that these are heterogeneous catalytic processes.

Acknowledgments

We thank the Department of Energy, Office of Basic Energy Sciences, Division of Chemical Sciences, for the support of this research.

References

1 . Hill, C. L. *Activation and Functionalization of Alkanes*, Ed.; Hill, C. L.; John Wiley & Sons, 1989, p 243.

2. Leising, R. A., Kim, J., Perez, M. A., Que, L. Jr. *J. Am. Chem. Soc.*, 115, 9524-9530 (1993)

3. Ganeshpure, P. A., Tembe, G. L., Satish, S., *J. Mol. Catal. A: Chemical*, 113, L423-425 (1996)

4. Barton, D. H. R., Beviere, S. D., Hill, D. R., *Tetrahedron* 50 (9), 2665-2670 (1994)

5. Corma, A., Esteve, P., Martinez, A., Valencia, S., *J. Catal.*, 152, 18-24 (1995)

6. Tateiwa, J., Horiuchi, H., Uemura, S., *J. Chem. Soc., Commnun.*, 2567-2568 (1994)

7. Suib, S. L. in *Recent Advances and New Horizons in Zeolite Science and Technology, Stud. Surf. Sci. Catal.*, 102, 47-74(1996)

8 Wang, J.-Y., Xia, G. G., Yin, Y.G., O'Young, C. L., Suib, S. L., *J. Catal.*, *in press (1998).*

Octahedral Layer (OL) and Octahedral Molecular Sieves (OMS). A Unique Class of Catalysts for the Oxidative Dehydrogenation of Cyclohexane

Jin-Yun Wang[1], Guan-Guang Xia[2], Yuan-Gen Yin[1], Steven L. Suib[1,2,3]*
[1]Department of Chemistry, [2]Institute of Material Science, [3]Department of Chemical Engineering, University of Connecticut, Storrs, CT 06268-4060

ABSTRACT

Manganese oxide based octahedral molecular sieves (OMS) with different structures and different doped metals have been tested as catalysts for the oxidative dehydrogenation of cyclohexane in the temperature range of 300°C to 450°C. All the OMS materials showed good activities with the one having the cryptomelane structure being most active. The products were mainly cyclohexene, pentanal, α-methyl tetrahydrogen furan (α-MTHF), and CO_x. There was no benzene detected in the products which distinguishes OMS materials from all other catalysts reported. The catalysts with the todorokite structure had higher selectivities and higher yields to useful products (cyclohexene, pentanal, α-MTHF) than other OMS materials. At 420°C with an oxygen to hydrocarbon ratio of 1.5 to 1, the conversion of cyclohexane was 39.6% with the selectivities of cyclohexene, 36.1%; α-MTHF, 10.7%; and pentanal, 19.5%.

I. Introduction

Oxidative dehydrogenation of cyclohexane has been intensively studied for several decades.[1] Various catalysts, such as activated alumina,[2] zeolites,[3-4] vanadate,[5] molybdena,[6] have been reported for this reaction. Zeolites as catalysts only produced benzene and combustion products such as CO_x. Transition metal oxide catalysts have produced some cyclohexene, but benzene and CO_x were always the main products, especially at high temperature or at high conversions of cyclohexane.

A new class of materials, octahedral layer and molecular sieves,[7,8] have shown very good activity toward alkane functionalization.[9] In pursuing their catalytic activity toward oxidative dehydrogenation of cyclohexane, three different manganese oxide materials[8] were used in this research including a layered material with the birnessite structure called OL-1; a microporous tunnel material with the todorokite structure called OMS-1; and a microporous tunnel material with the cryptomelane structure designated as OMS-2.

OL-1 is a layered material with a surface area (SA) of around 25 m²/g and average oxidation state (AOS) of about 3.5. OMS-1 possesses a 3 × 3 tunnel structure with a tunnel opening of about 6.9 Å (SA=20-30 m²/g, AOS=3.4-3.6), and OMS-2 possesses a 2 × 2 tunnel structure with a tunnel opening of about 4.6

Å (SA=70-80m^2/g, AOS≈3.9). The compositions of OL-1, OMS-1, and OMS-2 are Na$_4$Mn$_{14}$O$_{27}$.21H$_2$O, Mg$_{0.1-1.4}$Mn$_{6.3-6.4}$O$_{12}$.nH$_2$O, and KMn$_8$O$_{16}$.nH$_2$O respectively. When doped with a lower valent transition metal, these transition metal cations (M$^{2+/3+}$) can replace Mn$^{3+/4+}$ cations in the framework to form MO$_6$ units.

II. Experimental

2.1. Pretreatment of the Catalysts

The preparation of OL and OMS materials is discribed elsewhere.[8] All the catalysts used were sieved to 20-50 mesh. The catalyst was heated from room temperature to 150°C at a rate of 10°C/min in flowing air (10 mL/min) and stayed at 150°C for 30 minutes before air was switched to an appropriate oxygen containing gas. The system was then heated at a rate of 15°C/min to the desired reaction temperature and stabilized for 10 minutes before cyclohexane was introduced.

2.2. Catalytic Procedure

All reactions were studied under atmospheric pressure by a fixed bed quarter inch diameter U-type quartz reactor. The reaction temperature was monitored by a thermocouple and controlled by a CN2042 (OMEGA) temperature controller which can control the temperature to within ±1 K. The flow of oxygen containing gas was controlled by a Porter mass flow controller. Cyclohexane was introduced into the system with a Sage pump which was carefully calibrated.

Steady state was achieved after 90 minutes time on stream at 300°C, and it can be reached faster at higher temperature. Blank runs showed that the reactor wall, glass wool, and glass rod were inert to the reaction at temperatures below 400°C. Less than 3% conversion of cyclohexane was observed when the reaction temperature was 450°C. The reaction mixture was analyzed with a GC-5890 series II (HP). Two different columns were used: one was Supelcowax10 which was for analyzing hydrocarbons and oxygenated hydrocarbons; the other was Gaspro GSC (Alltech) which was used for analyzing CO$_x$, O$_2$, and N$_2$. The carbon balance was within 3%. All the reactions were done under 420°C (otherwise it will be specified) with 80 mg catalyst and a fixed rate of cyclohexane (0.726 mL/hr). All data were collected at steady state.

III. Results

3.1. Oxidative Dehydrogenation of Cyclohexane Catalyzed by Manganese Oxides with Different Structures (OL-1, OMS-1, and OMS-2).

Table 1 lists the results of the oxidative dehydrogenation of cyclohexane catalyzed by different OL and OMS materials with or without doped metals. The results indicate that the general order of activities of those materials and the selectivities toward pentanal are: OMS-2 > OMS-1 > OL-1. The order of selectivities for the formation of cyclohexene and α-MTHF, as well as the ability to suppress the combustion of hydrocarbons are OMS-1 > OL-1 > OMS-2.

Transition metal doped OMS materials have higher activities and tend to cause deep oxidation products (pentanal and CO_x) as compared to their non-doped counterparts. Other products besides those listed here are mainly lighter hydrocarbons (such as ethylene, propene, and butene, etc.) and small amounts of oxygenated hydrocarbons (acetyl aldehyde, propanyl aldehyde, etc.).

Table 1. Oxidative Dehydrogenation of Cyclohexane Catalyzed by Different OMS Materials at 420°C with air (12.6 mL/min)

Catalysts	(%) Conv.	(%) S. Cyclohexene	(%) S. α-MTHF	(%) S. Pentanal	(%) S. CO_x
OL-1	25.2	22.4	11.1	13.3	7.4
[Co]-OL-1	28.6	28.2	8.1	21.6	15.1
[Cu]-OL-1	31.4	29.3	13.5	13.3	14.0
OMS-1	28.9	38.2	11.6	15.7	8.6
[Co]-OMS-1	34.5	37.4	12.6	18.3	9.4
[Fe]-OMS-1	32.2	37.5	11.9	17.6	8.8
[Cu]-OMS-1	31.8	37.7	11.0	19.5	12.7
OMS-2	36.8	21.6	-	23.3	25.3
[Co]-OMS-2	40.2	20.2	-	20.2	32.4
[Ni]-OMS-2	43.5	16.7	-	22.2	35.6

3.2. Effects of Reaction Temperature.

Co doped OMS materials were chosen to examine temperature effects. Table 2 shows that the conversion of cyclohexane and the selectivity toward combustion (CO_x) increase with increase of reaction temperature for all the catalysts. The selectivities toward cyclohexene a nd α-MTHF decrease along

Table 2. Effects of Reaction Temperature on Oxydehydrogenation of Cyclohexane at 420°C with air (12.6 mL/min) catalyzed by Co Doped Catalysts

Catalysts	T. (°C)	(%) Conv.	(%)S. Cyclohexene	(%) S. α-MTHF	(%) S. Pentanal	(%) S. CO_x
[Co]-OL-1	360	22.6	30.3	10.4	16.2	11.7
	420	28.6	28.2	8.1	21.6	15.1
	450	32.4	25.7	6.6	23.5	17.8
[Co]-OMS-1	360	24.5	40.7	19.5	14.4	6.4
	420	34.5	37.4	12.6	18.3	9.4
	450	37.3	32.2	7.8	20.4	11.8
[Co]-OMS-2	360	30.7	24.0	4.1	38.7	24.6
	420	40.7	20.1	-	22.9	34.4
	450	45.8	13.9	-	16.8	39.9

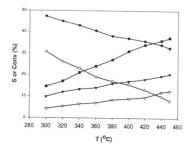

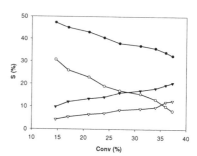

Fig. 1. The effects of temperature on cyclohexane conversion (■) and selectivities of cyclohexene (●), α-MTHF (○), pentanal (▼), and CO_x (∇) with [Co]-OMS-1 catalyst

Fig.2. The relationship of cyclohexane conversion and selectivity of cyclohexne (●), α-MTHF (○), pentanal (▼), and CO_x (∇) [Co]-OMS-1 catalyst

with increase of reaction temperature. The selectivities toward pentanal decrease with increase of reaction temperature for [Co]-OL-1 and [Co]-OMS-1 materials, but the opposite trend was observed for [Co]-OMS-2 materials

Detailed effects of temperature on the selectivity to different products for [Co]-OMS-1 as the catalyst are illustrated in Fig. 1. Within the temperature range studied, cyclohexene was always the main product even though it decreased with increasing temperature. The selectivity to α-MTHF decreased rapidly with increase of temperature, and the selectivities to pentanal and CO_x increased slowly in the temperature range studied.

The relationship between conversions and selectivities is shown in Fig. 2. The conversion to pentanal and CO_x increased with increase of conversion partially at the expense of conversion to cyclohexene and to α-MTHF. The conversion to cyclohexene was always the highest among others even when the cyclohexane conversion reached 40%.

3.3. Effects of Ratio of Oxygen to Hydrocarbon.

[Co]-OMS-1 was selected as a catalyst to study the effects of the ratio of oxygen to hydrocarbons on the oxidative dehydrogenation. To avoid mass transfer effects, the amount of cyclohexane and total flow of oxygen containing gas were kept the same (cyclohexane: 0.726 mL/hour, gas: 12.6 mL/min) while the concentration of oxygen in the oxygen containing gas was varied. The concentrations were 10, 20 and 30% and they corresponded to oxygen to hydrocarbons ratios of: 0.5:1.0, 1.0:1.0, and 1.5:1.0 respectively.

Table 3 shows that more oxygen present in the reaction mixture increases the conversion of cyclohexane, and the selectivities toward deep (%)oxidation products such as pentanal and CO_x. The selectivities toward cyclohexene and α-MTHF decrease as the oxygen concentration increases. No benzene was detected under all conditions studied in this work.

Table 3. Effects of Ratio of O_2 to Hydrocarbons for the Oxidative
Dehydrogenation of Cyclohexane at 420°C Catalyzed by [Co]-OMS-1

O_2/HC (molar)	(%) Conv.	(%) S. Cyclohexene	(%) S. α-MTHF	(%) S. Pentanal	(%) S. CO_x
0.5:1.0	25.7	39.2	13.8	16.4	6.7
1.0:1.0	34.5	37.4	12.6	18.3	9.4
1.5:1.0	39.6	36.1	10.7	19.5	11.1

IV. Discussion

4.1. Structure and Composition of the Catalysts.

OMS-2 materials are the most active but the least selective in the formation of cyclohexene and produce deep oxidation products (pentanal and CO_x) as shown in Table 1. This may be attributed to their higher surface area and higher AOS. The presence of magnesium and the 3 x 3 structure of OMS-1 may be reasons for differentiating OMS-1 from other materials. The basicity of MgO[10] may help the desorption of cyclohexene, a Lewis base, from the surface of the OMS-1 catalysts, and continuous oxidation or dehydrogenation may be minimized.

The doped transition metal ions, such as Fe^{3+}, Co^{2+}, Ni^{2+} and Cu^{2+}, have different oxidation states from $Mn^{3+/4+}$, and defects in the crystal structure may be introduced. These defects might contribute to the special activity of OMS-1 materials as well. The reason why the continuous dehydrogenation product benzene was not formed is unclear, and more work to study the structure and the surface properties (acidity, conductivity, redox, etc) of these catalysts needs to be done.

4.2. Reaction Temperature.

The selectivity to pentanal increased as the reaction temperature increased from 360°C to 450°C for OMS-1 and OL-1 materials, and this was different for OMS-2 materials whose selectivity to pentanal decreased as the temperature increased. This may be because pentanal is a relatively more stable intermediate than cyclohexene and α-MTHF. The stable products like pentanal will start to be oxidized either at higher temperatures for catalysts like OMS-1 and OL-1 or at lower temperatures for catalysts like OMS-2 materials. The results illustrated in Figs. 1 and 2 also indicate that all the hydrocarbons and oxygenated hydrocarbons are intermediates for the oxidation and that the degree of oxidation depends on reaction conditions.

4.3. Ratio of Oxygen to Hydrocarbons

The theoretical ratio of oxygen to hydrocarbons for the oxidative dehydrogenation of cyclohexane to form cyclohexene is 0.5. This is why selectivity to cyclohexene is higher at a ratio of 0.5 than at ratios of 1.0 and 1.5 under the same reaction temperatures. High oxygen content in the feed also helps to explain why pentanal is a more stable intermediate than cyclohexene and α-

MTHF because the selectivity to pentanal, like CO_x, increased with increase in the amount of oxygen.

V. Summary.

OMS materials with three different structures (OL-1, OMS-1, and OMS-2) were found to have very good activity for the oxidative dehydrogenation of cyclohexane. The unique characteristic of these catalysts is that no benzene was produced. OMS-1 materials showed higher total selectivities and total yields to useful products as compared to other materials. At similar levels of cyclohexane conversion, considerably less CO_x formation than most reported catalysts for oxidative dehydrogenation of cyclohexane has also been achieved.

VI. Acknowledgment.

We thank the Department of Energy, Office of Basic Energy Science, Division of Chemical Sciences for support of this research.

References:

1. Jouy, M., Balaceanu, J., *Proc. Int. Congr. Catal.*, 2nd, (Paris), 1, 645 (1960).
2. Uchida, A.; Nakazawa, T.; Oh-uchi, K.; Matsuda, S.;, *Ind. Eng. Chem. Prod. Res. Dev.*, 10 (2), 153 (1971).
3. Ben Taari, Y., *"Catalysis by Zeolites,"* *Stud. In Surf. Sci. Catal.*, 5, 167 (1980).
4. Coughlan, B.; Keane, M. A., *Catal. Lett.*, 4, 223(1990).
5. Michalakos, P. M.; Kung, M. C.; Jahan, I.; Kung, H. H., *J. Catal.* 140, 226 (1993).
6. Alyea, E. M.; Keane, M. K., *J. Catal.* 164, 28 (1996).
7. Shen, Y. F.; Zerger, P. R.; DeGuzman, R. N.; Suib, S. L.; McCurdy, L.; Potter, D. I.; O'Young, C. L., *Science*, 260, 511 (1993).
8. Suib, S. L. in *"Recent Advances and New Horizons in Zeolite Science and Technology,"* *Stud. in Surf. Sci. Catal.*, 102, 47 (1996).
9. Wang, J.-Y.; Xia, G. G.; Yin, Y. G.; Suib, S. L.; O'Young, C.-L., *J. Catal.*, In press (1998).
10. Tanabe. K. *Solid Acids and Bases*, Academic: New York, 1974.

Synthesis of Fluoroaniline Derivatives by Selective Palladium-Catalyzed Coupling of N-Methylpiperazine and Fluorohaloarenes

Stuart Hayden[a,b], John R. Sowa, Jr.[a,*]

[a]Department of Chemistry, Seton Hall University, South Orange, NJ 07079
[b]Labeled Compound Synthesis Department, Merck Research Laboratories,
P.O. Box 2000, Rahway, NJ 07065

Abstract

Selective coupling of fluorohaloarenes (halo = Cl, Br, I) with N-methylpiperazine is performed with the $Pd_2(dba)_3$/Binap/NaOtBu catalyst system to give fluoroaniline derivatives in >87 % isolated yield and >99 % selectivity.

Introduction

Recently there has been considerable interest in the use of homogeneous palladium catalysts for the coupling of amines and aryl halides. (1-4) These reactions are advantageous over traditional methods of preparing substituted aniline derivatives which include nucleophilic aromatic substitution and Ullmann coupling because of the high yields, mild conditions, and good functional group compatability. (1,2a,5) Also, the palladium-catalyzed reactions are effective with aryl halides containing electron donating as well as electron withdrawing substituents. (1-4) Although palladium-catalyzed coupling of amines and aryl halides has been extensively studied, only a few examples are reported where Pd catalysts are used for selective amination of mixed dihaloarenes to prepare chloroaniline (1b,3) and bromoaniline (1f) derivatives. Because of the ready commercial availability of fluorohaloarene starting materials, we envisioned that palladium coupling reactions could also be used to prepare fluoroaniline derivatives. This would expand the repertoire of methods for synthesis of fluoroanilines. In this paper, we report the first selective Pd-catalyzed amination of mixed fluorohaloarenes to yield fluoroaniline derivatives.

Experimental Section

Amination of 1-bromo-4-fluorobenzene. A three necked, 15 mL round-bottom flask (equipped with a condenser, septum adapter, and a J- Kem temperature probe) was charged with (*R*)-Binap (21 mg, 0.033 mmol), sodium *tert*-butoxide (185 mg, 1.90 mmol), N-methylpiperazine (221 mg, 2.20 mmol), 1-bromo-4-fluorobenzene (180 mg, 1.03 mmol), and tris(dibenzylidene-acetone)dipalladium (0) (10 mg, 0.02 mmol). Anhydrous toluene (3 mL) was added and the reaction was placed under an argon atmosphere at a flow rate between 50 to 100 mL/min. The reaction mixture was stirred and heated at 75 °C until completion (5 h) as indicated by GC analysis. After cooling to room temperature, the reaction mixture was diluted with methylene chloride (20 mL), filtered and concentrated by evaporation on a rotary evaporator (16 in Hg, 35 °C). The crude product was purified by flash chromatography using a Biotage Flash 40 unit with a silica gel cartridge (40g). The product was eluted with solvent mixture of CH_2Cl_2:EtOH:NH_4OH (25.0: 1.0:0.1) at a flow rate 20mL/min. The isolated product was first concentrated on a rotary evaporator (16 in Hg, 35 °C) to remove most of the solvent then by evaporating the remaining solvent at 40 °C (> 30 in Hg) for 10 min. The compound, 1-(4-fluorophenyl)-4-methylpiperazine (**3**), was obtained as an off-white solid (0.19 g) in 98 % yield.

Characterization data of fluoroaniline products:

1-(2-fluorophenyl)-4-methylpiperazine (**1**), ^1H NMR (CDCl$_3$, 400 MHz) δ 7.05-6.90 (m, 4H), 3.11(t, J = 4.4 Hz, 4H), 2.59 (t, J= 4.4 Hz, 4H), 2.34 (s, 3H); ^{13}C NMR (CDCl$_3$, 100 MHz), δ 156.85 (d, J_{CF} = 230 Hz), 125.49, 123.47 (d, J_{CF} = 7.6 Hz), 119.99 (d, J_{CF} = 3.0 Hz), 117.14 (d, J_{CF} = 20.5 Hz), 56.24, 51.53, 47.18.

1-(3-fluorophenyl)-4-methylpiperazine (**2**), ^1H NMR (CDCl$_3$, 400 MHz) δ 7.19 (dt, J_1 = 8.4 Hz, J_2 = 7.2 Hz, 1H), 6.68 (dd, J_1 = 8.0 Hz, J_2 = 2.0 Hz, 1H), 6.60 (dt, J_1 = 12.4 Hz, J_2 = 2.4 Hz, 1H), 6.53 (dt, J_1 = 8.0 Hz, J_2 = 2.2

Hz, 1H), 3.22 (t, J = 5.2 Hz, 4H), 2.57 (t, J = 5.2 Hz, 4H), 2.36 (s, 3H); ^{13}C NMR (CDCl$_3$, 100 MHz), δ 164.94 (d, J_{CF} = 242 Hz), 131.18 (d, J_{CF} = 9.8 Hz), 112.18, 106.88 (d, J_{CF} = 21.3 Hz), 103.74 (d, J_{CF} = 25.1 Hz), 56.00, 49.65, 47.20. Anal. Calcd for C$_{11}$F$_{15}$N$_2$: C, 68.02; H, 7.78. Found: C, 67.73; H, 8.05.

1-(4-fluorophenyl)-4-methylpiperazine (3), ^{1}H NMR (CDCl$_3$, 250 MHz) δ 7.0 - 6.8 (m, 4H), 3.2 (t, J = 5 Hz, 4H), 2.6 (t, J = 5 Hz, 4H), 2.4 (s, 3H); ^{13}C NMR (CDCl$_3$,100 MHz), δ 158.15 (d, J_{CF} = 230 Hz), 118.87 (d, J_{CF} = 7.6 Hz), 116.55 (d, J_{CF} = 22 Hz), 56.18, 51.15, 47.14. M.p. 36 - 38 °C. Anal. Calcd for C$_{11}$F$_{15}$N$_2$: C, 68.02; H, 7.78. Found: C, 67.80; H, 7.83.

Results and Discussion

Our initial studies attempted the coupling of N-methylpiperazine with *o*-bromofluorobenzene using a catalyst prepared from Pd$_2$(dba)$_3$ (dba = dibenzylideneacetone), 2 equivalents of tris(*o*-tolyl)phosphine, and sodium *tert*-butoxide as a co-reactant. (1c) However, no fluoroaniline products were observed over the range of temperatures (100 - 150 °C) and reaction times (24 h) explored. We thus turned to systems that incorporated bidentate phosphine ligands as studies by Buchwald (1) and Hartwig (2) indicated that catalysts containing bidentate phosphines such as Binap ((*R*) or (*S*) or racemic 2,2`-bis(diphenylphosphino)-1,1`-binapthyl), (1) and Dppf (1,1`-bis(diphenylphosphino)ferrocene) (2) showed superior reactivity. A convenient catalytic system (1a) is obtained by *in situ* combination of Pd$_2$(dba)$_3$ and Binap with a stoichiometric quantity of co-reactant, sodium *tert*-butoxide. Using this system, we successfully coupled N-methylpiperazine with *o*-bromofluorobenzene to give 1-(2-fluorophenyl)-4-methylpiperazine (1) in 89 % isolated yield in 5 h (eq 1).

In **Table 1**, we show the additional substrates that we have examined. In all cases the isolated yields are high (> 87 %) and selectivity for substitution of the nonfluoro-substituent over the fluoro-substituent is excellent (> 99 %). Because the fluoroaniline products are relatively volatile, removal of the solvent by evaporation on a rotary evaporator was performed at 35 - 40 °C and for only enough time to completely remove solvent. Reactions were carried out in toluene or without solvent with an excess of amine and as little as 2 mol % of $Pd_2(dba)_3$ catalyst.

In nucleophilic aromatic substitution reactions, fluoro-substituents are usually the most labile. (5b) Our studies demonstrate that the selectivity is reversed for the palladium-catalyzed reaction as the fluoro-substituent is the least labile halide. Less than 1 % of fluoro-substitution was observed by GC/MS under the harshest conditions necessary for this reaction, which was substitution of the chloro-substituent in *o*-chlorofluorobenzene (115 °C, 41 h). The high degree of selectivity is attributed to the strength of the aryl C-F bond and its resistance to oxidative addition by the palladium metal. Surprisingly, substitution of the chloro-substituent in these substrates (entry 5) occurs in high yield (91 %) when compared to the low yields (< 30 %) that are obtained with other aryl chlorides using the $Pd_2(dba)_3$/Binap/NaOtBu catalyst system. (1,6) While other studies have shown that the fluoro-substituent is inert under the conditions of palladium-catalyzed C-N coupling. (3c) Our study is the first to demonstrate that the fluoro-substituent also activates an aromatic ring to improve the overall reactivity of a chloro-substituent resulting in higher yields.

Conclusion

We report the first methodology for the catalytic synthesis of fluoroaniline derivatives using the recently discovered palladium-catalyzed amine coupling reaction. The procedure that we have developed is noteworthy for the high isolated yields (> 87 %) and high selectivity (> 99 %). In addition, this study adds to the breadth of knowledge about this new and important catalytic reaction.

Table 1. Reaction Data for Pd-catalyzed Coupling of N-Methylpiperazine with Fluorohaloarenes.

Entry	Fluorohaloarene	Solvent	Product[a]	Rxn Temp.	Rxn Time	Isolated Yield
1	o-BrC$_6$H$_4$F	toluene	1-(2-FC$_6$H$_4$)-4-Me-piperazine (1)	80 °C	5 h	89 %
2	m-BrC$_6$H$_4$F	toluene	1-(3-FC$_6$H$_4$)-4-Me-piperazine (2)	90 °C	7 h	97 %
3	p-BrC$_6$H$_4$F	toluene	1-(4-FC$_6$H$_4$)-4-Me-piperazine (3)	75 °C	5 h	98 %
4	o-IC$_6$H$_4$F	toluene	1-(2-FC$_6$H$_4$)-4-Me-piperazine (1)	85 °C	5 h	87 %
5	o-ClC$_6$H$_4$F	neat	1-(2-FC$_6$H$_4$)-4-Me-piperazine (1)	115 °C	42 h	91 %
6	o-ClC$_6$H$_4$F[b]	neat	no reaction	115 °C	>50 h	0 %[b]

[a]All fluoroaniline products were characterized by ^{1}H, ^{13}C NMR.
[b]Control sample, no catalyst used.

Acknowledgments

This research was supported by the Labeled Compound Synthesis Department of Merck Research Laboratories. We are grateful to Drs. Anthony Lu, David Melillo, Dennis Dean, Frank Fang, and Matthew Braun for many helpful discussions. We also thank Profs. Nicholas Snow and Robert Augustine at Seton Hall University for use of their GC/MS.

References and notes

†Dedicated to Br. Leo Michaels, Prof. of Chemistry, Manhattan College, for his many contributions to chemical research and education.

1. For leading references see: a) Wolfe, J. P.; Wagaw, S.; Buchwald, S. L. *J. Am. Chem. Soc.*, *118*, 7215 - 7216 (1996); b) Wolfe, J. P.; Buchwald, S. L. *J. Org. Chem.*, *61*, 1133 - 1135 (1996); c) Wagaw, S.; Rennels, R. A.; Buchwald, S. L. *J. Am. Chem. Soc.*, *119*, 8451 - 8458 (1997).

2. For leading references see: a) Hartwig, J. F. *Synlett*, 329-340 (1997); b) Driver, M. S.; Hartwig, *J. Am. Chem. Soc.*, *119*, 8232 - 8245 (1997); c) For a commentary see: Barta, N. S., Pearson, W. H. *Chemtracts - Org. Chem.*, *9*, 88 - 92 (1996).

3. a) Rossen, K., Pye, P. J.; Maliakal, A.; Volante, R. P. *J. Org. Chem.*, *62*, 6462 - 6463 (1997), b) Ward, Y. D.; Farina, V. *Tetrahedron Lett.*, *37*, 6993 - 6996 (1996), c) Hong, Y.; Tanoury, G. J.; Wilkinson, H. S.; Bakale, R. P.; Wald, S. A.; Senanayake, C. H. *Tetrahedron Lett.*, *38*, 5607 - 5610 (1997); d) Zhao, S.-H.; Miller, A. K.; Berger, J.; Flippin, L. A. *Tetrahedron Lett.*, *26*, 4463 - 4466 (1996).

4. Beller, M.; Riermeier, T. H.; Reisinger, C.-P.; Herrmann, W. A. *Tetrahedron Lett.*, *38*, 2073 - 2074 (1997).

5. a) March, J. *Advanced Organic Chemistry, 4th ed.*; Wiley: New York, 1992; Chap. 13; b) Paine, A. J. *J. Am. Chem. Soc.*, *109*, 1496 (1987).

6. Good yields for catalytic aminations of aryl chlorides using a nickel catalyst (Wolfe, J. P.; Buchwald, S. L. *J. Am. Chem. Soc.*, *119*, 6054 - 6058 (1997)) and a palladacycle catalyst (ref 4) have been recently reported.

New Ammoxidation Catalysts by Using Vanadyl(IV) Orthophosphate Precursor Compounds

Andreas Martin and Bernhard Lücke

Institut für Angewandte Chemie Berlin-Adlershof e.V.

Rudower Chaussee 5, D-12489 Berlin, Germany

Abstract

Novel $(VO)_3(PO_4)_2 \cdot 7\ H_2O$ and $(VO)_3(PO_4)_2 \cdot 9\ H_2O$ vanadyl(IV) ortho-phosphate hydrates (V/P = 1.5) were used as precursor compounds and transformed into highly active methylaromatics ammoxidation catalysts. The catalytic studies of the orthophosphate hydrate derived specimens showed a significant enhancement of the aromatic feed conversion rate in comparison to similar transformation products derived from $VOHPO_4 \cdot \frac{1}{2}\ H_2O$ precursor (V/P = 1) as well as a higher nitrile selectivity at similar conversion rates in comparison to conventional V_2O_5-containing catalyst generated by wet impregnation.

Introduction

In addition to their application as partial oxidation catalysts [e.g. ref. 1-4], vanadium phosphates (VPO) are also successfully employed as highly active and selective materials in the heterogeneous catalytic ammoxidation of substituted methylaromatics and methylheteroaromatics [5-7]. The well-known $(VO)_2P_2O_7$ that is generated by $VOHPO_4 \cdot \frac{1}{2}\ H_2O$ dehydration belongs to them, other catalysts are mainly formed under NH_3-containing feed during transformation periods, starting from different VPO precursor compounds [8].

Recently, it was found out that VPO's, being nearly inactive catalysts themselves like pure $(NH_4)_2(VO)_3(P_2O_7)_2$ will be active specimens if they carry vanadium oxide particles. Such solids can be formed by a one-step phase transformation of a number of VPO precursor compounds in NH_3-containing feed streams. Amongst other precursors, we have used $VOHPO_4 \cdot \frac{1}{2}\ H_2O$ (molar ratio V/P = 1) as precursor compound and found crystalline $(NH_4)_2(VO)_3(P_2O_7)_2$ (V/P = 0.75) by XRD [9]. The remained quarter of the original vanadium of the precursor compound forms a XRD-amorphous mixed-valent vanadium oxide phase (V_xO_y). This V_xO_y phase is responsible for the catalytic activity.

This paper reports on the usage of novel vanadyl(IV) orthophosphate $((VO)_3(PO_4)_2)$ as precursor (V/P = 1.5) compounds. The synthesis of $(VO)_3(PO_4)_2 \cdot 9\ H_2O$ was firstly described by Teller et al. [10]. More recently, Wolf succeeded in synthesizing a less water-containing product, $(VO)_3(PO_4)_2 \cdot 7\ H_2O$ [11]. The heterogeneous catalytic ammoxidation of toluene to benzonitrile was used as a model reaction. As-synthesized, pure $(NH_4)_2(VO)_3(P_2O_7)_2$

and the transformation products of $VOHPO_4 \cdot \frac{1}{2} H_2O$ were included for comparison.

Experimental

For the preparation of the $(VO)_3(PO_4)_2 \cdot 7 H_2O$ (VP_{7H}) precursor, H_3PO_4 (85%, 0.721 mol) and $(COOH)_2 \cdot 2 H_2O$ (0.824 mol) in 172 ml H_2O were heated up to 343 K. V_2O_5 (0.55 mol) was added in portions. This solution was evaporated at ca. 383 K during 24 h using a rotary evaporator. The water-unsoluble product was recovered by filtering and drying [11]. $(VO)_3(PO_4)_2 \cdot 9 H_2O$ (VP_{9H}) was synthesized in a similar procedure. $VOHPO_4 \cdot \frac{1}{2} H_2O$ ($VP_{\frac{1}{2}H}$) was used as precursor compound for comparison. The synthesis procedure has been previously reported, using an aqueous medium [12]. Furthermore, pure $(NH_4)_2(VO)_3(P_2O_7)_2$ (AVP_{syn}) was synthesized from V_2O_5 and $(NH_4)_2HPO_4$ in accordance with a method described in detail in ref. 9. Table 1 summarizes some characterization data of the precursor samples.

Table 1 Chemical analyses and vanadium valence states of the precursor materials

Samples	H (wt.%)	V (wt.%)	P (wt.%)	H : V : P molar ratio	V valence state
$VP_{\frac{1}{2}H}$	1.29	29.03	18.33	2.18 : 0.96 : 1	3.993
VP_{7H}	2.60	29.35	12.53	6.43 : 1.42 : 1	4.001
VP_{9H}	3.27	28.19	11.12	9.12 : 1.54 : 1	4.000

The precursor compounds ($VP_{\frac{1}{2}H}$), (VP_{7H}) and (VP_{9H}) were pelletized and crushed (1 - 1.25 mm) and then pretreated in an U-tube quartz-glass reactor either by heating up to 673 K under an NH_3-air-H_2O vapour flow (molar ratio = 1 : 7 : 5, total flow = 13 l/h) for 5 h (route A) or by heating up to 873 K under N_2 (7 l/h) for 50 h (route B).

The catalytic properties of the solids were determined during the ammoxidation of toluene to benzonitrile using a fixed bed U-tube quartz-glass reactor (1.5 ml catalyst volume). The following reaction conditions were applied: toluene : NH_3 : air : H_2O vapour = 1 : 4.5 : 32 : 24, atmospheric pressure, W/F = ca. 10 g h mol^{-1}. The catalytic runs were performed at ca. 573, 598 and 623 K. Toluene conversion and benzonitrile yield were followed by on-line capillary GC.

The results of a substantial characterization of the generated and used catalysts by XRD, FT-IR spectroscopy, ^{31}P-MAS-NMR spectroscopy and chemical analyses are described elsewhere [13].

Results and Discussion

The precursor materials were transformed during pretreatments in the presence of ammonia-containing gases (route A) as well as under nitrogen (route B), leading into materials those contain crystalline $(NH_4)_2(VO)_3(P_2O_7)_2$ (AVP) or $(VO)_2P_2O_7$ (VPP) specimens, respectively, and, in addition too, amorphous as well as partly crystalline vanadium oxides (V_xO_y). These oxides represent the molar vanadium excess of the precursor material in comparison to the defined, crystalline proportion of the transformation product.

Firstly, AVP-containing catalysts $([(NH_4)_2(VO)_3(P_2O_7)_2 + V_xO_y]$ $(AVP_{gen}))$ were generated during pretreatment route A. The catalysts will be designated as follows: AVP_{gen7H} (precursor material: $(VO)_3(PO_4)_2 \cdot 7\ H_2O$), AVP_{gen9H} (precursor material: $(VO)_3(PO_4)_2 \cdot 9\ H_2O$) and $AVP_{gen½H}$ (precursor material: $VOHPO_4 \cdot ½\ H_2O$). The AVP_{gen7H} and AVP_{gen9H} solids contain a higher proportion of V_xO_y in comparison to $AVP_{gen½H}$ as shown by chemical analyses and vanadium valence state determination (Table 2) but the V^{IV}/V^V ratio remains nearly the same (ca. 1). The following equations (eq. 1 and 2) should emphasize these experimental results.

$$4\ V^{IV}OHPO_4 \cdot ½\ H_2O\ (2\ NH_3 + O_2)\ \Rightarrow\ [(NH_4)_2(V^{IV}O)_3(P_2O_7)_2 + V^{IV/V}_xO_y] + 3\ H_2O \quad (1)$$

$$2\ (V^{IV}O)_3(PO_4)_2 \cdot 7\ H_2O\ (2\ NH_3 + O_2)\ \Rightarrow\ [(NH_4)_2(V^{IV}O)_3(P_2O_7)_2 + 3\ V^{IV/V}_xO_y] + 13\ H_2O \quad (2)$$

$VP_{½H}$ (V/P ratio = 1) is transformed into $AVP_{gen½H}$ and 75 % of the original vanadium should be find in the AVP phase, the rest must be located in the vanadium oxide proportion. Otherwise, the vanadyl(IV) orthophosphate hydrates VP_{7H} and VP_{9H} (V/P ratio = 1.5) are also transformed into AVP but only 50 % of the original vanadium amount can be find in the AVP phase, the other half remains in the V_xO_y phase.

The precursor compounds were also pretreated under nitrogen (route B). The high temperatures and the long-term pretreatment procedure were necessary because the crystalline vanadyl(IV) orthophosphates VP_{7H} and VP_{9H} rapidly dehydrate, forming amorphous, water-free $(VO)_3(PO_4)_2$ that recrystallizes very slow. $VP_{½H}$ was pretreated in this study in the same way to keep comparable reaction conditions. The phase transformation leads into vanadyl pyrophosphates $((VO)_2P_2O_7$ (VPP)) as proven by the XRD patterns. These solids will be described as follows: VPP_{7H} (precursor material: $(VO)_3(PO_4)_2 \cdot 7\ H_2O$), VPP_{9H} (precursor material $(VO)_3(PO_4)_2 \cdot 9\ H_2O$) and $VPP_{½H}$ (precursor material $VOHPO_4 \cdot ½\ H_2O$).

Equations 3 and 4 illustrate the described transformations and the formation of an additional portion of V_xO_y for example in the case of the VP_{7H} precursor transformation (eq. 4).

$$2 \ V^{IV}OHPO_4 \cdot \tfrac{1}{2} H_2O \ (N_2) \ \Rightarrow \ (V^{IV}O)_2P_2O_7 + 2 \ H_2O \tag{3}$$

$$(V^{IV}O)_3(PO_4)_2 \cdot 7 \ H_2O \ (N_2) \ \Rightarrow \ [(V^{IV}O)_2P_2O_7 + V^{IV}{}_xO_y] + 7 \ H_2O \tag{4}$$

These ideas are also reflected in the chemical analyses data (Table 2). $VPP_{\frac{1}{2}H}$ shows a vanadium valence state of nearly +4 because it consists of vanadyl pyrophosphate only. For example, VPP_{9Hant} (sample $[(V^{IV}O)_2P_2O_7 + V^{IV}{}_xO_y]$ obtained from $(VO)_3(PO_4)_2 \cdot 9 \ H_2O$) after nitrogen treatment) reveals a vanadium valence state of 4.029 that slightly differs from the basic material but it can be seen that VPP_{7H} and VPP_{9H} obtained after ammoxidation runs have significant V^V proportions in the V_xO_y phase. The VPP catalysts always exhibit hydrogen and nitrogen (Table 2), probably pointing to NH_4^+ species generated on the solid surface during ammoxidation runs by hydrolysis of some P-O-P, V-O-V or V-O-P links.

Table 2 Chemical analyses, BET surface areas and vanadium valence states of the solids derived after precursor pretreatments (route A: NH_3-air-H_2O vapour, 673 K, 5 h; route B: N_2, 873 K, 50 h) and ammoxidation runs

Samples	N (wt.%)	H (wt.%)	H : N mol. ratio	V (wt.%)	BET surface area / m² g⁻¹	V valence state
route A						
$AVP_{gen\frac{1}{2}H}$	4.95	1.28	3.62	29.17	2.95	4.111
AVP_{gen7H}	3.81	0.95	3.49	35.24	3.60	4.248
AVP_{gen9H}	3.99	0.98	3.44	34.75	2.70	4.237
route B						
$VPP_{\frac{1}{2}H}$	1.05	0.16		32.83	5.31	4.007
VPP_{7H}	0.08	0.03		38.23	1.09	4.209
VPP_{9H}	0.23	0.07		38.66	0.63	4.136
$VPP_{9Hant}{}^a$	-	-		38.68	n.d.	4.029

^{a} after N_2 treatment only

The results of the catalytic experiments during the ammoxidation of toluene using the AVP_{gen} catalysts are summarized in Figure 1. The solids possess noticeable activities at T > 550 K. The benzonitrile selectivity is very high ($\geq$ 95 %) and carbon oxides were formed only to a low extent. AVP_{gen7H} and AVP_{gen9H} reveal results, being similar to one another. Moreover, the results clearly demonstrate a significant increase of the toluene conversion on the samples derived from the orthophosphate precursors in comparison to the $VP_{\frac{1}{2}H}$ precursor. Furthermore, the decrease of the reaction selectivity to benzonitrile is not significantly enlarged as suspected due to a higher V^V amount. AVP_{syn} was used for comparison and it can be seen that it shows very poor catalytic activity only.

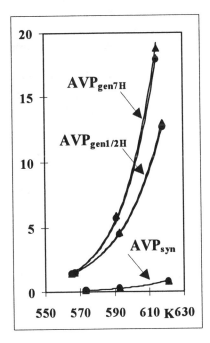

Fig. 1 Toluene conversion (mol%, ▲) and benzonitrile yield (mol%, ●) on AVP$_{syn}$, AVP$_{gen½H}$ and AVP$_{gen7H}$

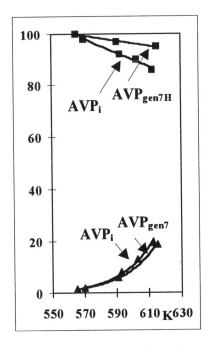

Fig. 2 Comparison of toluene conversion (mol%, ▲) and nitrile selectivity (%, ■) of AVP$_{gen7H}$ and AVP$_i$ (AVP$_{syn}$ impregnated with NH$_4$VO$_3$)

Interestingly, a comparison of the conversion and selectivity data of AVP$_{gen7H}$ and a catalyst obtained by impregnation of AVP$_{syn}$ with a defined amount of NH$_4$VO$_3$ (AVP$_i$) corresponding to the one of the V$_x$O$_y$-phase in AVP$_{gen½H}$ reveals that the nitrile selectivity on AVP$_i$ drastically drops despite a comparable toluene conversion (Fig. 2).

Similar results were obtained using the VPP pyrophosphate catalysts, i.e., toluene conversion rates are increased up to those, being comparable to the AVP$_{gen}$ catalysts by using the orthophosphate precursors. The toluene conversion as well as benzonitrile yields are rather low due to the low BET surface areas but conversions refering to the surface area are in the same order of magnitude.

The studies have shown that vanadyl orthophosphate hydrates $((VO)_3(PO_4)_2 \cdot n\ H_2O\ (n = 7;9))$ can be successfully transformed into vanadium oxide-containing ammonium vanadyl pyrophosphate as well as vanadyl pyrophosphate catalysts, depending on the pretreatment procedure. These catalysts reveal an enlarged catalytic activity in the ammoxidation of toluene, processing with nitrile selectivities remaining almost the same as for catalysts derived from the well known $VOHPO_4 \cdot \frac{1}{2}\ H_2O$ precursor. The enlarged catalytic activity is due to a higher proportion of V^{IV}/V^{V} mixed-valent vanadium oxide domains, existing on the catalyst surface as well as in the bulk and reaches values of known V_2O_5-containing catalysts obtained by wet impregnation.

Acknowledgement

The authors thank the Federal Ministry of Education, Science, Research and Technology of the FRG and the Berlin Senate Department for Science, Research and Culture (project 03C3005) for financial support.

References

1. E. Bordes, *Catal. Today*, **1**, 499 (1987).
2. *Vanadyl Pyrophosphate Catalysts*, ed. G. Centi, *Catal. Today*, **16 (1993)**.
3. M. Abon and J.-C. Volta, *Appl. Catal. A*, **157**, 173 (1997).
4. F. Cavani and F. Trifirò, *Appl. Catal. A*, **157**, 195 (1997).
5. B. Lücke and A. Martin, *Chem. Ind.* (Dekker), **62** (Catal. Org. React.), 479 (1995).
6. A. Martin and B. Lücke, *Chem. Ind.* (Dekker), **68** (Catal. Org. React.), 451 (1996).
7. I. Matsuura, *Stud. Surf. Sci. Catal.*, **72**, 247 (1992).
8. A. Martin, A. Brückner, Y. Zhang and B. Lücke, *Stud. Surf. Sci. Catal.*, **108**, 377 (1997).
9. Y. Zhang, A. Martin, G.-U. Wolf, S. Rabe, H. Worzala, B. Lücke, M. Meisel and K. Witke, *Chem. Mater.*, **8**, 1135 (1996).
10. R.G. Teller, P. Blum, E. Kostiner and J.A. Hriljac, *J. Solid State Chem.*, 1992, **97**, 10.
11. G.-U. Wolf, unpublished results.
12. H. Berndt, K. Büker, A. Martin, M. Meisel, A. Brückner and B. Lücke, *J. Chem. Soc., Faraday Trans.*, **91**, 725 (1995).
13. A. Martin, G.-U. Wolf, U. Steinike and B. Lücke; *J. Chem. Soc., Faraday Trans.*, submitted.

Index